10th Anniversary of *Coatings*: Invited Papers in "Surface Characterization, Deposition and Modification" Section

10th Anniversary of *Coatings*: Invited Papers in "Surface Characterization, Deposition and Modification" Section

Editor

Georgios Skordaris

Basel • Beijing • Wuhan • Barcelona • Belgrade • Novi Sad • Cluj • Manchester

Editor
Georgios Skordaris
Aristotle University of Thessaloniki
Thessaloniki
Greece

Editorial Office
MDPI
St. Alban-Anlage 66
4052 Basel, Switzerland

This is a reprint of articles from the Special Issue published online in the open access journal *Coatings* (ISSN 2079-6412) (available at: https://www.mdpi.com/journal/coatings/special_issues/10th_surf_charact_depos_modif).

For citation purposes, cite each article independently as indicated on the article page online and as indicated below:

Lastname, A.A.; Lastname, B.B. Article Title. *Journal Name* **Year**, *Volume Number*, Page Range.

ISBN 978-3-7258-0261-6 (Hbk)
ISBN 978-3-7258-0262-3 (PDF)
doi.org/10.3390/books978-3-7258-0262-3

© 2024 by the authors. Articles in this book are Open Access and distributed under the Creative Commons Attribution (CC BY) license. The book as a whole is distributed by MDPI under the terms and conditions of the Creative Commons Attribution-NonCommercial-NoDerivs (CC BY-NC-ND) license.

Contents

About the Editor ix

Preface xi

Ben D. Beake
Nano- and Micro-Scale Impact Testing of Hard Coatings: A Review
Reprinted from: *Coatings* 2022, 12, 793, doi:10.3390/coatings12060793 1

Yixue Duan, Gongchuan You, Zhe Zhu, Linfeng Lv, Xiaoqiao Liao, Xin He, et al.
Reconstructed NiCo Alloy Enables High-Rate Ni-Zn Microbattery with High Capacity
Reprinted from: *Coatings* 2023, 13, 603, doi:10.3390/coatings13030603 29

Zhenhua Chu, Haonan Shi, Fa Xu, Jingxiang Xu, Xingwei Zheng, Fang Wang, et al.
Study of the Corrosion Mechanism of Iron-Based Amorphous Composite Coating with
Alumina in Sulfate-Reducing Bacteria Solution
Reprinted from: *Coatings* 2022, 12, 1763, doi:10.3390/coatings12111763 37

**Georgios Skordaris, Konstantinos Vogiatzis, Leonidas Kakalis, Ioannis Mirisidis,
Vasiliki Paralidou and Soultana Paralidou**
Increasing the Life Span of Tools Applied in Cheese Cutting Machines via Appropriate
Micro-Blasting
Reprinted from: *Coatings* 2022, 12, 1343, doi:10.3390/coatings12091343 48

**Chih-Jui Chang, Chih-Wei Lai, Wei-Cheng Jiang, Yi-Syuan Li, Changsik Choi,
Hsin-Chieh Yu, et al.**
Fabrication and Characterization of P-Type Semiconducting Copper Oxide-Based Thin-Film
Photoelectrodes for Solar Water Splitting
Reprinted from: *Coatings* 2022, 12, 1206, doi:10.3390/coatings12081206 64

Daniel Kalús, Daniela Koudelková, Veronika Mučková, Martin Sokol and Mária Kurčová
Contribution to the Research and Development of Innovative Building Components with
Embedded Energy-Active Elements
Reprinted from: *Coatings* 2022, 12, 1021, doi:10.3390/coatings12071021 76

**Yuqing Zhang, Laura Vespignani, Maria Grazia Balzano, Leonardo Bellandi, Mara Camaiti,
Nadège Lubin-Germain and Antonella Salvini**
Low Fluorinated Oligoamides for Use as Wood Protective Coating
Reprinted from: *Coatings* 2022, 12, 927, doi:10.3390/coatings12070927 105

Yushan Ruan, Lineng Chen, Lianmeng Cui and Qinyou An
PPy-Modified Prussian Blue Cathode Materials for Low-Cost and Cycling-Stable Aqueous
Zinc-Based Hybrid Battery
Reprinted from: *Coatings* 2022, 12, 779, doi:10.3390/coatings12060779 130

Shuai-Shuai Lv, Yu Zhang, Hong-Jun Ni, Xing-Xing Wang, Wei-Yang Wu and Chun-Yu Lu
Effects of Additive and Roasting Processes on Nitrogen Removal from Aluminum Dross
Reprinted from: *Coatings* 2022, 12, 730, doi:10.3390/coatings12060730 142

Yaoting He, Jiafei Yan, Xuzhao He, Wenjian Weng and Kui Cheng
PLLA/Graphene Nanocomposites Membranes with Improved Biocompatibility and
Mechanical Properties
Reprinted from: *Coatings* 2022, 12, 718, doi:10.3390/coatings12060718 153

Bo Hu, Kang Han, Chunhua Han, Lishan Geng, Ming Li, Ping Hu and Xuanpeng Wang
Solid-Solution-Based Metal Coating Enables Highly Reversible, Dendrite-Free Aluminum Anode
Reprinted from: *Coatings* **2022**, *12*, 661, doi:10.3390/coatings12050661 . . . . **165**

Liguang Yang, Wensuo Ma, Fei Gao and Shiping Xi
Effect of EDM and Femtosecond-Laser Groove-Texture Collision Frequency on Tribological Properties of 0Cr17Ni7Al Stainless Steel
Reprinted from: *Coatings* **2022**, *12*, 611, doi:10.3390/coatings12050611 . . . . **175**

Charis Gryparis, Themis Krasoudaki and Pagona-Noni Maravelaki
Self-Cleaning Coatings for the Protection of Cementitious Materials: The Effect of Carbon Dot Content on the Enhancement of Catalytic Activity of TiO_2
Reprinted from: *Coatings* **2022**, *12*, 587, doi:10.3390/coatings12050587 . . . . **195**

Shuaishuai Lv, Hongjun Ni, Xingxing Wang, Wei Ni and Weiyang Wu
Effects of Hydrolysis Parameters on AlN Content in Aluminum Dross and Multivariate Nonlinear Regression Analysis
Reprinted from: *Coatings* **2022**, *12*, 552, doi:10.3390/coatings12050552 . . . . **213**

Chi-Yu Chu, Pei-Ying Lin, Jun-Sian Li, Rajendranath Kirankumar, Chen-Yu Tsai, Nan-Fu Chen, et al.
A Novel SERS Substrate Based on Discarded Oyster Shells for Rapid Detection of Organophosphorus Pesticide
Reprinted from: *Coatings* **2022**, *12*, 506, doi:10.3390/coatings12040506 . . . . **226**

Jun Su, Jiaye Ye, Zhenyu Qin and Lidong Sun
Polytetrafluoroethylene Modified Nafion Membranes by Magnetron Sputtering for Vanadium Redox Flow Batteries
Reprinted from: *Coatings* **2022**, *12*, 378, doi:10.3390/coatings12030378 . . . . **238**

Vicente Hernandez, Romina Romero, Sebastián Arias and David Contreras
A Novel Method for Calcium Carbonate Deposition in Wood That Increases Carbon Dioxide Concentration and Fire Resistance
Reprinted from: *Coatings* **2022**, *12*, 72, doi:10.3390/coatings12010072 . . . . **249**

Hongjun Ni, Weiyang Wu, Shuaishuai Lv, Xingxing Wang and Weijia Tang
Formulation of Non-Fired Bricks Made from Secondary Aluminum Ash
Reprinted from: *Coatings* **2022**, *12*, 2, doi:10.3390/coatings12010002 . . . . **261**

Emilio Márquez, Juan J. Ruíz-Pérez, Manuel Ballester, Almudena P. Márquez, Eduardo Blanco, Dorian Minkov, et al.
Optical Characterization of H-Free *a*-Si Layers Grown by rf-Magnetron Sputtering by Inverse Synthesis Using Matlab: Tauc–Lorentz–Urbach Parameterization
Reprinted from: *Coatings* **2021**, *11*, 1324, doi:10.3390/coatings11111324 . . . . **271**

Fani Stergioudi, Aikaterini Baxevani, Azarias Mavropoulos and Georgios Skordaris
Deposition of Super-Hydrophobic Silver Film on Copper Substrate and Evaluation of Its Corrosion Properties
Reprinted from: *Coatings* **2021**, *11*, 1299, doi:10.3390/coatings11111299 . . . . **297**

Martina Zuena, Ludovica Ruggiero, Giulia Caneva, Flavia Bartoli, Giancarlo Della Ventura, Maria Antonietta Ricci and Armida Sodo
Assessment of Stone Protective Coatings with a Novel Eco-Friendly Encapsulated Biocide
Reprinted from: *Coatings* **2021**, *11*, 1109, doi:10.3390/coatings11091109 . . . . **312**

Yu Zhang, Hongjun Ni, Shuaishuai Lv, Xingxing Wang, Songyuan Li and Jiaqiao Zhang
Preparation of Sintered Brick with Aluminum Dross and Optimization of Process Parameters
Reprinted from: *Coatings* **2021**, *11*, 1039, doi:10.3390/coatings11091039 325

Wenyan Gu, Rong Zhan, Rui Li, Jiaxin Liu and Jiaqiao Zhang
Preparation and Characterization of PU/PET Matrix Gradient Composites with
Microwave-Absorbing Function
Reprinted from: *Coatings* **2021**, *11*, 982, doi:10.3390/coatings11080982 339

Denisa Anca, Iuliana Stan, Mihai Chisamera, Iulian Riposan and Stelian Stan
Experimental Study Regarding the Possibility of Blocking the Diffusion of Sulfur at
Casting-Mold Interface in Ductile Iron Castings
Reprinted from: *Coatings* **2021**, *11*, 673, doi:10.3390/coatings11060673 355

Yuan Yin, Chen Li, Yinuo Yan, Weiwei Xiong, Jingke Ren and Wen Luo
MoS_2-Based Substrates for Surface-Enhanced Raman Scattering: Fundamentals, Progress and
Perspective
Reprinted from: *Coatings* **2022**, *12*, 360, doi:10.3390/coatings12030360 365

About the Editor

Georgios Skordaris

Dr.-Eng. Georgios Skordaris is Professor and Director of the Laboratory for Machine Tools and Manufacturing Engineering, School of Mechanical Engineering, Aristotle University of Thessaloniki (AUTH), Greece. His scientific interests are in manufacturing processes, coatings for cutting tools and machine elements, FEA modelling, superficial treatments, and the characterization of coated tool's strength, fatigue and adhesion properties. He has both published numerous papers in international peer-reviewed journals and presented at international conferences on these topics. He has been the scientific lead in many research projects supported by national and European Union research funds, as well as by well-established companies, in the fields of tools and coatings. He has served on several national and international committees and as an editorial member/guest editor of several reputed journals.

Preface

Celebrating the 10th anniversary of *Coatings* through a Special Issue focused on surface and interface engineering is a great way to highlight advancements in the field. This Special Issue provides a platform for researchers and experts to share their latest findings and innovations related to various coating deposition processes and surface treatments for attaining the enhanced protection and improved functionality of modified or treated surfaces under challenging conditions. By comprehensively characterizing the properties of coated or treated surfaces, researchers can gain insights into how these modifications impact material performance. This information is valuable for both advancing our understanding of surface engineering principles and guiding the practical applications of this research.

Georgios Skordaris
Editor

Review

Nano- and Micro-Scale Impact Testing of Hard Coatings: A Review

Ben D. Beake

Micro Materials Ltd., Willow House, Yale Business Village, Ellice Way, Wrexham LL13 7YL, UK; ben@micromaterials.co.uk; Tel.: +44-1978-261615

Abstract: In this review, the operating principles of the nano-impact test technique are described, compared and contrasted to micro- and macro-scale impact tests. Impact fatigue mechanisms are discussed, and the impact behaviour of three different industrially relevant coating systems has been investigated in detail. The coating systems are (i) ultra-thin hard carbon films on silicon, (ii) DLC on hardened tool steel and (iii) nitrides on WC-Co. The influence of the mechanical properties of the substrate and the load-carrying capacity (H^3/E^2) of the coating, the use of the test to simulate erosion, studies modelling the nano- and micro-impact test and performing nano- and micro-impact tests at elevated temperature are also discussed.

Keywords: nano-impact; micro-impact; fatigue; fracture

Citation: Beake, B.D. Nano- and Micro-Scale Impact Testing of Hard Coatings: A Review. *Coatings* **2022**, *12*, 793. https://doi.org/10.3390/coatings12060793

Academic Editor: Alessandro Latini

Received: 6 May 2022
Accepted: 6 June 2022
Published: 8 June 2022

Publisher's Note: MDPI stays neutral with regard to jurisdictional claims in published maps and institutional affiliations.

Copyright: © 2022 by the author. Licensee MDPI, Basel, Switzerland. This article is an open access article distributed under the terms and conditions of the Creative Commons Attribution (CC BY) license (https:// creativecommons.org/licenses/by/ 4.0/).

1. Introduction

Impact resistance is critical in many applications of coating systems involving highly loaded mechanical contact. These include automotive and aero-engine components and interrupted high-performance machining operations where intermittent high strain rate contact occurs [1–3]. In a diesel engine system, diamond-like carbon (DLC) coatings are deposited on many components in the powertrain, including fuel injectors, tappets, pistons, and piston rings, where they can be subjected to repetitive impacts in service [2]. In a gas turbine engine, high-temperature erosion of the thermal barrier coatings that protect the underlying superalloy turbine blades is a key factor lowering service life and restricting operating temperatures.

Cyclic impact tests are used as model tests for assessing coating durability under dynamic loading [4–6]. Bulk materials and coatings systems often undergo fatigue deformation mechanisms in the multi-cycle tests that are not observed in single-cycle tests [7,8]. In an impact test on a coated system, the test severity and positions of peak impact-induced stresses relative to the coating–substrate interface can be controlled by varying the impact energy and the geometry of the test probe. Fatigue mechanisms can vary with the ratio of coating thickness t to the indenter radius R [9,10] (t/R), so it can be very useful to perform impact tests with different contact sizes to obtain data over a range of t/R. Therefore, macro-scale, micro-scale and nano-scale impact tests have been developed. The differences between them and how these influence the observed behaviour is discussed in more detail in later sections. Deformation and failure mechanisms depend on applied load and indenter sharpness. t/R values are very low ($\approx$0.001) in macro-scale tests of thin physical vapour deposition (PVD) coatings with cemented carbide or hardened steel spherical indenters with 1–3 mm end radius. The peak von Mises stresses that result in plastic deformation are located deep into the substrate, and hence, the fatigue behaviour can be strongly influenced by the substrate properties [1,4–7,11]. Although they can be useful, macro-scale impact tests have some limitations. An alternative approach to determining coating fatigue resistance is to perform nano-scale impact tests at higher t/R with much sharper probes. These accelerated tests are of much shorter duration than macro-scale tests and subject coatings to more severe conditions that replicate the high stresses generated in actual operating conditions. The position of peak von Mises stresses relative to the coating-substrate interface is completely different in the nano- and macro-scale tests. Nano-impact tests are very sensitive to small differences in coating properties and have shown excellent correlation to coating performance in applications. In particular, there have been many studies on Al-rich (Ti,Al)N-based PVD coatings on cemented carbide that have shown strong correlation between the wear of coated tools in high-speed machining applications and the fracture resistance found in the nano-impact test [12–23].

Nano-impact testing utilises the depth-sensing capability of a multifunctional nanomechanical test system (NanoTest system, Micro Materials Ltd., Wrexham, UK) to perform impact testing at strain rates that are several orders of magnitude higher than those in quasi-static indentation tests [24–29]. Although nano-impact is the most common terminology, the technique was originally termed micro-impact [30] and has also been described as impact indentation or impulse impact. The small scale tests provide more localised assessment of impact resistance. They have potential advantages in high throughput, automation and surface sensitivity, so they are particularly suited to thin coatings/small volumes and in investigating the influence of nano/microstructure on performance.

To bridge the gap in t/R between the nano- and macro-ranges, the micro-impact test, involving higher loads and larger probe sizes than in nano-impact, has been developed as an instrumented accelerated test sensitive to coating and substrate together where stresses can be concentrated near interface(s) in the system [31–37]. In the micro-impact test, coating and substrate deformation is important, and coatings can be subjected to high bending stresses. The importance of the strain rate on the fatigue failure of coatings has been highlighted by Bouzakis and co-workers, with even only a relatively modest increase in strain rate decreasing the fatigue endurance limit of $Al_{0.6}Ti_{0.4}N$-coated WC-Co [38]. The high strain rate contact in nano- and micro-impact tests can provide closer simulation of the performance of coatings systems under highly loaded intermittent contact and the evolution of wear under these conditions than tests at a lower strain rate.

Originally envisaged as a test method primarily to assess the degradation of coatings to repeated localised stresses, the availability of single and multiple impact configurations, nano- and micro-scale load ranges and different indenter geometries have resulted in the development of a wide range of applications [39–59]. Applications of single impact tests have included (i) strain rate sensitivity [24,25,27,28], (ii) dynamic hardness [29,41,42], (iii) dynamic H/E [45], (iv) energy absorption [40] and (v) particle–matrix delamination [46]. Applications of repetitive impact tests have included (i) the evolution of dynamic hardness and debonding [51], (ii) erosive wear simulation by matching contact size [43], (iii) fracture toughness [56], and (iv) understanding how hierarchical structures influence damage tolerance in natural tough materials [47]. A single impact is effectively a high strain rate indentation test. Analysis methods for single-impact data have been developed using the approaches outlined by Schuh, van Vliet and others [24,29,41] to measure the dynamic hardness (impulse hardness) of the material, i.e., its effective hardness at high strain rate. The nano-impact test has also been used to assess toughness at a small scale [56]. Although it does not measure quasi-static fracture toughness K_{1c}, it can provide a quantitative assessment of resistance to fatigue fracture, or effective dynamic toughness, under repetitive loading. The power of the test as a reliable simulation tool is that in many cases, this is more representative of actual contact conditions in applications; i.e., wear resistance is controlled by a combination of load support and resistance to fracture rather than by coating hardness or toughness alone. To improve our understanding of coating system behaviour under repetitive impact, it has proved beneficial to also (i) develop analysis methods for quantifying deformation in single impacts [40,41], (ii) perform repetitive impacts on uncoated substrates [49], (iii) develop test metrics from single impacts, which can be used to detect the onset of fracture [50,52], and (iv) support conclusions with multi-sensing approaches such as acoustic emission monitoring [53,54].

In this review, the operating principles of the nano-impact test technique are described, and nano-impact tests are compared to micro- and macro-scale tests. Impact fatigue mechanisms are discussed, and the impact behaviour of three different industrially relevant coating systems—(i) ultra-thin hard carbon films on silicon, (ii) diamond-like carbon (DLC) on hardened tool steel, and (iii) PVD nitrides on cemented carbide—is investigated in detail. This is followed by sections describing the influence of the substrate mechanical properties and the load-carrying capacity (H^3/E^2) of the coating, the use of the test to simulate erosion, studies modelling the nano- and micro-impact test and lastly, performing nano- and micro-impact tests at elevated temperature.

2. Nano-Impact—Experimental Setup, Test Basics and Test Metrics

In the nano-impact test, a diamond indenter is withdrawn to a set distance from the sample surface and then rapidly accelerated to produce a high strain rate impact event. The depth-sensing capability of a commercial nanoindentation system (NanoTest, Micro Materials Ltd., Wrexham, UK) is used to monitor the degradation of surface from repeated localised stresses at high rates of strain, which are orders of magnitude higher than in normal (quasi-static) nanoindentation. The configuration is shown schematically in Figure 1a.

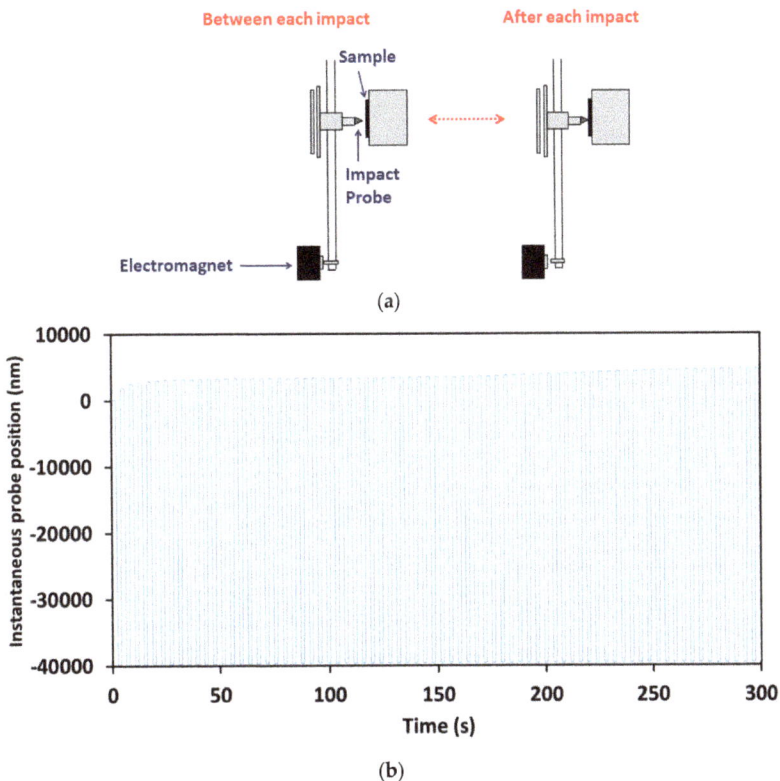

Figure 1. (a) Schematic representation of experimental configuration for repetitive nano-impact test. (b) Example of instantaneous probe position in a repetitive micro-impact test.

An initial surface contact by the impact probe under a minimum contact load determines the depth zero at the beginning of the nano-impact test experiment. The actuated (static) coil force is then applied, producing elastoplastic deformation by indentation. The corresponding initial indentation depth under load, h_0, which includes elastic and plastic deformation, is used to confirm that the depth zero is measured correctly and the test did not impact in an anomalous region of the surface. Repetitive contacts are produced by electromagnetic actuation where the impact probe is rapidly withdrawn from the surface (e.g., to 10 µm above the surface, as shown schematically on the left-hand side of Figure 1b) and then accelerated over this distance to impact the surface (right-hand side of Figure 1b), producing true high strain rate impact events (see also Figure 2) where the probe leaves the surface between each subsequent impact. The under-load impact depth, h, is always larger than h_0, as the dynamic impact force is significantly larger than the static impact load, due to inertia. Once the probe comes to rest, it is retracted, and with periodic actuation, the surface re-impacted at the same position at a set frequency, typically at 4 s intervals, to produce a cyclic impact test. The position of the impacting diamond probe under load is recorded throughout the test, allowing the progression of damage to be monitored cycle by cycle. An example is shown in Figure 1b.

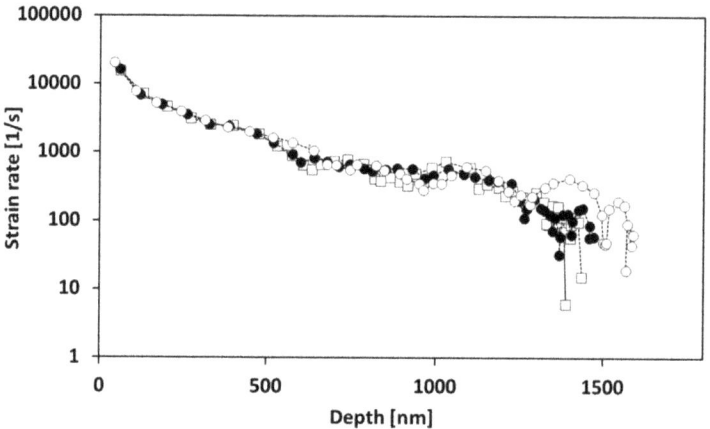

Figure 2. Strain rate in a nano-impact test on alumina with a cube corner indenter at 5 mN.

Several authors have shown that the strain rate at contact in the nano-impact test can be extremely high, typically in the region of 10^4–10^5 s^{-1} [26–28,55,57]. To illustrate this, Figure 2 shows how the strain rate varies with time during a single impact event on a bulk alumina sample when impacted by a cube corner diamond probe (three repeats are shown). Although the strain rate reduces after contact, it remains at a high level throughout the majority of the impact event.

Experimental parameters such as the test probe geometry, applied load, acceleration distance and the total number of impact cycles and their frequency are user-controlled in the nano-impact test to alter the severity of the test and its duration. A cube corner diamond indenter (with a small end radius of ≈50 nm) has been the most popular choice of impact probe, as its geometry produces high contact strain, which is beneficial in driving impact-induced fracture within a short test time. The applied load and accelerating distance control the impact energy delivered to the sample. Typical nano-impact test parameters that have been used for testing hard coatings are: (i) cube corner diamond impact probe, (ii) 90° impact angle, (iii) 25–150 mN applied load, (iv) 15 µm accelerating distance, (v) 0.25 Hz impact frequency, (vi) 300 s test duration (i.e., 75 impacts in total), (vii) 5–10 repeat tests at each load, and (viii) normal laboratory temperature (e.g., 22 °C). The general procedure for micro-impact tests is the same.

Qin and co-workers [29] have split the impact process into three stages: (i) acceleration, (ii) indentation, and (iii) rebound. High-resolution analysis of probe depth vs. time data is used to determine a range of metrics from single impacts including (i) coefficient of restitution (V_{in}/V_{out}) and (ii) from knowledge of the effective mass of the pendulum, fractional potential energy (KE_{in}/KE_{out}) [50]. Analysis of single nano-impacts [29] provided an estimate of the fraction of the impact energy transferred to the sample as ≈0.7 and the fraction lost through losses to the system, i.e., transmission into the pendulum, vibration, and air damping as ≈0.3.

The response of a material to repetitive contact in the nano-impact test depends on its ductility or brittleness. On a ductile material, there is a gradual increase in probe penetration depth. The rate of depth increase slows with continued impacts, particularly for strongly work-hardening materials. In contrast, for a brittle material, there are often several abrupt increases in probe depth during the test due to cohesive and/or interfacial failures. A typical example on a coating system is shown in Figure 3. At 50 mN, there was no clear failure, but at higher loads, the abrupt increases due to fracture are clear.

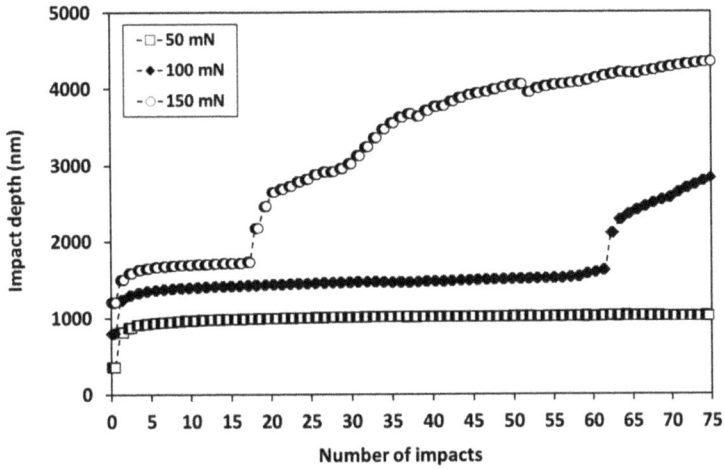

Figure 3. Variation in probe depth with number of impacts in nano-impact tests at 50–150 mN on $Ti_{0.25}Al_{0.65}Cr_{0.1}N$ PVD coating on cemented carbide with a sharp cube corner indenter.

Since it is trivial to set up multiple tests in an automated schedule, typically, multiple replicate tests are performed at different positions (e.g., in a grid array) on the sample surface to improve the statistical significance of the results. The impact resistance of different coatings can be assessed by the number of impacts required for failure to occur in 50% of the tests. Rebound impacts are essentially elastic [39] so that only the initial impact in each cycle is counted. Failure probability can be estimated by ranking the number of impacts-to-failure events in order of increasing fatigue resistance and then assigning a probability of failure to the nth ranked failure event in a total sample size of N, according to Equation (1), in an analogous approach to the treatment of distributions of failure stresses in Weibull statistics.

$$P(f) = n/(N+1) \quad (1)$$

By combining failure probability data at different loads, a plot of the number of impacts required for failure to occur in 50% of tests vs. the impact force can be obtained. Failure in the nano-impact test can be strongly load-dependent. As an example, Figure 4 illustrates how the failure probability changes with load and number of impacts for an 80 nm ta-C coating on Si when impacted by a spherical indenter with a 4.6 µm end radius [58].

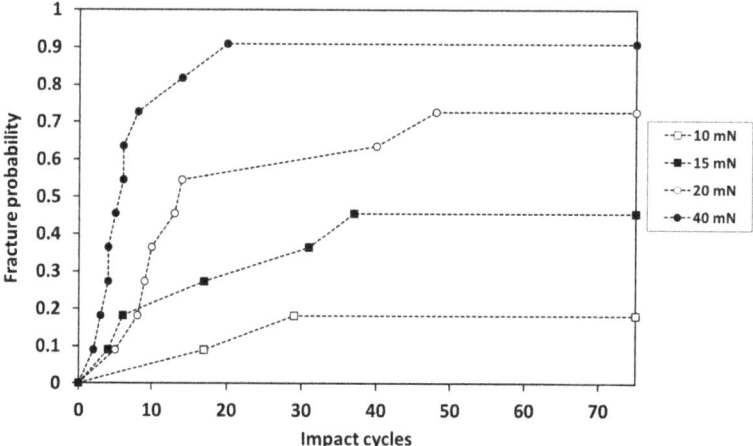

Figure 4. Failure probability vs. load for 80 nm ta-C on Si.

Several methods have been used to confirm that the abrupt changes in depth (as shown in Figure 3) are due to the onset of fracture. Jennett and Nunn [50] have used high-resolution analysis to monitor the change in fractional potential energy absorbed with continued impacting, showing a marked increase in energy absorbed for the impacts that resulted in abrupt increases in probe depth. In nano-impact tests on the bulk ceramic materials, alumina and partially stabilised zirconia, simultaneous acoustic emission (AE) detection has been used to reveal which impacts cause cracking [53,54]. Although there was a correlation between impacts that caused a large increase in depth being accompanied by bursts of AE, the in situ monitoring of AE revealed a more complex behaviour with crack systems developing over several impacts before a larger burst of AE for the impact resulting in material removal under the impact probe. Shi and co-workers [52] used high-resolution data acquisition of single nano-impacts on CrN to reveal changes in depth–time data when fracture occurred. In cyclic impact, more stochastic behaviour is observed from the onset of cracking. An indirect but practically useful indication of fracture is the onset of variability in depth vs. number of impacts in repeat tests [58]. When there was no fracture in any of the tests, the reproducibility in probe depth was typically very good. Higher variability begins once fracture occurs after a certain number of impacts in some tests but not others.

3. Comparison between Nano-, Micro- and Macro-Scale Impact Tests

The general procedure for nano- and micro-impact tests is the same. In micro-impact tests on hard coatings, the experimental parameters are typically the same except for accelerating distance, applied load and probe geometry. The accelerating distance is typically set at 40 µm so that differences in impact energy are obtained by altering the applied load (0.5–5 N). Sphero-conical diamond test probes with end radii of 8–100 µm have been most commonly used [36,49,59]. The impact energy is given by the product of the impulse force and accelerating distance. Since the accelerating distance is typically kept constant, it is common to report data in terms of the actuated impact force. Due to higher forces and accelerating distances, the energy supplied in micro-impact is typically from $\times 100$ to $\times 1000$ greater than in the nano-impact test, which enables spherical probes to be used effectively, causing fracture rapidly.

Typical experimental parameters in each type of test are summarised in Table 1. Although the principles behind nano-/micro- and macro-scale impact tests are common, there is a fundamental difference of approach in the nano-/micro- tests, which are depth-sensing, as the change in depth under load is monitored throughout the test with a capacitive sensor, and the macro-impact tests, which are not depth-sensing. Instead, coating durability in macro-scale impact tests has been assessed by post-test evaluations of damage such as crater volume [60] or failed-area ratio (defined as area of substrate exposure divided by the total contact area) [4,61]. Nano- and micro-impact tests are accelerated tests that are typically much shorter and probe coating system behaviour under more severe conditions where there is greater coating strain. Detailed information on the fatigue failure mechanisms in nano/micro-impact tests is obtained through setting up automated arrays of rapid repeat tests at different loads (e.g., 5–10 repeats per load) with cycle-by-cycle monitoring of the damage providing a precise measure of the number of impact cycles to coating failure in each test.

Table 1. Comparison between nano-, micro- and macro-scale impact test techniques (1).

-	Nano-Impact	Micro-Impact	Macro-Impact
Depth sensing	Y	Y	N
Accurate time-to-failure recorded	Y	Y	N
Test duration	1–60 min (2)	1–60 min (2)	Extended duration
Number of impact cycles	15–450	15–900	10^2–10^6
Test probe material	Diamond	Diamond	WC-Co, hardened steel, Si_3N_4
Test probe radius, R	≈50 nm	5–100 μm	1–3 mm
t/R	≈10	≈0.1	≈0.001
Sensitivity to coating mechanical properties	High	High	may be low
Sensitivity to adhesion	Medium	High	may be low
Automatic scheduling of multiple tests/tests on multiple samples	Y	Y	N
Applied load (N)	0.001–0.2	0.1–5	>>100
Accelerating distance (μm)	10–15	10–60	can remain in contact

(1) Macro-scale data taken from references [4,60–72]. (2) most commonly 5 min, i.e., 75 impacts.

The micro-impact test at higher t/R than macro-impact can be more sensitive to coating and substrate together, since stresses can be concentrated near interfaces in the coating system [31]. Beake, Liskiewicz and co-workers [31–37] have used this technique to investigate (i) impact resistance of hard carbon coatings on hardened tool steel [33–35,37] and (ii) PVD nitrides on WC-Co [31,32,36,37]. The impact energy in nano- and micro-impact tests is much lower than in the macro-scale tests, but critically, it is acting over a very small volume so that the resultant impact energy density is high. The size of the affected volume can be estimated by $2.4a \times \pi a^2$, where a is the contact radius and $2.4a$ is the depth of the primary indentation zone [73,74]. In a study of micro-impact of TiAlCrN/NbN coatings, the calculated energy densities when using $R = 8$ or 20 μm probes were ≈2–4 GJ/m^3, resulting in rapid fracture [36].

The potential advantages of studying coating fatigue resistance by nano- or micro-scale tests are the much shorter duration of the experiments compared to conventional, high-cycle macro-scale tests and impact-by-impact monitoring of the impact-induced deformation process that provides a precise record of the exact number of cycles to failure with detailed information on the fatigue failure mechanism. It is possible to use nano- or micro-impact testing to automatically build up complete S-N fatigue curves from single samples, enabling rapid screening to evaluate the performance of novel coating compositions and load-dependent deformation mechanisms to be evaluated.

4. Impact Fatigue Mechanisms

By altering the impact load, probe sharpness and test duration, it is possible to study fatigue mechanisms. In nano- and micro- scale impact tests, the instantaneous probe position is recorded throughout, so changing depth can be correlated with a post-test analysis of impact craters by SEM, with additional sub-surface information available from FIB cross-sections through the craters [75]. The initial resistance to crack nucleation and subsequent crack propagation under cyclic loading are studied by monitoring the evolution in probe depth.

When comparing coatings and their deformation mechanisms across wide force ranges and/or different probe geometries, it is common to plot depth vs. number of impacts from multiple tests together. However, small inflexions in probe depth that can occur at the onset of failure events may be obscured. As an alternate approach to compare coating behaviour at different applied load and/or with different probe geometries, the change in depth after the initial impact (i.e., $[h - h_0]$) with continued impact can provide a more useful indication of the damage progression [37,49,59]. This approach enables (i) effective comparison of nano- and micro-impact data with different probe geometries and (ii) convenient investigation of load-dependent behaviour at either length scale whilst retaining the same probe geometry. However, it removes the effects of initial load-dependent coating bending. An example of this is discussed in more detail in Section 6 (Figure 10).

Coating failure in a nano-impact test is usually accompanied by abrupt increases in probe depth, as shown in Figure 3. For some coating systems, the on-load probe depth has been found to decrease with continued impacts under certain conditions when spherical probes are used [51,55,76]. This backward depth evolution has been confirmed [55] by post-test AFM imaging of impact scars showing volume uplift. This uplift has been considered to be the results of interfacial delamination occurring without (or before) the accompanying fracture that results in the increase in depth. It is more commonly found at low load where stresses are relatively low and impact-induced plasticity is minimised. This behaviour indicates that under certain conditions, the impact test may be used to assess adhesion strength, particularly where debonding is induced without being preceded by appreciable plasticity.

Experimental studies with a range of different probe geometries have clearly shown that hard PVD coatings can display fatigue behaviour under cyclic loading. The location and extent of cracking observed depends on a range of factors including coating and substrate mechanical properties and coating thickness, and test conditions including the test probe geometry, applied load and number of cycles. Experimental studies of repetitive micro-impact by spherical indenters with end radii 17–20 μm and macro-scale tests with larger radius probes have reported crack formation at the top surface at the periphery of the contact where high tensile stresses exist. For example, Tarrés and co-workers [77] have studied the damage mechanism under cyclic loading of PVD TiN-coated hard metal substrate by a 1.25 mm radius WC-Ni spherical indenter at 200–900 N. This occurred through (i) nucleation of a surface circular crack after plastic deformation of the substrate, (ii) gradual crack growth down from the coating surface with increasing cycles through the coating thickness, and (iii) substrate cracking without any intermediate interface delamination. The critical loads for cracking under monotonic and cyclic loading were used to determine the fatigue sensitivity of the TiN coating. Spherical probe geometry has been generally preferred for investigating repetitive indentation/impact damage evolution and fatigue sensitivity [77], but sharper probe geometries such as Vickers, Berkovich or cube corner also can generate surface fatigue. For example, surface radial cracks which grow gradually extend from the impact zone with an increasing number of cycles and/or load. In cyclic Vickers indentation of hard coatings on tool steels, the observed crack morphology around the indent was found to depend on the H/E ratio of the coatings [78]. A quasi-plastic damage mode with radial cracks that increased in length under continued cyclic loading was found in coatings with relatively lower H/E [78].

Thin zirconia, alumina and zirconia-alumina bi-layer coatings deposited on glass have been studied by nano- and micro-impact testing [59] as a model brittle coating/brittle substrate system. Back-scattered SEM imaging revealed a range of load-dependent deformation mechanisms including (i) radial cracking without chipping/delamination, (ii) concentric ring cracks leading to chipping/delamination, (iii) chipping/delamination accompanied by spiral cracking outside the chipped/delaminated region and (iv) chipping/delamination accompanied by substrate fracture. A range of load-dependent mechanisms were also found in micro-impact testing of TiAlCrN/NbN nanomultilayered coatings on WC-Co [36]. Ring cracks precede radial cracks in indentation [79]. In spherical indentation and cyclic fatigue tests, circumferential cracking is commonly observed initially, transitioning to radial cracking and chipping at higher loads and/or longer cycling.

FIB milling has been used to study sub-surface crack networks developed by repetitive nano-impact testing. Zhang and co-workers [57] observed that lateral cracks developed in 10 µm TiN/Ti multilayers on Ti6Al4V. Chen and co-workers noted [75] that despite extensive spallation, there was no interfacial cracking in TiAlSiN and TiN coatings on hardened steel. Ma and co-workers used [80] X-FIB to show that degradation in columnar TiN and TiAlN-TiN bilayer coatings on steel subjected to indentation with a spherical indenter was predominantly by shear at columnar grain boundaries. Circumferential cracking outside the indentation zone was also observed. In cyclic loading of ≈1.5 µm thick TiN on 304 stainless steel with a $R = 5$ µm indenter; Cairney and co-workers [81] used FIB to show that the principal deformation mechanism appeared to be sliding along intercolumnar cracks. They proposed that fatigue occurred through a reduction in the shear stress at the column boundaries with repeated indentation. The reduction in shear stress resulted in greater load being transferred to the softer substrate, with a consequent increase in penetration depth [81]. Shi and co-workers [82] noted that in nano- and micro-impact tests on graphitic carbon coatings on stainless steel substrate, i.e., a substrate that does not provide as much load support as hardened tool steel, the failure location was not at the periphery of the crater but in the centre (underneath).

Abdollah and co-workers [83] proposed a three-stage impact deformation–wear transition map to describe wear evolution in impact tests of a DLC coating at 70–240 N with a 1 mm radius steel indenter: (i) initial steel substrate plastic deformation occurring without coating wear, (ii) suppression of substrate plasticity and (iii) coating wear. Boundaries between these regions depended [83] on the impact load and number of impact cycles. Under these conditions (low t/R), the presence of the thin hard coating barely influences the elastoplastic deformation of the steel substrate, and the mean contact pressure was close to 1.1 times the substrate yield stress. In micro-impact tests of graded a-C:H and a-C coatings on hardened M2 steel, the mean pressure during the test was calculated from the dynamic impact force and the contact area under load [33]. The contact pressure gradually reduces with each successive impact to reach the plateau contact pressure where the contact is effectively elastic. In tests at the micro-scale, the mean pressure in this region was somewhat controlled by substrate yield stress, although the harder coatings also carried some of the impact load.

The extent of plastic deformation and resultant coating bending and tensile stresses developed at the edge of the contact varied with both the applied load (hence impact energy) and the radius of the indenter. Increasing the applied load produces greater substrate plasticity with higher tensile stresses. Failure is more severe and occurs after fewer impacts.

Finite element analysis (FEA) has shown that varying the t/R ratio can alter the dominant failure mechanisms in single indentation tests through changes to the location of initial yielding [9,10]. When the ratio is very low, as in impact tests with probes with 1–3 mm radii, at low enough load, substrate plasticity can be reduced, and the mechanical properties of the coating do not influence substrate elasto-plastic deformation. Under these conditions, high-cycle coating (or substrate) fatigue may occur, with the highest tensile stresses being generated very close to the contact periphery and blistering inside the impact zone. In impact tests with low t/R, detailed investigation of the fatigue wear process revealed blistering and subsequent delamination of isolated regions [69,70]. Micro-scale impact tests have been performed over a range of t/R by changing the test probe radius to investigate the influence of the substrate on the coating degradation mechanism.

5. Coating Systems

5.1. Ultra-Thin Hard Carbon Films on Silicon—Influence of Probe Geometry

The high surface-to-volume ratio in Si-based Micro-electro-mechanical-system (MEMS) devices makes interfacial interactions a dominant factor in their wear and lifetime. Silicon is a brittle material with little or no conventional plasticity and low fracture toughness. It has highly complex mechanical and tribological behaviour with pressure-dependent phase transformations and lateral cracking observed in indentation and brittle fracture in a range of mechanical contacts [84–86]. Wear and stiction forces have limited the reliability of silicon-based MEMS when/if mechanical contact occurs [87–89]. The reliability of MEMS devices under severe shock conditions is an active research area, which has been reviewed by Peng and You [90]. Hard ultra-thin carbon films, including tetrahedral amorphous carbon (ta-C) coatings deposited by filtered cathodic vacuum arc (FCVA), have been developed for MEMS applications, and protective low friction coatings have been developed for micromachined components. The carbon coatings restrict silicon phase transformation under mechanical loading by providing additional load support, reducing the load reaching the substrate and spreading the deformation out over a wider area [91].

The behaviour of 5–80 nm FCVA ta-C on Si [58,92] and 100 nm sputtered DLC on Si [55] under repetitive impact loading has been studied in nano-impact tests with different impact probe geometry and applied load. Repetitive impact of the 100 nm DLC with a 10 μm end radius spheroconical indenter produced delamination and uplift at low load. FEA suggested that the blistering and delamination occurred when the maximum von Mises stresses were near the coating–substrate interface. To reduce the severity of the test, they also performed indentation fatigue tests where the probe did not leave the sample between tests [55]. Under these lower strain rate conditions, coating failure occurred after much longer fatigue cycles. The high strain rate in the nano-impact test, where each cycle is a true impact event, is more efficient at promoting coating failure.

In the nano-impact behaviour of 5, 20, 60 and 80 nm ta-C coatings deposited on silicon by FCVA using a well-worn Berkovich indenter at sub-mN forces [92], it was found that the 60 and 80 nm ta-C coatings failed clearly after only a few impacts. These coatings were less resistant to impact-induced damage than the underlying Si under these conditions. The on-load probe depths at the end of the test (including elastic deformation) therefore primarily reflect differences in coating thickness (so that higher depths were found on thicker coatings). Under the low-impact forces, the impact-induced stresses were not high enough to cause phase transformation or lateral cracking in the silicon substrate, but the fatigue process causes coating failure, as has also been reported in nano-fretting tests (reciprocating short track length) of the same coatings with higher loads, blunter probes (R = 5 and 37 μm), and lower contact pressures below that required for phase transformation (under ≈11 GPa) [91]. Goel and co-workers reported that in molecular dynamics simulations of very thin carbon coatings, they were able to resist nano-impact by reducing the contact pressure in the silicon substrate [93] to below that required for phase transformation.

Single and repetitive nano-impact tests with a R = 4.6 μm spherical diamond probe were performed over a range of loads on the 5 and 80 nm ta-C coatings and uncoated Si(100) to investigate how damage tolerance of silicon was modified by the presence of the ultra-thin coatings [58]. At low impact load, the deformation mechanisms involved coating damage with minimal permanent substrate damage, with delamination outside of the impact crater for the 80 nm coating, but this did not occur for the 5 nm coating. Substrate fracture occurred at higher loads through a failure mechanism involving initially plastic deformation/phase transformation during the first few impact cycles, with subsequent brittle fracture after the completed plastic deformation. In the tests on the ta-C coatings, the impact depth was lower, with more impacts required before substrate fracture than in tests on uncoated silicon, particularly for the 80 nm ta-C coating. This improvement appears to be related to their enhanced load support, which restricts the silicon phase transformation.

Shi and co-workers suggested [58] that delamination of the 80 nm coating might be an additional impact energy dissipation mechanism.

5.2. DLC on Hardened Tool Steel

Fatigue resistance of DLC coatings under highly loaded repetitive contact is required for their performance and use in demanding contact applications, e.g., in a diesel engine powertrain [2,94,95]. However, DLC coatings are susceptible to poor durability under severe loading conditions. Under these conditions, the performance of DLC coatings is limited by their resistance to contact damage [10,96], and typically, they perform poorly at higher load despite being hard and elastic. Nano-impact tests with sharp cube corner indenters and micro-impact tests with spherical probes (R = 17–20 μm) have been effective at highlighting differences in resistance to the contact damage of thin hard carbon coatings deposited on hardened steel. Studies have investigated [33–35,37,97,98] the role of coating mechanical properties and layer architecture on the fatigue resistance and the load dependence of the failure mechanisms under repetitive impact.

a-C:H coatings typically have shown high brittleness when subjected to repetitive impact [33–35,37,97,98]. Nano-impact tests on a 2.3 μm a-C:H (2 μm a-C:H with 300 nm Cr bond layer) with a cube corner diamond probe over a 3–15 mN load range revealed a strongly load-dependent coating lifetime [98]. Nano-impact tests were performed [98] on PACVD a-C:H coatings on hardened M2 steel with the same surface mechanical properties ($H \approx 22$ GPa, $E \approx 200$ GPa) but different coating architectures. The coatings were (i) 2.3 μm a-C:H, (ii) 3.0 μm a-C:H with TiN interlayer and (iii) 4.5 μm multilayered a-C:H. All three coatings were susceptible to rapid impact-induced cohesive fracture in tests with a cube corner indenter at 5 mN load. After the initial fracture, the damage tolerance of the coatings was dependent on their thickness, with thicker coatings providing better load support to the softer steel substrate and wearing at lower rate.

The load dependence of the impact response on compositionally graded 2.5 μm a-C (Graphit-iC from Teer Coatings) and 2.8 μm a-C:H (Dymon-iC from Teer Coatings) coatings on M42 tool steel has been studied in nano- and micro-impact tests [33,97,98]. Hydrogen-free a-C coatings produced by closed field unbalanced magnetron sputter ion plating (CFUBMSIP) are reported to have a predominantly sp^2 bonded graphitic structure resulting in high I_d/I_g, low stress and hence good adhesion [99]. The a-C coating is lower in hardness but stiffer, and it consequently has lower H/E and H^3/E^2 than the a-C:H coatings.

Final depth data from nano-impact tests are shown in Figure 5a and the failure probability vs. number of impacts at 5 mN are shown in Figure 5b. The harder graded a-C:H coating with higher sp^3/sp^2 bonded C was significantly less durable under fatigue loading than the softer graded a-C. For the graded a-C:H, fewer impacts were required until fracture (Figure 5b), and there was a greater change in depth on fracturing and a larger final depth at the end of the test (Figure 5a). At 1 mN, there was cohesive fracture within the coating and ring cracking at ≥ 5 mN. There was only minor cohesive cracking after ≈ 280 impacts at the same (or higher) impact forces for the graded a-C coating. In the micro-impact test, the graded a-C:H coating also showed greater susceptibility to cracking under repetitive loading. Raman spectra acquired from the centre of the impact craters showed an increase in the I_d/I_g ratio over the unworn surface for graded a-C:H due to cracking. The initially very high I_d/I_g ratio on graded a-C did not change after 75 impacts at 0.75–2 N, and it decreased only slightly after 300 impacts at 2 N [33].

Micro-impact tests with an 18 μm end radius diamond indenter have been performed at 0.5–2 N on an a-C:H, Si-doped DLC and W-doped DLC coatings on hardened steel [35]. Si-doped DLC showed the lowest resistance to repetitive impact. The a-C:H, which was the hardest and highest H^3/E^2 of the coatings studied, was also susceptible to fracture throughout the load range. The softer W-doped DLC was more impact-damage tolerant than the other coatings, despite having lower wear resistance in reciprocating sliding [35] and nano-fretting tests [100]. Although the W-doped DLC had a hard CrN sub-layer, this does not appear to be the main reason for its damage tolerance. McMaster and co-workers [34] noted that in nano- and micro-impact tests, a W-doped DLC without a CrN sub-layer also showed significantly enhanced damage tolerance compared to a-C:H and Si-doped DLC.

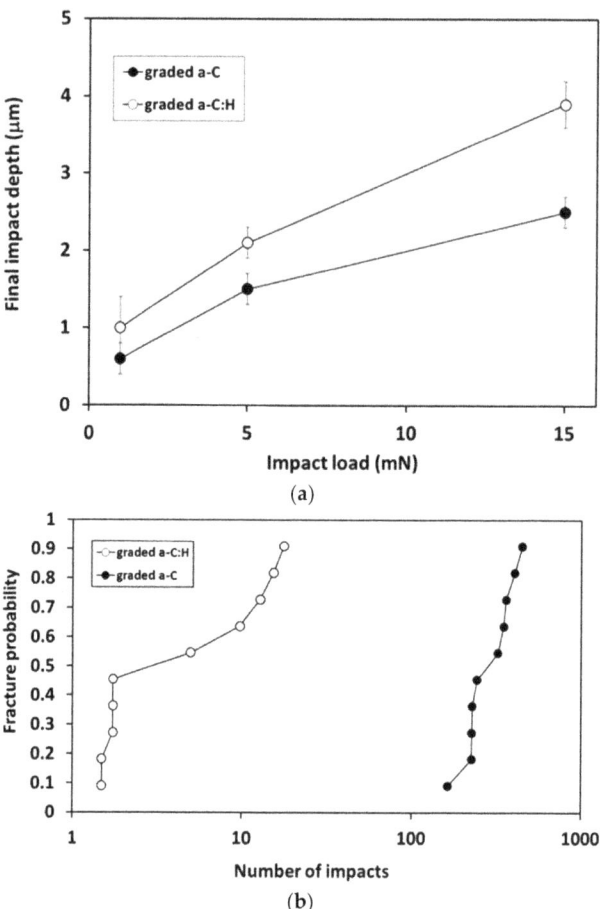

Figure 5. Nano-impact tests of graded a-C and a-C:H coatings on hardened M42 tool steel (**a**) Final depth at 1–15 mN, (**b**) Failure probability vs. number of impacts at 5 mN.

Studies of the impact performance of DLC coatings on hardened steel substrates in nano- [97,98], micro- [33–35,37] or macro-impact tests [101] have reported that coatings with lower hardness, H/E and H^3/E^2 were consistently significantly more impact resistant. In repetitive tests with 1.25 mm radius WC indenters, Ramírez and co-workers [102] found improved impact resistance for a soft W-doped carbon coating on cold-work steel than for a TiN coating on the same substrate. Under high load mechanical contact, where a combination of high load support and resistance to impact fatigue is required, an improved durability of coated components may be achieved by designing the coating system to combine these properties, rather than by increasing coating hardness alone, as this may be accompanied by brittle fracture and higher wear [33]. The combination of a coating with relatively lower H/E and a tough (i.e., damage tolerant) substrate appears beneficial for impact resistance [37].

5.3. Nitrides on WC-Co

Increasingly complex coating designs, with improved high-temperature oxidation resistance and other improved properties, have resulted in TiN being largely superceded in high-performance metal-cutting applications [103,104]. Ternary and quarternary coatings, microstructure control, layer architecture design (e.g., nano-multilayer coatings) and residual stress optimisation have been developed to increase cutting speeds and machine hard-to-cut materials economically.

Ternary nitrides have shown enhanced tool life over TiN-coated tools. Improved fracture resistance for (Ti,Al)N coatings compared to TiN has been reported in nano-impact tests [75,101]. High Al-fraction (Al = 0.52–0.67) coatings have been developed combining dense nanocrystalline or columnar microstructures, high oxidation resistance, good mechanical properties and low thermal conductivity at elevated temperature with the potential for self-adaptive behaviour through the formation of Al-rich tribo-films [12–23,101,103–116]. These Al-rich coatings typically outperform $Ti_{0.5}Al_{0.5}N$ in cutting tests and perform well in machining aerospace alloys such as titanium alloys [14,105], Ni-based superalloys (Inconel 718, Waspaloy, ME16) [14,16], and other difficult-to-machine materials including hardened steel [17,106–108], stainless steel [112] and super duplex stainless steel [110].

Despite the tribological complexity of high-speed metal cutting and the limitation of the nano-impact test to simulate the exact contact conditions, many studies have shown that a very strong correlation exists between fracture resistance in the nano-impact test and reduced wear of Al-rich (Ti,Al)N-based coatings on cemented carbide coated tools in high-speed machining [12–23,101,116]. In metal cuttings, many different wear mechanisms can be operative, and the resultant tool life is influenced by many factors besides coating mechanical properties. Studies where the (Ti,Al)N coating properties were modified without changing their microstructure, e.g., (i) through post-deposition micro-blasting or (ii) substrate bias during deposition, resulting in changes in tool life that have been directly correlated to the coating behaviour in the nano-impact test [3,18,21,117,118], show that rapid nano-impact tests are very useful as screening tests for coating optimisation in selecting potential coatings for cutting trials.

The earliest example where nano-impact tests were used as part of a comparative study with tool life data was by Fox-Rabinovich and co-workers [12]. They reported better performance of $Al_{0.7}Cr_{0.3}N$ than $Ti_{0.5}Al_{0.5}N$ in end milling of AISI 1040 steel, interrupted turning 42CrMo4V steel and deep hole drilling of hardened structural steel. Figure 6 shows nano-impact test data and cutting data in end milling 1040 structural steel. The better performance of $Al_{0.7}Cr_{0.3}N$ would not be possible to predict from room-temperature nanomechanical data, since this coating was softer with lower H/E and H^3/E^2 than the $Ti_{0.5}Al_{0.5}N$.

In nano-impact tests, $Al_{0.67}Ti_{0.33}N$ also showed significantly improved resistance to repetitive impact than $Ti_{0.5}Al_{0.5}N$ [13]. Fox-Rabinovich and co-workers reported [13] longer tool life for $Al_{0.67}Ti_{0.33}N$ than $Ti_{0.5}Al_{0.5}N$ in face milling of 1040 steel, end milling of 4340 steel and Ti6Al4V. Moderate improvement for more Al-rich (Ti,Al)N coatings compared to $Ti_{0.5}Al_{0.5}N$ has also been reported in face milling of low-carbon steel [112] and turning of medium carbon steel. Inspektor and Salvador [103] reported that with the increasing Al:Ti ratio, there was a gradual increase in the life of (Ti,Al)N-coated tools when face milling of 4140 steel. The higher Al-fraction coatings display multifunctional and adaptive behaviour in high-speed metal cutting, which results in improved tool life [12–23,101,103–121]. The coatings can more efficiently protect the tool from thermal softening through (i) more effective age-hardening by spinodal decomposition, (ii) lower thermal conductivity and brittleness at elevated temperature, and (iii) protective alumina-based tribo-films.

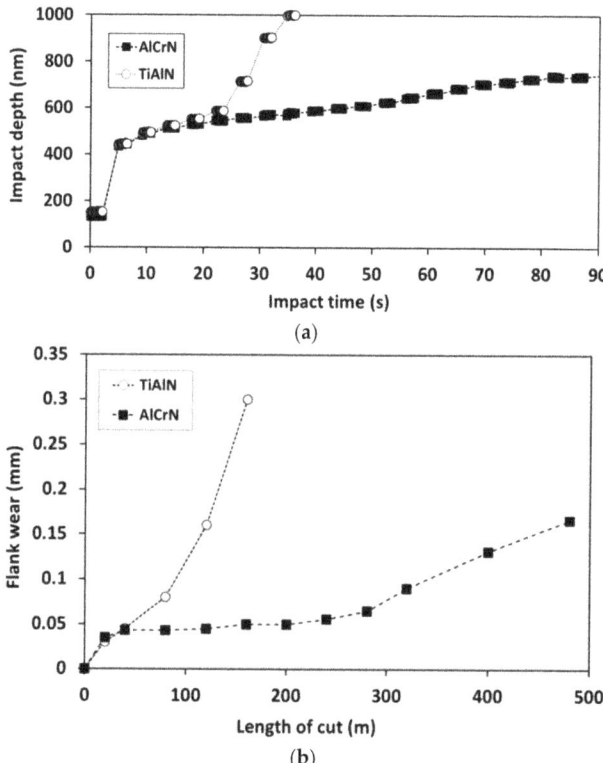

Figure 6. Comparative nano-impact and tool life data for $Al_{0.7}Cr_{0.3}N$ and $Ti_{0.5}Al_{0.5}N$ coatings on cemented carbide tool inserts (**a**) Nano-impact depth vs. time data, (**b**) cutting tool data in end milling AISI 1040 structural steel.

Nano-impact tests have also been used to study comparative fracture resistance in $Al_{0.67}Ti_{0.33}N$ and $Ti_{0.1}Al_{0.70}Cr_{0.2}N$ coatings [14,20]. $Ti_{0.1}Al_{0.70}Cr_{0.2}N$ has lower tool life than $Al_{0.67}Ti_{0.33}N$ in cutting the aerospace alloys Ti6Al4V and Waspaloy. In nano-impact tests, both coatings behave similarly on initial impact, but with repetitive impact, the TiAlCrN fractures dramatically resulting in much larger final impact depth [14,20].

Monolayer columnar coatings that have weak columnar boundaries which can act as lines of weakness for the development of through-thickness cracks that lead to extensive chipping often have lower durability. Designing coatings to be denser with additional interfaces has generally proved an effective strategy. Multilayer nitride coatings have shown improved performance in a wide range of tribological tests and machining applications [122–127]. Multilayer nitride coatings with high H^3/E^2 are discussed in more detail in Section 7.

Bouzakis and co-workers reported that varying the through-thickness multilayer microstructure of ≈8 µm thick $Al_{0.54}Ti_{0.46}N$ coatings by periodically stopping the deposition process to increase the number of layers from one to four enhanced their resistance to repetitive nano-impact and increased cutting life [18,19]. Stopping and restarting the coating deposition resulted in layering through abrupt changes in grain growth. They showed that a further increase in cutting life was achieved by increasing the number of interfaces by depositing a nanocomposite coating with approximately 600 alternating layers of 24 nm TiAlN and 3 nm TiN to the same 8 µm total thickness. There was a clear (inverse) correlation between the final impact depth and cutting life at short and long cutting edge entry durations reported in their data, as shown in Figure 7. A reduction in fracture resulted in lower final impact depth and longer tool life.

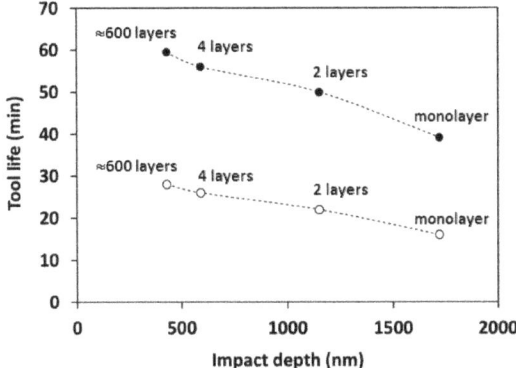

Figure 7. Relationship between impact depth and cutting life for 8 μm $Al_{0.54}Ti_{0.46}N$ coatings with differing microstructures (open circles = 0.3 ms entry time; closed circles = 5 ms entry time).

Nano-impact tests have been used to study the influence of compressive stresses developed during wet micro-blasting, with either angular Al_2O_3 or more spherical ZrO_2 grain materials, on the brittleness of 3.5 μm thick $Al_{0.54}Ti_{0.46}N$ coatings deposited on WC-Co [4,18,117,118]. Micro-blasting induces high compressive stresses within the coating which may reduce wear, although results are sensitive to the blasting pressure as well as the grain size and geometry of the abrasive materials. In addition to increasing coating hardness, the induced compressive stresses can result in increased coating brittleness. For a given micro-blasting condition and abrasive grain diameter, the abrasion with ZrO_2 was less intense than with Al_2O_3 due to the spherical nature of the ZrO_2 [4,18,117,118]. In optimising the wet micro-blasting conditions for improved cutting performance, Bouzakis and co-workers supplemented their cutting data with nanomechanical data and FEA [4,18,117,118].

High-impact resistance correlated with longer cutting tool life when milling AISI 4140 hardened steel (Figure 8). The trends in tool life with micro-blasting pressure were replicated in the nano-impact test. Maximum tool life and impact resistance were found at 0.2 MPa micro-blasting pressure. The relative ranking of cutting performance after micro-blasting with ZrO_2 and Al_2O_3 at a given pressure and the switch in relative performance above 0.2 MPa were reproduced in the nano-impact tests.

The influence of residual stress on the tool life of $Al_{0.55}Ti_{0.45}N$ coatings on cemented carbide in turning AISI 1045 steel has been studied by Skordaris and co-workers [21]. Coatings with different residual stress were obtained by depositing at different bias voltage (40, 65, 85 V), with higher bias voltages producing more compressively stress coatings. The coating deposited at 40V, which had the lowest compressive stress, was annealed to introduce tensile stress. An optimum level of compressive residual stress (-2.7 GPa) in the coating deposited at 65 V produced the best cutting performance. The improved performance for the coating with moderate compressive stress is consistent with other reports of too much compressive stress lowering durability [128]. There was a clear relationship between the final nano-impact depth and tool life, as shown in Figure 9. Skordaris and co-workers noted that coating fatigue occurred through over-stressing of the coating [21]. This contrasts to the situation in a macro-scale impact test where the coating is assumed to deform as a thin elastic plate. This may be another reason why the accelerated nano- and micro-impact tests correlate well with actual cutting behaviour.

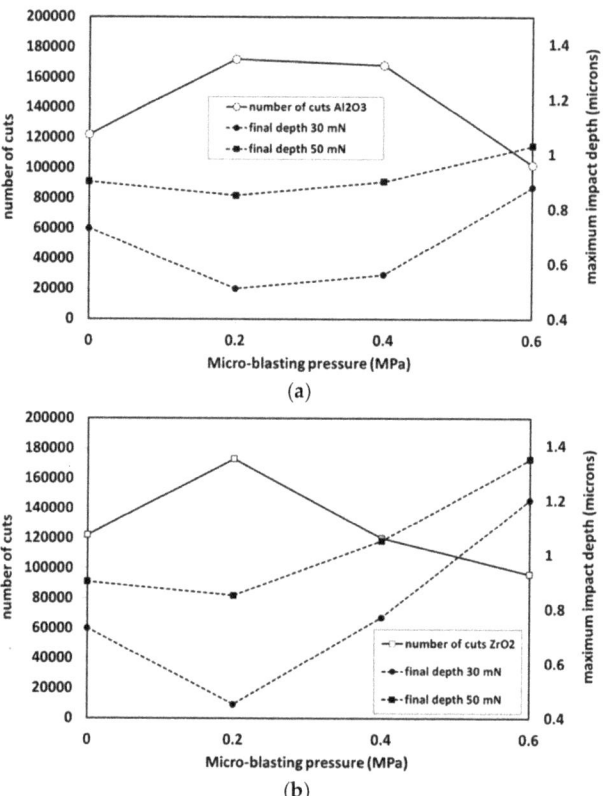

Figure 8. Correlation between maximum impact depth in tests at 30 or 50 mN and the number of cuts before tool failure at different micro-blasting pressure (**a**) Al$_2$O$_3$ abrasive (**b**) ZrO$_2$ abrasive.

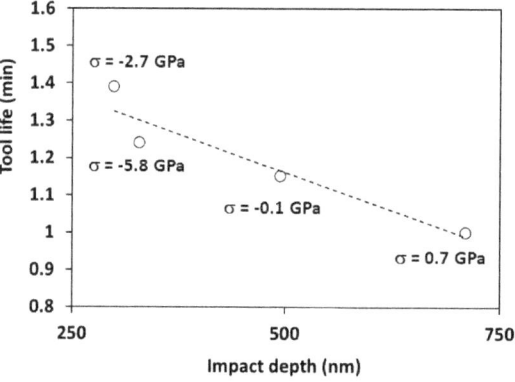

Figure 9. Influence of residual stress on tool life of Al$_{0.55}$Ti$_{0.45}$N coatings on cemented carbide in turning AISI 1045 steel and correlation to nano-impact behaviour.

Chowdhury and co-workers reported [116] that a 3 μm thick nano-multilayered TiAlCrSiYN/TiAlCrN with a 100 nm TiAlCrN interlayer, with optimised residual stress, showed better impact resistance than the other architectures they studied (either TiAlCrN or TiAlCrSiYN monolayers, TiAlCrSiYN/TiAlCrN nano-multilayers without an interlayer, or of lower thickness), which is consistent with its improved tool life in dry ball nose high speed milling of hardened H13 steel. Chowdhury and co-workers more recently investigated [23] the influence of varying TiAlCrN interlayer thickness (100, 300 or 500 nm) on the performance of nano-multilayered TiAlCrSiYN/TiAlCrN in dry high-speed milling of H13 steel. They found that there was a longer tool life with the 300 nm interlayer. In nano-impact tests, the coating with 300 nm interlayer was more resistant to spallation than the coatings with 100 or 500 nm interlayers, which is consistent with its longer tool life.

In a nano- and micro-impact comparative study of $Ti_{0.1}Al_{0.7}Cr_{0.2}N$, $Ti_{0.25}Al_{0.65}Cr_{0.1}N$ and $Al_{0.67}Ti_{0.33}N$ PVD coatings deposited on cemented carbide, the coatings exhibited strongly load-dependent fatigue behaviour in both nano-impact tests with a sharp cube corner indenter and in micro-impact tests with a $R = 17$ μm spherical indenter [31]. The relative ranking of the three coatings was the same in both tests. $Ti_{0.25}Al_{0.65}Cr_{0.1}N$, which had greater load-carrying capability due to slightly higher H^3/E^2 and greater thickness, performed best. There were differences in the impact fatigue mechanism in nano- and micro-scale impact tests due to the different stress distributions generated under repetitive nano-impact with a sharp cube corner probe failure occurred by chipping of the coating. In repetitive micro-impact with the blunter spherical probe, this coating chipping was also accompanied by debonding around the contact periphery and substrate fatigue.

The influence of t/R on the deformation behaviour has been studied in micro-impact tests on TiAlCrN/NbN nano-multilayer coatings on WC-Co by varying the sharpness of the diamond indenters, using $R = 8, 20, 100$ μm end radius probes [36]. With the 100 μm probe, there was no clear failure. Deformation with the 8 and 20 μm radius diamond probes was strongly load-dependent. At lower load, the dominant fracture behaviour was coating fracture through a three-stage process: (1) ring cracking, (2) radial cracking and (3) chipping. As the load increased, there was a transition to more substrate-dominated modes, and the impact stress field extended deeper into the WC-Co substrate, with less coating chipping and more carbide break-up.

6. Substrate Effects

In macro-scale impact tests using mm-sized WC indenters, the stresses for plasticity are far into the substrate, and hence, fatigue behaviour is influenced by substrate properties [1,129]. Knotek [1] noted that CrN coatings had better impact resistance when deposited on tool steel than on hard metal substrate due to stress relief by plastic deformation. The macro-scale impact wear of TiAlN and TiN coatings has been investigated by Yoon and co-workers on AISI D2 steel and WC-Co substrates [6]. There was lower crater volume and a longer number of cycles-to-fracture for the TiAlN when deposited on D2 steel. The more ductile tool steel substrate minimises the accumulation of elastic strain at high load. When deposited on WC-Co, the TiAlN was initially more resistant than the TiN, but with continued impacts, TiAlN exhibited pronounced brittle cracking, resulting in a dramatic increase in impact wear volume.

The WC-Co substrate is more impact fatigue resistant than the hard PVD coatings deposited on it. This can be seen in micro-impact tests at 300 mN with a $R = 8$ μm probe (Figure 10a). By showing data as depth increases ($h - h_0$), as in Figure 10b, it can be seen that the TiAlCrN/NbN coating had initially slightly better resistance than the substrate, but with continued impact, there was a transition to a more severe damage mechanism which was absent on the uncoated substrate under the same conditions. Bromark and co-workers [11] reported that the relative erosion resistance of uncoated and TiN-coated steels by SiC erodent at 20 m·s^{-1} was dependent on impingement angle, with improved erosion resistance for the uncoated steels than the TiN-coated steels found at a higher angle.

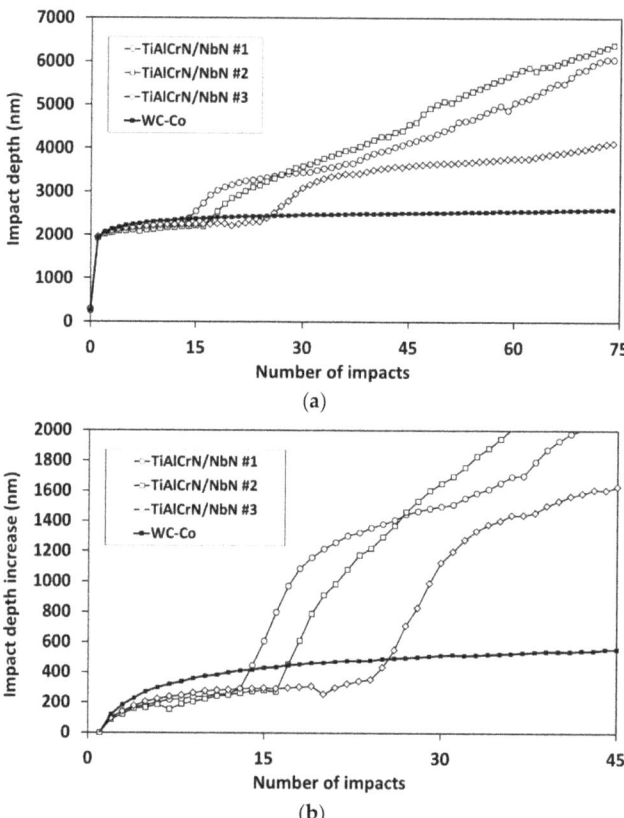

Figure 10. Micro-impact test data on TiAlCrN/NbN coated WC-Co (three repeat tests shown) and uncoated WC-Co: (**a**) impact depth vs. number of impacts, (**b**) increase in depth vs. number of impacts, for the first 45 impacts.

Beake and co-workers compared the micro-impact behaviour of carbon coatings on hardened tool steel and nitrides on cemented carbide substrates tested under the same conditions [37]. These authors showed that as the load in the micro-impact test increased, there was an increasing contribution of the properties of the substrate (specifically, its load support—influencing coating bending, and its ductility—influencing damage tolerance) to the coating system response whilst retaining high sensitivity to the coating properties. On cemented carbide substrates, coatings with higher H^3/E^2 performed well, although it was not possible to avoid lateral fracture at higher load. On hardened tool steel with its lower load support, the carbon coatings were subjected to higher bending strains. Under these severe conditions, carbon coatings with high H/E were too brittle and susceptible to extensive lateral fracture, but carbon-based coatings with more moderate hardness and relatively low H/E on hardened tool steel were more resistant to radial cracking and lateral fracture [33,35–37]. The damage tolerance of the coating systems at higher load was helped by the greater ductility of the hardened tool steel substrate.

With much softer substrates than hardened steel or cemented carbide, the reduced load support commonly results in the coating being broken through during the first few impacts of a nano- or micro-scale impact test. The abrupt depth step characteristic of brittle fracture is absent, and the depth vs. number of impacts behaviour is closer to that of a ductile material. Shi and co-workers reported the failure of graphite-like carbon coatings on stainless steel under the impact crater [82]. The coating is unable to accommodate the plastic deformation of the substrate and tensile stresses develop that result in cracking with repetitive contact. Yonezu and co-workers [130] reported that spiral cracks form within the indent crater for a 12 mm DLC on a 304 stainless steel. Mendibide and co-workers [131] purposely did not subject the steel substrate to a thermal anneal before cyclic fatigue testing with a R = 300 µm indenter. The coating was therefore subjected to greater strain. A total of 4000 impacts of 0.8 mJ energy were needed before delamination of multilayered TiN/CrN, which was much larger than the number of impacts needed on either monolayered TiN or CrN coatings (500 or 100 impacts, respectively). In macro-impact tests with 1.25 and 2.5 mm radius probes, Bouzakis [4] stated that superficial thin coating layers did not influence the failure initiation of the underlying coatings, i.e., lower sensitivity to coatings due to the larger radius probes.

7. Load Carrying Capacity, H^3/E^2

Contact mechanics shows that for a flat surface in elastic/plastic contact with a rigid ball of radius R, the yield pressure (P_y) is a function of H^3/E^2, as shown in Equation (2) [132]. At a given contact pressure, contact is more likely to be elastic for a surface with higher H^3/E^2. H^3/E^2 can be considered as a measure of the resistance to plastic deformation or the load-carrying capacity of a material.

$$P_y = 0.78\, R^2 (H^3/E^2) \qquad (2)$$

In a coated system, behaviour is more complex, since changing the radius of the test probe changes the location of initial yielding, with the subsequent fatigue behaviour being dependent on where this takes place. Although H^3/E^2 is still important, the microstructure and mechanical properties of a coating system are intimately linked so should not be considered in isolation.

Multilayered TiAlCrSiYN/TiAlCrN coatings have been developed showing adaptive behaviour and longer life when deposited on cemented carbide tools in high-speed machining. In nano-impact tests, multilayered TiAlCrSiYN/TiAlCrN with higher H^3/E^2 showed better impact resistance than monolayered TiAlCrSiYN coatings with lower H^3/E^2 [16,17,20]. In contrast, for hard carbon coatings on hardened tool steel, coatings with higher H^3/E^2 did not show enhanced impact resistance. Bousser and co-workers [133] observed that in Vickers indentation of 8–13 µm CrN and CrSiN coatings on stainless steel, the ratio of indentation depth to film thickness at which circular cracking occurred was *inversely* correlated with coating H^3/E^2. This implies it is the inability of the coating to accommodate the deformation of the soft steel substrate that drives the cracking process under these highly loaded conditions [133].

Multilayer coating design has also proved effective on other substrates providing less load support than WC-Co in tribological and impact tests. Chen and co-workers studied the response of multilayer TiAlSiN and monolayer TiN coatings on hardened tool steel in nano-impact tests [75]. Greater repetitive impact load was required for chipping in the multilayered TiAlSiN. This was due to a combination of microstructural advantage (less columnar with multilayer structure to aid crack deflection) and mechanical (higher H^3/E^2) advantage in comparison to the monolayered columnar TiN. On a non-hardened tool steel substrate, a multilayered TiN/CrN showed a much larger number of impacts to failure than monolayered TiN or CrN coatings [131]. Enhanced crack resistance of TiN/CrN on hardened tool steel was also reported by Roa and co-workers in indentation tests [123] through the interlayers restricting intergranular shear sliding.

The impact resistance of TiFeN and TiFeMoN coatings on silicon has been investigated [44,134]. It was found that increasing coating H^3/E^2 improved its resistance to single impact. At lower impact load, the ratio of impact depth to film thickness was low, and higher H^3/E^2 was able to prevent crack formation. However, under repetitive contact at high load, it was not possible to prevent substrate yield and fracture, and there was no benefit in increased H^3/E^2. In the absence of a toughening mechanism such as crack deflection from a multilayer microstructure, very high H^3/E^2 does not ensure impact resistance at high load when the coating is unable to protect the substrate from deforming. Durability and damage tolerance under cyclic loading requires resistance to both crack initiation and propagation.

8. Erosion Simulation

Shipway and Hutchings [135] have reviewed the use of erosion testing to evaluate coating durability, highlighting different methods employed to assess wear rate (erosion scar width, depth, mass loss). More recently, nano-impact tests have been used to assess coating durability under repetitive contact and simulate erosion testing [34,43]. Chen and co-workers have reported that the influence of thermal ageing on the solid particle erosion testing of columnar EB-PVD TBCs for aero-engines correlated with rapid nano-impact tests. The similar contact footprint in both types of test was highlighted [43]. Zhang and co-workers [57] have studied the influence of modulation period in 10 μm thick TiN/Ti multilayer coatings on Ti6Al4V on the damage mechanism in nano-impact tests using a well-worn Berkovich indenter as the impact probe. Interestingly, these authors found that when the period of modulation was reduced from micro (1000 nm) to nano- (60 nm), the impact resistance decreased. The well-defined interfaces and thicker Ti layers present in the coating with 1000 nm modulation period were able to effectively restrict lateral cracking. A clear correlation between DLC coating performance in nano- and micro-impact tests and resistance to sand erosion was reported by McMaster and co-workers [34].

9. Modelling Nano- and Micro-Impact

Bouzakis and co-workers have used FEA to model wear of TiAlN coatings in repetitive nano-impact tests with cube corner indenters [136–138]. They developed 3D-FEA and 2D axis-symmetric FEA models using ANSYS LS-DYNA software to simulate the damage progression. With the sharp cube corner probe geometry, the impact-induced stresses were more localized than with blunter probes; hence, failure initially proceeded by damage evolution under the indenter rather than cracking at the top surface at the contact periphery more commonly observed when spherical probes are used in micro-impact. The small contact size with the sharper cube corner probe enabled 3D-FEA simulation of the progressive damage.

Feng and co-workers [139] performed a numerical study of the fatigue behaviour of a model coating system composed of (i) TiN coating, (ii) a case-hardened diffusion zone with graded mechanical properties and (iii) H11 steel substrate under cyclic loading by a R = 300 μm indenter. Crack initiation and propagation under cyclic loading was simulated with an irreversible cohesive zone model, which enabled local degradation of the material properties with the increasing cycles to be incorporated into the model by a damage variable. A crack formed at the edge of the contact area between the indenter and coated surface during the first few loading cycles, which under further cyclic loading gradually progressed through the coating. The bending stress at the edge of contact area, caused by the plastic deformation of hardened case, influenced the crack initiation. A compressive stress, due to increasing indentation contact pressure during reloading, forced crack closure. Subsequent unloading released this compressive stress, causing crack reopening. The study indicated that the irreversible cohesive zone model could track crack propagation under cyclic loading; therefore, it has potential to predict the load-bearing capacity of coating systems under contact fatigue loading.

10. Elevated Temperature Impact Testing

Changes to coating and substrate mechanical properties at elevated temperature alter the location of and magnitude of the developed stresses in contact and the subsequent dominant mechanism. Temperature-dependent changes in deformation influence the damage tolerance, with plastic deformation generally prevailing over brittle fracture as a major damage mode. With this greater plasticity and reduced brittleness, it becomes more difficult to drive brittle fracture within the short test duration in a nano- or micro-impact test.

In high-speed machining applications, high stresses and temperatures are generated in the contact zone. In continuous high-speed cutting applications (e.g., turning), the temperatures can reach 700–1000 °C due to frictional heating [140,141]. In interrupted cutting of difficult-to-cut materials such as Ti6Al4V, they can be significantly lower, and at lower cutting speed, it may be well below 400 °C [105]. The correlation between coating performance in room-temperature impact tests and tool life commonly observed in practice implies that either coating brittleness is significantly more important than the hot hardness/load support in the application (i.e., cracking is life-limiting) or that relative differences in coating mechanical properties do not change as the temperature increases (i.e., coatings with higher H, H^3/E^2 at room temperature also maintain these advantages at elevated temperature), which may, at least in part, be a consequence of the relatively lower temperatures in milling operations compared to continuous turning.

Bouzakis and co-workers have developed a high-temperature macro-impact test capability utilising compressed inert gas heating to 600 °C [142–144]. These authors showed that coating failure required more load at higher temperatures, with the effect being particularly strong around 200 °C [143]. Performing nano- and micro-scale impact tests at elevated temperature is an alternative approach to simulate the high contact stresses and temperatures generated in high-speed interrupted contacts in milling.

Nano-impact tests up to 500 °C have been performed on 3 µm $Ti_{0.5}Al_{0.5}N$ and $Al_{0.67}Ti_{0.33}N$ deposited on H10A cemented carbide (6 wt.% Co) [13] and micro-impact tests up to 600 °C on 2 µm PVD TiAlSiN and nanomultilayered TiAlN/TiSiN coatings on P30 cemented carbide (10 wt.% Co) [32]. Alongside these tests, nanoindentation and micro-scratch tests were also performed over the same temperature range to understand how the coating properties change with increasing temperature.

In the nano-impact tests at 500 °C, there was lower susceptibility to coating fracture than at room temperature for both coatings, which is consistent with the significant softening shown in nanoindentation tests [13]. Due to the reduction in coating hardness, the impact stresses in the high-temperature tests were lower, which resulted in reduced fracture. $Al_{0.67}Ti_{0.33}N$ showed improved resistance to fracture in the elevated temperature nano-impact test and longer tool life than $Ti_{0.5}Al_{0.5}N$ in machining steels and Ti6Al4V [13,145]. Micro-impact tests at 25 °C on TiAlSiN and TiAlN/TiSiN displayed a brittle response with fatigue and a transition to more rapid wear after fracture [32]. There was a change in the dominant fatigue mechanism from fracture-dominated to more plasticity-dominated deformation at higher temperatures. Nanoindentation and micro-scratch tests at 600 °C indicated this was related to significant substrate—rather than coating—softening. In these nano- and micro-impact test studies [13,32], increasing temperature reduced the mechanical properties of the coating system, which resulted in reduced fracture in the impact tests. In the case of the $Ti_{0.5}Al_{0.5}N$ and $Al_{0.67}Ti_{0.33}N$ coatings deposited on a cemented carbide substrate with low Co fraction, this reduction in high-temperature hardness was primarily through coating softening. In contrast, for the TiAlSiN coatings, the reduction in coating properties was less severe, but there was more substrate softening due to the higher Co fraction in the cemented carbide used.

11. Outlook/Conclusions

The challenge in developing laboratory test methods for validating coating performance in industrial applications is to devise experimental tests that can simplify the complex contact conditions whilst retaining sufficient key features so they are practically useful [146–148]. High strain rate nano- and micro-impact tests can effectively overcome the limitations of quasi-static nanomechanical testing or impact/cyclic indentation testing at larger scale. A strong correlation between coating performance in the nano- and micro-impact tests and interrupted contact situations such as erosion or metal cutting has been reported in many studies. In this review, case studies on (i) ultra-thin hard carbon films on silicon, (ii) DLC on hardened tool steel and (iii) nitrides on WC-Co have been used to show how substrate load support and coating thickness influence how coating mechanical properties affect impact resistance. For nitrides on WC-Co, impact resistance can be enhanced by coating design with a combination of optimised mechanical properties and microstructure (e.g., high H^3/E^2 and multilayer structure). For carbon coatings on hardened tool steel, there is higher coating strain, and impact resistance is significantly worse for a-C:H coatings with high H^3/E^2 than softer a-C or WC/C coatings with lower H^3/E^2. For ultra-thin coatings on Si, phase transformation processes in the silicon substrate can be important. Substrate ductility and load support influence impact resistance in coated systems. Although stress fields generated in lower load nano-impacts with cube corner indenters or higher load micro-impacts with spherical probes are different in both cases, substrate properties cannot be ignored. Interestingly, for the coating systems studied so far, the relative ranking in nano- and micro-impact tests has been the same; i.e., a coating that performs well in nano-impact also performs well in micro-impact.

Both nano- and micro-impact tests have shown clear correlation with in-service performance and therefore have significant potential to be used standard tools in coating screening and optimization campaigns. Compared to macro-impact tests, they have many potential benefits including the ability to automatically test a large number of coatings in a short space of time. Statistical data are conveniently obtained through performing a large number of small contact size rapid repeat tests at different locations on the same sample.

Active research directions for future development of the nano- and micro-impact test techniques fall into two categories: (i) extending instrumentation capabilities and (ii) post-test characterization for more detailed analysis of the deformation mechanisms. These include modifications of the test setup to more closely simulate specific repetitive contact scenarios such as high-temperature erosion by increasing maximum test temperatures (e.g., to 900 °C) and performing angled impacts. Rueda-Ruiz and co-workers have recently shown [27,28] that the addition of an integrated load cell to enable the direct measurement of impact forces is a more effective approach in measuring high strain rate hardness in single impacts, and this will be extended to the study of repetitive impacts. Sub-surface damage can be assessed by cross-sectional FIB analysis [54]. FIB cross-sections of impact tests stopped after different numbers of cycles would be able to show locations of initial impact damage and crack propagation in detail. Such analysis could be supplemented with the modelling of repetitive impact, e.g., by adapting the approach in [139] to micro-scale impact tests with smaller radii probes to improve our fundamental understanding of the interrelationships between coating microstructure and mechanical properties and impact wear resistance.

Funding: The support of Innovate UK through project #100200751 is acknowledged.

Institutional Review Board Statement: Not applicable.

Informed Consent Statement: Not applicable.

Data Availability Statement: Not applicable.

Acknowledgments: The author would like to acknowledge the contributions of his colleagues at Micro Materials Ltd.: Jim Smith, Steve Goodes, Nick Pickford and Adrian Harris, and our many long-time research collaborators on nano- and micro-impact testing, in developing this test method and our understanding and the application of it in producing more impact-resistant coatings. A special thanks to Tomasz Liskiewicz (Manchester Metropolitan University, UK), Vlad Vishnyakov (University of Huddersfield, UK), German Fox-Rabinovich and Stephen Veldhuis (McMaster University, Canada), Jose Endrino (Nano4Energy, Spain), Luis Isern (Cranfield University, UK), Sam McMaster (Coventry University, UK), Radim Ctvrtlik and Jan Tomastik (University of Palacky, Czech Republic), Jian Chen (Southeast University, China), Xiangru Shi (Hohai University, China), Mario Rueda-Ruiz and Jon Molina (IMDEA, Spain).

Conflicts of Interest: The author declares no conflict of interest.

References

1. Knotek, O.; Bosserhoff, B.; Schrey, A.; Leyendecker, T.; Lemmer, O.; Esser, S. A new technique for testing the impact load of thin films: The coating impact test. *Surf. Coat. Technol.* **1992**, *54–55*, 102–107. [CrossRef]
2. Lawes, S.; Hainsworth, S.; Fitzpatrick, M. Impact wear testing of diamond-like carbon films for engine valve-tappet surfaces. *Wear* **2010**, *268*, 1303–1308. [CrossRef]
3. Bouzakis, K.-D.; Flocke, F.; Skordaris, G.; Bouzakis, E.; Geradis, S.; Katirtzoglou, G.; Makrimallakis, S. Influence of dry microblasting grain quality on wear behavior of TiAlN coated tools. *Wear* **2011**, *271*, 783–791. [CrossRef]
4. Bouzakis, K.-D.; Siganos, A.; Leyendecker, T.; Erkens, G. Thin hard coatings fracture propagation during the impact test. *Thin Solid Films* **2004**, *460*, 181–189. [CrossRef]
5. Bantle, R.; Matthews, A. Investigation into the impact wear behavior of ceramic coatings. *Surf. Coat. Technol.* **1995**, *74–75*, 857–868. [CrossRef]
6. Yoon, S.Y.; Yoon, S.-Y.; Chung, W.-S.; Kim, K.H. Impact-wear behaviors of TiN and Ti-Al-N coatings on AISI D2 steel and WC-Co substrates. *Surf. Coat. Technol.* **2004**, *177–178*, 645–650. [CrossRef]
7. Ramírez, G.; Mestra, A.; Casas, B.; Valls, I.; Martínez, R.; Bueno, R.; Góez, A.; Mateo, A.; Llanes, L. Influence of substrate microstructure on the contact fatigue strength of coated cold-work tool steels. *Surf. Coat. Technol.* **2012**, *206*, 3069–3081. [CrossRef]
8. Kim, D.K.; Jung, Y.-G.; Peterson, I.; Lawn, B. Cyclic fatigue of intrinsically brittle ceramics in contact with spheres. *Acta Mater.* **1999**, *47*, 4711–4725. [CrossRef]
9. Michler, J.; Blank, E. Analysis of coating fracture and substrate plasticity induced by spherical indentors: Diamond and diamond-like carbon layers on steel substrates. *Thin Solid Films* **2001**, *381*, 119–134. [CrossRef]
10. Bernoulli, D.; Wyss, A.; Raghavan, R.; Thorwarth, K.; Hauert, R.; Spolenak, R. Contact damage of hard and brittle thin films on ductile metallic substrates: An analysis of diamond-like carbon on titanium substrates. *J. Mater. Sci.* **2015**, *50*, 2779–2787. [CrossRef]
11. Bromark, M.; Hedenqvist, P.; Hogmark, S. The influence of substrate material on the erosion resistance of TiN coated steels. *Wear* **1995**, *186–187*, 189–194. [CrossRef]
12. Fox-Rabinovich, G.S.; Beake, B.D.; Veldhuis, S.C.; Endrino, J.L.; Parkinson, R.; Shuster, L.S.; Migranov, M.S. Impact of mechanical properties measured at room and elevated temperatures on wear resistance of cutting tools with TiAlN and AlCrN coatings. *Surf. Coat. Technol.* **2006**, *200*, 5738–5742. [CrossRef]
13. Beake, B.D.; Smith, J.F.; Gray, A.; Fox-Rabinovich, G.S.; Veldhuis, S.C.; Endrino, J.L. Investigating the correlation between nano-impact fracture resistance and hardness/modulus ratio from nanoindentation at 25–500 °C and the fracture resistance and lifetime of cutting tools with $Ti_{1-x}Al_xN$ (x = 0.5 and 0.67) PVD coatings in milling operations. *Surf. Coat. Technol.* **2007**, *201*, 4585.
14. Fox-Rabinovich, G.S.; Kovalev, A.I.; Aguirre, M.H.; Beake, B.D.; Yamamoto, K.; Veldhuis, S.C.; Endrino, J.L.; Wainstein, D.L.; Rashkovskiy, A.Y. Design and performance of AlTiN and TiAlCrN PVD coatings for machining of hard to cut materials. *Surf. Coat. Technol.* **2009**, *204*, 489–496. [CrossRef]
15. Beake, B.; Fox-Rabinovich, G.; Veldhuis, S.; Goodes, S. Coating optimisation for high-speed machining with advanced nanomechanical test methods. *Surf. Coat. Technol.* **2009**, *203*, 1919–1925. [CrossRef]
16. Fox-Rabinovich, G.S.; Beake, B.D.; Yamamoto, K.; Aguirre, M.H.; Veldhuis, S.C.; Dosbaeva, G.; Elfizy, A.; Biksa, A.; Shuster, L.S.; Rashkovskiy, A.Y. Structure, properties and wear performance of nano-multilayered TiAlCrSiYN/TiAlCrN coatings during machining of Ni-based aerospace superalloys. *Surf. Coat. Technol.* **2010**, *204*, 3698–3706. [CrossRef]
17. Fox-Rabinovich, G.; Yamamoto, K.; Beake, B.; Kovalev, A.; Aguirre, M.H.; Veldhuis, S.; Dosbaeva, G.; Wainstein, D.; Biksa, A.; Rashkovskiy, A. Emergent behavior of nano-multilayered coatings during dry high speed machining of hardened tool steels. *Surf. Coat. Technol.* **2010**, *204*, 3425–3435. [CrossRef]
18. Bouzakis, K.-D.; Michailidis, N.; Skordaris, G.; Bouzakis, E.; Biermann, D.; M'Saoubi, R. Cutting with coated tools: Coating technologies, characterization methods and performance optimization. *CRIP Ann. Manuf. Technol.* **2012**, *61*, 703–723. [CrossRef]
19. Skordaris, G.; Bouzakis, K.; Charalampous, P.; Bouzakis, E.; Paraskevopoulou, R.; Lemmer, O.; Bolz, S. Brittleness and fatigue effect of mono- and multi-layer PVD films on the cutting performance of coated cemented carbide inserts. *CIRP Ann. Manuf. Technol.* **2014**, *63*, 93. [CrossRef]
20. Beake, B.; Fox-Rabinovich, G. Progress in high temperature nanomechanical testing of coatings for optimising their performance in high speed machining. *Surf. Coat. Technol.* **2014**, *255*, 1021115. [CrossRef]
21. Skordaris, G.; Bouzakis, K.D.; Kotsanis, T.; Charalampous, P.; Bouzakis, E.; Breidenstein, B.; Bergmann, B.; Denkena, B. Effect of PVD film's residual stress on their mechanical properties, brittleness, adhesion and cutting performance of coated tools. *CIRP J. Manuf. Sci. Technol.* **2017**, *18*, 145–151. [CrossRef]
22. Bouzakis, K.D.; Skordaris, G.; Bouzakis, E.; Kotsanis, T.; Charalampous, P. A critical review of characteristic techniques for improving the cutting performance of coated tools. *J. Mach. Eng.* **2017**, *17*, 25–44.
23. Chowdhury, S.; Bose, B.; Yamamoto, K.; Veldhuis, S.C. Effect of interlayer thickness on nano-multilayer coating performance during high speed dry milling of H13 tool steel. *Coatings* **2019**, *9*, 737. [CrossRef]
24. Trelewicz, J.R.; Schuh, C.A. The Hall–Petch breakdown at high strain rates: Optimizing nanocrystalline grain size for impact applications. *Appl. Phys. Lett.* **2008**, *93*, 171916. [CrossRef]
25. Somekawa, H.; Schuh, C.A. High-strain-rate nanoindentation behavior of fine-grained magnesium alloys. *J. Mater. Res.* **2012**, *27*, 1295–1302. [CrossRef]

26. Wheeler, J.M. Nanoindentation under Dynamic Conditions. Ph.D. Thesis, University of Cambridge, Cambridge, UK, 2009.
27. Rueda-Ruiz, M.; Beake, B.D.; Molina-Aldareguia, J.M. New instrumentation and analysis methodology for nano-impact testing. *Mater. Des.* **2020**, *192*, 108715. [CrossRef]
28. Rueda-Ruiz, M.; Beake, B.D.; Molina-Aldareguia, J.M. Determination of rate dependent properties in cohesive frictional materials by instrumented indentation. *JOM* **2022**, *74*, 2206–2219. [CrossRef]
29. Qin, L.; Li, H.; Shi, X.; Beake, B.D.; Xiao, L.; Smith, J.F.; Sun, Z.; Chen, J. Investigation on dynamic hardness and high strain rate indentation size effects in aluminium (110) using nano-impact. *Mech. Mater.* **2019**, *133*, 55–62. [CrossRef]
30. Beake, B.D.; Goodes, S.R.; Smith, J.F. Micro-impact testing: A new technique for evaluating fracture toughness, adhesion, erosive wear resistance and dynamic hardness. *Surf. Eng.* **2001**, *17*, 187. [CrossRef]
31. Beake, B.D.; Isern, L.; Endrino, J.L.; Fox-Rabinovich, G.S. Micro-impact testing of AlTiN and TiAlCrN coatings. *Wear* **2019**, *418–419*, 102–110. [CrossRef]
32. Beake, B.D.; Bird, A.; Isern, L.; Endrino, J.; Jiang, F. Elevated temperature micro-impact testing of TiAlSiN coatings produced by physical vapour deposition. *Thin Solid Films* **2019**, *688*, 137358. [CrossRef]
33. Beake, B.D.; Liskiewicz, T.W.; Bird, A.; Shi, X. Micro-scale impact testing—A new approach to studying fatigue resistance in hard carbon coatings. *Tribol. Int.* **2020**, *149*, 105732. [CrossRef]
34. McMaster, S.J.; Liskiewicz, T.W.; Neville, A.; Beake, B.D. Probing fatigue resistance in multi-layer DLC coatings by micro- and nano-impact: Correlation to erosion tests. *Surf. Coat. Technol.* **2020**, *402*, 126319. [CrossRef]
35. Beake, B.D.; McMaster, S.J.; Liskiewicz, T.W.; Neville, A. Influence of Si- and W- doping on micro-scale reciprocating wear and impact performance of DLC coatings on hardened steel. *Tribol. Int.* **2021**, *160*, 107063. [CrossRef]
36. Beake, B.; Bergdoll, L.; Isern, L.; Endrino, J.; Fox-Rabinovich, G.; Veldhuis, S. Influence of probe geometry in micro-scale impact testing of nano-multilayered TiAlCrN/NbN coatings deposited on WC-Co. *Int. J. Refract. Met. Hard Mater.* **2021**, *95*, 105441. [CrossRef]
37. Beake, B.; Isern, L.; Endrino, J.; Liskiewicz, T.; Shi, X. Micro-scale impact resistance of coatings on hardened tool steel and cemented carbide. *Mater. Lett.* **2021**, *284*, 129009. [CrossRef]
38. Bouzakis, K.; Maliaris, G.; Makrimallakis, S. Strain rate effect on the fatigue failure of thin PVD coatings: An investigation by novel impact tester with adjustable repetitive force. *Int. J. Fatigue* **2012**, *44*, 89–97. [CrossRef]
39. Wheeler, J.; Dean, J.; Clyne, T. Nano-impact indentation for high strain rate testing: The influence of rebound impacts. *Extrem. Mech. Lett.* **2019**, *26*, 35–39. [CrossRef]
40. Constantinides, G.; Tweedie, C.A.; Holbrook, D.M.; Barragan, P.; Smith, J.F.; Van Vliet, K.J. Quantifying deformation and energy dissipation of polymeric surfaces under localized impact. *Mater. Sci. Eng. A* **2008**, *489*, 403–412. [CrossRef]
41. Constantinides, G.; Tweedie, C.A.; Savva, N.; Smith, J.F.; Van Vliet, K.J. Quantitative impact testing of energy dissipation at surfaces. *Exp. Mech.* **2009**, *49*, 511–522. [CrossRef]
42. Zehnder, C.; Peltzer, J.-N.; Gibson, J.S.-L.; Korte-Kerzel, S. High strain rate testing at the nano-scale: A proposed methodology for impact nanoindentation. *Mater. Des.* **2018**, *151*, 17–28. [CrossRef]
43. Chen, J.; Beake, B.D.; Wellman, R.G.; Nicholls, J.R.; Dong, H. An investigation into the correlation between nano-impact resistance and erosion performance of EB-PVD thermal barrier coatings on thermal ageing. *Surf. Coat. Technol.* **2012**, *206*, 4992–4998. [CrossRef]
44. Zhang, X.; Zhang, S.; Beake, B.D. Toughness Evaluation of Thin Hard Coatings and Films. In *Thin Films and Coatings: Toughening and Toughness Characterisation*; Zhang, S., Ed.; CRC Press: Boca Raton, FL, USA, 2015; Chapter 2; pp. 48–121.
45. Arreguin-Zavala, J.; Milligan, J.; Davies, M.I.; Goodes, S.R.; Brochu, M. Characterization of nanostructured and ultrafine-grain aluminum-silicon claddings using the nanoimpact indentation technique. *JOM* **2013**, *65*, 763–768. [CrossRef]
46. Prakash, C.; Gunduz, I.E.; Oskay, C.; Tomar, V. Effect of interface chemistry and strain rate on particle-matrix delamination in an energetic material. *Eng. Fract. Mech.* **2018**, *191*, 46–64. [CrossRef]
47. Huang, W.; Shishehbor, M.; Guarín-Zapata, N.; Kirchhofer, N.D.; Li, J.; Cruz, L.; Wang, T.; Bhowmick, S.; Stauffer, D.; Manimunda, P.; et al. A natural impact-resistant bicontinuous composite nanoparticle coating. *Nat. Mater.* **2020**, *19*, 1236–1243. [CrossRef] [PubMed]
48. Cinca, N.; Beake, B.D.; Harris, A.J.; Tarrés, E. Micro-scale impact testing on cemented carbide. *Int. J. Refract. Met. Hard Mater.* **2019**, *84*, 105045. [CrossRef]
49. Beake, B.; Isern, L.; Harris, A.; Endrino, J. Probe geometry and surface roughness effects in microscale impact testing of WC-Co. *Mater. Manuf. Proc.* **2020**, *35*, 836–844. [CrossRef]
50. Jennett, N.M.; Nunn, J. High resolution measurement of dynamic (nano) indentation impact energy: A step towards the determination of indentation fracture resistance. *Philos. Mag.* **2011**, *91*, 1200–1220. [CrossRef]
51. Wheeler, J.; Gunner, A. Analysis of failure modes under nano-impact fatigue of coatings via high-speed sampling. *Surf. Coat. Technol.* **2013**, *232*, 264–268. [CrossRef]
52. Shi, X.; Li, H.; Beake, B.D.; Bao, M.; Liskiewicz, T.W.; Sun, Z.; Chen, J. Dynamic fracture of CrN coating by highly-resolved nano-impact. *Surf. Coat. Technol.* **2020**, *383*, 125288. [CrossRef]
53. Ctvrtlik, R.; Tomastik, J.; Vaclavek, L.; Beake, B.D.; Harris, A.J.; Martin, A.S.; Hanak, M.; Abrham, P. High-resolution acoustic emission monitoring in nanomechanics. *JOM* **2019**, *71*, 3358–3367. [CrossRef]

54. Beake, B.D.; Ctvrtlik, R.; Harris, A.J.; Martin, A.S.; Vaclavek, L.; Manak, J.; Ranc, V. High frequency acoustic emission monitoring in nano-impact of alumina and partially stabilised zirconia. *Mater. Sci. Eng. A* **2020**, *780*, 139159. [CrossRef]
55. Faisal, N.; Ahmed, R.; Goel, S.; Fu, Y. Influence of test methodology and probe geometry on nanoscale fatigue failure of diamond-like carbon film. *Surf. Coat. Technol.* **2014**, *242*, 42–53. [CrossRef]
56. Frutos, E.; González-Carrasco, J.L.; Polcar, T. Repetitive nano-impact tests as a new tool to measure fracture toughness in brittle materials. *J. Eur. Ceram. Soc.* **2016**, *36*, 3235–3243. [CrossRef]
57. Zhang, H.; Li, Z.; He, W.; Ma, C.; Chen, J.; Liao, B.; Li, Y. Damage mechanisms evolution of TiN/Ti multilayer films with different modulation periods in cyclic impact conditions. *Appl. Surf. Sci.* **2021**, *540*, 148366. [CrossRef]
58. Shi, X.; Beake, B.D.; Liskiewicz, T.W.; Chen, J.; Sun, Z. Failure mechanism and protective role of ultrathin ta-C films on Si (100) during cyclic nano-impact. *Surf. Coat. Technol.* **2019**, *364*, 32–42. [CrossRef]
59. Beake, B.D.; Isern, L.; Bhattacharyya, D.; Endrino, J.L.; Lawson, K.; Walker, T. Nano- and micro-scale impact testing of zirconia, alumina and zirconia-alumina duplex optical coatings on glass. *Wear* **2020**, *462–463*, 203499. [CrossRef]
60. Cassar, G.; Banfield, S.; Wilson, J.A.-B.; Housden, J.; Matthews, A.; Leyland, A. Impact wear resistance of plasma diffusion treated and duplex treated/PVD-coated Ti–6Al–4V alloy. *Surf. Coat. Technol.* **2012**, *206*, 2645–2654. [CrossRef]
61. Zha, X.; Jiang, F.; Xu, X. Investigating the high frequency fatigue failure mechanisms of mono and multilayer PVD coatings by the cyclic impact tests. *Surf. Coat. Technol.* **2018**, *344*, 689–701. [CrossRef]
62. Antonov, M.; Hussainova, I.; Sergejev, F.; Kulu, P.; Gregor, A. Assessment of gradient and nanogradient PVD coatings under erosive, abrasive and impact wear conditions. *Wear* **2009**, *267*, 898–906. [CrossRef]
63. Lamri, S.; Langlade, C.; Kermouche, G. Damage phenomena of thin hard coatings submitted to repeated impacts: Influence of the substrate and film properties. *Mater. Sci. Eng. A* **2013**, *560*, 296–305. [CrossRef]
64. Batista, J.C.A.; Godoy, C.; Matthews, A. Impact testing of duplex and non-duplex (Ti,Al)N and Cr–N PVD coatings. *Surf. Coat. Technol.* **2003**, *163–164*, 353–361. [CrossRef]
65. Mo, J.L.; Zhu, M.H.; Leyland, A.; Matthews, A. Impact wear and abrasion resistance of CrN, AlCrN and AlTiN PVD coatings. *Surf. Coat. Technol.* **2013**, *215*, 170–177. [CrossRef]
66. Treutler, C.P.O. Industrial use of plasma-deposited coatings for components of automotive fuel injection systems. *Surf. Coat. Technol.* **2005**, *200*, 1969–1975. [CrossRef]
67. Bao, M.D.; Zhu, X.D.; He, J.W. Evaluation of the toughness of hard coatings. *Surf. Eng.* **2006**, *22*, 11–14. [CrossRef]
68. Zhu, X.; Dou, H.; Ban, Z.; Liu, Y.; He, J. Repeated impact test for characterisation of hard coatings. *Surf. Coat. Technol.* **2007**, *201*, 5493–5497. [CrossRef]
69. Ledrappier, F.; Langlade, C.; Vannes, A.-B.; Gachon, Y. Damage phenomena observed on PVD coatings subjected to repeated impact tests. *Plasma Process. Polym.* **2007**, *4*, 5835–5839. [CrossRef]
70. Ledrappier, F.; Langlade, C.; Gachon, Y.; Vannes, A.-B. Blistering and spalling of thin hard coatings submitted to repeated impacts. *Surf. Coat. Technol.* **2008**, *202*, 1789–1796. [CrossRef]
71. Qiu, L.; Zhu, X.; Lu, S.; He, G.; Xu, K. Quantitative evaluation of bonding strength for hard coatings by interfacial fatigue strength under cyclic indentation. *Surf. Coat. Technol.* **2017**, *315*, 303–313. [CrossRef]
72. Qiu, L.; Zhu, X.; He, G.; Xu, K. The repeated spherical indentation test: An efficient way to evaluate the adhesion of hard coatings. *Surf. Eng.* **2016**, *32*, 578–584. [CrossRef]
73. Kalidindi, S.; Pathak, S. Determination of the effective zero-point and the extraction of spherical nanoindentation stress–strain curves. *Acta Mater.* **2008**, *56*, 3523–3532. [CrossRef]
74. Pathak, S.; Kalidindi, S.R. Spherical nanoindentation stress–strain curves. *Mater. Sci. Eng. R* **2015**, *91*, 1–36. [CrossRef]
75. Chen, J.; Ji, R.; Khan, R.H.U.; Li, X.; Beake, B.D.; Dong, H. Effects of mechanical properties and layer structure on the cyclic loading of TiN-based coatings. *Surf. Coat. Technol.* **2011**, *206*, 522–529. [CrossRef]
76. Beake, B.; Goodes, S.; Smith, J.; Madani, R.; Rego, C.; Cherry, R.; Wagner, T. Investigating the fracture resistance and adhesion of DLC films with micro-impact testing, Diam. *Relat. Mater.* **2002**, *11*, 1606. [CrossRef]
77. Tarrés, E.; Ramírez, G.; Gaillard, Y.; Jiménez-Piqué, E.; Llanes, L. Contact fatigue behavior of PVD-coated hardmetals. *Int. J. Refract. Met. Hard Mater.* **2009**, *27*, 323–341. [CrossRef]
78. Sivitski, A.; Gregor, A.; Saarna, M.; Kulu, P.; Sergejev, F. Application of the indentation method for cracking resistance evaluation of hard coatings on tool steels. *Est. J. Eng.* **2009**, *15*, 309–317. [CrossRef]
79. Abudaia, F.; Evans, J.; Shaw, B. Spherical indentation fatigue cracking. *Mater. Sci. Eng. A* **2005**, *391*, 181–187. [CrossRef]
80. Ma, L.; Cairney, J.; Hoffman, M.; Munroe, P. Deformation and fracture of TiN and TiAlN coatings on a steel substrate during nanoindentation. *Surf. Coat. Technol.* **2006**, *200*, 3518–3526. [CrossRef]
81. Cairney, J.; Tsukano, R.; Hoffman, M.; Yang, M. Degradation of TiN coatings under cyclic loading. *Acta Mater.* **2004**, *52*, 3229–3237. [CrossRef]
82. Shi, X.; Chen, J.; Beake, B.D.; Liskiewicz, T.W.; Wang, Z. Dynamic contact behavior of graphite-like carbon films on ductile substrate under nano/micro-scale impact. *Surf. Coat. Technol.* **2021**, *422*, 127515. [CrossRef]
83. Abdollah, M.F.B.; Yamaguchi, Y.; Akao, T.; Inayoshi, N.; Miyamoto, N.; Tokoroyama, T.; Umehara, N. Deformation-wear transition map of DLC coating under cyclic impact loading. *Wear* **2012**, *274–275*, 435–441. [CrossRef]
84. Cook, R.F. Strength and sharp contact fracture of silicon. *J. Mater. Sci.* **2006**, *41*, 841. [CrossRef]

85. Bhowmick, S.; Cha, H.; Jung, Y.-G.; Lawn, B.R. Fatigue and debris generation at indentation-induced cracks in silicon. *Acta Mater.* **2009**, *57*, 582–589. [CrossRef]
86. Domnich, V.; Gogotsi, Y. Phase transformations in silicon under contact loading. *Rev. Adv. Mater. Sci.* **2002**, *3*, 1–36.
87. Williams, J.A.; Le, H.R. Tribology and MEMS. *J. Phys. D Appl. Phys.* **2006**, *39*, R201–R214. [CrossRef]
88. Tanner, D.; Miller, W.; Eaton, W.; Irwin, L.; Peterson, K.; Dugger, M.; Senft, D.; Smith, N.; Tangyunyong, P.; Miller, S. The effect of frequency on the lifetime of a surface micromachined microengine driving a load. In Proceedings of the 1998 IEEE International Reliability Physics Symposium, Reno, NV, USA, 30 March–2 April 1998; pp. 26–35.
89. Ku, I.S.Y.; Reddyhoff, T.; Holmes, A.S.; Spikes, H.A. Spikes, wear of silicon surfaces in MEMS. *Wear* **2011**, *271*, 1050–1058. [CrossRef]
90. Peng, T.; You, Z. Reliability of MEMS in shock environments: 2000–2020. *Micromachines* **2021**, *12*, 1275. [CrossRef]
91. Beake, B.D.; Davies, M.I.; Liskiewicz, T.W.; Vishnyakov, V.M.; Goodes, S.R. Nano-scratch, nanoindentation and fretting tests of 5–80 nm ta-C films on Si (100). *Wear* **2013**, *301*, 575. [CrossRef]
92. Beake, B.D.; Lau, S.P.; Smith, J.F. Evaluating the fracture properties and fatigue wear of tetrahedral amorphous carbon films on silicon by nano-impact testing. *Surf. Coat. Technol.* **2004**, *177–178*, 611–615. [CrossRef]
93. Goel, S.; Agrawal, A.; Faisal, N.H. Can a carbon nano-coating resist metallic phase transformation in silicon substrate during nanoimpact? *Wear* **2014**, *315*, 38–41. [CrossRef]
94. Erdemir, A.; Donnet, C. Tribology of diamond-like carbon films: Recent progress and future prospects. *J. Phys. D Appl. Phys.* **2006**, *39*, R311–R327. [CrossRef]
95. Gåhlin, R.; Larsson, M.; Hedenqvist, P. Me-C:H coatings in motor vehicles. *Wear* **2001**, *249*, 302–309. [CrossRef]
96. Bernoulli, D.; Rico, A.; Wyss, A.; Thorwarth, K.; Best, J.P.; Hauert, R.; Spolenak, R. Improved contact damage resistance of hydrogenated diamond-like carbon (DLC) with a ductile α-Ta interlayer. *Diam. Relat. Mater.* **2015**, *58*, 78–83. [CrossRef]
97. Beake, B.D.; Smith, J.F. Nano-impact testing—Aan effective tool for assessing the resistance of advanced wear-resistant coatings to fatigue failure and delamination. *Surf. Coat. Technol.* **2004**, *188–189*, 594. [CrossRef]
98. Beake, B.D. Evaluation of the fracture resistance of DLC coatings on tool steel under dynamic loading. *Surf. Coat. Technol.* **2005**, *198*, 90. [CrossRef]
99. Camino, D.; Jones, A.; Mercs, D.; Teer, D. High performance sputtered carbon coatings for wear resistant applications. *Vacuum* **1999**, *52*, 125. [CrossRef]
100. Liskiewicz, T.W.; Beake, B.D.; Schwarzer, N.; Davies, M.I. Short note on improved integration of mechanical testing in predictive wear models. *Surf. Coat. Technol.* **2013**, *237*, 212–215. [CrossRef]
101. Mandal, P.; Beake, B.D.; Paul, S. Effect of deposition parameters on TiAlN coating using pulsed DC CFUBMS. *Surf. Eng.* **2013**, *29*, 287. [CrossRef]
102. Ramírez, G.; Jiménez-Piqué, E.; Mestra, A.; Vilaseca, M.; Casellas, D.; Llanes, L. A comparative study of the contact fatigue behavior and associated damage micromechanisms of TiN- and WC:H-coated cold-work tool steel. *Tribol. Int.* **2015**, *88*, 263–270. [CrossRef]
103. Inspektor, A.; Salvador, P.A. Architecture of PVD coatings for metalcutting applications: A review. *Surf. Coat. Technol.* **2014**, *257*, 138–153. [CrossRef]
104. Fox-Rabinovich, G.S.; Yamamoto, K.; Kovalev, A.I. Chapter 10—Synergistic alloying of self-adaptive wear-resistant coatings. In *Self-Organization during Friction: Advanced Surface Engineered Materials and Systems Design*; Fox-Rabinovich, G.S., Totten, G., Eds.; CRC Press LLC.: Boca Raton, FL, USA, 2007; pp. 297–334.
105. Michailidis, N. Variations in the cutting performance of PVD-coated tools in milling Ti6Al4V, explained through temperature-dependent coating properties. *Surf. Coat. Technol.* **2016**, *304*, 325–329. [CrossRef]
106. Beake, B.; Ning, L.; Gey, C.; Veldhuis, S.; Kornberg, A.; Weaver, A.; Khanna, M.; Fox-Rabinovich, G. Wear performance of different PVD coatings during wet end milling of H13 tool steel. *Surf. Coat. Technol.* **2015**, *279*, 118–125. [CrossRef]
107. Fox-Rabinovich, G.S.; Veldhuis, S.C.; Dosbaeva, G.K.; Yamamoto, K.; Kovalev, A.I.; Wainstein, D.L.; Gershman, I.S.; Shuster, L.S.; Beake, B.D. Nanocrystalline coating design for extreme applications based on the concept of complex adaptive behavior. *J. Appl. Phys.* **2008**, *103*, 083510. [CrossRef]
108. Bouzakis, K.-D.; Hadjiyiannis, S.; Skordaris, G.; Mirisidis, I.; Michailidis, N.; Efstathiou, K.; Pavlidou, E.; Erkens, G.; Cremer, R.; Rambadt, S.; et al. The effect of coating thickness, mechanical strength and hardness properties on the milling performance of PVD coated cemented carbide inserts. *Surf. Coat. Technol.* **2004**, *177–178*, 657–664. [CrossRef]
109. Endrino, J.; Fox-Rabinovich, G.; Gey, C. Hard AlTiN, AlCrN PVD coatings for machining of austenitic stainless steel. *Surf. Coat. Technol.* **2006**, *200*, 6840–6845. [CrossRef]
110. Ahmed, Y.S.; Paiva, J.M.; Covelli, D.; Veldhuis, S.C. Investigation of coated cutting tool performance during machining of super duplex stainless steels through 3D wear evaluations. *Coatings* **2017**, *7*, 127. [CrossRef]
111. Kalss, W.; Reiter, A.; Derflinger, V.; Gey, C.; Endrino, J. Modern coatings in high performance cutting applications. *Int. J. Refract. Met. Hard Mater.* **2006**, *24*, 399–404. [CrossRef]
112. Hörling, A.; Hultman, L.; Odén, M.; Sjölén, J.; Karlsson, L. Mechanical properties and machining performance of $Ti_{1-x}Al_xN$ coated cutting tools. *Surf. Coat. Technol.* **2005**, *191*, 384. [CrossRef]

113. Erkens, G.; Cremer, R.; Hamoudi, T.; Bouzakis, K.-D.; Mirisidis, I.; Hadjiyiannis, S.; Skordaris, G.; Asimakopoulos, A.; Kombogiannis, S.; Anastopoulos, J.; et al. Properties and performance of high aluminium containing (Ti,Al)N based supernitride coatings in innovative cutting applications. *Surf. Coat. Technol.* **2004**, *177–178*, 727–734. [CrossRef]
114. Mosquera, A.; Mera, L.; Fox-Rabinovich, G.S.; Martínez, R.; Azkona, I.; Endrino, J.L. Advantages of nanoimpact fracture testing in studying the mechanical behavior of CrAl(Si)N coatings. *Nanosci. Nanotechnol. Lett.* **2010**, *2*, 352–356. [CrossRef]
115. Mosquera, A.; Mera, L.; Fox-Rabinovich, G.S.; Martínez, R.; Azkona, I.; Endrino, J.L. Statistical analysis of nanoimpact testing of hard CrAl(Si)N coatings. *Mater. Res. Soc. Symp. Proc.* **2011**, *1339*, 1113390502. [CrossRef]
116. Chowdhury, S.; Beake, B.D.; Yamamoto, K.; Bose, B.; Aguirre, M.; Fox-Rabinovich, G.S.; Veldhuis, S.C. Improvement of wear performance of nano-multilayer PVD coatings under dry hard end milling conditions based on their architectural development. *Coatings* **2018**, *8*, 59. [CrossRef]
117. Bouzakis, K.D.; Bouzakis, E.; Skordaris, G.; Makrimallakis, S.; Tsouknidas, A.; Katirtzoglou, G.; Gerardis, S. Effect of PVD films wet micro-blasting by various Al_2O_3 grain sizes on the wear behavior of coated tools. *Surf. Coat. Technol.* **2011**, *205*, S128–S132. [CrossRef]
118. Bouzakis, K.D.; Skordaris, G.; Bouzakis, E.; Tsouknidas, A.; Makrimallakis, S.; Gerardis, S.; Katirtzoglou, G. Optimisation of wet micro-blasting on PVD films with various grain materials for improving the coated tools' cutting performance. *CIRP Ann. Manuf. Technol.* **2011**, *60*, 587–590. [CrossRef]
119. Fox-Rabinovich, G.S.; Endrino, J.L.; Agguire, M.H.; Beake, B.D.; Veldhuis, S.C.; Kovalev, A.I.; Gershman, I.S.; Yamamoto, K.; Losset, Y.; Wainstein, D.L.; et al. Mechanism of adaptability for the nano-structured TiAlCrSiYN-based hard physical vapor deposition coatings under extreme frictional conditions. *J. Appl. Phys.* **2012**, *111*, 064306. [CrossRef]
120. Fox-Rabinovich, G.S.; Yamamoto, K.; Beake, B.D.; Gershman, I.S.; Kovalev, A.I.; Veldhuis, S.C.; Aguirre, M.H.; Dosbaeva, G.; Endrino, J.L. Hierarchical adaptive nanostructured PVD coatings for extreme tribological applications: The quest for nonequilibrium states and emergent behavior. *Sci. Technol. Adv. Mater.* **2012**, *13*, 043001. [CrossRef]
121. Fox-Rabinovich, G.S.; Gershman, I.S.; Veldhuis, S. Thin-Film PVD coating metamaterials exhibiting similarities to natural processes under extreme tribological conditions. *Nanomaterials* **2020**, *10*, 1720. [CrossRef]
122. Zhang, Q.; Xu, Y.; Zhang, T.; Wu, Z.; Wang, Q. Tribological properties, oxidation resistance and turning performance of AlTiN/AlCrSiN multilayer coatings by arc ion plating. *Surf. Coat. Technol.* **2018**, *356*, 1–10. [CrossRef]
123. Roa, J.J.; Jiménez-Pique, E.; Martínez, R.; Ramírez, G.; Tarragó, J.J.; Rodríquez, R.; Llanes, L. Contact damage and fracture micromechanisms of multi-layered TiN/CrN coatings at micro- and nano-length scales. *Thin Solid Films* **2014**, *571*, 308–315. [CrossRef]
124. Chang, Y.-Y.; Wu, C.-J. Mechanical properties and impact resistance of multi-layered TiAlN/ZrN coatings. *Surf. Coat. Technol.* **2013**, *231*, 62–66. [CrossRef]
125. Ou, Y.X.; Lin, J.; Che, H.L.; Sproul, W.D.; Moore, J.J.; Lei, M.K. Mechanical and tribological properties of CrN/TiN multilayer coatings deposited by pulsed dc magnetron sputtering. *Surf. Coat. Technol.* **2015**, *276*, 152–159. [CrossRef]
126. Vereschaka, A.A.; Grigoriev, S.N. Study of cracking mechanisms in multi-layered composite nano-structured coatings. *Wear* **2017**, *378–379*, 43–57. [CrossRef]
127. Vereschaka, A.; Tabakov, V.; Grigoriev, S.; Sitnikov, N.; Milovich, F.; Andreev, N.; Sotova, C.; Kutina, N. Investigation of the influence of the thickness of nanolayers in wear-resistant layers of Ti-TiN-(Ti,Cr,Al)N coating on destruction in the cutting and wear of carbide cutting tools. *Surf. Coat. Technol.* **2020**, *385*, 125402. [CrossRef]
128. Wang, Y.X.; Zhang, S. Toward hard yet tough ceramic coatings. *Surf. Coat. Technol.* **2014**, *258*, 1–16. [CrossRef]
129. Yang, J.; Botero, C.A.; Cornu, N.; Ramírez, G.; Mestra, A.; Llanes, L. Mechanical response under contact loads of, AlCrN-coated tool materials. *IOP Conf. Ser. Mater. Sci. Eng.* **2013**, *48*, 012003. [CrossRef]
130. Yonezu, A.; Liu, L.; Chen, X. Analysis on spiral crack in thick diamond-like carbon films subjected to spherical contact loading. *Mater. Sci. Eng. A* **2008**, *496*, 67–76. [CrossRef]
131. Mendibide, C.; Steyer, P.; Fontaine, J.; Goudeau, P. Improvement of the tribological behavior of PVD nanostratified TiN/CrN coatings—An explanation. *Surf. Coat. Technol.* **2006**, *201*, 4119–4124. [CrossRef]
132. Johnson, K.L. *Contact Mechanics*; Cambridge University Press: London, UK, 1985; p. 464. ISBN 0521347963.
133. Bousser, E.; Benkahoul, M.; Martinu, L.; Klemberg-Sapieha, J.E. Effect of microstructure on the erosion resistance of Cr-Si-N coatings. *Surf. Coat. Technol.* **2008**, *203*, 776–780. [CrossRef]
134. Beake, B.D.; Vishnyakov, V.; Colligon, J.S. Nano-impact testing of TiFeN and TiFeMoN films for dynamic toughness evaluation. *J. Phys. D Appl. Phys.* **2011**, *44*, 085301. [CrossRef]
135. Shipway, P.; Hutchings, I. Measurement of coating durability by solid particle erosion. *Surf. Coat. Technol.* **1995**, *71*, 1–8. [CrossRef]
136. Skordaris, G.; Bouzakis, K.; Charalampous, P. A dynamic FEM simulation of the nano-impact test on mono- or multi-layered PVD coatings considering their graded strength properties determined by experimental-analytical procedures. *Surf. Coat. Technol.* **2015**, *265*, 53–61. [CrossRef]
137. Bouzakis, K.; Gerardis, S.; Skordaris, G.; Bouzakis, E. Nano-impact test on a TiAlN PVD coating and correlation between experimental and FEM results. *Surf. Coat. Technol.* **2011**, *206*, 1936–1940. [CrossRef]
138. Skordaris, G.; Bouzakis, K.; Charalampous, P. A critical review of FEM models to simulate the nano-impact test on PVD coatings. *MATEC Web Conf.* **2018**, *188*, 04017. [CrossRef]

139. Feng, J.; Qin, Y.; Liskiewicz, T.W.; Beake, B.D.; Wang, S. Crack propagation of a thin hard coating under cyclic loading: Irreversible cohesive zone model. *Surf. Coat. Technol.* **2021**, *426*, 127776. [CrossRef]
140. Naskar, A.; Chattopadhyay, A. Investigation on flank wear mechanism of CVD and PVD hard coatings in high speed dry turning of low and high carbon steel. *Wear* **2018**, *396–397*, 98–106. [CrossRef]
141. Santhanam, A.; Quinto, D.; Grab, G. Comparison of the steel-milling performance of carbide inserts with MTCVD and PVD TiCN coatings. *Int. J. Refract. Metal. Hard Mater.* **1996**, *14*, 31–40. [CrossRef]
142. Bouzakis, K.-D.; Pappa, M.; Bouzakis, E.; Skordaris, G.; Gerardis, S. Correlation between PVD coating strength properties and impact resistance at ambient and elevated temperatures. *Surf. Coat. Technol.* **2010**, *205*, 1481. [CrossRef]
143. Bouzakis, K.-D.; Lili, E.; Sampris, A.; Michailidis, N.; Pavlidou, E.; Skordaris, G. Impact test on PVD coatings, at elevated temperatures. In *THE Coatings, Proceedings of the 5th International Conference and EUREKA Brokerage Event, Kallithea, Greece, 5–7 October 2005*; Bouzakis, K.-D., Denkena, B., Toenshoff, H.-K., Geiger, M., Eds.; Ziti Publications: Thessaloniki, Greece, 2005; pp. 311–321.
144. Bouzakis, E. Fatigue endurance assessment of DLC coatings on high-speed steels at ambient and elevated temperatures by repetitive impact tests. *Coatings* **2020**, *10*, 547. [CrossRef]
145. Beake, B.; Endrino, J.; Kimpton, C.; Fox-Rabinovich, G.; Veldhuis, S. Elevated temperature repetitive micro-scratch testing of AlCrN, TiAlN and AlTiN PVD coatings. *Int. J. Refract. Met. Hard Mater.* **2017**, *69*, 215–226. [CrossRef]
146. Matthews, A.; Franklin, S.; Holmberg, K. Tribological coatings: Contact mechanisms and selection. *J. Phys. D Appl. Phys.* **2007**, *40*, 5463–5475. [CrossRef]
147. Beake, B.D.; Vishnyakov, V.M.; Liskiewicz, T.W. Integrated nanomechanical characterisation of hard coatings. In *Protective Thin Coatings Technology*; Zhang, S., Ting, J.-M., Wu, W.-Y., Eds.; CRC Press: Boca Raton, FL, USA, 2021; pp. 95–140.
148. Beake, B.D. The influence of the H/E ratio on wear resistance of coating systems—Insights from small-scale testing. *Surf. Coat. Technol.* **2022**, 128272. [CrossRef]

Article

Reconstructed NiCo Alloy Enables High-Rate Ni-Zn Microbattery with High Capacity

Yixue Duan [1,2,†], Gongchuan You [2,†], Zhe Zhu [2], Linfeng Lv [1,2], Xiaoqiao Liao [2], Xin He [2], Kai Yang [2], Ruiqi Song [2], Peng Tian [2,*] and Liang He [1,2,3,*]

1 State Key Laboratory of Advanced Technology for Materials Synthesis and Processing, Wuhan University of Technology, Wuhan 430070, China
2 School of Mechanical Engineering, Sichuan University, Chengdu 610065, China
3 Med+X Center for Manufacturing, West China Hospital, Sichuan University, Chengdu 610041, China
* Correspondence: tianpeng@scu.edu.cn (P.T.); hel20@scu.edu.cn (L.H.)
† These authors contributed equally to this work.

Abstract: Miniaturized powering devices with both sufficient capacity as well as fast charging capability are anticipated to support microelectronics with multi-functions. However, most reported miniaturized energy storage devices only display limited performances around capacity or rate performance, and it remains challenging to develop high-rate microdevices with large capacities. Herein, a reconstructed NiCo alloy is proposed as a promising microcathode for a Ni-Zn microbattery with a high-rate performance and large capacity. With the reconstructed layer compactly adhered on the metal substrate, the activated NiCo alloy demonstrates an excellent conductivity close to metals. Meanwhile, the abundant alloying defect contributes to a relatively higher reconstruction depth up to 20 nm. Both the superior electron transport and the higher reaction depth facilitate the simultaneous excellent performance in the reaction rate and capacity. As a consequence, the microcathode achieves a large capacity up to 1.51 mAh cm^{-2}, as well as an excellent rate performance with a capacity retention of 82.9% when the current density is expanded to 100 mA cm^{-2}. More surprisingly, such excellent performance can shift towards the full Ni-Zn microbattery, and the fast-charging capability based on large capacity can stably maintain 7000 cycles. This unique strategy of reconstructed NiCo alloy microcathode provides a new direction for the construction of high-performance output units.

Keywords: Ni-Zn microbattery; NiCo alloy; reconstruction; high rate; high capacity

1. Introduction

With the boom of the Internet of Things (IoT) in the fields of integrated systems and flexible electronics, there forms a growing demand for high-performance miniaturized energy storage devices, including microbatteries (MBs) and microsupercapacitors (MSCs) [1–5]. Typically, MSCs are featured with having the merits of a long cycling durability and a high-rate capability, while also having the drawbacks of self-discharge issues and a low energy supply, which severely restrict their practical applications. To the contrary, MBs with high energy densities could well maintain a stable voltage output for a long period in powering advanced miniaturized devices [6–10]. Among the ever-developed MBs, aqueous Ni-Zn MBs hold the most application anticipation for their relatively high voltage output (~1.8 V), fast reaction kinetics, and abundant resources. There have been reports about Ni-Zn batteries with high energy densities through dense Ni cathodes [11] and 3D Zn anodes [12]. However, the development of Ni-Zn MBs is neither sufficient in the capacity supply nor in the reaction rate.

To address the above issues of Ni-Zn MBs associated with capacity supply and reaction rate, many reports have proposed powder techniques with various structures and composition designs, or deposition techniques with active materials on current collectors [13–15]. However, these traditional techniques suffer from the limited reaction transport when a

thick electrode is applied [3]. Generally, when more active materials are constructed in a given electrode, it is hard to guarantee that the reactivity is always maintained at a high level. This effectiveness decline could be attributed to the accompanied higher reaction impedance and longer diffusion distance, thus resulting in a worse power performance when more energy is provided. Additionally, even though the design of the thin-layer reaction with a fine structure is effective at reactivity [16], the contact resistance of active materials and current collectors cannot be ignored if a high-rate current is applied. Recently, there have been reports about the concept of reconstruction based on the nickel element in catalysis, which could effectively provide a high performance of catalyzers. Generally, as an activation strategy, reconstruction refers to the in situ transformation of pre-fabricated materials through oxidization/reduction reactions under alkaline electrochemical environments [17–19]. This approach is essentially a rearrangement of the atoms on the atomic layer of the crystal surface, leading to a change in the two-dimensional structure of the surface layer, and the utilizing depth of the substrate is very shallow (~10 nm). A reconstruction strategy can also be used as a reference for material activation in the field of energy storage. For example, in our previous work, we firstly adopted the strategy of reconstruction to obtain in situ active $Ni(OH)_2/NiOOH$ on a nanoporous Ni substrate for high-rate Ni-Zn MB [20]. The reconstructed nickel microcathode was featured as having an ultrahigh reactivity as well as an excellent rate performance. However, an inadequate Ni deposition and a shallow reconstruction layer correspondingly led to low production of active materials, which cannot provide satisfactory capacity. To this end, how to construct a high-rate Ni-Zn MB with a large capacity remains challenging.

Herein, a NiCo alloy was proposed as the pre-fabricated material for reconstruction, and the constructed Ni-Zn MB delivered both a high-rate performance and large capacity. Initially, we co-deposited the NiCo alloy hierarchical porous structure by virtue of a bubble template. The NiCo alloy skeleton is larger and more porous than that of pure nickel, demonstrating an enhanced surface area for reconstruction activation. Then, during the reconstruction process in aqueous alkali, the abundant alloying defect contributes to a relatively higher reconstruction depth of up to 20 nm. Therefore, more $Ni(OH)_2/NiOOH$ components are produced while the superior electron transport is still well maintained. Meanwhile, benefiting from the naturally formed double hydroxide in the reconstructed layer, the structure deformation during the proton insertion/extraction is greatly eased, resulting in an enhanced cycling durability. As a consequence, both high-rate performance and large capacity are ensured for the reconstructed microcathode on the premise of electrochemical stability. The reaction capacity is realized to 1.51 mAh cm^{-2}, and the current density can be expanded up to 100 mA cm^{-2}. Lastly, the cycling retention is more than 80% in 7000 cycles. The final fabricated Ni-Zn MB inherits this excellent electrochemical performance with a slightly decreased capacity of 1.39 mAh cm^{-2}, an enhanced rate current of 200 mA cm^{-2}, and an almost maintained cycling durability in 7000 cycles.

2. Materials and Methods

2.1. Preparation of Reconstructed NiCo Alloy Microcathode

The CHI 760E electrochemical workstation (CH Instruments, Shanghai, China) was utilized for the electrochemical operation. In brief, the customized miniaturized nickel plate (effective area: 0.2 cm × 0.5 cm) served as the working electrode with a saturated calomel electrode as the reference electrode, and a Pt plate as the counter electrode. The deposition process for the NiCo alloy was conducted at a cathodic potential of 6 V in the solution of 2 M NaCl, 2 M NH_4Cl, 0.1 M $NiCl_2$, and x M $CoCl_2$ (the variable x refers to 0, 0.015, 0.03, and 0.045). After washing the microelectrodes with the deionized water a few times, the mentioned reconstruction processes were switched to an alkaline solution (1 M KOH) with a Hg/HgO electrode as the reference electrode instead. The typical reconstruction method of CV refers to a cyclic sweep within the potential range of 0.2–0.6 V at a scan rate of 10 mV s^{-1}. Furthermore, according to the previous report, the saturated reconstruction cycles are within 750 cycles, and, herein, we followed the procedures.

Safety Statement: During the electrodeposition process, the chlorine and hydrogen will be released. Finally, the experiment is suggested to be conducted in a fume hood.

2.2. Preparation of Zinc Microanode

The zinc microelectrode was obtained via the electrochemical process as well. The same customized miniaturized nickel plate was selected as the working electrode and an ordinary zinc plate worked as the counter electrode. The electrodeposition was conducted at a constant potential of −1.5 V for 30 min in a solution of 6 M ZnO-saturated solution of KOH.

2.3. Assembly of Ni-Zn MB

Firstly, 500 mg of sodium polyacrylate was slowly added into a 20 mL solution of 6 M ZnO-saturated solution of KOH. Then, after stirring for 30 min, the transparent gel electrolyte was formed. Two films tailored from plastic bag were selected as the package materials. Firstly, three edges of the two films were sealed via a sealer machine. Afterward, the reconstructed NiCo alloy microcathode and Zn microanode were put inside the micro pocket and separated via two thin cardboards. After transferring the gel electrolyte into the middle and removing the separated cardboards, the final edge was sealed and the packaged Ni-Zn MB was obtained.

2.4. Structure and Composition Characterizations

The morphologies and microstructural and component characteristics of the samples were measured using FE-SEM (ZEISS Gemini 300, Carl Zeiss AG, Oberkochen, Germany), transmission electron microscopy with a voltage of 200 kV (TEM, F200X, Thermo Fisher Scientific, Waltham, MA, USA). The chemical states and atomic structure information were investigated by XPS (Thermo Scientific K-Alpha, Thermo Fisher Scientific, Waltham, MA, USA).

2.5. Electrochemical Measurements

Cyclic voltammetry (CV) and galvanostatic charge/discharge (GCD) tests were all conducted on a CHI 760E. The single microelectrode was tested in a three-electrode system and the Ni-Zn MB was packed for the test. All the tests related with single NiCo alloy microelectrodes were conducted in 1 M KOH solutions while the Ni-Zn packed MB was in the gel electrolyte of the 6 M ZnO-saturated solution of KOH. The specific capacity (C) of the single microelectrode or assembled MB was calculated by the formulas $C = It/A$, where I is the discharge current (mA cm^{-2}) and A is the geometric area of the microelectrode for the operation.

3. Results and Discussion

We prepared four microelectrodes (denoted as Ni, NiCo0.15, NiCo0.3, and NiCo0.45) by varying the amount of Co^{2+} in the deposition process. As shown in Figure 1, the deposited pure Ni is densely featured with uniform small micro pores (mostly around 5 μm) as a result of the accompanied release of gas. When the Co^{2+} was added, it could be clearly observed that the skeleton of the deposited alloy and the pores became larger, and this trend tended to expand with the increase in added Co^{2+}. This is possibly due to the provided alloy deposition environment and the initial formed alloy substrate with a high reactivity [21], which lowered the reaction barrier for both metal depositions and hydrogen evolution. As a result, the deposition of metals as well as the gas release became severe at the same given potential, and a large metallic skeleton with micro pores and cracks was finally obtained. Meanwhile, we also combined energy dispersive spectrometer to analyze the 4 SEM images, and the resulted Ni/Co ratios were 1:0, 1:0.18, 1:0.35, and 1:0.51, respectively. This ratio result reflects that the deposition potential of Co is slightly lower than that of Ni and the ratio of deposited alloy is close to that of the original added ions.

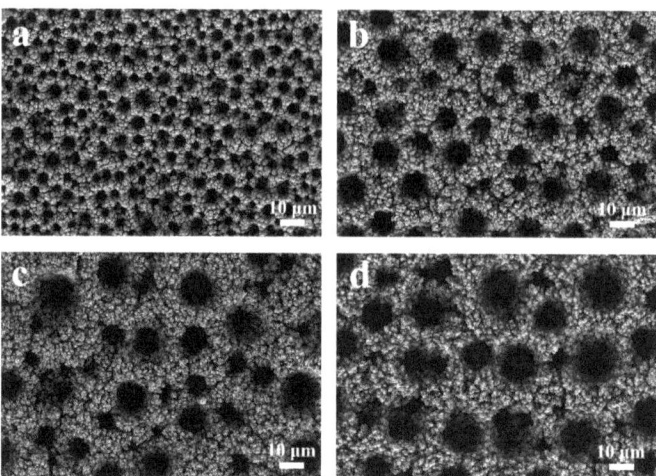

Figure 1. SEM images of prepared (**a**) Ni, (**b**) NiCo0.15, (**c**) NiCo0.3, and (**d**) NiCo0.45 microelectrodes.

To analyze the composition of the reconstructed layer, the TEM and high-resolution TEM (HRTEM) were investigated. As could be observed in Figure 2a, the contrast in the TEM images revealed that the depth of the activated layer was up to 20 nm, and the subsequent redox reactions were maintained since compact connection with the metallic substrate. The d spacings about the reconstructed lattice fringes were 0.244 and 0.232 nm, corresponding to both (101) planes of $Co(OH)_2$ (JCPDS No. 001-0357) and $Ni(OH)_2$ (JCPDS No. 003-0177), indicating that the deposited alloy was partly reconstructed. For the inner layer, the d spacings of 0.177 and 0.204 nm clearly referred to the (200) plane of Ni (JCPDS No. 001-1258) and the (111) plane of Co (JCPDS No. 001-1259), respectively (Figure 2b). We then carried out the XPS measurement to clarify the chemical environment of the reconstructed alloy microelectrode. As shown in Figure 2c, the signals of $Ni(OH)_2$ (855.75 and 873.3 eV) and NiOOH (856.8 and 874.45 eV) existed in the reconstructed layer of pure Ni [22], indicating that the final reconstructed products contained the nickel hydroxide and its oxidized derivative. Furthermore, affected by the alloying effect, the electron cloud migration left a slight negative shift (approximately 0.4 eV) for the reconstructed NiCo alloy [23]. As shown in Figure 2d, when it comes to the Co 2p signals, both $Co(OH)_2$ (781 and 796.36 eV) and CoOOH (782.3 and 797.9 eV) could be identified [24], similar to that of nickel. All of the results confirmed the hydroxides product after the reconstruction treatment.

Around the electrochemical performance associated with the enhanced capacity, we compared the discharge curves of four reconstructed microelectrode whose deposition times were set to be 60 s. As shown in Figure 3a, with the increase in the Co content, the related alloy microelectrode displayed a higher capacity. Compared with pure nickel hydroxide, the slight addition of cobalt hydroxide means more active sites and an optimized diffusion polarization, thus increasing the discharge plateau even though more active materials were provided. However, when the Co content further increased to 0.3 or 0.45, the ohmic resistance arising from the increased load of active materials may affect the reaction polarization [25,26], generating a negative effect on the discharge plateau. Here, it should be noted that the increased capacity was not only attributed to the enhanced reconstruction depth, but also from the increased deposition when Co ions were involved in the reduction reactions. Finally, when more metallic structures were loaded via extending the deposition time, the optimal mixture ratio was not NiCo0.45, as its structure would become unstable when deposition time reached 120 s. Instead, the NiCo0.3 ensured a stable deposition time of up to 180 s. The related discharge curves under various current densities and the cycling performance of such microelectrode are shown in Figure 3b,c, respectively. The discharge capacities were 1.51, 1.49, 1.45, 1.37, and 1.25 mAh cm^{-2} at the current densities of 5, 10, 20, 50, and 100 mA cm^{-2}, respectively. To evaluate the stability of the high-rate performance, a cycling test at the current density of 100 mA cm^{-2} was conducted for both microelectrodes of NiCo0.3 and pure Ni (180 s). Having benefited from the stabilizing effect of double hydroxides [4,27], the NiCo0.3 microelectrode displayed a capacity retention of 83.36% after 7000 cycles, while that of pure Ni microelectrode dropped to 48.83% after 3000 cycles. These results confirmed that the reconstructed alloy (NiCo0.3) is a promising cathode with a high capacity, high rate, and high cycling durability.

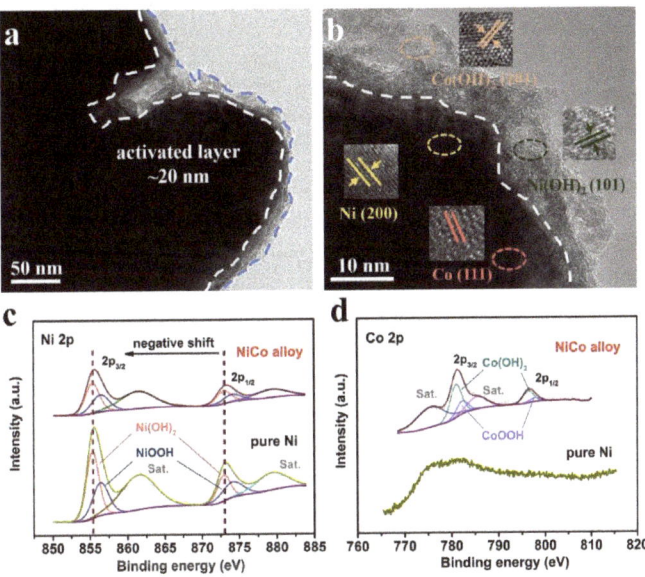

Figure 2. (**a**) TEM image and (**b**) HRTEM image of reconstructed NiCo alloy, and XPS (**c**) Ni 2p and (**d**) Co 2p signals for both pure Ni and NiCo alloys after reconstruction.

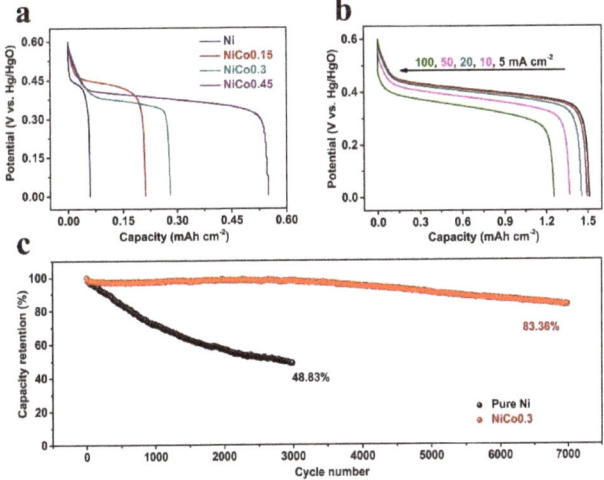

Figure 3. Discharge curves of (**a**) four microelectrodes at the current density of 10 mA cm^{-2} and (**b**) the NiCo0.3 microelectrode under various current densities. (**c**) The cycling performance of pure Ni and NiCo0.3 microelectrodes.

The assembly process of Ni-Zn MB is shown in Figure 4a. To further demonstrate the practical performance of this microcathode in assembled Ni-Zn MBs, CV curves, galvanostatic discharge profiles, and cycling test were conducted. As shown in Figure 4b, there was no obvious change in the CV shape when the scan rate increased from 1 to 10 mV s^{-1}, indicating a good electrochemical reversibility. According to previous research about double hydroxides, the redox peak can be ascribed to the following electrochemical reaction: Zn + CoOOH + NiOOH + 2 KOH + 2 H$_2$O $\rightleftharpoons$ K$_2$[Zn(OH)$_4$] + Co(OH)$_2$ + Ni(OH)$_2$ [28,29]. For the rate performance of this Ni-Zn MB, the increased electrolyte concentration (6 M) further expanded the rate current to 200 mA cm^{-2}. The discharge capacities were 1.39, 1.37, 1.36, 1.32, and 1.23 mAh cm^{-2} at the current densities of 10, 20, 50, 100, and 200 mA cm^{-2}, respectively (Figure 4c). Similar to the exploration of the mentioned stability of the fast charge/discharge capability, the long-term cycling durability of the Ni-Zn MB was investigated at 200 mA cm^{-2}. As observed in Figure 4d, the assembled MB delivered an excellent cycling performance with a capacity retention of 92.0% in 7000 cycles along with a nearly 100% Coulombic efficiency. All these performances clearly indicate the successful assembly of the Ni-Zn MB with the reconstructed NiCo0.3 microcathode, and that the high-rate performance with a large capacity and the cycling durability is advanced in the ever-reported MBs, which is promising for practical applications in microelectronics [8,15,30].

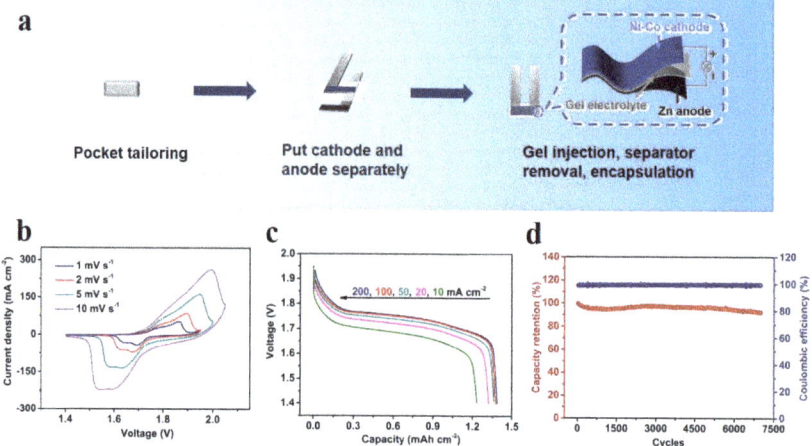

Figure 4. Assembly process and electrochemical performance of the Ni-Zn MB with reconstructed NiCo0.3 microcathode: (**a**) schematic diagram of assembly process, (**b**) CV curves at various scan rates, (**c**) discharge curves under various current densities and (**d**) cycling results.

4. Conclusions

In summary, a high-rate Ni-Zn microbattery with a large capacity was realized through the reconstruction approach for the NiCo alloy. By virtue of the bubble template, a highly porous metallic structure was obtained with various Ni-Co ratios in the preparation. When the Co content in the NiCo alloy increased, the higher alloy content facilitated the deposition reactions, leading to an enhanced metal loading and surface area. Additionally, benefiting from the highly active NiCo alloy structure, a deeper reconstruction of up to 20 nm was achieved, resulting in a higher capacity when compared with pure nickel. Meanwhile, the formed reconstructed layer contained double hydroxides which possessed enhanced electrostatic forces between the hydroxides when compared with the pure nickel hydroxide, hence realizing a more stable cycling performance. As a result, the final assembled Ni-Zn MB with reconstructed alloy microelectrode achieved a high capacity of up to 1.39 mAh cm^{-2}, an ultrahigh rate performance with a capacity retention of 88.5% when the current density increased to 200 mA cm^{-2}, and an excellent cycling stability with over 90% retention after 7000 fast-charging cycles. This optimization design from the perspective of substrate metal provides an effective strategy for developing MBs with large capacities and high-rate performances simultaneously.

Author Contributions: Conceptualization, Y.D. and G.Y.; methodology, L.L.; software, Z.Z.; validation, Y.D., G.Y., L.H. and P.T.; formal analysis, X.L.; investigation, X.H.; resources, K.Y.; data curation, Y.D.; writing—original draft preparation, Y.D. and G.Y.; writing—review and editing, Y.D. and G.Y.; visualization, R.S.; supervision, L.H. and P.T.; project administration, L.H. and P.T.; funding acquisition, L.H. and P.T. All authors have read and agreed to the published version of the manuscript.

Funding: This research was supported by the Fundamental Research Funds for the Central Universities (No. 20822041F4045), and the Science and Technology Project of Yibin Sanjiang New Area (No. 2023SJXQSXZJ003).

Institutional Review Board Statement: Not applicable.

Informed Consent Statement: Not applicable.

Data Availability Statement: Not applicable.

Conflicts of Interest: The authors declare no conflict of interest.

References

1. Yang, Y.; Gao, W. Wearable and flexible electronics for continuous molecular monitoring. *Chem. Soc. Rev.* **2019**, *48*, 1465–1491. [CrossRef] [PubMed]
2. Zhu, Z.; Kan, R.; Hu, S.; He, L.; Hong, X.; Tang, H.; Luo, W. Recent Advances in High-Performance Microbatteries: Construction, Application, and Perspective. *Small* **2020**, *16*, 2003251. [CrossRef] [PubMed]
3. Hao, Z.; Xu, L.; Liu, Q.; Yang, W.; Liao, X.; Meng, J.; Hong, X.; He, L.; Mai, L. On-chip Ni-Zn microbattery based on hierarchical ordered porous Ni@Ni(OH)$_2$ microelectrode with ultrafast ion and electron transport kinetics. *Adv. Funct. Mater.* **2019**, *29*, 1808470. [CrossRef]
4. Wang, Y.; Hong, X.; Guo, Y.; Zhao, Y.; Liao, X.; Liu, X.; Li, Q.; He, L.; Mai, L. Wearable Textile-Based Co-Zn Alkaline Microbattery with High Energy Density and Excellent Reliability. *Small* **2020**, *16*, 2000293. [CrossRef] [PubMed]
5. Duan, Y.; You, G.; Sun, K.; Zhu, Z.; Liao, X.; Lv, L.; Tang, H.; Xu, B.; He, L. Advances in wearable textile-based micro energy storage devices: Structuring, application and perspective. *Nanoscale Adv.* **2021**, *3*, 6271–6293. [CrossRef] [PubMed]
6. Pikul, J.H.; Gang Zhang, H.; Cho, J.; Braun, P.V.; King, W.P. High-power lithium ion microbatteries from interdigitated three-dimensional bicontinuous nanoporous electrodes. *Nat. Commun.* **2013**, *4*, 1732. [CrossRef]
7. Kyeremateng, N.A.; Hahn, R. Attainable energy density of microbatteries. *ACS Energy Lett.* **2018**, *3*, 1172–1175. [CrossRef]
8. Sun, G.; Jin, X.; Yang, H.; Gao, J.; Qu, L. An aqueous Zn-MnO$_2$ rechargeable microbattery. *J. Mater. Chem. A* **2018**, *6*, 10926–10931. [CrossRef]
9. Li, Y.; Zhu, M.; Bandari, V.; Karnaushenko, D.D.; Karnaushenko, D.; Zhu, F.; Schmidt, O.G. On-chip batteries for dust-sized computers. *Adv. Energy Mater.* **2022**, *12*, 2103641. [CrossRef]
10. Sun, P.; Li, X.; Shao, J.; Braun, P.V. High-Performance Packaged 3D Lithium-Ion Microbatteries Fabricated Using Imprint Lithography. *Adv. Mater.* **2021**, *33*, 2006229. [CrossRef] [PubMed]
11. Zhou, W.; Zhu, D.; He, J.; Li, J.; Chen, H.; Chen, Y.; Chao, D. A scalable top-down strategy toward practical metrics of Ni-Zn aqueous batteries with total energy densities of 165 W h kg^{-1} and 506 W h L^{-1}. *Energy Environ. Sci.* **2020**, *13*, 4157–4167. [CrossRef]
12. Parker, J.F.; Chervin, C.N.; Pala, I.R.; Machler, M.; Burz, M.F.; Long, J.W.; Rolison, D.R. Rechargeable nickel-3D zinc batteries: An energy-dense, safer alternative to lithium-ion. *Science* **2017**, *356*, 415–418. [CrossRef] [PubMed]
13. Qu, Z.; Zhu, M.; Yin, Y.; Huang, Y.; Tang, H.; Ge, J.; Li, Y.; Karnaushenko, D.D.; Karnaushenko, D.; Schmidt, O.G. A Sub-Square-Millimeter Microbattery with Milliampere-Hour-Level Footprint Capacity. *Adv. Energy Mater.* **2022**, *12*, 2200714. [CrossRef]
14. Hallot, M.; Roussel, P.; Lethien, C. Sputtered LiNi$_{0.5}$Mn$_{1.5}$O$_4$ thin films for lithium-ion microbatteries. *ACS Appl. Energy Mater.* **2021**, *4*, 3101–3109. [CrossRef]
15. Wen, W.; Wu, J.-M.; Jiang, Y.-Z.; Bai, J.-Q.; Lai, L.-L. Titanium dioxide nanotrees for high-capacity lithium-ion microbatteries. *J. Mater. Chem. A* **2016**, *4*, 10593–10600. [CrossRef]
16. Li, Y.-Q.; Shi, H.; Wang, S.-B.; Zhou, Y.-T.; Wen, Z.; Lang, X.-Y.; Jiang, Q. Dual-phase nanostructuring of layered metal oxides for high-performance aqueous rechargeable potassium ion microbatteries. *Nat. Commun.* **2019**, *10*, 4292. [CrossRef]
17. Liu, X.; Meng, J.; Ni, K.; Guo, R.; Xia, F.; Xie, J.; Li, X.; Wen, B.; Wu, P.; Li, M.; et al. Complete Reconstruction of Hydrate Pre-Catalysts for Ultrastable Water Electrolysis in Industrial-Concentration Alkali Media. *Cell Rep. Phys. Sci.* **2022**, *1*, 100241. [CrossRef]
18. Liu, X.; Meng, J.; Zhu, J.; Huang, M.; Wen, B.; Guo, R.; Mai, L. Comprehensive understandings into complete reconstruction of precatalysts: Synthesis, applications, and characterizations. *Adv. Mater.* **2021**, *33*, 2007344. [CrossRef]

19. Liu, X.; Ni, K.; Wen, B.; Guo, R.; Niu, C.; Meng, J.; Li, Q.; Wu, P.; Zhu, Y.; Wu, X.; et al. Deep reconstruction of nickel-based precatalysts for water oxidation catalysis. *ACS Energy Lett.* **2019**, *4*, 2585–2592. [CrossRef]
20. Zhu, Z.; Kan, R.; Wu, P.; Ma, Y.; Wang, Z.; Yu, R.; Liao, X.; Wu, J.; He, L.; Hu, S. A Durable Ni-Zn Microbattery with Ultrahigh-Rate Capability Enabled by In Situ Reconstructed Nanoporous Nickel with Epitaxial Phase. *Small* **2021**, *17*, 2103136. [CrossRef]
21. Zou, P.; Li, J.; Zhang, Y.; Liang, C.; Yang, C.; Fan, H.J. Magnetic-field-induced rapid synthesis of defect-enriched Ni-Co nanowire membrane as highly efficient hydrogen evolution electrocatalyst. *Nano Energy* **2018**, *51*, 349–357. [CrossRef]
22. Ren, X.; Wei, C.; Sun, Y.; Liu, X.; Meng, F.; Meng, X.; Sun, S.; Xi, S.; Du, Y.; Bi, Z.; et al. Constructing an adaptive heterojunction as a highly active catalyst for the oxygen evolution reaction. *Adv. Mater.* **2020**, *32*, 2001292. [CrossRef]
23. Zou, W.; Li, Q.; Zhu, Z.; Du, L.; Cai, X.; Chen, Y.; Zhang, G.; Hu, S.; Gong, F.; Xu, L.; et al. Electron cloud migration effect-induced lithiophobicity/lithiophilicity transformation for dendrite-free lithium metal anodes. *Nanoscale* **2021**, *13*, 3027–3035. [CrossRef]
24. Gao, P.; Zeng, Y.; Tang, P.; Wang, Z.; Yang, J.; Hu, A.; Liu, J. Understanding the synergistic effects and structural evolution of $Co(OH)_2$ and Co_3O_4 toward boosting electrochemical charge storage. *Adv. Funct. Mater.* **2022**, *32*, 2108644. [CrossRef]
25. Yang, C.; Xin, S.; Mai, L.; You, Y. Materials Design for High-Safety Sodium-Ion Battery. *Adv. Energy Mater.* **2021**, *11*, 2000974. [CrossRef]
26. Chen, H.; Wang, J.; Zhao, Y.; Zhang, J.; Cao, C. Electrochemical performance of Zn-substituted Ni $(OH)_2$ for alkaline rechargeable batteries. *J. Solid State Electrochem.* **2005**, *9*, 421–428. [CrossRef]
27. Guo, Y.; Hong, X.; Wang, Y.; Li, Q.; Meng, J.; Dai, R.; Liu, X.; He, L.; Mai, L. Multicomponent hierarchical Cu-doped NiCo-LDH/CuO double arrays for ultralong-life hybrid fiber supercapacitor. *Adv. Funct. Mater.* **2019**, *29*, 1809004. [CrossRef]
28. Huang, M.; Li, M.; Niu, C.; Li, Q.; Mai, L. Recent Advances in Rational Electrode Designs for High-Performance Alkaline Rechargeable Batteries. *Adv. Funct. Mater.* **2019**, *29*, 1807847. [CrossRef]
29. Gong, M.; Li, Y.; Zhang, H.; Zhang, B.; Zhou, W.; Feng, J.; Wang, H.; Liang, Y.; Fan, Z.; Liu, J.; et al. Ultrafast high-capacity NiZn battery with NiAlCo-layered double hydroxide. *Energy Environ. Sci.* **2014**, *7*, 2025–2032. [CrossRef]
30. Jiang, K.; Zhou, Z.; Wen, X.; Weng, Q. Fabrications of High-Performance Planar Zinc-Ion Microbatteries by Engraved Soft Templates. *Small* **2021**, *17*, 2007389. [CrossRef]

Disclaimer/Publisher's Note: The statements, opinions and data contained in all publications are solely those of the individual author(s) and contributor(s) and not of MDPI and/or the editor(s). MDPI and/or the editor(s) disclaim responsibility for any injury to people or property resulting from any ideas, methods, instructions or products referred to in the content.

Article

Study of the Corrosion Mechanism of Iron-Based Amorphous Composite Coating with Alumina in Sulfate-Reducing Bacteria Solution

Zhenhua Chu [1,*], Haonan Shi [1], Fa Xu [1], Jingxiang Xu [1], Xingwei Zheng [2], Fang Wang [1], Zheng Zhang [1] and Qingsong Hu [1]

[1] Department of Mechanical Engineering, College of Engineering, Shanghai Ocean University, Shanghai 201306, China
[2] College of Science, Donghua University, Shanghai 201620, China
* Correspondence: zhchu@shou.edu.cn

Abstract: In this work, a composite coating composed of iron-based amorphous material and alumina mixed with 13 wt.% titanium oxide (AT13) ceramic was successfully fabricated by High Velocity Air-fuel Flame Spray (HVAF). The corrosion process of the composite coating in Sulfate-Reducing Bacteria (SRB) solution for 31 d was investigated by Electrochemical Impedance Spectroscopy (EIS). The corrosion morphologies and corrosion products were tested by X-ray photoelectron spectroscopy. The corrosion mechanism can be divided into two stages: microbial adhesion and biofilm failure. The microbial adhesion on the surface of the composite coating improved the formation of biofilm, which improved the corrosion resistance. On the other hand, the SRB metabolic process in the biofilm accelerated the formation of corrosion products, which resulted in the failure of the biofilm and thus the composite coating was re-exposed in the corrosion solution.

Keywords: iron-based amorphous coating; ceramic; corrosion; SRB

1. Introduction

The marine environment is a complex corrosion environment. In addition to the corrosion damage by seawater medium, the widespread microorganisms in the marine environment can also affect the corrosion behavior of alloys [1,2]. Generally, the service failure of metal materials can be caused by microbiologically influenced corrosion (MIC), due to microorganisms and their metabolic activities [3–6]. This is an important type of corrosion, causing the failure of marine engineering materials. The SRB is one kind of well-known anaerobic bacterium existing widely in the deep sea and causing MIC. Researchers investigated the polarization resistance of stainless steel in a sterile solution and SRB solution, and the results indicated that the corrosion rate of the coupons in SRB solution was 10 times higher than that in sterile medium [7,8]. This is caused by the uneven adsorption of SRB biofilm on the surface of metal materials [9]. Therefore, the corrosion mechanism caused by SRB and protection technologies for severe corrosion environments require further study.

Coating technology is regarded as one of the effective approaches for the protection of steels from corrosion [10–12]. Water-borne coatings have been used widely in the last decade due to the provisions of low volatile organic compound emissions [13,14]. However, some problems also need to be noted, such as toxicity, carcinogenicity and the pollution of biocides, accompanying the release of antibacterial agents to the environment [15,16]. On the other hand, the interaction between the abrasion caused by sea mud and corrosion will accelerate the failure of materials. Thus, it is promising to fabricate a novel coating with high abrasion and corrosion resistance.

The high strength, high hardness and superior corrosion resistance of iron-based amorphous alloys have attracted the attention of researchers [17–20]. However, the poor glassy formation ability of amorphous alloy restricts its applications. In order to avoid the disadvantages of iron-based amorphous alloys, we can fabricate amorphous coatings using thermal spraying technology [21]. Zhou et al. [22] prepared a dense iron-based amorphous coating by high-speed oxygen fuel spraying. The electrochemical test results showed that the coating exhibited excellent corrosion resistance in 3.5% sodium chloride, 1 N hydrogen chloride and 1 N sulfuric acid solution. Chu [23,24] fabricated iron-based amorphous composite coatings with AT13 and titanium nitride with high corrosion resistance and wear resistance. However, the corrosion resistance of amorphous composite coating in SRB solution needs to be further studied.

In the present work, the corrosion resistance of iron-based amorphous coatings with AT13 in SRB solution was studied systematically. The electrochemical characteristics of iron-based amorphous composite coating and EIS were investigated. The corrosion mechanism caused by SRB was proposed.

2. Materials and Experimental Process

$Fe_{54}Cr_{25}Mo_{17}C_2B_2$ alloy amorphous powders were produced by high pressure Ar gas atomization. Then, sprayed AT13 ceramic powders (20–40 μm) and amorphous powders were mixed by a mechanical mixing machine for 4 h. The sprayed powders are shown in Figure 1. Mild steel (0.45 wt.% C) was selected as the substrate with a size of 10 mm × 10 mm × 12 mm. HVAF was adopted to fabricate iron-based amorphous coatings and composite coatings. The parameters of the spraying process by HVAF are summarized as shown in Table 1.

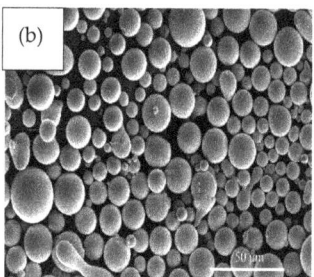

Figure 1. Sprayed powder morphologies: (**a**) AT13; (**b**) Fe-based amorphous powders.

Table 1. Spraying parameters of the HVAF process.

Coating by HVAF	Parameter
Spray distance (mm)	180
Air pressure (MPa)	0.54
Fuel 1 press (MPa)	0.48
Fuel 2 press (MPa)	0.26
Powder delivery rate (rpm)	3

The morphologies of coatings were observed by scanning electron microscopy (SEM, S4800, Hitachi, Tokyo, Japan). X-ray diffraction (XRD, Bruker D8 Focus, Billerica, MA, USA) was adopted to analyze the microstructures of coatings. A fluorescence microscope was used to observe the adhesion of microorganisms on the coating surface. Porosity data were determined according to the results of the SEM photos and Image-Plus software. More than 6 images were chosen.

Electrochemical tests were performed by a three-electrode cell, including a saturated calomel electrode (SCE), a graphite electrode as the reference and an auxiliary electrode. Specimens for the corrosion test were closely sealed with epoxy resin, leaving only an end-surface with a surface area of 1 × 1 cm^2 exposed for testing. A Tafel plot was created at a potential sweep rate of 0.5 mVs^{-1} from −100 mV to 1500 mV in SRB solution, which was open to air after immersing the specimens for an hour. In addition, EIS was examined in SRB solution. The impedance plots were interpreted on the basic of the equivalent circuit using a suitable fitting procedure by Echem Analyst. After EIS measurement, the corroded surface was examined by SEM. The corrosion products also were investigated by X-ray photoelectron spectroscopy.

In this study, the SRB were purchased from Beina Chuanglian Biotechnology Co., Ltd., Beijing, China. SRB medium included 0.98 g $MgSO_4$, 1.0 g yeast extract, 1.0 g NH_4Cl, 1.0 g Na_2SO_4, 0.5 g K_2HPO_4, 0.1 g $CaCl_2 \cdot 2H_2O$, 1.0 mg Resazurin, 0.5 g $FeSO_4 \cdot 7H_2O$, 0.1 g Na-thioglycolate, and 1 L deionized water. The PH of the medium was adjusted to 7.8 ± 0.2 by NaOH. The growth curve of SRB was measured by an ultraviolet spectrophotometer; see Figure 2. It can be seen from the figure that 0~5 h was the slow period, and 6~16h was the logarithmic growth period. The stable period occurred after 16 h. In this experiment, all samples were soaked in SRB solution when in the stable growth period. Meanwhile, fresh bacterial liquid was replaced every three days to ensure the activity of bacteria.

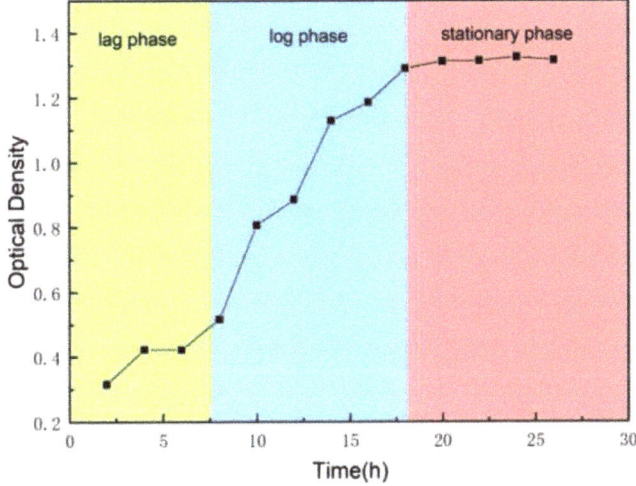

Figure 2. The growth curve of SRB.

3. Results

The as-sprayed iron-based amorphous coating and composite coatings with various contents of AT13 were tested by X-ray diffraction, and the XRD patterns are shown in Figure 3a. For the iron-based amorphous coating, there was a broad diffraction bump at the angle of 2θ = 30–43°. This represented the amorphous phase. For the composite coatings, the Bragg peaks were associated with AT13 crystalline phases, which indicates that the composite coating was composed with amorphous and AT13. The iron-based amorphous composite coating could be successfully prepared by HVAF technology.

The cross-section of the composite coating with iron-based amorphous and 15 wt.% AT13 is shown in Figure 3b. It is found that the as-sprayed coating is closely bonded to the substrate and the AT13 particles are distributed homogeneously; the composite coatings are dense structures. All of coatings had a thickness of about 300 μm.

The porosities of coatings were calculated based on the SEM photos. The statistical results are shown in Figure 4b. With the addition of AT13 into iron-based amorphous coating, the porosity is reduced. Meanwhile, the smallest porosity is obtained for the composite coating with 15% AT13.

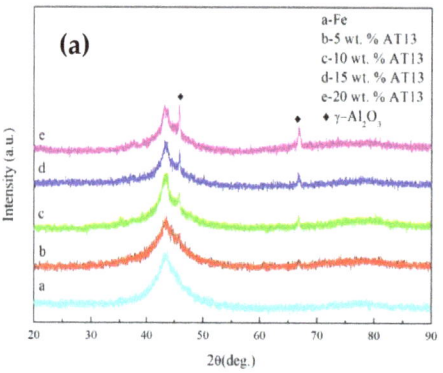 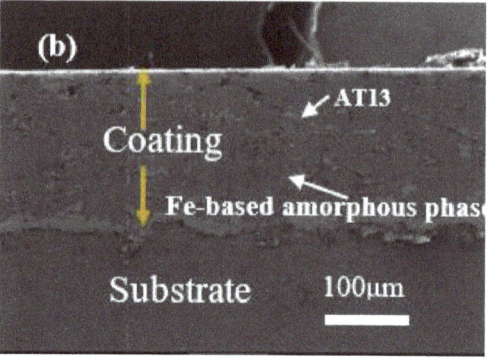

Figure 3. (a) XRD patterns of composite coatings with various content of AT13; (b) surface morphology of iron-based amorphous coating with 15 wt.% AT13.

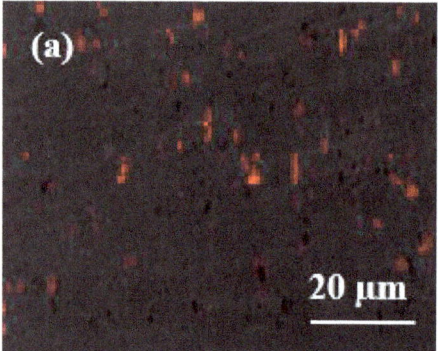 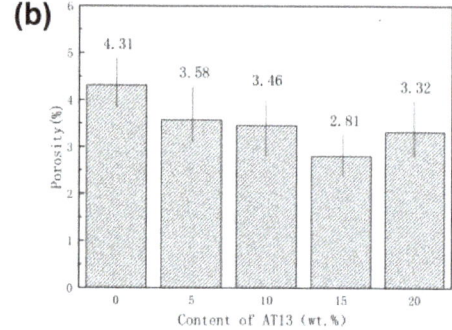

Figure 4. (a) SEM photo of coatings for calculation of porosity and (b) statistical porosities of coatings.

Firstly, the potendiodynamic polarization curves of different samples with various content AT13 composite coatings were investigated as shown in Figure 5. Some electrochemical parameters, such as corrosion potential (E_{corr}), corrosion current density (i_{corr}), transpassive current density (i_{pass}), transpassive potential (E_{tr}) and corrosion rate, are listed in Table 2. It clearly shows that the corrosion resistance of composite coatings is improved with the introduction of AT13. The E_{corr} of the composite coating is larger than that of iron-based amorphous coating, and the i_{corr} of the composite coating is lower than that of iron-based amorphous coating. Meanwhile, it is worthy to note that the composite coating with 15 wt.% AT13 has the largest corrosion potential and transpassive potential. This indicates that the best corrosion resistance was obtained for the composite coating with 15 wt.% AT13.

Furthermore, EIS measurement was adopted to evaluate the corrosion failure process, and the immersion experiments in SRB solution combined with EIS measurements were carried out to analyze the corrosion behavior of the coating. Figure 6 illustrates the EIS plots of the coatings with 15 wt.% AT13 compared with iron-based amorphous coatings immersed in SRB solution for 31 d. The Nyquist plots are shown in Figure 6a,c. It is shown that the capacitive arc decreases firstly in a small range for 1d. It is considered that the corrosive solution gradually penetrated into the coating from the pores of the surface. This indicates the corrosion resistance is reduced at the beginning.

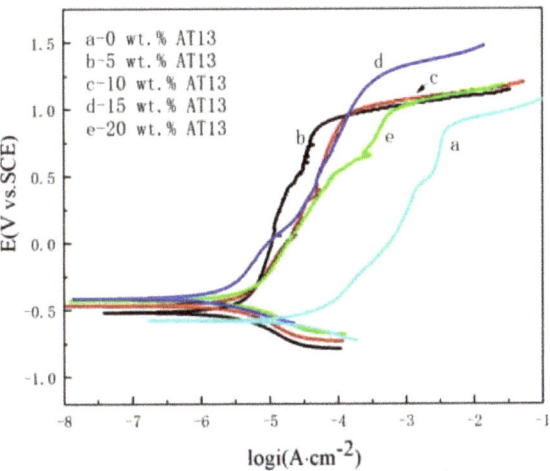

Figure 5. Potentiodynamic polarization curves of different samples with various contents of AT13.

Table 2. Electrochemical parameters obtained from potentiodynamic polarization curves of coatings.

Content of AT13/wt.%	E_{corr}/mV	i_{corr}/A cm^{-2}	i_{pass}/A cm^{-2}	E_{tr}/mV	Corrosion Rate/mpy
0	−580	5.14×10^{-5}	5.42×10^{-3}	867	25.06
5	−519	7.50×10^{-6}	6.23×10^{-5}	905	6.47
10	−467	5.01×10^{-6}	1.12×10^{-4}	993	5.49
15	−411	1.75×10^{-6}	5.37×10^{-4}	1236	2.07
20	−430	4.06×10^{-6}	6.94×10^{-4}	1012	4.60

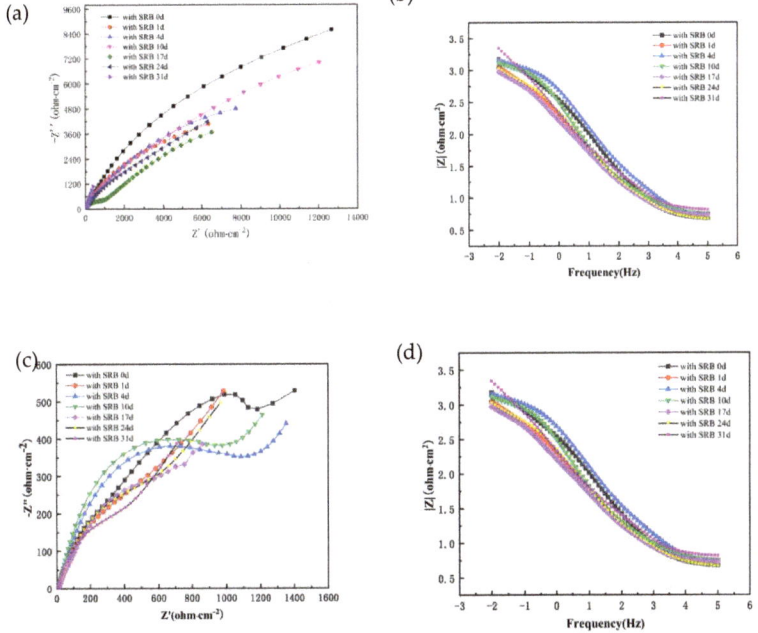

Figure 6. Electrochemical impedance spectrogram of Fe-based amorphous coating soaked in SRB solution: (**a**) Nyquist plot; (**b**) Bode plot; and composite coating with 15 wt.% AT 13 soaked in SRB solution: (**c**) Nyquist plot; (**d**) Bode plot.

However, the radius of the capacitive reactance arc for both the iron-based amorphous coating and the composite coating increased after being immersed for 4 d and 10 d. Meanwhile, for the composite coating with AT13, the radius of the capacitive reactance arc was larger than that of the initial stage. The largest radius was obtained when the soak time was 10 d. Generally, the large radius of the capacitive reactance arc means high corrosion resistance. Thus, it is suggested that the corrosion resistance is improved after immersion for 10 d. The arc radius of high-frequency capacitive reactance is related to the charge transfer resistance and corrosion product film. The increase of capacitive reactance arc radius indicates that the charge transfer resistance is increased, which means the protective ability of the corrosion product film is enhanced. On the other hand, it is worthy to note that Warburg impedance can be observed in the composite coating. this is caused by the diffusion of corrosive ions in solution. Meanwhile, the Bode impedance plots of the composite coating show two continuous semi-circle shapes for 4 d and 10 d (as shown in Figure 6d). This means that there are two time constants. One time constant is in the high frequency range related to the capacitance impedance of the coating. The other constant is the Warburg impedance caused by the reaction metabolic process of SRB.

As immersion time increased, the radius of capacitance arc was reduced gradually after 17 d. The smallest radius of the capacitance arc was obtained for the iron-based amorphous coating. This indicates more defects are formed in the coating due to corrosion. The other time constant in the Bode plot in the low frequency range is related to charge transfer resistance and a double-layer capacitance between the solution/coating interface.

According to the Bode plot, the EIS curves are fitted by the proposed equivalent circuit. There are two stages for the corrosive process. For the initial stage, the equivalent circuit model (Figure 7a) is used, which is composed of the resistance of electrolyte solution (R_s), the CPE of coating (Q_c) in parallel with the resistance of coating (R_{ct}), and inductive reactance W [25–27]. Model B (Figure 7b) is applied to analyze the later corrosion stage of the composite coatings. For model B, CPE-cf represents the characteristics of the external film layer and film resistance (R_f). The EIS fitting parameters of various components are summarized in Table 3. The Rct is decreased firstly and then increased. Finally, it is reduced. This indicates the initial increase of corrosion resistance. The biofilm is broken, so the electrolyte solution penetrates into the coating from the pores.

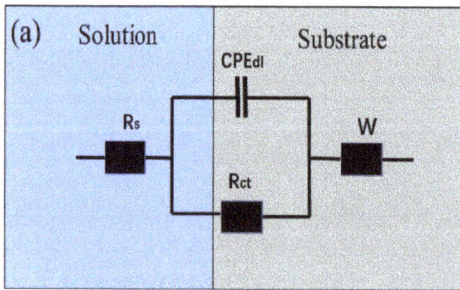 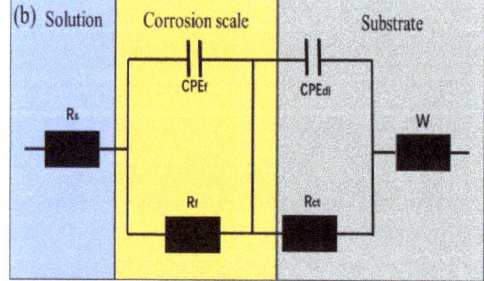

Figure 7. Schematic diagram of the equivalent circuit: (**a**) the pre-corrosion stage of the coating (0~10 d); (**b**) the corrosion stage of the coating (17~31 d).

Table 3. Summary of the EIS fitting parameters for various components.

Time (day)	R_s ($\Omega \cdot cm^{-2}$)	R_{ct} ($\Omega \cdot cm^{-2}$)	CPE_{dl}	R_f ($\Omega \cdot cm^{-2}$)	CPE_f	Goodness of Fit
0	4.56	950.1	8.58×10^{-4}	/	/	2.20×10^{-3}
1	5.036	402.1	2.07×10^{-3}	/	/	1.13×10^{-3}
4	3.845	838.3	5.05×10^{-4}	/	/	1.08×10^{-3}
10	5.026	680.4	7.94×10^{-4}	/	/	1.09×10^{-3}
17	4.992	177.1	8.80×10^{-3}	588.7	3.14×10^{-3}	3.19×10^{-4}
24	4.492	133.4	4.23×10^{-4}	489.8	1.46×10^{-3}	4.04×10^{-5}
31	2.781	101.6	1.2×10^{-1}	274.8	2.04×10^{-3}	6.47×10^{-4}

According to the results of the EIS, the corrosion process can be divided into two stages: the pre-corrosion stage and the corrosion stage. When the sample is immersed in the SRB solution for 1d, the adhesion of some microorganisms on the surface can be observed, as marked in Figure 8a. With the increased immersion time, the microbial film is formed with the metabolic process. So, the Warburg impedance appears in the EIS results due to the transfer of charge. On the other hand, the microbial film protects the substrate from corrosion. Therefore, the capacitive reactance arc radius is increased from 1 d to 10 d.

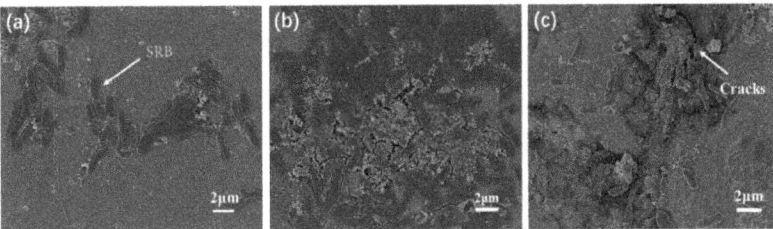

Figure 8. Corrosion morphology of AT13 composite coating: (**a**) soak for 1d; (**b**) soak for 17 days; (**c**) soak for 31 days.

However, when the immersion time exceeds 17 d, the broken microbial film can be observed as shown in Figure 8b. The corrosion resistance is reduced. Meanwhile, the capacitance is formed between the microbial film and the composite coating. The corrosive solution penetrates into the composite coating. When the immersion time is 31 d, the microbial film is peeled, and the crack on the composite coating is observed, as shown in Figure 8c.

Meanwhile, the corrosive surface of the composite coating was also observed by fluorescence microscope after 1 d, 17 d and 31 d, as shown in Figure 9. The green fluorescence represents the adhesion of SRB on the surface. It can be found that there are some SRBs on the coating surface when it is immersed for one day. With the increase of immersion time, more and more SRBs appear on the surface of the coating. When the sample is immersed for 31 d, some colonies are formed on the surface of the coating.

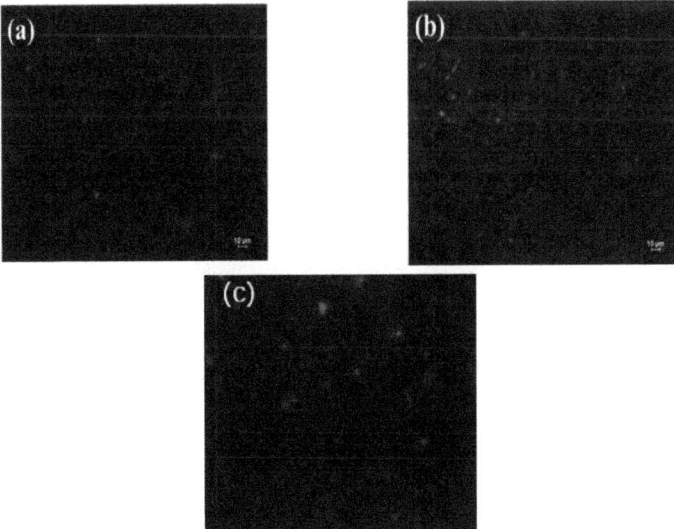

Figure 9. Adhesion of SRB on the surface of the composite coating after immersion for (**a**) 1 d, (**b**) 17 d, (**c**) 31 d.

The corrosion products were analyzed by X-ray photoelectron spectroscopy (XPS). Figure 10a shows the main distribution of sulfur ions in the corrosion products. S comes from the metabolite of SRB bacteria. The sulfur element is composed of four peaks, mainly SO_4^{2-} and S^{2-}. The peak area of S^{2-} is larger than that of SO_4^{2-}. According to the BCSR theory, SRB will consume SO_4^{2-}, and the final product is HS^-; the reactions are as follows [25,26]:

$$H_2O \rightarrow 2H^+ + OH^- \tag{1}$$

$$H^+ \xrightarrow{Hydrogenase} [H] \tag{2}$$

$$SO_4^{2-} + 8[H] \xrightarrow{SRB} 4H_2O + S^{2-} \tag{3}$$

where [H] acts as an electron transfer medium, which is generated by the cathode reaction. During the metabolic process of the SRB, the film is formed on the surface of the composite coating.

However, the metabolic process enhances the acidity covering the coating at the same time. This results in the corrosion of the coatings. The Fe element is transferred to Fe^{2+} and Fe^{3+} as follows:

$$Fe \rightarrow Fe^{2+} + 2e^- \tag{4}$$

$$Fe \rightarrow Fe^{3+} + 3e^- \tag{5}$$

$$Fe^{3+} + 3OH^+ \rightleftharpoons Fe(OH)_3 \tag{6}$$

$$Fe^{2+} + 2OH^+ \rightleftharpoons Fe(OH)_2 \tag{7}$$

$$Fe^{2+} + S^{2-} \rightarrow 2FeS \tag{8}$$

FeS is formed with the metabolic product of the SRB. The XPS spectrum of Fe, $Fe(OH)_3$ and FeS are mainly formed as shown in Figure 10b.

Meanwhile, there are three peaks in the XPS spectrum of the Al element, as shown in Figure 9c. They are Al_2O_3, Al_2S_3 and $Al(OH)_3$. The reactions are as follows:

$$Al^{3+} + S^{2-} \rightarrow Al_2S_3 \tag{9}$$

$$Al^{3+} + 3OH^+ \rightarrow Al(OH)_3 \tag{10}$$

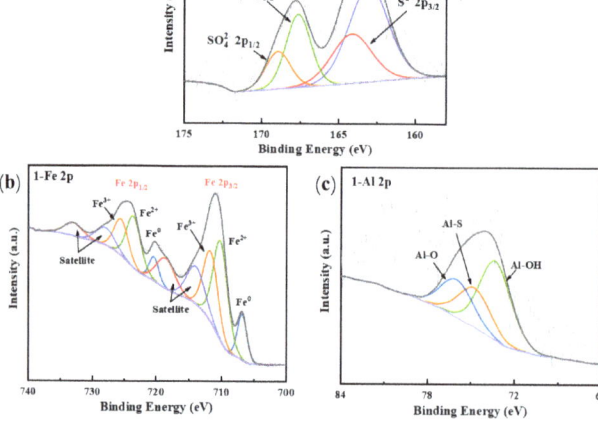

Figure 10. XPS atlas of corrosion product elements: (**a**) S; (**b**) Fe; (**c**) Al.

4. Discussion

At present, researchers have proposed various theories to explain the corrosion mechanism of SRB [27–29]. One of them is the cathode depolarization theory (CDT), which reported that the corrosion is caused by utilizing a hydrogenase to consume the hydrogen produced by cathodic reactions [30]. Another one is the mechanism of sulfides produced by the metabolism of SRB [31]. The biocatalytic cathodic sulfate reduction (BCSR) theory was proposed and developed by Gu [32] and Xu et al. [33]. The studies showed that SRB can induce MIC by acting as the acceptor of the electrons produced by anodic reactions (metal dissolution), which generally prevails under the conditions lack of carbon source.

According to above results, the corrosion mechanism is proposed. The schematic of the corrosion mechanism is shown in Figure 11. In the SRB environment, [H] is as a medium for electron transfer, so the corrosion of SRB causes a large amount of H to leave the composite surface, accelerating the cathode reaction. At the same time, the anode reaction remains stable, resulting in a large number of electrons being consumed on the composite surface. This indicates that the reaction through which SRB consumes [H] further promotes the metabolic activity of SRB. Meanwhile, the biofilm is formed, which can effectively block the contact between corrosive factors in the solution. However, when the corrosion products in the film cannot be metabolized, this leads to the deterioration of the film environment. At the same time, the lack of nutrients in the film will also cause the bacteria in the biofilm to start to take electrons from the composite coating. Then, with the acidity increase, the SRB corrosion produces porous FeS, which accelerates the absorption of H and the cathodic depolarization process, accelerating the corrosion rate. Finally, the film cracks due to the change of the internal environment, re-exposing the composite coating in the solution.

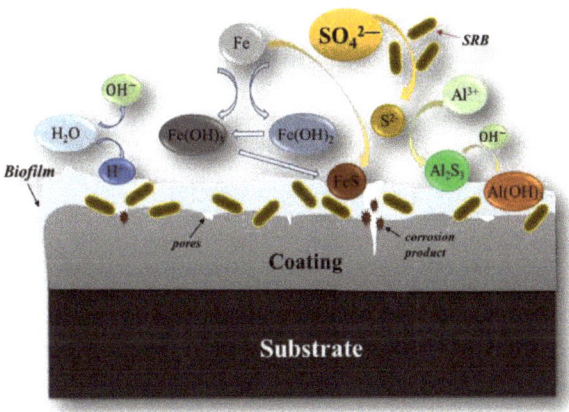

Figure 11. The schematic of the corrosion mechanism.

5. Conclusions

The composite coatings with iron-based amorphous and AT13 were prepared by HVAF. The corrosion process in SRB bacteria was studied by EIS. The corrosion products were analyzed by SEM and XPS. Based on the results, the corrosion mechanism was proposed. The SRB corrosion process can be divided into two stages: one is microbial adhesion, which is accompanied with SRB metabolic process and biofilm is formed. The other one is failure of the biofilm. With the acidity increase, the corrosion produces porous FeS and Al(OH)$_3$ increase, which accelerates the absorption of H and the cathodic depolarization process, accelerating the corrosion rate. Finally, the biofilm is destroyed and the composite coating is re-exposed in the solution.

Author Contributions: Formal analysis, J.X., X.Z. and Z.Z.; Investigation, Z.C. and Q.H.; Data curation, H.S., F.X. and F.W.; Writing—original draft preparation, Z.C. and F.X.; Writing—review and editing, Z.C. and H.S. All authors have read and agreed to the published version of the manuscript.

Funding: The present work was supported by the National Nature Science Foundation of China (Grant No. 51872072). The authors would like to express their gratitude for the support from the Fishery Engineering and Equipment Innovation Team of Shanghai High-level Local University, the State Key Laboratory for Mechanical Behavior of Materials, the program for Shanghai Collaborative In-novation Center for Cultivating Elite Breeds and Green-culture of Aquaculture animals (No. 2021-KJ-02-12) and the Natural Science Foundation of Shanghai (No. 20ZR1424000).

Institutional Review Board Statement: Not applicable.

Informed Consent Statement: Not applicable.

Data Availability Statement: Not applicable.

Conflicts of Interest: The authors declare no conflict of interest.

References

1. Enning, D.; Venzlaff, H.; Garrelfs, J.; Dinh, H.T.; Meyer, V.; Mayrhofer, K.; Hassel, A.W.; Stratmann, M.; Widdel, F. Marine sulfate-reducing bacteria cause serious corrosion of iron under electroconductive biogenic mineral crust. *Environ. Microbiol.* **2012**, *14*, 1772–1787. [CrossRef] [PubMed]
2. Liu, B.; Sun, M.H.; Lu, F.Y.; Du, C.W.; Li, X.G. Study of biofilm-influenced corrosion on X80 pipeline steel by a nitrate-reducing bacterium, bacillus cereus, in artificial Beijing soil. *Colloids Surf. B Biointerfaces* **2021**, *197*, 111356. [CrossRef] [PubMed]
3. Wang, D.; Liu, J.; Jia, R.; Dou, W.; Kumseranee, S.; Punpruk, S.; Li, X.; Gu, T. Distinguishing two different microbiologically influenced corrosion (MIC) mechanisms using an electron mediator and hydrogen evolution detection. *Corros. Sci.* **2020**, *177*, 108993. [CrossRef]
4. Dai, X.; Wang, H.; Ju, L.-K.; Cheng, G.; Cong, H.; Newby, B.M.Z. Corrosion of aluminum alloy 2024 caused by Aspergillus niger. *Int. Biodeterior. Biodegrad.* **2016**, *115*, 1–10. [CrossRef]
5. Kalnaowakul, P.; Xu, D.; Rodchanarowan, A. Accelerated corrosion of 316L stainless steel caused by Shewanella algae biofilms. *ACS Appl. Bio Mater.* **2020**, *3*, 2185–2192. [CrossRef]
6. Guo, Z.; Pan, S.; Liu, T.; Zhao, Q.; Wang, Y.; Guo, N.; Chang, X.; Liu, T.; Dong, Y.; Yin, Y. Bacillus subtilis inhibits Vibrio natriegens-induced corrosion via biomineralization in seawater. *Front. Microbiol.* **2019**, *10*, 1111. [CrossRef]
7. Yang, J.X.; Zhao, P.; Sun, C.; Xu, J. Effect of sulfate-reducing bacteria on the crevice corrosion behavior of Q235 steel. *J. Chin. Soc. Corros. Prot.* **2012**, *32*, 54–58. [CrossRef]
8. Cui, L.Y.; Liu, Z.Y.; Xu, D.K.; Hu, P.; Li, X.G. The study of microbiologically influenced corrosion of 2205 duplex stainless steel based on high-resolution characterization. *Corros. Sci.* **2020**, *174*, 108842. [CrossRef]
9. Castaneda, H.; Benetton, X.D. SRB-biofilm influence in active corrosion sites formed at the steel-electrolyte interface when exposed to artificial seawater conditions. *Corros. Sci.* **2008**, *50*, 1169–1183. [CrossRef]
10. Lin, B.L.; Lu, J.T.; Kong, G. Effect of molybdate post-sealing on the corrosion resistance of zinc phosphate coatings on hot-dip galvanized steel. *Corros. Sci.* **2008**, *50*, 962–967. [CrossRef]
11. Kartsonakis, I.; Balaskas, A.; Koumoulos, E.; Charitidis, C.; Kordas, G. Incorporation of ceramic nanocontainers into epoxy coatings for the corrosion protection of hot dip galvanized steel. *Corros. Sci.* **2012**, *57*, 30–41. [CrossRef]
12. Li, Y. Formation of nano-crystalline corrosion products on Zn-Al alloy coating exposed to seawater. *Corros. Sci.* **2001**, *43*, 1793–1800. [CrossRef]
13. Zhai, X.; Myamina, M.; Duan, J.; Hou, B. Microbial corrosion resistance of galvanized coatings with 4, 5-dichloro-2-n-octyl-4-isothiazolin-3-one as a biocidal ingredient in electrolytes. *Corros. Sci.* **2013**, *72*, 99–107. [CrossRef]
14. Maia, F.; Silva, A.P.; Fernandes, S.; Cunha, A.; Almeida, A.; Tedim, J.; Zheludkevich, M.L.; Ferreira, M.G.S. Incorporation of biocides in nanocapsules for protective coatings used in maritime applications. *Chem. Eng. J.* **2015**, *270*, 150–157. [CrossRef]
15. Zhuk, I.; Jariwala, F.; Attygalle, A.B.; Wu, Y.; Libera, M.R.; Sukhishvili, S.A. Self-defensive layer-by-layer films with bacteria-triggered antibiotic release. *ACS Nano* **2014**, *8*, 7733–7745. [CrossRef]
16. Taubes, G. The Bacteria Fight Back, American Association for the Advancement of Science. *Science* **2008**, *321*, 356–361. [CrossRef]
17. Chu, Z.; Deng, W.; Zheng, X.; Zhou, Y.; Zhang, C.; Xu, J.; Gao, L. Corrosion Mechanism of Plasma-Sprayed Fe-Based Amorphous Coatings with High Corrosion Resistance. *J. Therm. Spray Technol.* **2020**, *29*, 1111–1118. [CrossRef]
18. Huang, G.; Qu, L.; Lu, Y.; Wang, Y.; Li, H.; Qin, Z.; Lu, X. Corrosion resistance improvement of 45 steel by Fe-based amorphous coating. *Vacuum* **2018**, *153*, 39–42. [CrossRef]
19. Luo, Q.; Sun, Y.J.; Jiao, J.; Wu, Y.X.; Qu, S.J.; Shen, J. Formation and tribological behavior of AC-HVAF-sprayed nonferromagnetic Fe-based amorphous coatings. *Surf. Coat. Technol.* **2018**, *334*, 253–260. [CrossRef]
20. Shang, X.; Zhang, C.; Xv, T.; Wang, C.; Lu, K. Synergistic effect of carbide and amorphous phase on mechanical property and corrosion resistance of laser-clad Fe-based amorphous coatings. *Mater. Chem. Phys.* **2021**, *263*, 124407. [CrossRef]

21. Zhang, C.; Chu, Z.; Wei, F.; Yang, Y.; Dong, Y.; Huang, D.; Wang, L. Optimizing process and the properties of the sprayed Fe-based metallic glassy coating by plasma spraying. *Surf. Coat. Technol.* **2017**, *319*, 1–5. [CrossRef]
22. Zhou, Z.; Wang, L.; Wang, F.C.; Zhang, H.F.; Liu, Y.B.; Xu, S.H. Formation and corrosion behavior of Fe-based amorphous metallic coatings by HVOF thermal spraying. *Surf. Coat. Technol.* **2009**, *204*, 563–570. [CrossRef]
23. Chu, Z.; Zheng, X.; Zhang, C.; Xu, J.; Gao, L. Study the effect of AT13 addition on the properties of AT13/Fe-based amorphous composite coatings. *Surf. Coat. Technol.* **2019**, *379*, 125053–125060. [CrossRef]
24. Chu, Z.; Wei, F.; Zheng, X.; Zhang, C.; Yang, Y. Microstructure and properties of TiN/Fe-based amorphous composite coatings fabricated by reactive plasma spraying. *J. Alloys Compd.* **2019**, *785*, 206–213. [CrossRef]
25. Ali, N.; Zada, A.; Zahid, M.; Ismail, A.; Rafiq, M.; Riaz, A.; Khan, A. Enhanced photodegradation of methylene blue with alkaline and transition-metal ferrite nanophotocatalysts under direct sun light irradiation. *J. Chin. Chem. Soc.* **2019**, *66*, 402–408. [CrossRef]
26. Yasmeen, H.; Zada, A.; Liu, S. Surface plasmon resonance electron channeled through amorphous aluminum oxide bridged ZnO coupled g-C3N4 significantly promotes charge separation for pollutants degradation under visible light. *J. Photochem. Photobiol. A Chem.* **2020**, *400*, 112681. [CrossRef]
27. Zada, A.; Qu, Y.; Ali, S.; Sun, N.; Lu, H.; Yan, R.; Zhang, X.; Jing, L. Improved visible-light activities for degrading pollutants on TiO$_2$/g-C$_3$N$_4$ nanocomposites by decorating SPR Au nanoparticles and 2,4-dichlorophenol decomposition path. *J. Hazard. Mater.* **2018**, *342*, 715–723. [CrossRef]
28. Liu, H.; Cheng, Y.F. Mechanism of microbiologically influenced corrosion of X52 pipeline steel in a wet soil containing sulfate-reduced bacteria. *Electrochim. Acta* **2017**, *253*, 368–378. [CrossRef]
29. Dou, W.; Ru, J.; Peng, J.; Liu, J.; Chen, S.; Gu, T. Investigation of the mechanism and characteristics of copper corrosion by sulfate reducing bacteria. *Corros. Sci.* **2018**, *144*, S0010938X17322916. [CrossRef]
30. Anandkumar, B.; George, R.P.; Maruthamuthu, S.; Parvathavarthini, N.; Mudali, U.K. Corrosion characteristics of sulfate-reducing bacteria (SRB) and the role of molecular biology in SRB studies: An overview. *Corros. Rev.* **2016**, *34*, 41–63. [CrossRef]
31. Marciales, A.; Peralta, Y.; Haile, T.; Crosby, T.; Wolodko, J. Mechanistic microbiologically influenced corrosion modeling—A review. *Corros. Sci.* **2019**, *146*, 99–111. [CrossRef]
32. Gu, T.Y. New Understandings of biocorrosion mechanisms and their classifications. *J. Microb. Biochem. Technol.* **2012**, *4*, 3–6. [CrossRef]
33. Xu, D.K.; Gu, T.Y. Carbon source starvation triggered more aggressive corrosion against carbon steel by the desulfovibrio vulgaris biofilm. *Int. Biodeter. Biodegr.* **2014**, *91*, 74–81. [CrossRef]

Article

Increasing the Life Span of Tools Applied in Cheese Cutting Machines via Appropriate Micro-Blasting

Georgios Skordaris [1,*], Konstantinos Vogiatzis [1], Leonidas Kakalis [1], Ioannis Mirisidis [2], Vasiliki Paralidou [2] and Soultana Paralidou [2]

1 Laboratory for Machine Tools and Manufacturing Engineering, Mechanical Engineering Department, Aristotle University of Thessaloniki, 54124 Thessaloniki, Greece
2 Paralidis S.A., 8th Km Thessaloniki-Katerini, Kalochori, 57009 Thessaloniki, Greece
* Correspondence: skordaris@meng.auth.gr; Tel.: +30-2310-996-027

Abstract: The potential to increase the life span of tools applied in cheese cutting machines is of great importance, considering their cost and the risk of fragmented metallic parts of the tool being inserted into the cheese. Such tools are commonly manufactured using stainless steel 405 and are subjected to dynamic loads during their operation, leading to fatigue failure. An efficient method to improve the fatigue properties of such tools is the application of micro-blasting. In this work, for the first time, an experimental–analytical methodology was developed for determining optimum micro-blasting conditions and ascertaining a preventive replacement of the tool before its extensive fracture. This methodology is based on the construction of a pneumatic system for the precise cutting of cheese and simultaneous force measurements. Additionally, the entire cheese-cutting process is simulated by appropriate FEA modeling. According to the attained results, micro-blasting on steel tools significantly improves the resistance against dynamic loads, whilst the number of impacts that a tool can withstand until fatigue fracture is more than three times larger. Via the developed methodology, a preventive replacement of the tool can be conducted, avoiding the risk of a sudden tool failure. The proposed methodology can be applied to different tool geometries and materials.

Keywords: cheese cutting; tools; micro-blasting; fatigue

Citation: Skordaris, G.; Vogiatzis, K.; Kakalis, L.; Mirisidis, I.; Paralidou, V.; Paralidou, S. Increasing the Life Span of Tools Applied in Cheese Cutting Machines via Appropriate Micro-Blasting. *Coatings* 2022, 12, 1343. https://doi.org/10.3390/coatings12091343

Academic Editor: Jinyang Xu

Received: 24 August 2022
Accepted: 9 September 2022
Published: 15 September 2022

Publisher's Note: MDPI stays neutral with regard to jurisdictional claims in published maps and institutional affiliations.

Copyright: © 2022 by the authors. Licensee MDPI, Basel, Switzerland. This article is an open access article distributed under the terms and conditions of the Creative Commons Attribution (CC BY) license (https:// creativecommons.org/licenses/by/ 4.0/).

1. Introduction

The production of standard pieces of cheese concerning weight is a complicated process, and its interruption due to the failure of the cutting tool entails prolonged production time due to worn tool replacements and increased cost. In such a production process, the tools are commonly made of stainless steel, and they are subjected to dynamic loads during the cheese-cutting process. In this way, fatigue failure is the prevailing wear phenomenon, significantly reducing tool life [1,2]. Various experimental techniques supported by analytical models have been developed in the past for evaluating the fatigue behavior of coated and uncoated steel parts possessing different geometries [3–6]. Moreover, the perpendicular impact test is an efficient method for characterizing the material's fatigue endurance [7]. By conducting this test, the required force after one million impacts for the initiation of failure is determined. Based on the appropriate FEA model, the maximum equivalent stress developed in the material at the fatigue threshold force during its loading and the remaining one due to the plastic deformation is calculated [7]. As a consequence, the Smith-like diagram, as well as the Woehler diagram of the material, can be determined.

The success of products strongly depends on the properties of the outermost layer, which can be significantly enhanced by appropriate surface treatments and suitable coatings. In the past decade, the field of surface engineering has gained a leading role in materials science and engineering and attracted great scientific and research interest with a view to producing cost-effective and advanced multifunctional materials. In this way, the effect of various surface treatments, such as sandblasting, shot peening, etc., applied to different materials on their mechanical properties were extensively investigated by conducting well-established experimental procedures [8–10]. Among the examined parameters of shot-peening was the application of semi-random or regular shot peening [11], different pressures [9], etc. Furthermore, theoretical investigations were conducted to examine the effects of shot distance and impact sequence on the residual stress field in shot-peening [12]. Based on these studies, the improvement of the mechanical and fatigue properties of different materials after shot peening was revealed [13,14]. Moreover, in the tool industry, micro-blasting has been registered as an efficient method for increasing the life of coated tools [15–17]. However, the applied micro-blasting conditions have to be carefully selected since this process can lead to an augmentation of the cutting-edge radius, resulting in reduced cutting tool ability [18,19]. In the frame of the conducted research, the potential of an effective application of micro-blasting on tools applied for cheese cutting was investigated.

More specifically, various dry micro-blasting conditions were applied to the tool in order to attain a similar magnitude of cutting-edge roundness to the pristine one. An experimental–analytical methodology was developed for predicting the number of impacts that the tool can receive during cheese cutting before its fatigue fracture initiation. In this context, the mechanical properties of the untreated and micro-blasted tools were determined by nanoindentations and perpendicular impact tests coupled with appropriate FEA simulations. An appropriate device was designed and developed based on a pneumatic system for measuring the required forces for cheese cutting using micro-blasted and untreated tools. Finally, 3D-FEA models were developed using ANSYS software for simulating the cheese-cutting process and thus to determine the developed equivalent stresses using the calculated forces as input data. It must be pointed out that the modeling of the cutting processes has been denoted as a valuable tool for studying and predicting the tool life [1,20,21]. Based on the developed methodology, a preventive replacement of the tool can be conducted, avoiding the risk of a sudden tool failure.

2. Materials and Methods

2.1. The Used Tools and the Employed Devices

The geometry of the tools employed for cheese cutting is illustrated in Figure 1. The tools were made of annealed stainless steel 405. The radius of the cutting edge was measured using white light scanning by a 3D confocal system mSURF of NANOFOCUS AG [1], and amounted to 10 μm. Dry micro-blasting was carried out on tools using sharp-edged Al_2O_3 or spherical ZrO_2 grains, as illustrated in the same figure. The average grain size was 10 μm. Due to the different grains' geometry, the cutting edge topomorphy was expected to be variously affected.

The dry micro-blasting process was conducted using the WIWOX DI12SF located in the Laboratory for Machine Tools and Manufacturing Engineering of the Aristotle University of Thessaloniki (see Figure 2). All the micro-blasting parameters were kept constant except for the pressure. More specifically, the micro-blasting duration was equal to 4 s, the distance between the blasting nozzle and the tools was set to 100 mm, and the applied pressure amounted to 0.1 or 0.2 MPa. For determining the mechanical properties of the employed materials, nanoidentations were conducted by a FISCHERSCOPE H100 device (Helmut Fischer GmbH, Sindelfingen, Germany). The fatigue properties of the untreated and micro-blasted tools were assessed by perpendicular impact tests using an impact tester designed and manufactured by the Laboratory for Machine Tools and Manufacturing Engineering of the Aristotle University of Thessaloniki in collaboration with the company Impact-BZ (London, UK) [22]. The employed ceramic ball had a diameter of 5 mm. The applied time-dependent force signal is shown in Figure 2. Three-dimensional measurement facilities of the confocal microscope SURF of NANOFOCUS AG (Oberhausen, Germany) were used for evaluating the impact imprints. ANSYS 2021 R1 software was used for simulating the nanoindentation and impact tests and thus to determine the tool's mechanical properties [1].

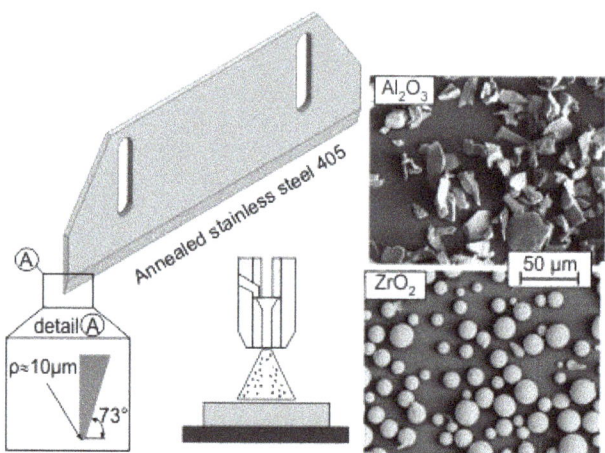

Figure 1. The applied tool geometry for cheese cutting and the used grains for micro-blasting.

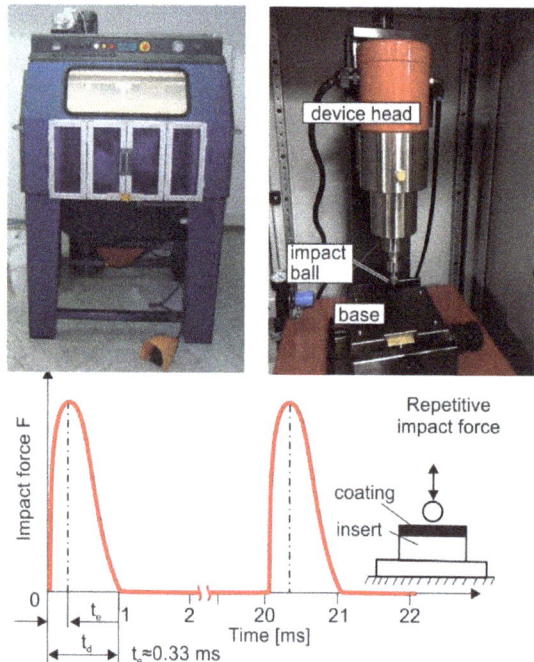

Figure 2. The employed devices for conducting micro-blasting and impact experiments as well as the applied force signal during impact test.

2.2. The Developed Device for Cheese Cutting

For conducting cutting experiments on cheese possessing different hardness, an experimental device was designed and manufactured by the Laboratory for Machine Tools and Manufacturing Engineering (LMTME) of the Mechanical Engineering Department at the

Aristotle University of Thessaloniki (see Figure 3). The cutting mechanism consists of a cylindrical shaped piston that uses air pressure in order to move the cutting knife up and down. For stabilizing the device, two high-rigidity beams were used, in which there is the possibility of moving the piston mechanism according to the needs of the experiment. The maximum knife's displacement is approximately 160 mm. In this way, cheese with different geometries can be cut. Between the piston and the cutting knife, a force measurement sensor was set in order to directly record the cutting force of the cheese. The operation of the developed device was controlled and monitored by a suitably developed algorithm using commercial computer software (Labview 8.6), thus enabling its fully automated operation. More specifically, the analog signal of the measurement device was turned into digital through an AT converter. The digital control, the data processing, the recording, and the results presentation were achieved through a developed algorithm using "Labview" software (8.6). Typical registered results, such as force, displacement, and velocity versus the experimental time taken from the developed software are presented at the bottom of Figure 3.

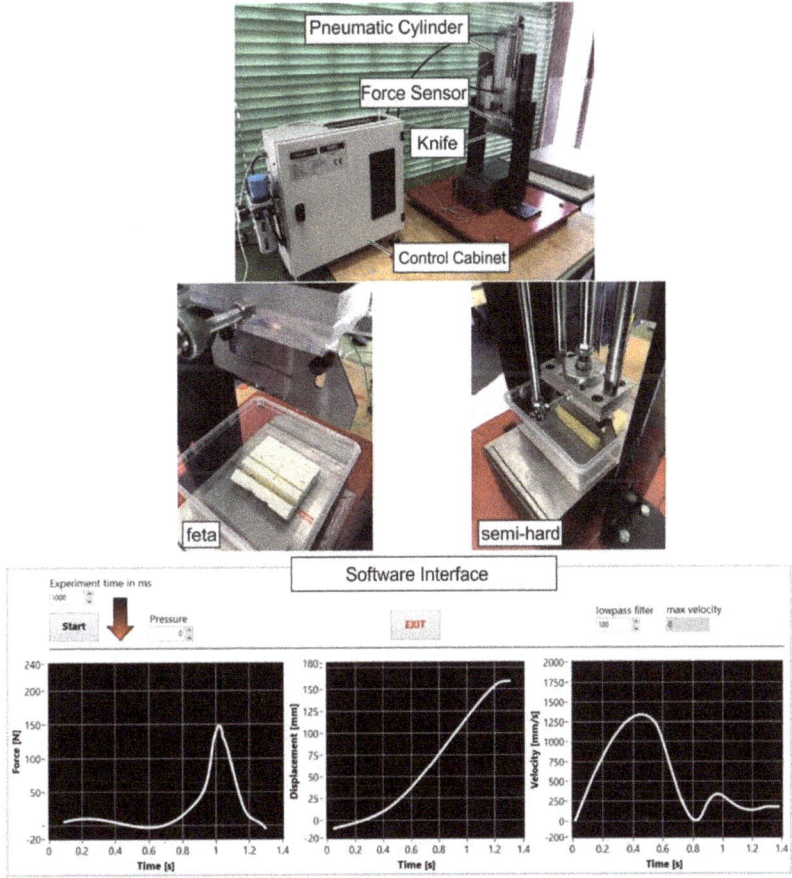

Figure 3. The developed device for measuring the cutting forces during cutting cheese experiments.

2.3. The Developed 3D-FEA Models for Simulating Cheese Cutting

Three-dimensional FEA models were developed using ANSYS 2021 R1 software for simulating the cheese-cutting process when untreated tools and micro-blasted ones were employed, as shown in Figure 4. The kinematic hardening rule was applied in the developed FEA model due to the fact that it leads to a rapid convergence in the corresponding FEM calculations [23].

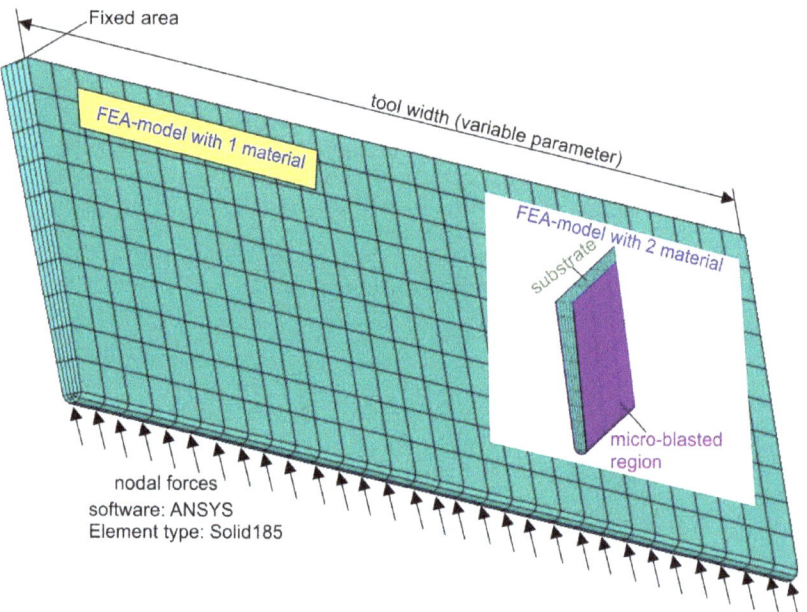

Figure 4. The developed FEA models for simulating cheese cutting experiments.

The kinematic hardening assumes that the yield surface remains constant in size and the surface translates in stress space with progressive yielding, whereas the Besseling model is used [24], also called the sub-layer or overlay model, to characterize the material behavior. In the case of the micro-blasted tool, a second material area was considered to possess the mechanical properties and the width of the deformed region of a micro-blasted tool, as will be described in the next session. A rigid film–substrate interface was considered in the case of the coated tool. The geometry of the tool was similar to that one depicted in Figure 1. The width of the tool was equal to the contact region between the cheese and the tool. Convergence studies were conducted to determine the optimal mesh density and attain a mesh-independent grid. Solid elements possessing a pyramid geometry were used for generating the meshed volume. In order to simulate the cheese cutting, perpendicular nodal forces were applied to the cutting-edge region until the calculated reaction loads in the fixed upper area were equal to the measured one. The mechanical properties of the employed materials were set accordingly to the extracted results shown in the next sessions. The materials' mechanical behavior was simulated as an isotropic using multilinear laws. With the aid of the developed FEA models, the developed maximum stresses corresponding to certain impact forces can be calculated in the case of untreated tools and micro-blasted ones.

3. Results

3.1. Hardness Characterization of the Untreated and Micro-Blasted Tools

Nanoindenation measurements were carried out to characterize the hardness of the examined materials using a Berkovich diamond indenter. The maximum indentation force was selected to be 15 mN in order to capture the effect of the micro-blasting process on superficial hardness modifications. The load–displacement diagrams of the untreated and micro-blasted tools at various pressures are illustrated in Figure 5a. To exclude the roughness effect on results accuracy, 30 measurements per nanoindentation were conducted.

In this way, the moving average of the indentation depth versus the indentation force was stabilized [1]. As observed in both micro-blasting grain cases, an increase in the pressure was associated with a reduction of the maximum indentation depth. The latter is an indication that there is a hardness augmentation after micro-blasting. This fact can be explained by considering the induced residual stresses in the material structure after micro-blasting and the resulting superficial material deformation. The effect of the different employed grains on the material hardness is more clearly visible in Figure 5b, where the course of maximum indentation force versus the micro-blasting pressure is shown. According to the attained results, the registered maximum indentation depths were comparably larger in the case of ZrO_2 grains under the same conditions. More specifically, the pristine maximum indentation depth amounted to 230 nm. After micro-blasting, the attained indentation depths at a pressure of 0.1 MPa were equal to 222 nm and 218 nm when Al_2O_3 and ZrO_2 were employed, respectively. Moreover, the increase in material hardness at higher micro-blasting pressure was not so intense since there was a limit to the induced residual stresses that a material can receive prior to its fracture.

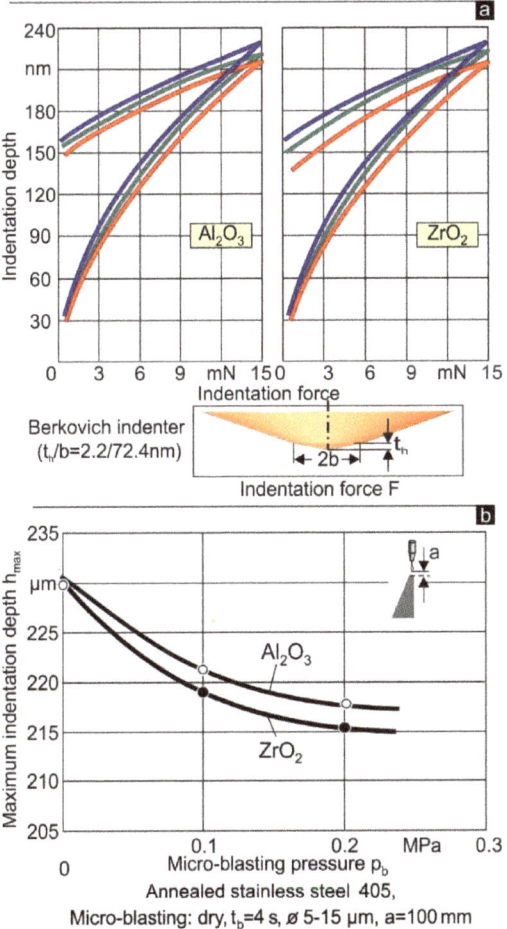

Figure 5. (a) Nanoindentation results of the untreated and micro-blasted tools; (b) maximum attained indentation depths after micro-blasting at various pressures.

In order to interpret the previous shown results, roughness measurements were conducted on variously micro-blasted tools. A characteristic roughness magnitude for showing the effect of different employed grains on surface integrity was the maximum peak to valley height Rt. Since Rt shows the height difference between the highest mountain and lowest valley within the measured range, potential superficial cracks created due to the grains' impact can be captured. The related results are shown in Figure 6. Each bar shown in this figure represents the mean value of 10 measurements. As can be observed in Figure 6, the roughness Rt became more intense when Al_2O_3 grains were used. This can be explained by considering the geometries of the used grains. Due to their sharp edges, a part of the initial kinetic energy of the Al_2O_3 grains was consumed to micro-chipping on the material surface, and in this way, the roughness increased. On the contrary, in the case of ZrO_2 grains, a bigger portion of the initial kinetic energy was used for material deformation and as a consequence, larger indentation depths were attained.

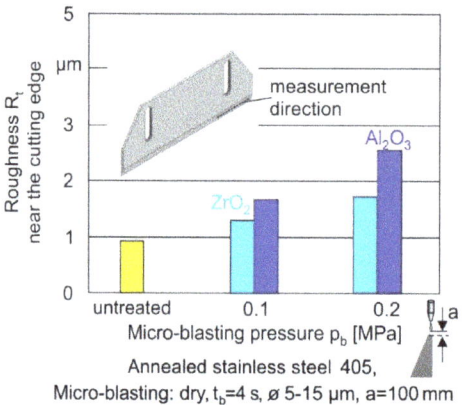

Figure 6. Roughness measurements on variously micro-blasted tools.

3.2. Cutting Edge Radius of the Untreated and Micro-Blasted Tools

A crucial issue for the effective application of tools in a cheese-cutting machine is the magnitude of their cutting-edge radii. This is attributed to the fact that the sharpness of the tool plays a dominant role in maintaining the surface integrity of the cheese after its cutting, which is desired to be as smooth as it can be. Although micro-blasting has been reported to be an efficient method for improving material hardness, it causes an enlargement of the cutting-edge radius [18]. In order to determine the variation of the cutting-edge radius of the micro-blasted tools, confocal microscopy measurements were carried out. The related results are shown in Figure 7. More specifically, it was necessary to register successive cross-sections of the cutting edges and to determine the average values (see Figure 7). The pristine cutting-edge radius of the untreated tools amounted to 10 μm. The conduct of micro-blasting resulted in an enlargement of the cutting-edge radius. Only in the case of Al_2O_3 grains and when the micro-blasting amounted to 0.1 MPa did this magnitude remain almost invariable. Based on these results, an optimum micro-blasting pressure of 0.1 MPa with Al_2O_3 grains was selected for improving the tool life.

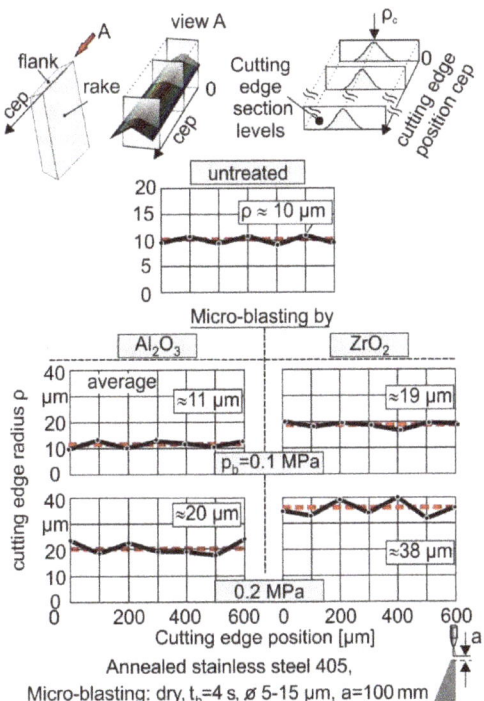

Figure 7. Measurements of cutting-edge radii via light scans along the cutting edge.

3.3. Mechanical Properties of the Untreated and Micro-Blasted Tools

For determining the mechanical properties of the untreated and micro-blasted tools at a pressure of 0.1 MPa, the FEM-based procedure concerning the stepwise simulation of the nanoindentation procedure, based on the SSCUBONI algorithm, was employed [1]. It has to be pointed out that only in these two cases the cutting-edge radii were almost equal, and the tools preserved their sharpness. In this way, using the nanoindentation results diagrams shown in Figure 5a, the input data to the developed algorithm was created, and the mechanical properties of the investigated materials were extracted. These results are illustrated in Figure 8. As observed, the elasticity modulus remained unaffected by the micro-blasting process, whereas the yield and rupture stress grew. Generally, micro-blasting results in a gradation of the strength properties [25]. In the described investigations, for simplifying the related calculations, it was assumed that evenly distributed strength properties develop up to a certain depth from the film surface. By applying the technique described in ref. [25], the occurring equivalent stress distribution during micro-blasting can be determined, and the width of the shadowed area associated with the plastically deformed region can be calculated. These data were employed in the developed FEA model, simulating the cheese-cutting process for determining the maximum developed stresses.

Furthermore, perpendicular impact tests were conducted at various impact loads in order to detect the critical ones for fatigue damage initiation in the case of untreated and micro-blasted tools. Related results in terms of the remaining imprint depths at various exercised loads after 10^6 impacts on annealed stainless steel 405 and the micro-blasted ones at a pressure of 0.1 MPa are shown in the diagram of Figure 9. To avoid the interpretation of surfaces asperity damages as material damage, the double of the tool's arithmetic roughness Ra was considered as a criterion for material fracture. According to the attained results, the critical impact loads for exceeding the above-mentioned criterion were 25 N and 50 N for the annealed stainless steel 405 and the micro-blasted ones, respectively. Characteristic impact imprints at various loads in the case of stainless steel 405 and the micro-blasted ones, scanned by white light via confocal microscopy, are shown in the bottom part of the figure.

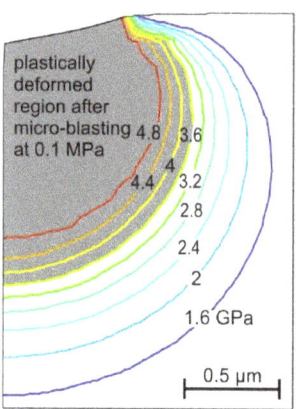

Figure 8. Mechanical properties of the untreated and micro-blasted tool at 0.1 MPa and the width of the deformed region after micro-blasting at 0.1 MPa.

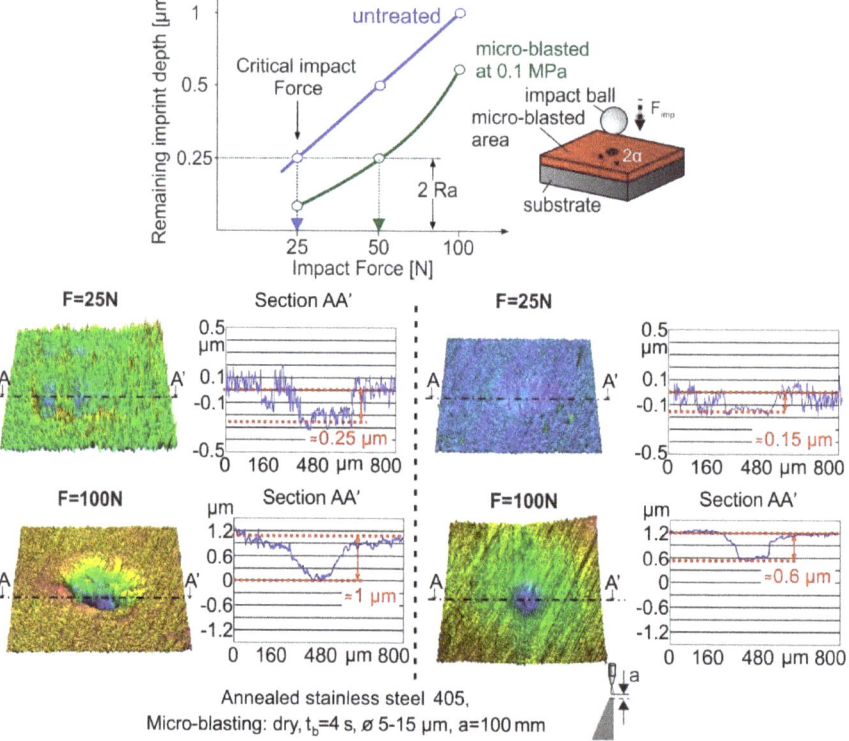

Figure 9. Perpendicular impact test results at various loads of the untreated and micro-blasted tool at 0.1 MPa.

Based on a FEM simulation of the impact process [1] and considering the results shown in Figure 9, the Smith-like diagrams and the Woehler diagrams of an annealed stainless steel 405 and the micro-blasted ones were determined and are illustrated in Figure 10. The Smith diagram presents the maximum load alterations versus its mean value for a fatigue-safe operation of stressed material. The Woehler diagram illustrates the fatigue-safe maximum stress versus the number of repetitive loads alternating from zero to a maximum value. Since the developed stresses during the operation of the cheese cutting tool alternate from zero up to a maximum value, the latter diagram can be used for predicting the number of impacts that the tool can receive during cheese cutting before its fatigue fracture initiation.

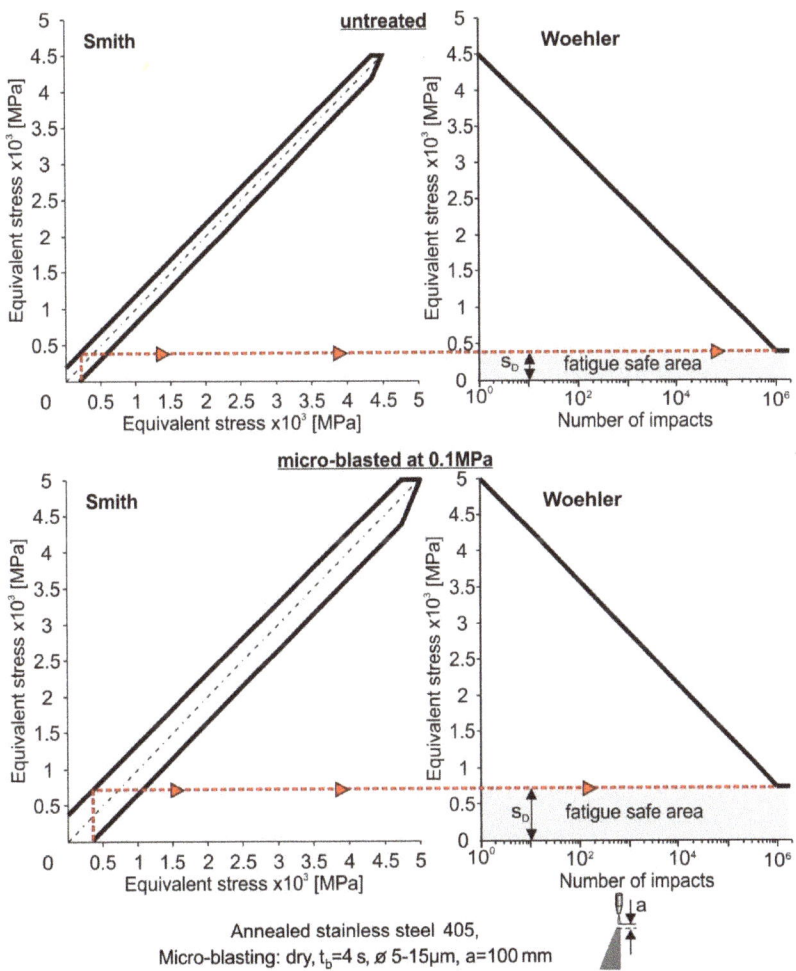

Figure 10. Determined Smith-like diagrams and Woehler diagrams of an annealed stainless steel 405 and the micro-blasted at 0.1 MPa.

3.4. Cutting Results

The experiments were carried out on two different types of cheese concerning its hardness (feta cheese and semi-hard cheese), as illustrated in Figure 11. The effect of cheese hardness on the developed forces is clearly visible. Every curve represents the mean value of approximately 20 experiments. For the semi-hard cheese, a cutting force of about 220 N is required, while for the feta cheese, only 100 N is needed. According to the conducted force measurements of cheeses with different heights, the maximum developed force remains practically invariable from the cheese geometry, and it depends only upon its hardness. Moreover, the knife's displacement and the velocity of the piston were measured, as shown in the bottom part of the figure. It is obvious that the measurements are almost identical for both types of cheese. The whole cutting process for a knife displacement of 160 mm lasts approximately less than 250 ms. Thus, dynamic loads are developed on the cutting edge of the knife during the repetitive cuts, leading to the initiation of fatigue failure after a certain number of cuts.

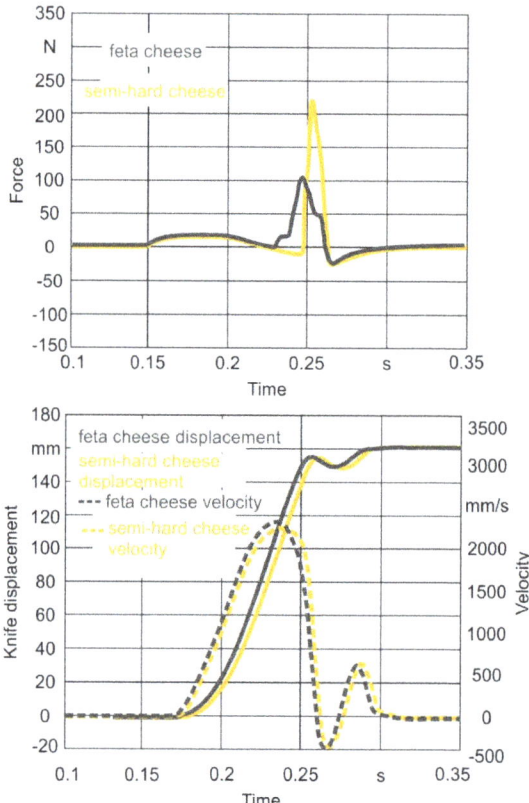

Figure 11. Measured forces as well as knife displacement and velocity in the case of cutting cheese with different hardness.

3.5. FEA Results during Cheese Cutting

By employing the aforementioned FEA models (see Figure 4), the developed equivalent stress fields in the tools during cheese cutting associated with the most intense loading case of semi-hard cheese were determined. The related results for both uncoated and micro-blasted tools are shown in Figure 12. The maximum stress in both cases amounts to 0.81 GPa. This stress is developed in the right part of the tool. This fact can be explained by the cutting-edge geometry (see Figure 1). More specifically, in this region, due to its geometry, higher bending stresses are developed, resulting in a more intense tool loading. Although the maximum equivalent stress does not exceed the material yield stress in both cases, it is higher than the critical fatigue endurance stress. As a consequence, a fatigue damage initiation is expected after approximately 200,000 impacts in the case of the uncoated tool (see Figure 13). Moreover, the application of micro-blasting at 0.1 MPa results in a significant augmentation of the number of impacts that the tool can withstand up to the fatigue fracture initiation. As can be observed in Figure 13, the first coating fracture occurred after approximately 700,000 impacts.

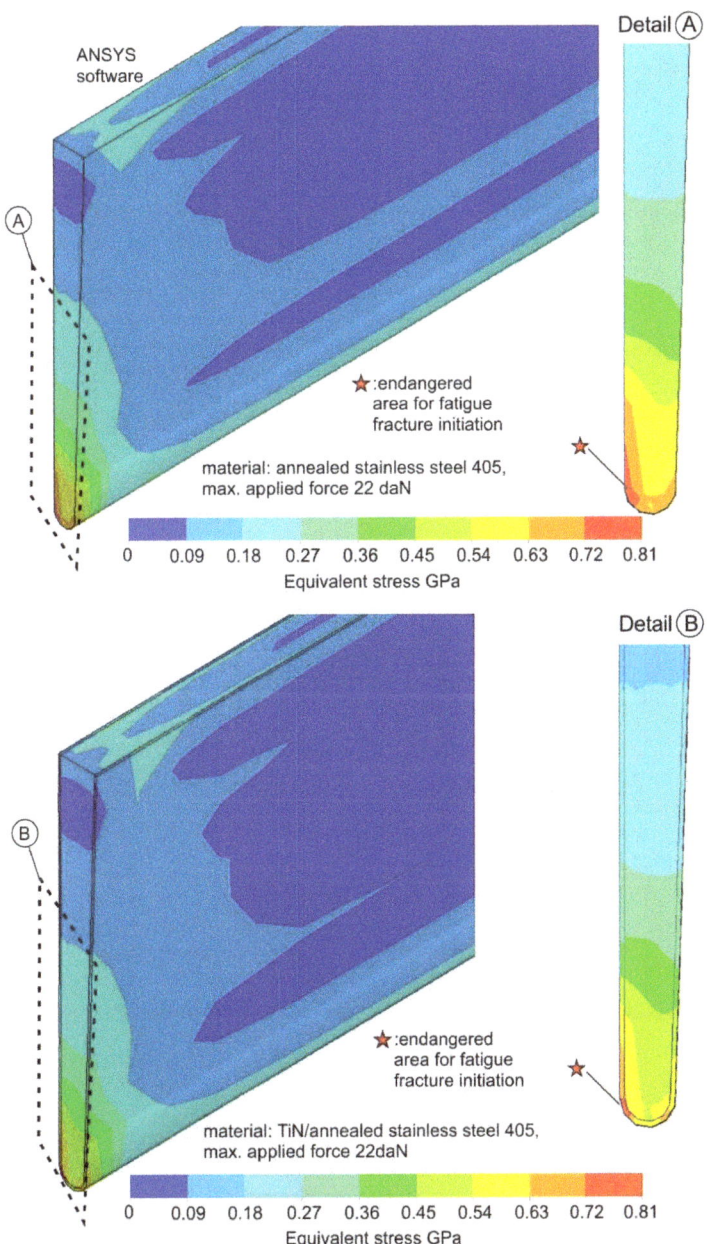

Figure 12. Developed equivalent stress fields in the case of annealed stainless steel 405 and the micro-blasted ones at a pressure of 0.1 MPa during cutting semi-hard cheese.

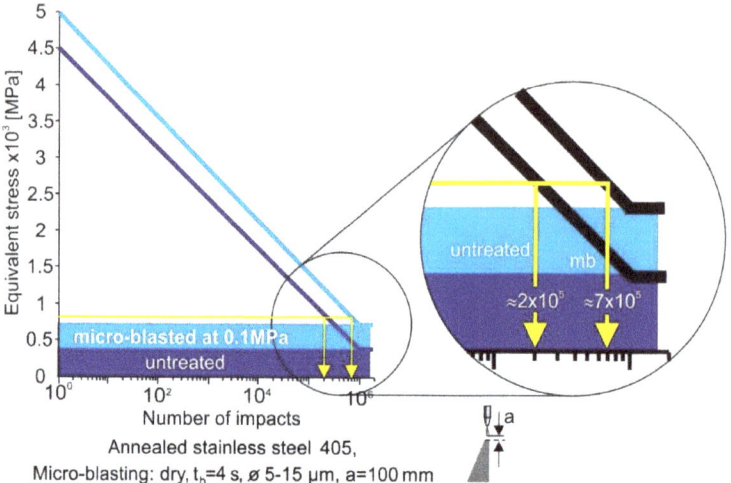

Figure 13. Prediction of the number of impacts up to the first fatigue fracture in the case of annealed stainless steel 405 and the micro-blasted ones at a pressure of 0.1 MPa during cutting semi-hard cheese.

4. Discussion

In the food industry, the production of standard pieces of cheese concerning its weight is a complicated process with high importance. In this context, cheese packages less than a certain weight are not accepted, whereas overweight packages are associated with an additional cost for industries. Moreover, the interruption of this process due to the failure of the cutting tool entails prolonged production time due to the replacement of worn tools and increased cost. In such a production process, the tools are commonly made of stainless steel, and they are subjected to dynamic loads during the cheese-cutting process. In this way, fatigue failure is the prevailing wear phenomenon, significantly reducing the tool life.

The importance of micro-blasting as a technique to increase the life span of tools has been registered in the past [18,19]. This process has as a target to induce residual stresses in the material structure and, in this way, to increase the mechanical properties [14]. As a consequence, the fatigue properties are also expected to be improved. However, this process has to be carefully applied concerning the selected conditions since it can modify the geometry of the cutting edge. In our previous publication [18], dry micro-blasting was conducted on TiAlN coatings deposited on cemented carbide inserts using Al_2O_3 or ZrO_2 grains. These coatings were characterized by comparably superior mechanical properties. As a result, larger micro-blasting pressures must be applied to deform the coating structure and thus induce residual stresses. In the case of the coated tools described in ref. [18], an enlargement of the cutting-edge radius takes place as the micro-blasting pressure increases. A larger cutting-edge radius leads to a reduction of the developed stresses on the coated tool during cutting and to a simultaneous tool life increase [26]. However, special care must be taken to avoid substrate revelation due to coating abrasion in the cutting-edge region during micro-blasting. In the case of annealed stainless steel 405, the applied conditions must appropriately change to attain sufficient increases in the mechanical characteristics without enlarging the cutting-edge radius. This must occur since the sharpness of the tool plays a dominant role in ensuring the surface integrity of the cheese after its cutting, making it as smooth as it can be.

In this paper, for the first time, an experimental-analytical methodology was developed for selecting optimum micro-blasting conditions dependent upon the tool geometry and properties and for determining the number of impacts that the tool can withstand until the first fatigue fracture. Via the developed methodology, preventive replacement of the tool can be conducted, avoiding the risk of sudden tool failure. This methodology was applied in the case tools employed in cheese cutting. To verify the developed methodology's capability to predict the number of impacts for fatigue damage initiation of the tool, cutting experiments were conducted using a semi-hard cheese as the workpiece material (see Figure 14). The tool possesses the geometry as illustrated in Figure 1, and it is made of annealed stainless steel 405. According to attained results, after 200,000 impacts, there was no sign of failure, and fatigue cracks appeared after the conduct of further 10,000 impacts, resulting in tool material removal. These results are in good agreement with the calculated ones. Thus, the capability of the developed methodology to predict the critical number of impacts for tool replacement is verified. The issue of the tool replacement prior to its fatigue fracture is very important. Despite the cheese cleanness process after the production of standard pieces of cheese concerning weight, the whole process became more secure since any possibility of fragmented metallic parts of the tool being inserted into the cheese was avoided. Another significant issue in the market is the surface integrity of the cheese region that must be cut. By assuring an unworn cutting-edge geometry of the tool, the surface integrity of the cheese is expected to be improved, and wasted material during cutting is predicted to be minimized.

Figure 14. Characteristic images of an unworn and worn tool after cheese cutting.

5. Conclusions

In the present work, the potential to enlarge the life span of tools applied in a cheese cutting machine via micro-blasting was investigated. Optimum micro-blasting conditions were detected, considering an almost invariable cutting-edge radius as a criterion. Moreover, a methodology was developed for predicting the number of impacts that the tool can receive during cheese cutting before its fatigue fracture initiation. This methodology was based on the construction of a pneumatic system for the precise cutting of cheese and the simultaneous measurement of the forces developed, as well as on appropriate FEA modeling. The following basic conclusions can be summarized:

- The conduct of dry micro-blasting at a pressure of 0.1 MPa using Al_2O_3 grains can increase the number of impacts that the tool can withstand, more than three times, until the fatigue fracture initiation.
- Experimental cutting investigations on semi-hard cheese using an untreated tool verify the capability of the developed methodology to predict the critical number of impacts for fatigue damage imitation.
- The proposed methodology can be applied to different tool geometries and materials since a parametric FEA model was developed.
- The issue of the tool replacement prior to its fatigue fracture is very important. Despite the cheese cleanness process after the production of standard pieces of cheese concerning weight, the whole process became more secure since any possibility of damaged parts being inserted into the cheese was avoided. Moreover, by assuring an unworn cutting-edge geometry of the tool, the surface integrity of the cheese was expected to be improved, and wasted material during cutting was minimized.

Author Contributions: Conceptualization, G.S.; methodology, G.S.; software, G.S., I.M., K.V. and L.K.; validation, G.S, K.V., L.K., I.M., V.P. and S.P.; formal analysis, G.S., I.M., K.V., and L.K.; investigation, G.S, K.V., L.K., I.M., V.P. and S.P.; data curation, G.S. and S.P.; writing—original draft preparation, G.S.; writing—review and editing, G.S.; visualization, G.S.; supervision, G.S.; project administration, G.S. and S.P.; All authors have read and agreed to the published version of the manuscript.

Funding: This research was carried out as part of the project «DESIGN, DEVELOPMENT & OPTIMIZATION OF AN INNOVATIVE CHEESE CUTTING MACHINE» (Project code: KMP6-0077379) under the framework of the Action «Investment Plans of Innovation» of the Operational Program «Central Macedonia 2014 2020», that is co-funded by the European Regional Development Fund and Greece.

Institutional Review Board Statement: Not applicable.

Informed Consent Statement: Not applicable.

Data Availability Statement: Not applicable.

Conflicts of Interest: The authors declare no conflict of interest.

References

1. Bouzakis, K.-D.; Michailidis, N.; Skordaris, G.; Bouzakis, E.; Biermann, D.; M'Saoubi, R. Cutting with coated tools: Coating technologies, characterization methods and performance optimization. *CIRP Ann. Manuf. Technol.* **2012**, *61*, 703–723. [CrossRef]
2. Xiaowei, L.; Hanlin, W.; Liuyang, F.; Huiyong, B. Low-cycle fatigue behavior for stainless-clad 304 + Q235B bimetallic steel. *Int. J. Fatigue* **2022**, *159*, 106831.
3. Knotek, O.; Bosserhoff, B.; Schrey, A.; Leyendecker, T.; Lemmer, O.; Esser, S. A New Technique for Testing the Impact Load of Thin Films: The Coating Impact Test. *Surf. Coat. Technol.* **1992**, *54–55*, 102–107. [CrossRef]
4. Gothivarekar, S.; Coppieters, S.; Talemi, R.; Debruyne, D. Effect of bending process on the fatigue behaviour of high strength steel. *J. Constr. Steel Res.* **2021**, *182*, 106662. [CrossRef]
5. Bin, Z.; Ali, H.; Xiaoman, Z.; Jikui, F.; Shao, S.; Khonsari, M.M.; Guo, S.; Meng, W.J. On the failure mechanisms of Cr-coated 316 stainless steel in bending fatigue tests. *Int. J. Fatigue* **2020**, *139*, 105733.
6. Guodong, W.; Yafei, M.; Zhongzhao, G.; Hanbing, B.; Lei, W.; Jianren, Z. Fatigue life assessment of high-strength steel wires: Beach marks test and numerical investigation. *Constr. Build. Mater.* **2022**, *323*, 126534.
7. Skordaris, G.; Bouzakis, K.-D.; Charalampous, P.; Kotsanis, T.; Bouzakis, E.; Bejjani, R. Bias voltage effect on the mechanical properties, adhesion and milling performance of PVD films on cemented carbide inserts. *Wear* **2018**, *404–405*, 50–61. [CrossRef]
8. Yu, W.; Yin, Y.; Zhou, J.; Xu, Q.; Feng, X.; Nan, H.; Zuo, J.; Wang, X.; Ding, X. Surface Condition Evolution and Fatigue Evaluation after Different Surface Processes for TiAl$_{47}$Cr$_2$Nb$_2$ Alloy. *Materials* **2022**, *15*, 5491. [CrossRef]
9. Avcu, Y.Y.; Gönül, B.; Yetik, O.; Sönmez, F.; Cengiz, A.; Guney, M.; Avcu, E. Modification of Surface and Subsurface Properties of AA1050 Alloy by Shot Peening. *Materials* **2021**, *14*, 6575. [CrossRef]
10. Rautio, T.; Jaskari, M.; Gundgire, T.; Iso-Junno, T.; Vippola, M.; Järvenpää, A. The Effect of Severe Shot Peening on Fatigue Life of Laser Powder Bed Fusion Manufactured 316L Stainless Steel. *Materials* **2022**, *15*, 3517. [CrossRef]
11. Matuszak, J.; Zaleski, K.; Skoczylas, A.; Ciecielag, K.; Kecik, K. Influence of Semi-Random and Regular Shot Peening on Selected Surface Layer Properties of Aluminum Alloy. *Materials* **2021**, *14*, 7620. [CrossRef] [PubMed]
12. Wang, Z.; Shi, M.; Gan, J.; Wang, X.; Yang, Y.; Ren, X. The Effects of Shot Distance and Impact Sequence on the Residual Stress Field in Shot Peening Finite Element Model. *Metals* **2021**, *11*, 462. [CrossRef]
13. Efe, Y.; Karademir, I.; Husem, F.; Maleki, E.; Karimbaev, R.; Amanov, A.; Unal, O. Enhancement in microstructural and mechanical performance of AA7075 aluminum alloy via severe shot peening and ultrasonic nanocrystal surface modification. *Appl. Surf. Sci.* **2020**, *528*, 146922. [CrossRef]
14. Chung, Y.-H.; Chen, T.-C.; Lee, H.-B.; Tsay, L.-W. Effect of Micro-Shot Peening on the Fatigue Performance of AISI 304 Stainless Steel. *Metals* **2021**, *11*, 1408. [CrossRef]
15. Barbatti, C.; Garcia, J.; Pitonak, R.; Pinto, H.; Kostka, A.; Di Prinzio, A.; Staia, M.H.; Pyzalla, A.R. Influence of micro-blasting on the microstructure and residual stresses of CVD j-Al$_2$O$_3$ coatings. *Surf. Coat. Technol.* **2009**, *203*, 3708–3717. [CrossRef]
16. Jacob, A.; Gangopadhyay, S.; Satapathy, A.; Mantry, S.; Jha, B.B. Influences of micro-blasting as surface treatment technique on properties and performance of AlTiN coated tools. *J. Manuf. Process.* **2017**, *29*, 407–418. [CrossRef]
17. Gadge, M.; Lohar, G.; Chinchanikar, S. A review on micro-blasting as surface treatment technique for improved cutting tool performance. *Mater. Today Proc.* **2022**, *64*, 725–730. [CrossRef]
18. Bouzakis, K.-D.; Klocke, F.; Skordaris, G.; Bouzakis, E.; Gerardis, S.; Katirtzoglou, G.; Makrimallakis, S. Influence of dry micro-blasting grain quality on wear behaviour of TiAlN coated tools. *Wear* **2011**, *271*, 783–791. [CrossRef]
19. Bouzakis, K.-D.; Tsouknidas, A.; Skordaris, G.; Bouzakis, E.; Makrimallakis, S.; Gerardis, S.; Katirtzoglou, G. Optimization of wet or dry microblasting on PVD films by various Al$_2$O$_3$ grain sizes for improving the coated tools' cutting performance. *Tribol. Ind.* **2011**, *33*, 49–56.

20. Astakhov, V.P. Cutting Force Modeling Genesis, State of the Art, and Development. In *Mechanical and Industrial Engineering: Historical Aspects and Future Directions*; Davim, J.P., Ed.; Springer Nature: Cham, Switzerland, 2022; pp. 39–93.
21. Astakhov, V.P. Authentication of FEM in Metal Cutting, Chapter 1. In *Finite Element Method in Manufacturing Processes*; Davim, J.P., Ed.; Wiley: Hoboken, NJ, USA, 2011; pp. 1–43.
22. Impact-BZ Ltd. Available online: www.impact-bz.com (accessed on 21 August 2022).
23. Bouzakis, K.-D.; Michailidis, N.; Hadjiyiannis, S.; Skordaris, G.; Erkens, G. A continuous FEM simulation of the nanoindentation to determine actual indenter tip geometries, material elastoplastic deformation laws and universal hardness. *Z. Fuer Met./Mater. Res. Adv. Tech.* **2002**, *93*, 862–869.
24. *ASTM STP 889*; American Society for Testing and Materials: Philadelphia, PA, USA, 1986; pp. 72–89.
25. Bouzakis, K.-D.; Skordaris, G.; Klocke, F.; Bouzakis, E. A FEM-based Analytical–Experimental Method for Determining Strength Properties Gradation in Coatings after Micro-blasting. *Surf. Coat. Technol.* **2009**, *203*, 2946–2953. [CrossRef]
26. Bouzakis, K.-D.; Michailidis, N.; Skordaris, G.; Kombogiannis, S.; Hadjiyiannis, S.; Efstathiou, K.; Pavlidou, E.; Erkens, G.; Rambadt, S.; Wirth, I. Optimisation of the cutting edge roundness and its manufacturing procedures of cemented carbide inserts, to improve their milling performance after a PVD coating deposition. *Surf. Coat. Technol.* **2003**, *163–164*, 625–630. [CrossRef]

Article

Fabrication and Characterization of P-Type Semiconducting Copper Oxide-Based Thin-Film Photoelectrodes for Solar Water Splitting

Chih-Jui Chang [1,†], Chih-Wei Lai [1,†], Wei-Cheng Jiang [1], Yi-Syuan Li [1], Changsik Choi [2,*], Hsin-Chieh Yu [1,*], Shean-Jen Chen [1] and YongMan Choi [1,*]

[1] College of Photonics, National Yang Ming Chiao Tung University, Tainan 71150, Taiwan
[2] Clean Energy Conversion Research Center, Institute for Advanced Engineering, Yongin 17180, Korea
* Correspondence: cschoi@iae.re.kr (C.C.); hcyu@nycu.edu.tw (H.-C.Y.); ymchoi@nycu.edu.tw (Y.C.)
† These authors contributed equally to this work.

Abstract: Solar light-driven hydrogen by photocatalytic water splitting over a semiconductor photoelectrode has been considered a promising green energy carrier. P-type semiconducting copper oxides (Cu_2O and CuO) have attracted remarkable attention as an efficient photocathode for photoelectrochemical (PEC) water splitting because of their high solar absorptivity and optical band gaps. In this study, CuO thin films were prepared using the sol-gel spin coating method to investigate the effects of aging time and layer dependency. Electrodeposition was also applied to fabricate Cu_2O thin films. Cu_2O thin films annealed at 300 °C are a hetero-phase system composed of Cu_2O and CuO, while those at 400 °C are fully oxidized to CuO. Thin films are characterized using atomic force microscopy (AFM), scanning electron microscopy (SEM), ultraviolet-visible spectroscopy (UV-VIS), Fourier transform infrared spectroscopy (FTIR), spectroscopic ellipsometry (SE), X-ray diffraction (XRD), X-ray photoelectron spectroscopy (XPS), and Raman microscopy. The hetero-phase thin films increase the photoconversion efficiency compared to Cu_2O. Fully oxidized thin films annealed at 400 °C exhibit a higher efficiency than the hetero-phase thin film. We also verified that CuO thin films fabricated using electrodeposition show slightly higher efficiency than the spin coating method. The highest photocurrent of 1.1 mA/cm^2 at 0.10 V versus RHE was measured for the fully oxidized CuO thin film under one-sun AM1.5G illumination. This study demonstrates a practical method to fabricate durable thin films with efficient optical and photocatalytic properties.

Keywords: p-type semiconductor; copper oxide; thin film; sol-gel deposition; electrodeposition; photoelectrochemical water splitting

Citation: Chang, C.-J.; Lai, C.-W.; Jiang, W.-C.; Li, Y.-S.; Choi, C.; Yu, H.-C.; Chen, S.-J.; Choi, Y. Fabrication and Characterization of P-Type Semiconducting Copper Oxide-Based Thin-Film Photoelectrodes for Solar Water Splitting. *Coatings* 2022, 12, 1206. https://doi.org/10.3390/coatings12081206

Academic Editors: Giorgos Skordaris and Fabio Palumbo

Received: 29 June 2022
Accepted: 16 August 2022
Published: 17 August 2022

Publisher's Note: MDPI stays neutral with regard to jurisdictional claims in published maps and institutional affiliations.

Copyright: © 2022 by the authors. Licensee MDPI, Basel, Switzerland. This article is an open access article distributed under the terms and conditions of the Creative Commons Attribution (CC BY) license (https://creativecommons.org/licenses/by/4.0/).

1. Introduction

Hydrogen generation using solar energy through water electrolysis with solar-powered electricity and photocatalytic solar water splitting is vital to fulfill the pipe dream of the hydrogen economy, due to global demand for renewable energy [1]. Among solar energy conversion methods, photoelectrochemical (PEC) water splitting is considered one of the most pivotal and sustainable approaches to tackle the global energy crisis. It dissociates water molecules directly into hydrogen and oxygen using sunlight and semiconducting PEC materials [2]. Even if it is sustainable without producing greenhouse gases, developing novel PEC materials with excellent efficiencies, durability, and cost-effectiveness is imperative for their commercialization [3]. PEC water splitting occurs sequentially through the production of electron-hole pairs due to absorbing sunlight, the charge drift and diffusion, the separation of charge carriers, and the reduction or oxidation of water on the surfaces with the charge carriers [4]. Semiconducting PEC materials must meet several requirements to split water molecules [5]. For example, the potential of the conduction band (CB) gap must be lower than the reaction potential of reduced water, and that of the valence band (VB) gap higher than the reaction potential of oxidized water.

The photoelectrodes for large-scale applications may be cost-effective metal oxides [6] among various semiconductor materials of metal oxides (TiO_2 and WO_3), metal sulfides (CdS and WS_2), and metal selenides (WSe_2 and ZnSe) [7]. Their ionic character originated from the hybrid orbitals of transition metal cations and oxygen (O 2p) could lead to a large band gap [5]. Furthermore, the VB's localized p-orbital characteristic causes the low mobility of electron holes in conventional metal oxides, resulting in no durable photoanode device producing a photocurrent greater than 10 mA/cm^2 with adequate photovoltage coupled with a photocathode [8]. In addition, metal oxides often experience short charge diffusion and slow charge transfer, owing to the formation of sluggish and self-trapped charge carriers [9]. Therefore, the search for novel, highly efficient, and cost-effective metal oxide semiconductors with a narrower band gap and a longer carrier diffusion length has attracted wide attention [10]. However, it remains a great challenge to propose high-performance materials for photocatalytic water splitting, for example, by regulating electronic band structures and creating the direct Z-scheme heterojunction (i.e., Co_9S_8/CdS [11], ZnO/CdS [12], and TiO_2/CdS [13]) and the p-n heterojunction (i.e., $B-C_3N_4$/$Mo-BiVO_4$ [14], Cu_2O/CeO_2 [15], and CuO/ZnO [16]). To the best of our knowledge, no candidate material meets all requirements with a high enough energy conversion efficiency [7].

Among semiconducting photoelectrode materials, copper oxides (Cu_xO; x = 1 or 2) have received remarkable attention due to their outstanding properties and numerous applications for solar-to-hydrogen conversion from water [17,18]. Copper oxides are naturally a p-type semiconductor [19] with high solar absorptivity, low thermal radiation, and low energy band gaps (Cu_2O: 2.1–2.6 eV and CuO: 1.3–2.1 eV) [20]. The band gaps also depend on sample preparation approaches and surface morphologies. Due to their high natural abundance, copper oxides can also be used for cost-effective solar cells [21]. Cu_2O-based solar cells with a band gap of 2.0 eV were reported to potentially reach a theoretical maximum efficiency of 20% [19,22] according to the Shockley-Queisser limit (SQL) [23]. Moreover, copper oxide-based photoelectrodes can be applied for photochemical CO_2 reduction [24]. Regardless of these advantages and the versatile applications of copper oxides as excellent solar absorbing materials, a grand challenge may be to rationally minimize the strong electron-hole recombination of copper oxides [22]. Various reports also address obtaining the optimal diffusion length of minority charge carriers and thicknesses of thin films [22,25]. Therefore, it may be crucial to effectively separate electron-hole pairs to improve the efficiency of photocatalytic solar water splitting using copper oxide-based photocathodes. To date, as summarized in Table S1, the fabrication of copper oxide-based photocathodes applied to PEC water splitting is a tremendous challenge that requires technological advances [26], since the efficiency (e.g., 0.10–1.75%) [17,27–37] depends on many factors, such as substrates, pH values, and surface modification. For the preparation of high-performance photoelectrode materials, metal oxide thin films can be prepared by numerous methods, including the sol-gel method, chemical vapor deposition (CVD), pulsed laser deposition (PLD), and electrodeposition. Among them, the sol-gel spin coating method is more cost-effective, faster, and higher-yield than physical methods (i.e., CVD and PLD). Its process comprises the preparation of a precursor solution, aging the solution for conversion of sol in the liquid to gel, depositing the sol-gel on a substrate by spin coating, and heat treatment of thin films. Even if the sol-gel approach is fast and straightforward compared to other thin-film deposition methods, the performance of thin films is influenced by several factors, including sol aging time and temperatures, sol concentrations, chelating agents, pre- and post-heating temperatures, and spin coating speed and time. Among the factors for preparing high-quality thin films, optimizing sol aging time and heat-treatment parameters may be essential for fabricating nanostructured thin films and controlling optical band gaps. Accordingly, this study first examined the effects of sol aging time, heat treatment, and layer dependency on p-type CuO thin films fabricated using the sol-gel spin coating method, followed by photocatalytic solar water splitting. CuO thin films were also prepared using electrodeposition with thermal oxidation to compare with the sol-gel spin coating method. We demonstrated the effect of the substrates on photocatalytic efficiencies using indium tin oxide (ITO or tin-doped indium oxide) and fluorine-doped tin oxide (FTO) coated glass substrates since their properties are affected by thermal treatment [38,39].

2. Experimental Details

2.1. Thin-Film Preparation

The sol-gel thin-film deposition method [40] was applied on ITO (~120 nm) coated glass substrates to examine the effects of CuO sol using p-type CuO thin films. ITO substrates were sequentially cleaned in an ultrasonic bath of acetone, isopropyl alcohol (IPA), ethanol, and deionized (DI) water for 5 min, followed by drying in an oven and ozone treatment for 10 min. For spin coating, we prepared a precursor solution by adding 1 mL of ethanolamine to 15 mL of IPA under stirring, followed by mixing with 0.68 g of copper (II) acetate. After the complete dissolution of the copper acetate, a solution of 0.2 g of ethylene glycol was added. The precursor solution was stirred for 24 h in air to reach complete mixing at 1500 rpm. Impurities were removed using a 0.45 μm syringe filter [41]. Then, the sol solution was aged at room temperature for 0, 12, 24, 36, and 48 h with magnetic stirring. The prefiltered 150 μL sol solution was placed on an ITO substrate at 2000 rpm for 30 s using a spin coater in a fume hood. The spin-coated thin films were then dried at 120 °C on a hot plate for 2 min, and was pre-annealed at 350 °C for 5 min to remove precursor chemicals. All thin-film samples were subjected to a final annealing process at 400 °C on a hot plate for 2 h in air. To examine layer dependency, the pre-annealing process at 350 °C for 5 min was repeated twice, four times, six times, and eight times, followed by the final annealing process at 400 °C for 2 h. Cu_2O thin films were also fabricated on FTO (~400 nm) coated glass substrates by electrodeposition. In this study, FTO substrates were applied for electrodeposition because of their excellent chemical resistance [38]. It was performed using the chronoamperometric (CA) method at −0.36 V versus Ag/AgCl (3 M NaCl) for 20 min in a solution of 3 M lactic acid and 0.4 M $CuSO_4$ (pH = 9) [29] added to 40 mL of DI water. The pH value was obtained using 5 M NaOH, while the entire electrodeposition process was conducted with magnetic stirring at 100 rpm. A potentiostat (BioLogic, SP-150) was used to supply a constant current during the electrodeposition process at room temperature using a coiled platinum wire (0.5 mm diameter) as a counter electrode. To obtain hetero-phase CuO/Cu_2O and pure CuO thin films, Cu_2O thin films were thermally oxidized at 300 °C and 400 °C, respectively, for 1 h in air, after cleaning the Cu_2O thin films using DI water and drying in an oven at 70 °C.

2.2. Characterization of Thin-Film Samples

Atomic force microscopy (AFM, Bruker Innova, MA, USA), with the tapping mode, and scanning electron microscopy (SEM, Zeiss Auriga, Oberkochen, Germany), at an accelerating voltage of 5.0 kV and a working distance of 5.2 mm after gold coating, were employed to examine the surface morphologies of thin films. Band gaps of thin films were measured using ultraviolet-visible spectroscopy (UV-VIS, Shimadzu UV1800, Kyoto, Japan), while Fourier transform infrared spectroscopy (FTIR, Thermo Nicolet iS-10, MA, USA) was used to investigate the temperature-dependent variation of precursor chemicals. Furthermore, we characterized the optical constants using variable angle spectroscopic ellipsometry (VASE, J.A.Woollam α-SE, Lincoln, NE, USA) to examine the thickness of CuO thin films by measuring at three different angles of 65°, 70°, and 75°. X-ray diffraction (XRD, Brucker D8 DISCOVER, Billerica, MA, USA) using Cu $K_α$ radiation (λ = 0.154184 nm), an X-ray photoelectron spectrometer (XPS, Thermo Scientific K-Alpha, Waltham, MA, USA) equipped with mono-chromated Al Kα radiation and a flood gun, and Raman microscopy (CL Technology UniDRON-S, New Taipei, Taiwan) with a 532 nm excitation wavelength were applied to characterize the temperature-dependent effects of thin films grown on FTO. PEC measurements were carried out with a typical three-electrode system using

Ag/AgCl (3 M NaCl) and a coiled platinum wire as reference and counter electrodes, respectively, and the working electrode of copper oxide thin films with an active area of 1.0 cm^2 in 0.2 M Na$_2$SO$_4$. The Nernst equation was applied to convert the potentials (V versus Ag/AgCl) to the reversible hydrogen electrode (RHE) scale (V versus RHE) [42]. The applied bias photon-to-current conversion efficiency (ABPE, %) was used to calculate the photoconversion efficiency of thin films using ABPE = [J$_P$(1.23 − V$_b$)/P$_{light}$]$_{AM1.5G}$, where J$_P$ is the photocurrent density measured in mA/cm^2, V$_b$ is the applied bias in V, and P$_{light}$ is the power density of incident light (AM1.5G 100 mW/cm^2 illumination, Enlitech SS-X, Kaohsiung, Taiwan) [43,44].

3. Results and Discussion

3.1. Sol Aging Time

The effects of the aging time of the CuO sol solution from 0 to 48 h were examined to understand the influence of structural and optical properties of CuO thin films. The AFM images of CuO thin films are displayed in Figure 1a–e. The AFM images in 3 μm × 3 μm were used to quantify the root mean square (RMS) roughness of thin-film surfaces, which could be calculated from the cross-sectional profile. As illustrated in Figure 1f, the RMS roughness of CuO thin films prepared at different aging times (0, 12, 24, 36, and 48 h) corresponded to 35.5, 20.5, 15.7, 12.8, and 12.5 nm, respectively. It demonstrated that the surface of CuO thin films became smoothened by surface diffusion and were stable after 36 h. Based on the stable RMS roughness, we assumed that copper oxide grains became homogeneous after annealing because of the stable CuO sol solution. Furthermore, we measured the thickness of CuO thin films using spectroscopic ellipsometry (SE). We first verified the thickness of ITO measured using SEM (~120 nm) with the Cauchy model [45] used for transparent thin films. Then, the Tauc-Lorenz method was applied to examine non-transparent CuO thin films [46]. Figure 1f shows the variation of the thickness of CuO thin films by changing the aging time of the sol. It was found that the film thickness was stabilized as the surface roughness became constant after 36 h, similar to the previous study [47]. Then, we conducted absorption measurements of CuO thin films using UV-VIS spectroscopy. We observed that longer aging times (0 and 12 h) showed less absorbance, confirming that the RMS roughness was correlated with absorbance [47]. The Tauc plot analysis [48,49] with a direct transition ($\gamma = \frac{1}{2}$) resulted in the band gap of ~1.99 eV [20] using $(\alpha h v)^{1/\gamma} = B(hv - E_g)$, where α, h, v, B, and E$_g$ were the absorption coefficient, the Plank constant, the photon energy, a constant, and the energy band gap, respectively. It verified that the aging time negligibly affected the optical property of the stabilization process. Our detailed examination showed that 36 h may be imperative to fabricate stable thin films. Therefore, in this study, we used the aging time of 36 h to prepare the sol-gel CuO thin films. As summarized in Figure 2a, FTIR spectra confirmed the complete removal of precursor chemicals and the formation of metal oxide phases after 300 °C. After heating as-deposited thin films, the characteristic peaks of the precursor chemical of copper(II) acetate disappeared [50], while the absorption peak at around 587 cm^{-1} assigned to the vibration of Cu(II)-O [51] appeared.

3.2. Characterization of Layer-Dependence Effects of Spin-Coated CuO Thin Films

The layer dependency of CuO thin films was performed to determine a reasonable thickness compared to that fabricated by electrodeposition. As mentioned above, the sol-gel thin-film deposition method [40] was applied on ITO substrates. As described above, two, four, six, and eight-layer CuO films were prepared. After each layer deposition, pre-annealing was conducted at 350 °C for 5 min, while final thin-film samples were annealed at 400 °C for 2 h. In addition to the surface morphology analysis, the band gap was obtained via the Tauc plot analysis using the absorbance spectra of each CuO thin-film layer. Figure 2b,c show the absorption spectra and averaged energy band gaps of CuO thin films with different layers. The absorption intensity increased as the thickness of the thin films became thicker, indicating that the transmittance decreased accordingly. To improve the accuracy of the optical energy gap measurements, due to the inhomogeneity of the thin films, we measured each thin-film sample three times, resulting in 1.83, 1.73, 1.62, and 1.49 eV, respectively. The growth of grains caused a decrease in the number of grain boundaries. This led to an increase in the absorption intensity and a shift of the absorption edge to higher wavelengths. Consequently, the energy gap decreased [52]. Shown in Figure 3a–d are AFM-based images to examine the surface morphology of the layer dependency. As shown in Figure 3e, the RMS roughness analysis illustrated a linear relationship with the number of layers of CuO thin films (8.9, 15.1, 15.6, and 20.4 nm versus two, four, six, and eight layers, respectively). Furthermore, we observed that the thickness of the CuO thin film using ellipsometry with the Tauc-Lorentz method linearly increased with the growth of CuO layers (Figure 3e, 97.8, 123.9, 180.4, and 274.2 nm versus two, four, six, and eight layers, respectively). The layer-dependence study suggested that eight-layer CuO thin films under the experimental conditions were optimal for further analyses to avoid too low transmittance of CuO thin films.

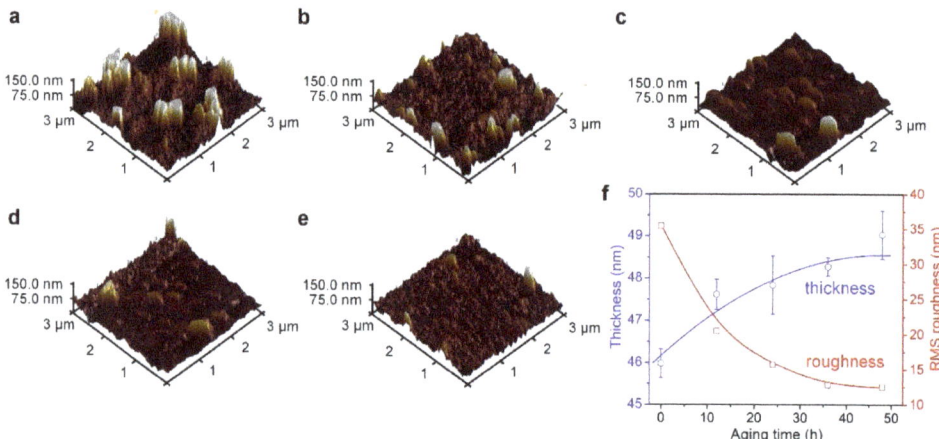

Figure 1. AFM images of CuO thin films deposited on an ITO substrate obtained with the tapping mode AFM for the aging time of (**a**) 0, (**b**) 12, (**c**) 24, (**d**) 36, and (**e**) 48 h. Image areas are 3 μm × 3 μm. (**f**) Variation of the RMS roughness as a function of aging time and the thickness of CuO thin films measured using SE.

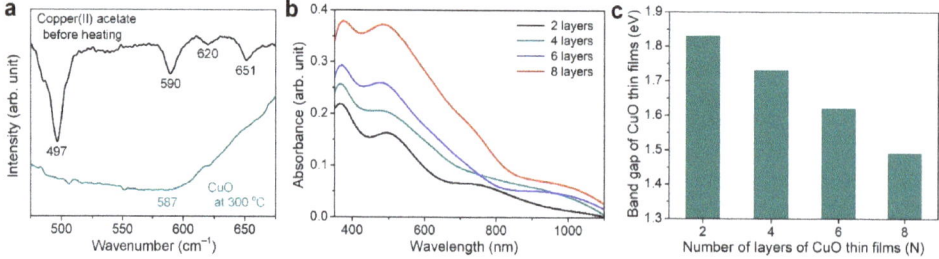

Figure 2. (**a**) FTIR spectra of the sol-gel spin-coated thin film before and after annealing at 300 °C in air. (**b**) Absorbance spectra and (**c**) optical band gaps with different layers of CuO thin films.

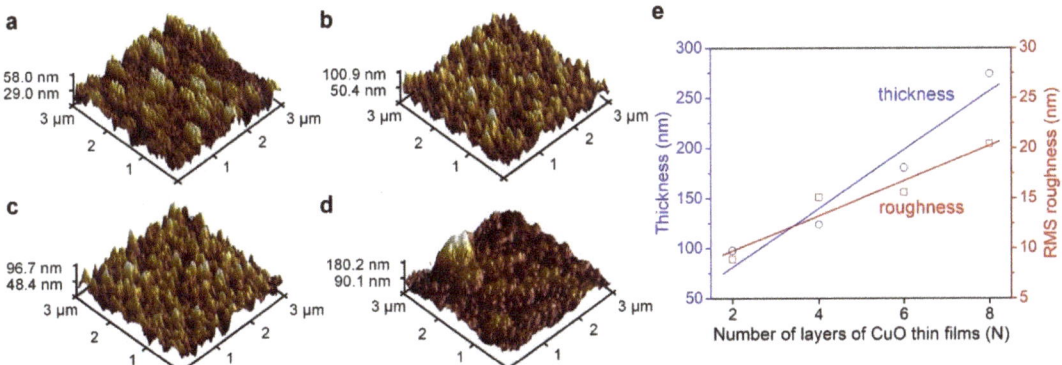

Figure 3. AFM images of (**a**) two, (**b**) four, (**c**) six, and (**d**) eight-layer CuO thin films deposited on ITO after the annealing process at 400 °C for 2 h. Image areas are 3 µm × 3 µm. (**e**) Variation of the RMS roughness and the thickness of CuO thin films measured using SE.

3.3. Preparation and Characterization of Copper Oxide Thin Films Using Electrodeposition

As summarized in Figure 4, we carried out SEM measurements to examine whether the surfaces of thin films, using spin coating and electrodeposition, were dense enough for PEC measurements. As an initiative, we first examined the PEC performance of CuO/ITO and CuO/FTO prepared using the sol-gel spin-coating method and annealed at 400 °C. As shown in Figure S1, the ITO-based CuO thin film showed a lower PEC performance than the FTO-based thin film, due to the change of the electrical property at >300 °C [38,39]. Accordingly, we applied the FTO substrate to examine the photoconversion efficiency in this study. We prepared yellowish-red Cu_2O thin films using electrodeposition at room temperature to fabricate hetero-phase CuO/Cu_2O and CuO films on FTO. As shown in Figure 4a,b, we observed that the thickness of electrodeposition-based thin films was approximately 100 nm. In addition, it demonstrated that the Cu_2O thin film was well prepared without cracks, and the thermally oxidized samples became denser with smaller grains. More characterizations were carried out to examine how Cu_2O was transformed into CuO after annealing. As shown in Figure 4c, the spin-coated CuO surfaces consisted of nanostructured and homogeneous grains, leading to a high surface roughness compared to Cu_2O-based samples. We assumed that these surface morphologies would influence the PEC performance because nanostructured surfaces significantly affect optical properties [53], making nanostructured photoelectrodes more effective than conventional bulk films. SEM confirmed that the thickness of the spin-coated CuO thin film on FTO was ~257 nm, which was in line with that deposited on ITO using SE (Figure 3e, ~274 nm). We conducted surface characterizations to understand the chemical composition of thin films using XRD, Raman microscopy, and XPS. As shown in Figure 5a, the two major diffraction peaks at 36.7° and 42.4° corresponding to (111) and (200) planes were attributed to Cu_2O (JCPDS file No. 05-0667) [54,55]. The spin-coated CuO (JCPDS file No. 78–0428) thin film was identified by those at 33.5°, 35.5°, and 38.6°, corresponding to (110), (002), and (111) planes [55,56]. After thermal oxidation at 300 °C, the thin film was characterized by a mixture of Cu_2O and CuO. Then, after thermal oxidation at 400 °C, the XRD patterns became identical to those of CuO, confirming the complete oxidation of Cu_2O to CuO at 400 °C under the experimental conditions. However, as the intensity of the XRD pattern was not remarkable, Raman spectroscopy was used to characterize the surfaces to complement the XRD analysis, clarifying the change of thin-film-based copper oxides. Shown in Figure 5b are Raman spectra for thin-film copper oxides deposited on FTO using spin coating and electrodeposition before and after annealing at 300 °C and 400 °C scanned in the spectral region of 100 to 800 cm^{-1}. The characteristic Raman bands for CuO at 292 cm^{-1}

and Cu$_2$O at 213, 529, and 633 cm^{-1} agreed with those in the literature [57]. Like the XRD measurements, we verified that the Cu$_2$O-based thin-film sample annealed at 400 °C to become CuO, while the sample that annealed at 300 °C exhibited the characteristic peaks of Cu$_2$O and CuO. It revealed that the Cu$_2$O thin films annealed at 300 °C could be considered a hetero-phase system composed of CuO and Cu$_2$O. To identify the elemental compositions and oxidation states of CuO and Cu$_2$O thin-film samples prepared by the sol-gel spin coating and electrodeposition methods, XPS measurements were performed [58,59]. The survey spectra of Cu$_2$O and CuO deposited on FTO are shown in Figure S2, where copper and oxygen elements are characterized. The C 1s peak at 284.8 eV was applied to calibrate the XPS spectra of Cu$_2$O and CuO [60,61]. As shown in Figure 5c, the main peaks of the Cu$_2$O thin film at 932.8 eV and 952.5 eV (Cu$^+$) and those of the CuO thin film at 933.3 eV and 953.6 eV (Cu^{2+}) corresponded to Cu 2p$_{3/2}$ and Cu 2p$_{1/2}$, respectively [62]. Furthermore, the shake-up satellite peaks were detected in the region of 940.0–943.5 eV and at 962.2 eV [63,64]. Figure 5d shows the XPS analysis of the O 1s peaks for Cu$_2$O and CuO. The peak at 529.7 eV observed from the thermal oxidation sample was identified as lattice oxygen in CuO, while that at 530.3 eV was assigned to lattice oxygen in Cu$_2$O [65]. The optical properties of copper oxide thin films were also examined by the band gap change in the range of 300–1100 nm. All thin films exhibited higher absorption in the UV region (350 to 390 nm) than in the visible region (Figure 6a). The absorption of the Cu$_2$O film was greatly enhanced in the range of 400–800 nm after annealing. As shown in Figure 6b, the 2.45 eV band gap of Cu$_2$O aligned with those in the literature [20]. After annealing, as anticipated, its band gap shifted down to that of CuO (i.e., 1.61 eV (annealed at 300 °C), 1.51 eV (annealed at 400 °C), and 1.49 eV (spin-coated CuO)) [20].

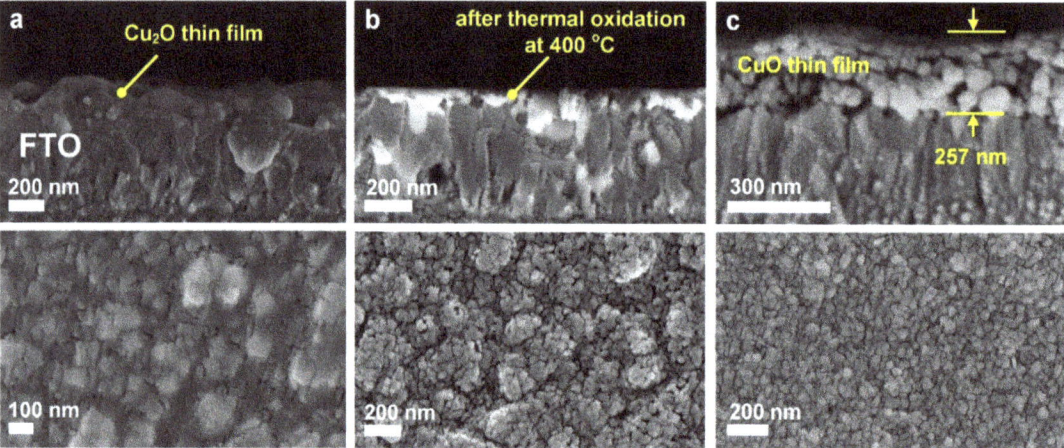

Figure 4. SEM images of cross-sectional and top views for (**a**) the as-prepared Cu$_2$O thin film by electrodeposition and (**b**) the thermally oxidized sample at 400 °C after electrodeposition. (**c**) The spin-coated eight-layer CuO thin film annealed at 400 °C. Thin films are deposited on FTO.

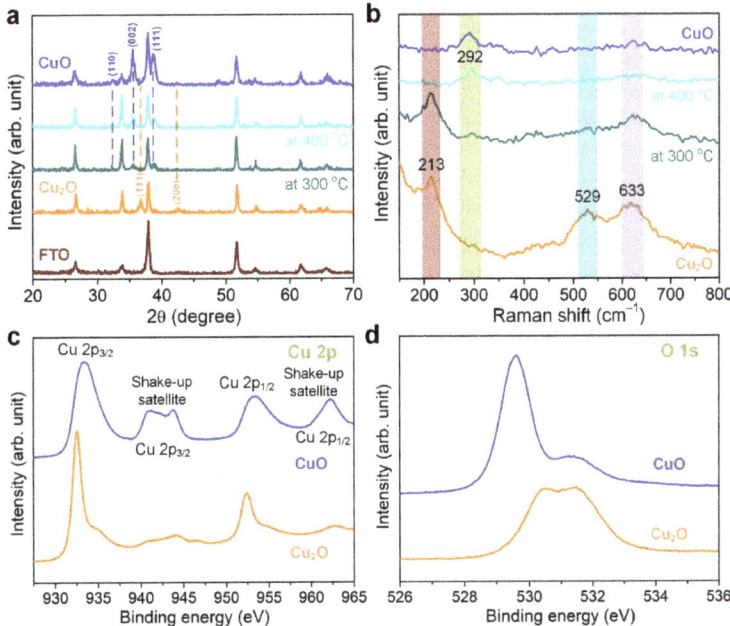

Figure 5. (a) XRD patterns and (b) Raman spectra of as-prepared Cu_2O and after thermal oxidation at 300 °C and 400 °C and CuO fabricated by the sol-gel spin coating method on FTO. XPS spectra of (c) Cu 2p and (d) O 1s for Cu_2O and CuO on FTO. The survey spectra are shown in Figure S1.

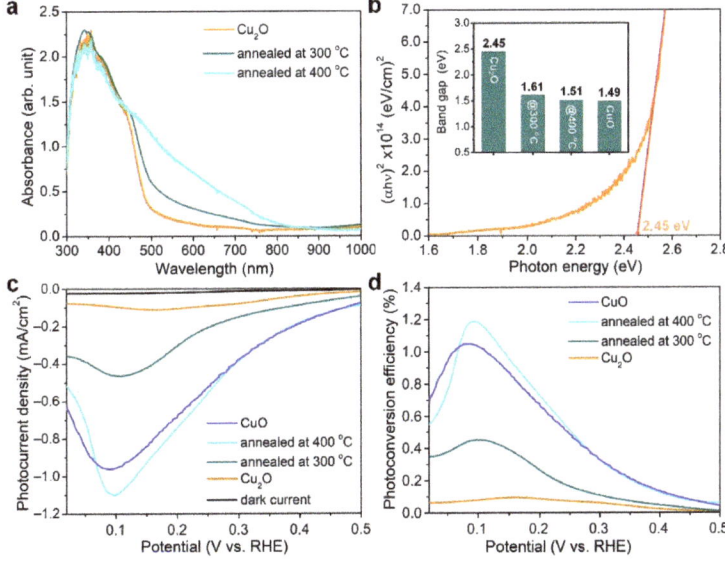

Figure 6. (a) Absorption spectra of Cu_2O and thermally oxidized at 300 °C and 400 °C, and (b) Tauc plot for the Cu_2O thin film deposited on FTO. The inset shows a comparison of the band gaps of different samples. (c) Photocurrent densities and (d) photoconversion efficiencies of Cu_2O and annealed thin films at 300 °C and 400 °C and the spin-coated CuO thin film on FTO.

3.4. Photocatalytic Solar Water Splitting Using P-Type Semiconducting Copper Oxide Thin Films

After characterizing photoelectrode materials, as summarized in Figure 6c,d, we conducted PEC water splitting experiments, demonstrating the onset potential of photoelectrodes at ~0.45 V [29,66]. Similar to the previous study [29], the photoconversion efficiency of Cu_2O prepared by electrodeposition was 0.09%, while thermally oxidized electrode materials at 300 °C and 400 °C showed enhanced efficiencies of 0.45% and 1.19%, respectively (Figure 6d), compared to Cu_2O. We also measured the photoconversion efficiency of the spin-coated CuO thin film, and its efficiency of 1.05% was comparable to that annealed at 400 °C. As described above, the nanostructured surface of thin films might influence the efficiency. As we characterized that Cu_2O samples annealed at 400 °C were identified as CuO, we hypothesized that electrodeposition-based electrode materials might improve photocurrent collection, owing to a better interface between the nanostructured photoelectrodes and substrates, leading to less energy loss during transport. In summary, hetero-phase thin films (Cu_2O and CuO) boosted PEC performance compared to the Cu_2O film, while a fully oxidized sample at 400 °C exhibited better performance. A rational design of high-efficiency copper oxide-based photoelectrodes might be critical to understanding the mechanisms, since the enhancement of photoconversion efficiencies through hydrogen evolution was related to the carrier concentrations of Cu_2O and CuO (3.1×10^{17} cm^{-3} and 2.4×10^{18} cm^{-3}, respectively) [29].

4. Conclusions

Cu_2O thin films were successfully fabricated using electrodeposition, while CuO thin films were also prepared using the sol-gel spin coating method. We studied the effects of aging time and layer dependency based on surface morphologies and optical property variation, suggesting an optimal eight-layer CuO thin film (~270 nm) for further experiments. XRD and Raman spectroscopic measurements confirmed that Cu_2O thin films annealed at 300 °C are a hetero-phase system composed of Cu_2O and CuO. It revealed that >400 °C annealing is essential for fabricating pure CuO. The surface morphology examination using SEM and AFM observed the nanostructured formation from thermal oxidation, suggesting a good junction between electrode materials and the substrate. Consequently, it may affect photoconversion efficiencies. The Tauc plot was successfully applied to measure the energy band gaps of thin films, which were consistent with those in the literature (Cu_2O: 2.45 eV and CuO: 1.49–1.51 eV). The hetero-phase thin film had a slightly larger band gap of 1.61 eV than CuO. We verified that CuO thin films fabricated via electrodeposition exhibit a slightly higher efficiency than the sol-gel spin coating method (1.19% versus 1.05%). A systematic thickness-dependent study on Cu_2O thin films, by adjusting preparation conditions (i.e., pH and temperatures of precursor solutions and annealing temperatures), may influence PEC performance by generating heterojunction layers [18,29]. This study would offer pivotal information to achieve the rational design of highly efficient and cost-effective PEC water splitting and CO_2 seawater splitting using different fabrication approaches for copper oxide-based photoelectrodes.

Supplementary Materials: The following supporting information can be downloaded at: https://www.mdpi.com/article/10.3390/coatings12081206/s1, Table S1: Compilation of representative copper oxide-based photocathodes used for PEC water splitting, Figure S1: Comparison of photoconversion efficiencies of CuO deposited on FTO and ITO using the sol-gel spin-coating method, Figure S2: XPS survey spectrum for (a) Cu_2O and (b) CuO deposited on FTO.

Author Contributions: Conceptualization, Y.C.; data curation, C.-J.C. and C.-W.L.; investigation, C.-J.C., C.-W.L., W.-C.J. and Y.-S.L.; methodology, C.-J.C. and C.-W.L.; project administration, C.C., H.-C.Y. and Y.C.; software, C.-W.L.; supervision, Y.C.; visualization, C.-J.C. and C.-W.L.; writing—original draft preparation, C.-J.C. and C.-W.L.; writing—review and editing, C.C., H.-C.Y., S.-J.C. and Y.C. All authors have read and agreed to the published version of the manuscript.

Funding: We acknowledge the Ministry of Science and Technology of Taiwan (MOST Grant No. 110-2221-E-A49-017-MY3 and 109-2221-E-009-146). C.C. thanks the Ministry of SMEs and Startups (Project No. P0016558), Korea.

Institutional Review Board Statement: Not applicable.

Informed Consent Statement: Not applicable.

Data Availability Statement: Not applicable.

Acknowledgments: C.-J.C. and C.-W.L. thank Cong-You Lin and Tsai-Te Wang for technical support.

Conflicts of Interest: The authors declare no conflict of interest.

References

1. Moniz, S.J.; Shevlin, S.A.; Martin, D.J.; Guo, Z.-X.; Tang, J. Visible-light driven heterojunction photocatalysts for water splitting—A critical review. *Energy Environ. Sci.* **2015**, *8*, 731–759. [CrossRef]
2. Montoya, J.H.; Seitz, L.C.; Chakthranont, P.; Vojvodic, A.; Jaramillo, T.F.; Nørskov, J.K. Materials for solar fuels and chemicals. *Nat. Mater.* **2017**, *16*, 70–81. [CrossRef] [PubMed]
3. Dias, P.; Mendes, A. Hydrogen production from photoelectrochemical water splitting. In *Encyclopedia of Sustainability Science and Technology*; Meyers, R.A., Ed.; Springer: New York, NY, USA, 2017; pp. 1–52.
4. Hisatomi, T.; Kubota, J.; Domen, K. Recent advances in semiconductors for photocatalytic and photoelectrochemical water splitting. *Chem. Soc. Rev.* **2014**, *43*, 7520–7535. [CrossRef] [PubMed]
5. Yang, Y.; Niu, S.; Han, D.; Liu, T.; Wang, G.; Li, Y. Progress in developing metal oxide nanomaterials for photoelectrochemical water splitting. *Adv. Energy Mater.* **2017**, *7*, 1700555. [CrossRef]
6. Díez-García, M.I.; Gómez, R. Progress in ternary metal oxides as photocathodes for water splitting cells: Optimization strategies. *Sol. RRL* **2022**, *6*, 2100871. [CrossRef]
7. Saraswat, S.K.; Rodene, D.D.; Gupta, R.B. Recent advancements in semiconductor materials for photoelectrochemical water splitting for hydrogen production using visible light. *Renew. Sustain. Energ. Rev.* **2018**, *89*, 228–248. [CrossRef]
8. Liu, C.; Dasgupta, N.P.; Yang, P. Semiconductor nanowires for artificial photosynthesis. *Chem. Mater.* **2014**, *26*, 415–422. [CrossRef]
9. Rettie, A.J.E.; Chemelewski, W.D.; Emin, D.; Mullins, C.B. Unravelling small-polaron transport in metal oxide photoelectrodes. *J. Phys. Chem. Lett.* **2016**, *7*, 471–479. [CrossRef]
10. Jin, H.; Guo, C.; Liu, X.; Liu, J.; Vasileff, A.; Jiao, Y.; Zheng, Y.; Qiao, S.-Z. Emerging two-dimensional nanomaterials for electrocatalysis. *Chem. Rev.* **2018**, *118*, 6337–6408. [CrossRef]
11. Feng, C.; Chen, Z.; Jing, J.; Sun, M.; Han, J.; Fang, K.; Li, W. Synergistic effect of hierarchical structure and Z-scheme heterojunction constructed by CdS nanoparticles and nanoflower-structured Co_9S_8 with significantly enhanced photocatalytic hydrogen production performance. *J. Photochem. Photobiol.* **2021**, *409*, 113160. [CrossRef]
12. Wang, S.; Zhu, B.; Liu, M.; Zhang, L.; Yu, J.; Zhou, M. Direct Z-scheme ZnO/CdS hierarchical photocatalyst for enhanced photocatalytic H_2-production activity. *Appl. Catal. B Environ.* **2019**, *243*, 19–26. [CrossRef]
13. Meng, A.; Zhu, B.; Zhong, B.; Zhang, L.; Cheng, B. Direct Z-scheme TiO_2/CdS hierarchical photocatalyst for enhanced photocatalytic H_2-production activity. *Appl. Surf. Sci.* **2017**, *422*, 518–527. [CrossRef]
14. Ye, K.-H.; Li, H.; Huang, D.; Xiao, S.; Qiu, W.; Li, M.; Hu, Y.; Mai, W.; Ji, H.; Yang, S. Enhancing photoelectrochemical water splitting by combining work function tuning and heterojunction engineering. *Nat. Commun.* **2019**, *10*, 3687. [CrossRef]
15. Wang, Y.; Tang, G.; Wu, Y.; Zhao, J.; Zhang, H.; Zhou, M. Cu_2O/CeO_2 Photoelectrochemical water splitting: A nanocomposite with an efficient interfacial transmission path under the Co-action of p-n heterojunction and micro-mesocrystals. *Chem. Eur. J.* **2022**, *28*, e202103459.
16. Zhang, K.; Cai, W.-F.; Shi, J.-W.; Chen, Q.-Y. Architecture lattice-matched cauliflower-like CuO/ZnO p–n heterojunction toward efficient water splitting. *J. Chem. Technol. Biotechnol.* **2022**, *97*, 914–923. [CrossRef]
17. Chiang, C.-Y.; Aroh, K.; Franson, N.; Satsangi, V.R.; Dass, S.; Ehrman, S. Copper oxide nanoparticle made by flame spray pyrolysis for photoelectrochemical water splitting–Part II. Photoelectrochemical study. *Int. J. Hydrogen Energy* **2011**, *36*, 15519–15526. [CrossRef]
18. Jeong, D.; Jo, W.; Jeong, J.; Kim, T.; Han, S.; Son, M.-K.; Jung, H. Characterization of Cu_2O/CuO heterostructure photocathode by tailoring CuO thickness for photoelectrochemical water splitting. *RSC Adv.* **2022**, *12*, 2632–2640. [CrossRef]
19. Meyer, B.K.; Polity, A.; Reppin, D.; Becker, M.; Hering, P.; Klar, P.J.; Sander, T.; Reindl, C.; Benz, J.; Eickhoff, M.; et al. Binary copper oxide semiconductors: From materials towards devices. *Phys. Status Solidi B* **2012**, *249*, 1487–1509. [CrossRef]
20. Murali, D.S.; Kumar, S.; Choudhary, R.J.; Wadikar, A.D.; Jain, M.K.; Subrahmanyam, A. Synthesis of Cu_2O from CuO thin films: Optical and electrical properties. *AIP Adv.* **2015**, *5*, 047143. [CrossRef]
21. Hussain, S.; Cao, C.; Usman, Z.; Nabi, G.; Butt, F.K.; Mahmood, K.; Ali, A.; Arshad, M.I.; Amin, N. Effect of films morphology on the performance of Cu_2O PEC solar cells. *Optik* **2018**, *172*, 72–78. [CrossRef]
22. Kunturu, P.P.; Huskens, J. Efficient solar water splitting photocathodes comprising a copper oxide heterostructure protected by a thin carbon layer. *ACS Appl. Energy Mater.* **2019**, *2*, 7850–7860. [CrossRef]

23. Shockley, W.; Queisser, H.J. Detailed balance limit of efficiency of p-n junction solar cells. *J. App. Phys.* **1961**, *32*, 510–519. [CrossRef]
24. de Brito, J.F.; Araujo, A.R.; Rajeshwar, K.; Zanoni, M.V.B. Photoelectrochemical reduction of CO_2 on Cu/Cu_2O films: Product distribution and pH effects. *Chem. Eng. J.* **2015**, *264*, 302–309. [CrossRef]
25. Liu, Y.; Turley, H.K.; Tumbleston, J.R.; Samulski, E.T.; Lopez, R. Minority carrier transport length of electrodeposited Cu_2O in ZnO/Cu_2O heterojunction solar cells. *Appl. Phys. Lett.* **2011**, *98*, 162105. [CrossRef]
26. Paracchino, A.; Mathews, N.; Hisatomi, T.; Stefik, M.; Tilley, S.D.; Grätzel, M. Ultrathin films on copper(I) oxide water splitting photocathodes: A study on performance and stability. *Energy Environ. Sci.* **2012**, *5*, 8673–8681. [CrossRef]
27. Xing, H.; Lei, E.; Guo, Z.; Zhao, D.; Liu, Z. Enhancement in the charge transport and photocorrosion stability of CuO photocathode: The synergistic effect of spatially separated dual-cocatalysts and p-n heterojunction. *Chem. Eng. J.* **2020**, *394*, 124907. [CrossRef]
28. Jian, J.; Kumar, R.; Sun, J. Cu_2O/ZnO p–n junction decorated with NiO_x as a protective layer and cocatalyst for enhanced photoelectrochemical water splitting. *ACS Appl. Energy Mater.* **2020**, *3*, 10408–10414. [CrossRef]
29. Yang, Y.; Xu, D.; Wu, Q.; Diao, P. Cu_2O/CuO bilayered composite as a high-efficiency photocathode for photoelectrochemical hydrogen evolution reaction. *Sci. Rep.* **2016**, *6*, 35158. [CrossRef]
30. Bae, H.; Burungale, V.; Na, W.; Rho, H.; Kang, S.H.; Ryu, S.-W.; Ha, J.-S. Nanostructured CuO with a thin g-C_3N_4 layer as a highly efficient photocathode for solar water splitting. *RSC Adv.* **2021**, *11*, 16083–16089. [CrossRef]
31. Mary, A.S.; Murugan, C.; Pandikumar, A. Uplifting the charge carrier separation and migration in Co-doped $CuBi_2O_4/TiO_2$ pn heterojunction photocathode for enhanced photoelectrocatalytic water splitting. *J. Colloid Interface Sci.* **2022**, *608*, 2482–2492. [CrossRef]
32. Baek, S.K.; Kim, J.S.; Yun, Y.D.; Kim, Y.B.; Cho, H.K. Cuprous/Cupric heterojunction photocathodes with optimal phase transition interface via preferred orientation and precise oxidation. *ACS Sustain. Chem. Eng.* **2018**, *6*, 10364–10373. [CrossRef]
33. Ma, M.; Lei, E.; Zhao, D.; Xin, Y.; Wu, X.; Meng, Y.; Liu, Z. The p-n heterojunction of $BiVO_4/Cu_2O$ was decorated by plasma Ag NPs for efficient photoelectrochemical degradation of Rhodamine B. *Colloids Surf. A Physicochem. Eng. Asp.* **2022**, *633*, 127834. [CrossRef]
34. Hossain, R.; Nekouei, R.K.; Al Mahmood, A.; Sahajwalla, V. Value-added fabrication of NiO-doped CuO nanoflakes from waste flexible printed circuit board for advanced photocatalytic application. *Sci. Rep.* **2022**, *12*, 12171. [CrossRef] [PubMed]
35. Chen, Y.-C.; Dong, P.-H.; Hsu, Y.-K. Defective indium tin oxide forms an ohmic back contact to an n-type Cu_2O photoanode to accelerate charge-transfer kinetics for enhanced low-bias photoelectrochemical water splitting. *ACS Appl. Mater. Interfaces* **2021**, *13*, 38375–38383. [CrossRef]
36. Zhong, X.; Song, Y.; Cui, A.; Mu, X.; Li, L.; Han, L.; Shan, G.; Liu, H. Adenine-functionalized graphene oxide as a charge transfer layer to enhance activity and stability of Cu_2O photocathode for CO_2 reduction reaction. *Appl. Surf. Sci.* **2022**, *591*, 153197. [CrossRef]
37. Kalanur, S.S.; Lee, Y.J.; Seo, H. Enhanced and stable photoelectrochemical H_2 production using a engineered nano multijunction with Cu_2O photocathode. *Mater. Today Chem.* **2022**, *26*, 101031. [CrossRef]
38. Zardetto, V.; Brown, T.M.; Reale, A.; Di Carlo, A. Substrates for flexible electronics: A practical investigation on the electrical, film flexibility, optical, temperature, and solvent resistance properties. *J. Polym. Sci. B Polym. Phys.* **2011**, *49*, 638–648. [CrossRef]
39. Li, F.; Chen, C.; Tan, F.; Li, C.; Yue, G.; Shen, L.; Zhang, W. Semitransparent inverted polymer solar cells employing a sol-gel-derived TiO_2 electron-selective layer on FTO and $MoO_3/Ag/MoO_3$ transparent electrode. *Nanoscale Res. Lett.* **2014**, *9*, 579. [CrossRef]
40. Oral, A.Y.; Menşur, E.; Aslan, M.H.; Başaran, E. The preparation of copper(II) oxide thin films and the study of their microstructures and optical properties. *Mater. Chem. Phys.* **2004**, *83*, 140–144. [CrossRef]
41. Lim, Y.-F.; Chua, C.S.; Lee, C.J.J.; Chi, D. Sol-gel deposited Cu_2O and CuO thin films for photocatalytic water splitting. *Phys. Chem. Chem. Phys.* **2014**, *16*, 25928–25934. [CrossRef]
42. Wang, L.; Lee, C.-Y.; Schmuki, P. Solar water splitting: Preserving the beneficial small feature size in porous α-Fe_2O_3 photoelectrodes during annealing. *J. Mater. Chem. A* **2013**, *1*, 212–215. [CrossRef]
43. Chu, S.; Li, W.; Yan, Y.; Hamann, T.; Shih, I.; Wang, D.; Mi, Z. Roadmap on solar water splitting: Current status and future prospects. *Nano Futures* **2017**, *1*, 022001. [CrossRef]
44. Masoumi, Z.; Tayebi, M.; Lee, B.-K. Ultrasonication-assisted liquid-phase exfoliation enhances photoelectrochemical performance in α-Fe_2O_3/MoS_2 photoanode. *Ultrason. Sonochem.* **2021**, *72*, 105403. [CrossRef] [PubMed]
45. Elmas, S.; Korkmaz, Ş.; Pat, S. Optical characterization of deposited ITO thin films on glass and PET substrates. *Appl. Surf. Sci.* **2013**, *276*, 641–645. [CrossRef]
46. Trzcinski, M.; Antończak, A.; Domanowski, P.; Kustra, M.; Wachowiak, W.; Naparty, M.; Hiller, T.; Bukaluk, A.; Wronkowska, A. Characterisation of coloured $TiO_x/Ti/$glass systems. *Appl. Surf. Sci.* **2014**, *322*, 209–214.
47. Li, Y.M.; Xu, L.-h.; Li, X.; Shen, X.Q.; Wang, A. Effect of aging time of ZnO sol on the structural and optical properties of ZnO thin films prepared by so-gel method. *Appl. Surf. Sci.* **2010**, *256*, 4543–4547. [CrossRef]
48. Jacob, S.S.K.; Kulandaisamy, I.; Raj, I.L.P.; Abdeltawab, A.A.; Mohammady, S.Z.; Ubaidullah, M. Improved optoelectronic properties of spray pyrolysis coated Zn doped Cu_2O thin films for photodetector applications. *Opt. Mater.* **2021**, *116*, 111086. [CrossRef]

49. Tauc, J.; Grigorovici, R.; Vancu, A. Optical properties and electronic structure of amorphous germanium. *Phys. Status Solidi B* **1966**, *15*, 627–637. [CrossRef]
50. Singhal, A.; Pai, M.R.; Rao, R.; Pillai, K.T.; Lieberwirth, I.; Tyagi, A.K. Copper(I) oxide nanocrystals—One step synthesis, characterization, formation mechanism, and photocatalytic properties. *Eur. J. Inorg. Chem.* **2013**, *2013*, 2640–2651. [CrossRef]
51. Zhang, Y.C.; Tang, J.Y.; Wang, G.L.; Zhang, M.; Hu, X.Y. Facile synthesis of submicron Cu_2O and CuO crystallites from a solid metallorganic molecular precursor. *J. Cryst. Growth* **2006**, *294*, 278–282. [CrossRef]
52. Li, J.; Kolekar, S.; Ghorbani-Asl, M.; Lehnert, T.; Biskupek, J.; Kaiser, U.; Krasheninnikov, A.V.; Batzill, M. Layer-dependent band gaps of platinum dichalcogenides. *ACS Nano* **2021**, *15*, 13249–13267. [CrossRef]
53. Al-Enizi, A.M.; Shaikh, S.F.; Tamboli, A.M.; Marium, A.; Ijaz, M.F.; Ubaidullah, M.; Moydeen Abdulhameed, M.; Ekar, S.U. Hybrid ZnO flowers-rods nanostructure for improved photodetection compared to standalone flowers and rods. *Coatings* **2021**, *11*, 1464. [CrossRef]
54. Prabu, R.D.; Valanarasu, S.; Ganesh, V.; Shkir, M.; Kathalingam, A.; AlFaify, S. Effect of spray pressure on optical, electrical and solar cell efficiency of novel Cu_2O thin films. *Surf. Coat. Technol.* **2018**, *347*, 164–172. [CrossRef]
55. Panzeri, G.; Cristina, M.; Jagadeesh, M.S.; Bussetti, G.; Magagnin, L. Modification of large area Cu_2O/CuO photocathode with CuS non-noble catalyst for improved photocurrent and stability. *Sci. Rep.* **2020**, *10*, 18730. [CrossRef]
56. Moumen, A.; Hartiti, B.; Comini, E.; El Khalidi, Z.; Arachchige, H.M.M.M.; Fadili, S.; Thevenin, P. Preparation and characterization of nanostructured CuO thin films using spray pyrolysis technique. *Superlattices Microstruct.* **2019**, *127*, 2–10. [CrossRef]
57. Balık, M.; Bulut, V.; Erdogan, I.Y. Optical, structural and phase transition properties of Cu_2O, CuO and Cu_2O/CuO: Their photoelectrochemical sensor applications. *Int. J. Hydrogen Energy* **2019**, *44*, 18744–18755. [CrossRef]
58. Ghodselahi, T.; Vesaghi, M.; Shafiekhani, A.; Baghizadeh, A.; Lameii, M. XPS study of the Cu@Cu_2O core-shell nanoparticles. *Appl. Surf. Sci.* **2008**, *255*, 2730–2734. [CrossRef]
59. Poulston, S.; Parlett, P.; Stone, P.; Bowker, M. Surface oxidation and reduction of CuO and Cu_2O studied using XPS and XAES. *Surf. Interface Anal.* **1996**, *24*, 811–820. [CrossRef]
60. Li, J.P.H.; Zhou, X.; Pang, Y.; Zhu, L.; Vovk, E.I.; Cong, L.; van Bavel, A.P.; Li, S.; Yang, Y. Understanding of binding energy calibration in XPS of lanthanum oxide by in situ treatment. *Phys. Chem. Chem. Phys.* **2019**, *21*, 22351–22358. [CrossRef]
61. Greczynski, G.; Hultman, L. Compromising science by ignorant instrument calibration—Need to revisit half a century of published XPS data. *Angew. Chem. Int. Ed.* **2020**, *59*, 5002–5006. [CrossRef]
62. Hiraba, H.; Koizumi, H.; Kodaira, A.; Nogawa, T.; Yoneyama, T.; Matsumura, H. Influence of oxidation of copper on shear bond strength to an acrylic resin using an organic sulfur compound. *Materials* **2020**, *13*, 2092. [CrossRef]
63. Svintsitskiy, D.A.; Kardash, T.Y.; Stonkus, O.A.; Slavinskaya, E.M.; Stadnichenko, A.I.; Koscheev, S.V.; Chupakhin, A.P.; Boronin, A.I. In Situ XRD, XPS, TEM, and TPR study of highly active in CO oxidation CuO nanopowders. *J. Phys. Chem. C* **2013**, *117*, 14588–14599. [CrossRef]
64. Dubale, A.A.; Pan, C.-J.; Tamirat, A.G.; Chen, H.-M.; Su, W.-N.; Chen, C.-H.; Rick, J.; Ayele, D.W.; Aragaw, B.A.; Lee, J.-F.; et al. Heterostructured Cu_2O/CuO decorated with nickel as a highly efficient photocathode for photoelectrochemical water reduction. *J. Mater. Chem. A* **2015**, *3*, 12482–12499. [CrossRef]
65. Martin, L.; Martinez, H.; Poinot, D.; Pecquenard, B.; Le Cras, F. Comprehensive X-ray photoelectron spectroscopy study of the conversion reaction mechanism of CuO in lithiated thin film electrodes. *J. Phys. Chem. C* **2013**, *117*, 4421–4430. [CrossRef]
66. Kyesmen, P.I.; Nombona, N.; Diale, M. A Promising three-step heat treatment process for preparing CuO films for photocatalytic hydrogen evolution from water. *ACS Omega* **2021**, *6*, 33398–33408. [CrossRef]

Article

Contribution to the Research and Development of Innovative Building Components with Embedded Energy-Active Elements

Daniel Kalús *, Daniela Koudelková, Veronika Mučková, Martin Sokol and Mária Kurčová

Faculty of Civil Engineering, Slovak University of Technology, Radlinského 11, 81005 Bratislava, Slovakia; daniela.koudelkova@stuba.sk (D.K.); veve.muckova@gmail.com (V.M.); martin.sokol@stuba.sk (M.S.); maria.kurcova@stuba.sk (M.K.)
* Correspondence: daniel.kalus@stuba.sk; Tel.: +421-2-328-88-661

Abstract: The research described in this study focuses on the innovation and optimization of building envelope panels with integrated energy-active elements in the thermal barrier function. It is closely related to developing and implementing the prototype prefabricated house IDA I with combined building-energy systems using renewable energy sources. We were inspired by the patented ®ISOMAX panel and system, which we have been researching and innovating for a long time. The thermal barrier has the function of eliminating heat loss/gain through the building envelope. By controlling the heat/cold transfer in the thermal barrier, it is possible to eliminate the thickness of the thermal insulation of the building envelope and thus achieve an equivalent thermal resistance of the building structure that is equal to the standard required value. The technical solution of the ISOMAX panel also brings, besides the use of the thermal barrier function, the function of heat/cold accumulation in the load-bearing part of the building envelope. Our research aimed to design and develop a panel for which the construction would be optimal in terms of thermal barrier operation and heat/cold accumulation. As the production panels in the lost formwork of expanded polystyrene (according to the patented system) proved to be too complicated and time consuming, and often showed shortcomings from a structural point of view, the next goal was to design a new, statically reliable panel construction with integrated energy-active elements and a time-saving, cost-effective, unified production directly in the panel factory. In order to develop and design an innovative panel with integrated energy-active elements, we analyzed the composition of the original panel and designed the composition of the innovative panel. We created mathematical–physical models of both panels and analyzed their energy potential. By induction and an analog form of formation, we designed the innovative panel. Based on the synthesis of the knowledge obtained from the scientific analysis and the transformation of this data, most of the building components and all the panels with integrated energy-active elements were manufactured directly in the prefabrication plant. Subsequently, the prototype of the prefabricated house IDA I was realized. The novelty of our innovative building envelope panel solution lies in the panel's design, which has a heat loss/gain that is 2.6 times lower compared to the ISOMAX panel.

Keywords: active energy elements (EAE); active thermal protection (ATP); thermal barrier (TB); renewable energy sources (RES); heat/cold accumulation

Citation: Kalús, D.; Koudelková, D.; Mučková, V.; Sokol, M.; Kurčová, M. Contribution to the Research and Development of Innovative Building Components with Embedded Energy-Active Elements. *Coatings* **2022**, *12*, 1021. https://doi.org/10.3390/coatings12071021

Academic Editor: Giorgos Skordaris

Received: 22 June 2022
Accepted: 14 July 2022
Published: 19 July 2022

Publisher's Note: MDPI stays neutral with regard to jurisdictional claims in published maps and institutional affiliations.

Copyright: © 2022 by the authors. Licensee MDPI, Basel, Switzerland. This article is an open access article distributed under the terms and conditions of the Creative Commons Attribution (CC BY) license (https://creativecommons.org/licenses/by/4.0/).

1. Introduction

Energy self-sufficiency and security are among the priorities of all governments. Scientists worldwide are looking for solutions to halt climate change on our planet and reduce dependence on fossil fuels. Our research focuses on innovating and optimizing energy systems that use renewable energy and waste heat from technologies. Building structures that have an internal energy source with active thermal protection (ATP) represent technical solutions with a high potential to use environmentally friendly energy sources.

Active thermal protection is a dynamic process that is characterized by the building of structures with integrated active elements which are themselves characterized by one or more functions in different modes of energy systems operations. The energy functions of ATP are: thermal barrier, large-scale radiant low-temperature heating/high-temperature cooling, heat/cool storage, solar and ambient energy capture, and heat/cool heat recovery.

The research described in this study focuses on the innovation and optimization of building envelope panels with integrated energy-active elements in the thermal barrier function. The field of combined building-energy systems has been researched in our department for a long time, since approximately 2004. The contribution to the research and development of innovative building components with embedded energy-active elements is closely related to developing and implementing the prototype prefabricated house, IDA I (the name of the prototype prefabricated house was derived from the first name of the wife of the client of the development and implementation of the building, Ida, in accordance with the work contract), with combined building-energy systems using renewable energy sources (RES)(Figure 1).

Figure 1. A view of the construction of the prototype prefabricated house IDA I [1].

We were inspired by the patented ®ISOMAX panel and system, [2], which we have been researching and innovating for a long time. The thermal barrier has the function of eliminating heat loss/gain through the building envelope. By controlling the heat/cold transfer in the thermal barrier, it is possible to eliminate the thickness of the thermal insulation of the building envelope and thus achieve an equivalent thermal resistance for the building structure that is equal to the standard required value.

Based on a request from the practice, we were approached by AQUA IDA Slovakia, s. r. o. in 2005. (Currently Paneláreň Vrakuňa, a. s.)—the owner of the license of the building technology with the name and trademark ®ISOMAX [2]—to develop a prototype prefabricated house called IDA I. In accordance with the work contract HZ 04-309-05 between the customer and the Department of Building Services of the Faculty of Civil Engineering of the STU in Bratislava, a prototype of the panel house IDA I was developed, designed, and implemented in 2005–2006. It currently serves as an administrative building for the joint-stock company Paneláreň Vrakuňa (responsible researcher: Kalús, D.) [1].

The technical solution of the ISOMAX panel also brings, besides the use of the thermal barrier function, the function of heat/cold accumulation in the load-bearing part of the building envelope. Our research aimed to design and develop a panel for which the construction would be optimal in terms of thermal barrier operation and heat/cold accumulation. The novelty of our innovative building envelope panel solution lies in the panel's design, which has a 2.6 times lower heat loss/gain compared to the ISOMAX panel from the load-bearing part of the panels, which accumulates heat/cool for controlled operation of the thermal barrier. So, in addition to a higher equivalent thermal resistance compared to the ISOMAX panel, our innovative building envelope panel has significantly lower requirements for the operation of the circulation pumps. This result lowers the building's energy intensity and is both economically efficient and environmentally friendly.

The production panels in lost formwork of expanded polystyrene, according to the patented system, proved too complicated and time-consuming (Figure 2). Because the panels often showed shortcomings from a structural point of view, the next goal was to design a new, statically reliable panel construction method that had integrated energy-active elements and a timesaving, cost-effective, unified production directly in the panel factory. The sponsor's priority for the applied research was that as many of the components of the IDA I prototype prefabricated house as possible should be mass-produced in prefabrication.

Figure 2. View of ISOMAX panel production from lost formwork [2].

In this study, we used the following methods of scientific work:
(1) Analysis and synthesis of knowledge in the field of active thermal protection and thermal barrier (Section 2);
(2) Description of the basic calculation equations for thermal barrier sizing (Section 3.1);
(3) Analysis of the ISOMAX panel composition and the development of a mathematical–physical model (Section 3.2);
(4) Analysis of the composition of the innovative envelope panel solution and development of the mathematical–physical model, (Section 3.3);
(5) Development and design of innovative panels and the details of panel joining (Section 3.4);
(6) Inductive and analog forms of formation of the innovative panel with integrated energy-active elements (Section 4.1);
(7) Analysis of the energy potential of the retrofitted panel with a thermal barrier compared to the original panel (Section 4.2);
(8) Synthesis of the knowledge obtained from the scientific analysis and the transformation of the data into the design and implementation of the IDA I prefabricated house prototype (Section 4.3).

Finally, we defined the objectives of our further research in Section 5.

2. Overview of Studies Dealing with Active Thermal Protection and Thermal Barrier

Lehmann et al., (2007) [3]: The authors examined the possibilities of use and functionality of thermally activated building systems (TABS). For the analysis of thermal comfort factors, maximum permissible thermal gains in the room, and cooling of the building mass for a typical administrative building, the building simulation program TRNSYS was used, in which the RC modeling approach for TABS was gradually developed. Based on the results, it can be concluded that, depending on the maximum permissible daily amplitude of the room temperature, it is possible to use the modules to control typical heat gain profiles with a peak load of up to approx. 50 W/m^2 of floor area. However, the design of the panels will, in most cases, be decisive for transitional periods with already high solar gains and a still-limited comfort range, thus limiting the maximum load to lower values. The results also showed that room-side processes are hardly affected by supply-side processes. In the case of cooling, this makes it possible to extend the re-cooling period of the fabric to 24 h a day with correspondingly "high" supply temperatures and a reduction in the maximum load by up to 50%. The stated results can indicate whether the slats can be used in a specific building and provide relevant parameters for dimensioning the slats.

Gwerder et al., (2008) [4]: The article outlined the basic thermal models, assumptions, and gives a procedure and example of the application of the calculation method for TABS. The integration of the building structure as TABS for energy storage has been shown to be energy efficient and economically viable for cooling and heating buildings. However, control remained an issue that needed to be improved. This paper outlined a method to enable both sizing and automated control of TABS with automatic switching between cooling/heating modes for variable comfort criteria.

Rijksen et al., (2010) [5]: The study aimed to establish general guidelines for the required cooling performance of an office building using TABS. On-site measurements aimed at obtaining the required cooling performance of the entire building as well as individual zones were carried out. In addition, indoor room environmental parameters and TABS surface temperatures were measured. The measured data were used to analyze the predictive performance of the simulation model. To obtain general guidelines for the required cooling performance of a standard office building, whole building simulations were used to determine the impact of different window sizes as well as variable internal heat gains. The required cooling performance was compared with the cooling performance of a system without energy balancing (e.g., cooled ceiling panels). Research has shown that TABS can reduce the cooling performance of a radiator by up to 50%. The results presented in this paper can be used for guidance in the first phase of the design process. The results focused on temperate climates and were derived for the climatic conditions in the Netherlands.

Kowalczuk and Krzaczek (2010) [6]: The authors focused on analyzing the thermal performance and stability of the thermal barrier (TB). The research was carried out on a 3D FE model (three-dimensional finite-element model) of a prefabricated outer wall component that had a built-in TB formed by a polypropylene U-pipe system with a flowing liquid. The temperature of the TB was fixed at 17 °C, although it varied only to a small extent during the year. The FE analysis was supported by a new SVC (scheduling variable controller) control system implemented in FORTRAN to simulate real working conditions. The FE simulations showed that the optimal mass flow range is 0.05–1.0 kg/s. No significant influence of TB placement on the thermal behavior of the external prefabricated component was found. The optimal design of the outer wall with a TB is a multilayer structure composed of at least three layers: an outer insulating layer, a massive core layer with the TB U-pipe system, and an inner layer material of low thermal conductivity. The TB reduces the heating and cooling requirements by at least three times compared to a traditional exterior wall without a TB. The TB temperature, almost fixed at 17 °C, radically reduces the risk of water vapor condensation. It allows for designing insulation layers on the inside of the exterior walls. The TB technique can thus be successfully implemented both in designing new buildings and in the thermal modernization of existing buildings.

Lehmann et al., (2011) [7]: The authors investigated the influence of a control strategy, a hydronic circuit, and a (cold) generation system on the energy efficiency of TABS. Based on a case simulation study for a typical Central European office building, they found that TABS with separate zoned return water systems can achieve energy savings of around 20–30% of heating and cooling demand compared to conventional zoned return water. With the intermittent pulse width modulation (PWM) system control strategy, it is possible to reduce the electrical energy consumption of circulating water pumps by more than 50% compared to continuous operation. In terms of cooling generation for TABS, free cooling with a wet cooling tower has been shown to be most effective when the cooling source is outside air. Variants with mechanical chillers show 30–50% higher electrical energy requirements for the production and distribution of cold, even though their running times are much shorter than the cooling towers' operating times.

Stojanovic et al., (2014) [8]: The study focused on the possibilities of increasing the energy efficiency of buildings when using integrated thermally activated systems that use geothermal energy. It analyzed the effect of temperature, as well as the type and thickness of other materials of the nontransparent part of the building façade, on energy efficiency, as this type of element strongly depends on these parameters. The building energy demand for heating was obtained for a real apartment building in Serbia using EnergyPlus software. The building with all the necessary inputs for the simulation was modeled in Google SketchUp using the Open Studio Plug-in. The obtained results were compared with the measured energy consumption for heating. The results showed that thermally activated building systems are a good way to increase the energy efficiency of a building and that, by applying certain temperatures within this element, a low-energy house standard can be achieved.

Babiak et al., (2015) [9]: A study compared TABS with traditional convection air conditioning and fan coil systems in office buildings in France, Lyon. TABS is a combined heating and cooling system with piping embedded in structural concrete slabs or walls of multistory buildings. An evaluation project had been performed by appointing a building energy simulation consultant. The thermal indoor climate and energy modeling were conducted, and heating/cooling load profiles were provided for selected systems. The HVAC (heating, ventilation, and air conditioning) schemes were selected based on what was commonly specified in France for similar projects. Based on the thermal modeling results, the consultant created schemes for each method of mechanical service. The cost amount was obtained from various sources. A life cycle cost analysis provides an assessment of different methods of heating, cooling, and ventilation. The cost data and building energy simulations showed that TABS reduces the total cost of the building by a significant 16–27% compared to other air handling equipment. The results also showed that selecting TABS for the HVAC scheme will improve the indoor thermal energy environment (class B PMV index ranges from 22% to 24% more working hours in the two selected rooms). In conclusion, TABS has proven to be adaptable and cost-effective for French conditions.

Yu et al., (2016) [10]: The authors proposed a novel idea of using an MTC (minitubular capillary network) with low-grade thermal water for the thermal activation of conventional walls. In contrast to the general indoor water embedded design, this innovation applies MTC at the boundary between the space and the outdoor environment and brings the thermal energy closer to the load. It can significantly relax the constraints on water temperature and facilitate the direct use of low-grade renewable energy in buildings without conversion. To investigate the thermodynamic performance of the wall, a transient dynamic model was developed for the cases with and without the MTC thermal layer. Simulated results showed that the embedded MTC wall could significantly change the thermodynamics of the wall, from balancing environmental influences to indirectly cooling the space. This wall can activate the wall, can effectively stabilize the internal surface temperature, compensate for heat gains, and supply cooling energy to the space in summer.

Kisilewicz et al., (2019) [11]: In the paper, the authors discussed how an active insulation system can replace commonly used standard passive insulation systems. The research was carried out in an experimental apartment building in the city of Nyiregyház in Hungary, which was equipped with an innovative system that had a direct connection from the ground heat exchanger to the wall heat exchanger. The results were obtained using the method of experimental measurements of selected physical quantities and dynamic simulation. Initial research results concluded that active thermal insulation significantly improves the insulation parameters of the outer wall. In the analyzed periods, the total amount of heat loss through the perimeter walls decreased from 53% in February to 81% in November. The equivalent thermal transmittance U_{eq} of the analyzed wall was dependent on the local climatic conditions and was 0.047 W/(m^2·K) in November and 0.11 W/(m^2·K) in March, while the standard heat transfer value was 0.282 W/(m^2·K). These research results should be the basis for implementing an innovative system in buildings with almost zero energy consumption.

Figiel and Leciej-Pirczewska (2020) [12]: the authors studied the effect of an active thermal barrier on heat loss and CO_2 emissions. The analysis was performed for a family house in a temperate climate based on parameters taken from one of the Polish meteorological databases. The calculations were carried out using current procedures for assessing the energy efficiency of the building. General calculations of energy demand for heating, cooling, and ventilation were based on methods from CEN standards. Research has shown that the application of a thermal barrier has the effect of reducing heat loss from the inside. Placing the barrier close to the exterior is beneficial and installing it during a thermal renovation can provide significant energy improvements and be more environmentally friendly.

Kalús et al., (2021) [13]: The article focused on describing an innovative solution and application of active thermal protection (ATP) of buildings using thermal insulation panels, with the active regulation of heat transfer in the form of a contact insulation system. Thermal insulation panels are part of a prefabricated light outer shell, which, together with a system of low-temperature heating and high-temperature cooling, creates an internal environment. The energy source is usually renewable energy sources or technological waste heat. Research and development of an innovative facade system with ATP is in the phase of computer simulations and laboratory preparation for measuring thermal insulation panels' parameters with different energy function combinations. The paper presented the theoretical assumptions, calculation procedure, and parametric study of three basic design solutions of combined energy wall systems in the function of low-temperature radiant heating and high-temperature radiant cooling. The most significant limitation of implementing this technology in practice is that thermal insulation panels with ATP are not yet certified and produced by any manufacturer. We recommend further research, especially towards multifunctional thermal insulation panels with ATP in low-temperature radiant heating/high-temperature radiant cooling modes, but also as a solar energy absorber and ambient energy in cooperation with heat pumps.

Kalús et al., (2021) [14]: The study aimed to evaluate TBs in terms of energy demand, economic efficiency, and environmental friendliness by comparing the use of a classic perimeter wall with the required thickness of thermal insulation (that meets the normative requirements for thermal resistance) and a perimeter wall with an integrated TB (significantly eliminating the thermal insulation thickness). The use of a thermal barrier was evaluated using a number of economic indicators. Economic indicator one compared the cost of heat delivered to the TB in a building with significantly eliminated insulation with the saved cost of insulation at standard thickness. Economic indicator two compared the cost of heat delivered to the TB in a building with significantly eliminated thermal insulation with the potential gain from the sale of the usable area of the building gained relative to the area with the normative thickness of thermal insulation. Economic indicator three compared the cost of heat delivered to the TB in a building with significantly eliminated insulation with the cost of grey energy at the normative insulation thickness. On the basis of a parametric study based on theoretical assumptions, it can be concluded that the thermal barrier represents a very promising and effective solution in terms of evaluating economic indicators one to three, which are even more significant if heat from renewable energy sources (RES) or waste heat is used for TB.

Research in the field of active thermal protection and thermally activated building systems in connection with renewable energy sources have been carried out by many other scientists and their colleagues, including as Maruyama et al., (1989) [15], Olesen et al., (2006) [16], Krecké et al., (2007) [17], Gwerder et al., (2009) [18], Xie et al., (2012) [19], Doležel (2014) [20], Ibrahim et al., (2021) [21], and Kalús et al., (2010) [22].

3. Innovated Thermal Barrier Panel Compared to the Patented ISOMAX Panel

In Section 3.1, we present the initial calculation relations for the sizing of the thermal barrier. In Section 3.2, we analyze the composition of the patented ISOMAX panel. We present the results of a parametric study of the temperatures in the different layers of the building structure and graphically illustrate the progression of temperatures from the interior to the exterior in winter and summer. We analyze the energy potential of the patented ISOMAX panel. In Section 3.3, we analyze the design of an innovative thermal barrier panel and also present the results of a parametric study of the temperatures in the different layers of the building structure and graphically illustrate the progression of temperatures from the interior to the exterior in winter and summer. We analyze the energy potential of the upgraded panel. In Section 3.4, we give the basic details of joining the panels.

3.1. Initial Calculation Relations for Thermal Barrier Dimensioning

The idea of using a thermal barrier to eliminate heat loss/heat gain throughout the building structure is based on the knowledge of the temperatures between the layers—specifically between the static load-bearing layer and thermal insulation layers of the building structure. When calculating the thermal resistance R (($m^2 \cdot K$)/W) and the heat transfer coefficient U (W/($m^2 \cdot K$)) of a multilayer building structure, the temperatures between the layers are calculated.

The thermal resistance of the jth structure is calculated by:

$$R_i = \frac{d_j}{\lambda_j}, \tag{1}$$

where:

R_i is the thermal resistance of the jth layer of the structure (($m^2 \cdot K$)/W),
d_j is the thickness of the jth layer of the structure (m), and
λ_j is the coefficient of thermal conductivity of the jth layer of the structure (W/(m K)) [23].

The thermal resistance of a multilayer structure shall be calculated:

$$R_c = \sum R_i, \tag{2}$$

$$R = R_{si} + R_c + R_{se}, \tag{3}$$

where:

R_j is the thermal resistance of the jth layer of the structure (($m^2 \cdot K$)/W),
R_c is the total thermal resistance of the structure (($m^2 \cdot K$)/W),
R_{si} is the thermal resistance to heat transfer at the internal surface of the structure (($m^2 \cdot K$)/W),
R_{se} is the thermal resistance to heat transfer at the external surface of the structure (($m^2 \cdot K$)/W), and
R is the thermal resistance of the structure (($m^2 \cdot K$)/W), [23].

The value of the heat transfer coefficient of a multilayer structure is calculated as follows:

$$U = \frac{1}{R_{si} + R + R_{se}}, \tag{4}$$

where:

U is the heat transfer coefficient of the structure (($m^2 \cdot K$)/W),
R_{si} is the thermal resistance to heat transfer at the internal surface of the structure (($m^2 \cdot K$)/W),
R is the thermal resistance of the structure (($m^2 \cdot K$)/W), and
R_{se} is the thermal resistance to heat transfer at the external surface of the structure (($m^2 \cdot K$)/W), [23].

The temperature in the jth layer of the structure is calculated:

$$\theta j = \theta i - U \times (\theta i - \theta_e) \times (R_{si} + \sum R_j), \tag{5}$$

where:

θj is the temperature in the jth layer of the structure (°C),
θi is the internal design temperature (°C),
θ_e is the outdoor design temperature in winter (°C),
U is the heat transfer coefficient of the structure ((m²·K)/W),
R_{si} is the thermal resistance to heat transfer at the internal surface of the structure ((m²·K)/W), and
$\sum R_j$ is the sum of thermal resistances of the jth layers of the structure ((m²·K)/W) [23].

The external wall prefabricated component is assumed to be a 3D system (Figures 4 and 7). The heat transfer equation for transient conditions in a Cartesian coordinate system is as follows:

$$C_p \rho \frac{\partial T}{\partial t} = \lambda \left(\frac{\partial^2 T}{\partial x} + \frac{\partial^2 T}{\partial y} + \frac{\partial^2 T}{\partial z} \right), \tag{6}$$

where:

C_p is the specific heat under constant pressure (kJ/kg K),
p is the density of the wall layer material (kJ/m³)],
T is the temperature (°C), and
t is the time (s) [24].

Suppose the outer wall at constant temperature T_i separates the inner zones from the ambient conditions—it is possible to define boundary conditions on the S_i and S_e surfaces using Newton's law. When using the convection/radiative heat transfer coefficient, it is possible to determine the inner S_i surface's convection and radiative heat transfer rates. Based on these facts, the boundary conditions after the modification of Equation (6) on the S_i surface are given as follows:

$$\lambda \left. \frac{\partial T(t)}{\partial x} \right|_{S_i} = h_i [T_{Fi}(t) - T_i] \tag{7}$$

where:

λ is the thermal conductivity (W/m K),
h_i is the convective/radiative heat transfer coefficient on the internal surface (W/m²·K),
$T_{Fi}(t)$ is the internal surface temperature (°C), and
T_i is the internal air temperature (°C) [24].

For the heat exchange between the external surface S_e and the external environment, convection and radiation are considered separately. The convection heat transfer coefficient defines convection, and the solar temperature defines radiation. The actual heat transfer mechanism between the roof or wall surface and the outside air is replaced by a fictitious solar temperature T_e, which provides the same heat transfer rate. Considering variable ambient conditions, the boundary conditions at the external surface S_e can be defined as follows:

$$\lambda \left. \frac{\partial T}{\partial x} \right|_{S_e} = h_e(t)[T_e(t) - T_{Fe}(t)] \tag{8}$$

where:

$h_e(t)$ is the convective heat transfer coefficient on the external surface (W/m²·K),
$T_e(t)$ is the solar temperature (°C), and
$T_{Fe}(t)$ is the external surface temperature (°C) [24].

The boundary conditions on the adiabatic surfaces S_{a1} and S_{a2} are defined as follows:

$$q(t) | S_{a1} = 0, \tag{9}$$

and

$$q(t) | S_{a2} = 0, \tag{10}$$

where:

$q(t)$ is the heat flux normal to the surface (W/m²), and
S_{a1} and S_{a2} are adiabatic surfaces (m²) [24].

3.2. The Composition of the Patented ISOMAX Panel

Building envelope panels, according to the ISOMAX system [2], consist of a lost formwork on the interior and exterior side—slabs of expanded polystyrene with a thickness of 75 mm and a load-bearing static reinforced concrete part with a thickness of 150 mm, in which PP-20/2 tubes are embedded at an axial distance of 100 to 250 mm from each other, forming the circuits of the thermal barrier (Figure 3). The wall constructed in this way functions not only as a thermal barrier, but also accumulates heat and cold in the mass of the load-bearing part where a large-capacity reservoir is formed.

Figure 3. View of the lost formwork of the ISOMAX panel and the principal location of the thermal barrier [2].

In our research, in addition to troubleshooting the ISOMAX panels for on-site implementation, we also focused on the energy analysis of the thermal barrier function and the heat/cold accumulation potential in the mass of the load-bearing reinforced concrete wall. For the purpose of the parametric study, a mathematical–physical model of the wall ISOMAX was created (Figure 4). The basic physical properties of the building materials comprising the ISOMAX panel—thickness, bulk density, thermal conductivity coefficient, specific heat capacity, and diffusion resistance factor—are tabulated in the mathematical–physical models included in Figure 4.

The function of the thermal barrier is to reduce heat loss/gain through the external building envelope. In terms of the calculation procedure described in Section 3.1, the thermal resistance of the building structure R ((m²·K)/W), the thermal transmittance of the building structure U (W/(m²·K)), the thermal transmittance of the thermal insulation on the exterior side U_{TIe} (W/(m²·K)), and the temperature θ_{TB} (°C) in the thermal barrier layer for different thicknesses of the thermal insulation on the exterior side are determined. Table 1 shows the values for the heating period for an indoor temperature of $\theta_i = +20$ °C and an outdoor design temperature of $\theta_e = -11$ °C. Table 2 shows the values for the summer period for an indoor temperature of $\theta_i = +26$ °C and an outdoor temperature of $\theta_e = +34$ °C. Suppose the external thermal insulation thickness remains constant, and a heat transfer fluid is supplied to the thermal barrier to heat/cool the layer. In that case, the building structure has an equivalent thermal resistance $R_{equivalent}$ ((m²·K)/W) corresponding to the temperature in the thermal barrier layer (see Tables 1 and 2).

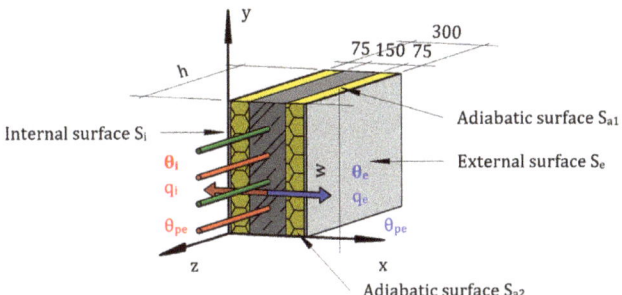

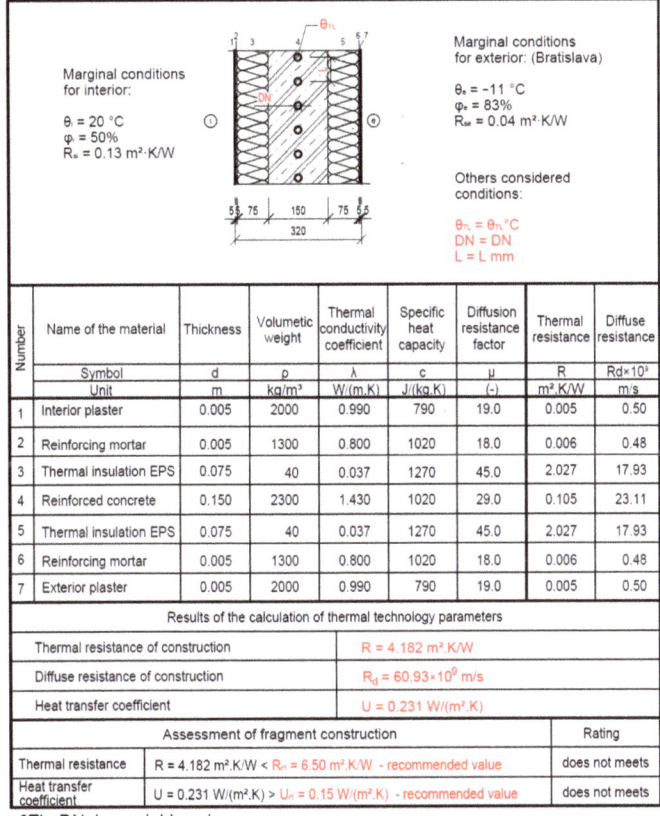

Figure 4. Mathematical–physical model: thermal properties and evaluation of the ISOMAX perimeter wall fragment. q_i—heat flow towards the interior (W/m^2), q_e—heat flow towards the exterior (W/m^2), θ_i—internal calculation temperature (°C), θ_e—outdoor calculation temperature (°C), θ_{pi}—interior surface temperature (°C), θ_{pe}—exterior surface temperature (°C), θ_{TL}—temperature of the heating medium (°C), φ_i—design relative humidity of the indoor air (%), φ_e—design relative humidity of the outdoor air (%), R_{se}—thermal resistance to heat transfer at the external surface of the structure ((m^2·K)/W), R_{si}—thermal resistance to heat transfer at the internal surface of the structure ((m^2·K)/W), i—interior, e—exterior, DN—pipe dimension, L—pipe pitch (mm), h—fragment length (m), w—height of the fragment (m).

Table 1. Results of calculation of physical variables during the heating season.

Heating by ISOMAX 75–150-d_2												
d_2 (mm)	75	100	125	150	175	200	225	250	300	400	500	750	1000
R (m²·K/W)	4.182	4.857	5.533	6.209	6.884	7.560	8.236	8.911	10.263	12.965	15.668	22.425	29.182
U (W/(m²·K))	0.231	0.199	0.175	0.157	0.142	0.129	0.119	0.110	0.096	0.076	0.063	0.044	0.034
U_{TI} (m²·K/W))	0.481	0.363	0.292	0.244	0.209	0.183	0.163	0.147	0.123	0.092	0.074	0.049	0.037
θ_m (°C)	3.91	5.98	7.64	8.95	10.01	10.88	11.62	12.24	13.25	14.64	15.55	16.88	17.60

Table 2. Results of calculation of physical quantities during the cooling season.

Cooling by ISOMAX 75–150-d_2												
d_2 (mm)	75	100	125	150	175	200	225	250	300	400	500	750	1000
R (m²·K/W)	4.182	4.857	5.533	6.209	6.884	7.560	8.236	8.911	10.263	12.965	15.668	22.425	29.182
U (W/(m²·K))	0.231	0.199	0.175	0.157	0.142	0.129	0.119	0.110	0.096	0.076	0.063	0.044	0.034
U_{TI} (m²·K/W))	0.481	0.363	0.292	0.244	0.209	0.183	0.163	0.147	0.123	0.092	0.074	0.049	0.037
θ_m (°C)	30.17	29.62	29.19	28.85	28.58	28.35	28.16	28.00	27.74	27.38	27.15	26.80	26.62

Figure 5 shows the temperature evolution in each layer of the ISOMAX panel during the heating and cooling seasons.

For the ISOMAX panel, the total thermal resistance R = 4.182 ((m²·K)/W) and the heat transfer coefficient U = 0.231 (W/(m²·K)). The heat transfer coefficient of the thermal insulation on the exterior side of the external wall is $U_{TI,e}$ = 0.481 (W/(m²·K)). If we reach a temperature of +20 °C in the thermal barrier layer in winter, the specific heat loss from the thermal barrier to the exterior at a mean temperature of the heat transfer medium of +20 °C would be q = 0.481 × (20 − (−11)) = 14.911 W/m². If a temperature of +26 °C is reached in the thermal barrier layer in summer, the specific heat gain to the thermal barrier from the exterior at a mean temperature of the heat transfer medium of +26 °C will be q = 0.481 × (26 − 34)) = −3.848 W/m².

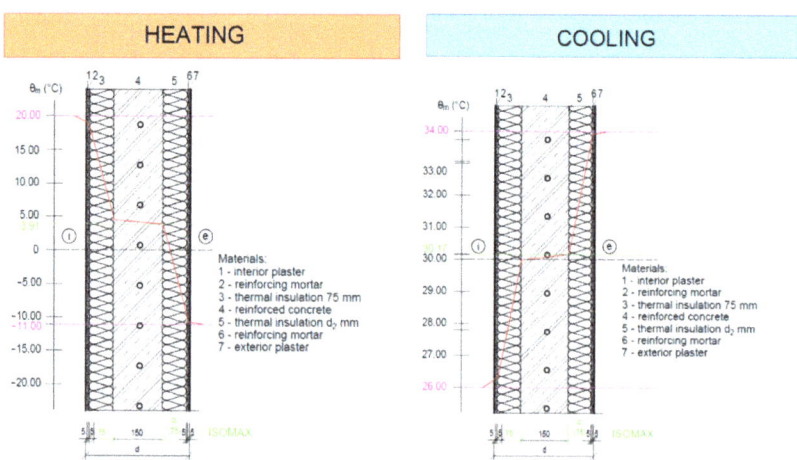

Figure 5. Temperatures in individual layers of the ISOMAX panel, i—interior, e—exterior.

3.3. Analysis of the Design of an Innovative Panel with a Thermal Barrier

The design of the innovative panel with a thermal barrier was part of the overall upgraded solution of the ISOMAX system, which resulted in the design and realization of the prototype prefabricated house IDA I (Figure 6). The perimeter panels were designed with thermal insulation made of expanded polystyrene on the interior side with a thickness of 100 mm, and on the exterior side with a thickness of 200 mm, based on the energy analysis and the requirements for reducing the energy demand for heating. The supporting static reinforced concrete part of the panels, in which PP-20/2 tubes were embedded at an axial distance of 100 to 250 mm from each other, forming the thermal barrier circuits, was left at 150 mm thickness. Figure 7 shows a mathematical–physical model of an innovative thermal barrier perimeter panel solution.

Figure 6. Prototype of prefabricated house IDA I—view of the innovative panels with built-in thermal barrier and construction of the first floor [1].

By analogy with the ISOMAX panel, according to the calculation procedure described in Section 3.1, the thermal resistance of the building structure R ($(m^2 \cdot K)/W$), the heat transfer coefficient of the building structure U ($W/(m^2 \cdot K)$), the heat transfer coefficient of the thermal insulation on the outside U_{TIe} ($W/(m^2 \cdot K)$), and the temperature θ_{TB} (°C) in the thermal insulation layer were determined for the upgraded panel for different thicknesses of the thermal insulation on the outside. Table 3 shows the values for the heating period for an indoor temperature of θ_i = +20 °C and an outdoor design temperature of θ_e = −11 °C. Table 4 shows the values for the summer period for an indoor temperature of θ_i = +26 °C and an outdoor temperature of θ_e = +34 °C. If the thickness of the external thermal insulation remains constant and a heat transfer fluid is supplied to the thermal barrier to heat/cool the layer, the building structure has an equivalent thermal resistance $R_{equivalent}$ ($(m^2 \cdot K)/W$) corresponding to the temperature in the thermal barrier layer (see Tables 3 and 4).

The basic physical properties of the building materials comprising the innovative panel—thickness, bulk density, thermal conductivity coefficient, specific heat capacity, and diffusion resistance factor—are tabulated in the mathematical–physical models included in Figure 7.

For the upgraded panel, the total thermal resistance R = 8.236 ($(m^2 \cdot K)/W$) and the heat transfer coefficient U = 0.119 ($W/(m^2 \cdot K)$). The heat transfer coefficient of the thermal insulation on the exterior side of the external wall is $U_{TI,e}$ = 0.183 ($W/(m^2 \cdot K)$). If we reach a temperature of +20 °C in the thermal barrier layer in winter, the specific heat loss from the thermal barrier to the exterior at a mean temperature of the heat transfer medium of +20 °C would be q = 0.183 × (20 − (−11)) = 5.673 W/m². If a temperature of +26 °C is reached in the thermal barrier layer in summer, the specific heat gain to the thermal barrier from the exterior at a mean temperature of the heat transfer medium of +26 °C will be q = 0.183 × (26 − 34)) = −1.464 W/m². Figure 8 shows the temperature evolution in the different layers of the upgraded panel during the heating and cooling season.

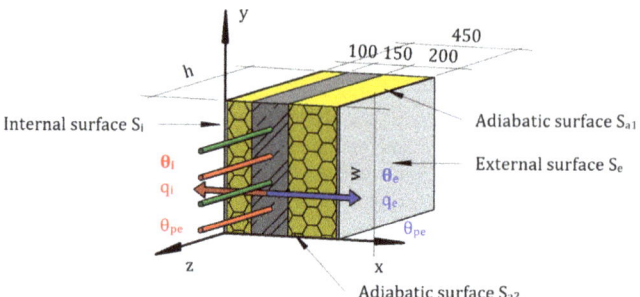
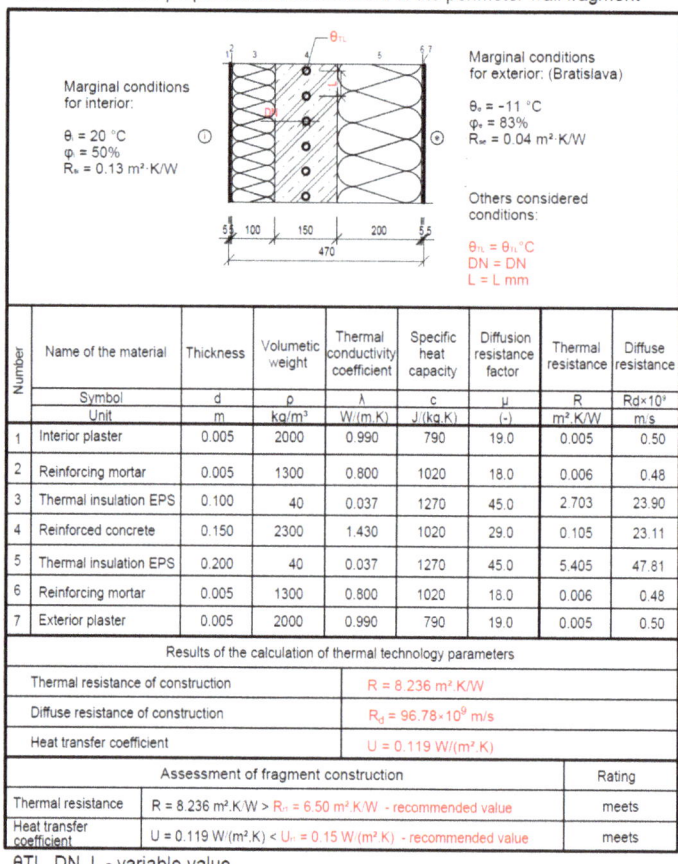

Figure 7. Mathematical–physical model: composition of an upgraded thermal insulation envelope panel. q_i—heat flow towards the interior (W/m^2), q_e—heat flow towards the exterior (W/m^2), θ_i—internal calculation temperature (°C), θ_e—outdoor calculation temperature (°C), θ_{pi}—interior surface temperature (°C), θ_{pe}—exterior surface temperature (°C), θ_{TL}—temperature of the heating medium (°C), φ_i—design relative humidity of the indoor air (%), φ_e—design relative humidity of the outdoor air (%), R_{se}—thermal resistance to heat transfer at the external surface of the structure ((m^2·K)/W), R_{si}—thermal resistance to heat transfer at the internal surface of the structure ((m^2·K)/W), i—interior, e—exterior, DN—pipe dimension, L—pipe pitch (mm), h—fragment length (m), w—height of the fragment (m).

Table 3. Results of calculation of physical variables during the heating season.

	Heating by Innovated Construction 100–150-d_2											
d_2 (mm)	75	100	125	150	175	200	225	250	300	400	500	750	1000
R (m^2·K/W)	4.857	5.533	6.209	6.884	7.560	8.236	8.911	9.587	10.938	13.641	16.344	23.100	29.857
U (W/(m^2·K))	0.199	0.175	0.157	0.142	0.129	0.119	0.110	0.102	0.090	0.072	0.061	0.043	0.033
U_{TI} (m^2·K/W))	0.481	0.363	0.292	0.244	0.209	0.183	0.163	0.147	0.123	0.092	0.074	0.049	0.037
θ_m (°C)	1.82	3.97	5.67	7.04	8.17	9.05	9.93	10.63	11.77	13.38	14.46	16.07	16.96

Table 4. Results of calculation of physical variables during the cooling season.

	Cooling by Innovated Construction 100–150-d_2											
d_2 (mm)	75	100	125	150	175	200	225	250	300	400	500	750	1000
R (m^2·K/W)	4.857	5.533	6.209	6.884	7.560	8.236	8.911	9.587	10.938	13.641	16.344	23.100	29.857
U (W/(m^2·K))	0.199	0.175	0.157	0.142	0.129	0.119	0.110	0.102	0.090	0.072	0.061	0.043	0.033
U_{TI} (m^2·K/W))	0.481	0.363	0.292	0.244	0.209	0.183	0.163	0.147	0.123	0.092	0.074	0.049	0.037
θ_m (°C)	30.69	30.14	29.7	29.34	29.05	28.8	28.6	28.42	28.12	27.71	27.43	27.01	26.79

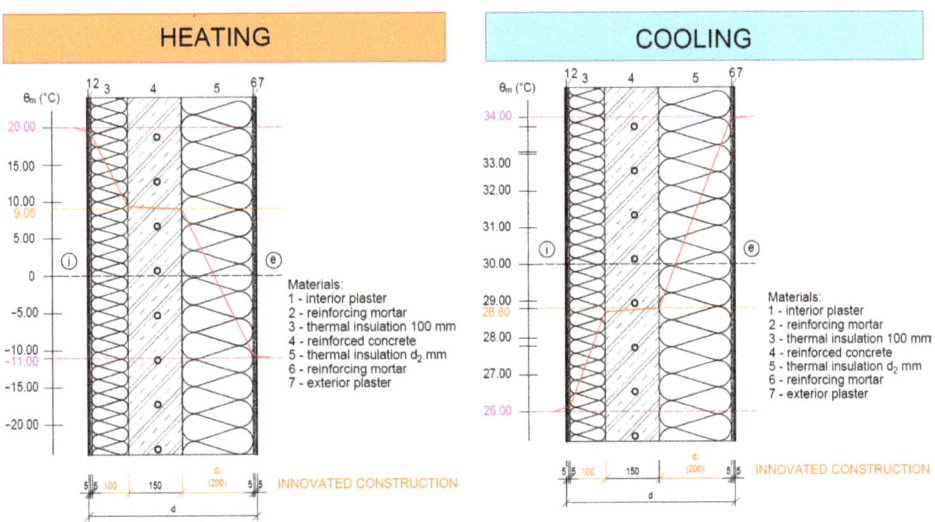

Figure 8. Temperatures in individual layers of the upgraded panel, i—interior, e—exterior.

This means that the construction of the perimeter panel that was designed by us shows approximately a 2.6 times lower specific heat loss from the thermal barrier to the exterior at a mean temperature of the heat transfer medium of 20 °C than the wall of the ISOMAX system [2].

3.4. Details on Connecting Panels during Construction

Our research was not focused on the static and technological design of the panel joints; therefore, we only presented the principal details of the panel joints (Figures 9 and 10). There are several variants for joining reinforced concrete panels. In designing the IDA I prototype panel house, the structural engineer and the panel plant technologist designed the standard joints at that time.

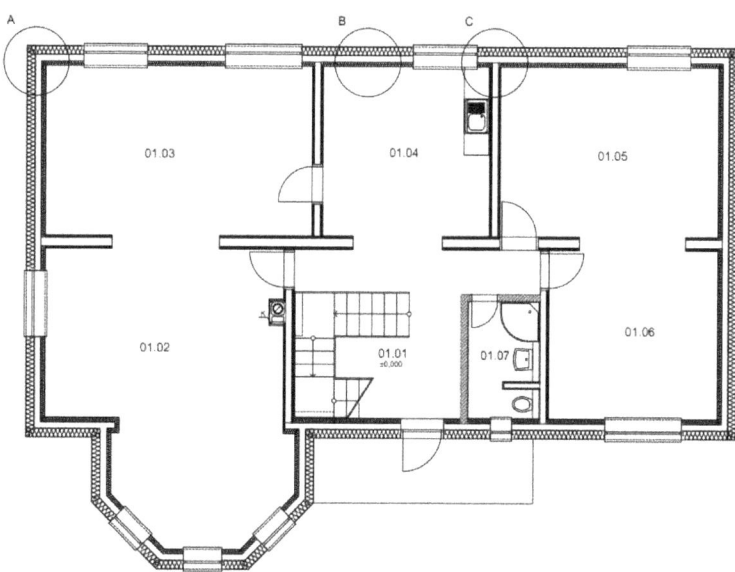

Figure 9. First floor plan showing the location of details A, B, and C; K—chimney for fireplace; the digit +0.000 indicates the ground floor elevation.

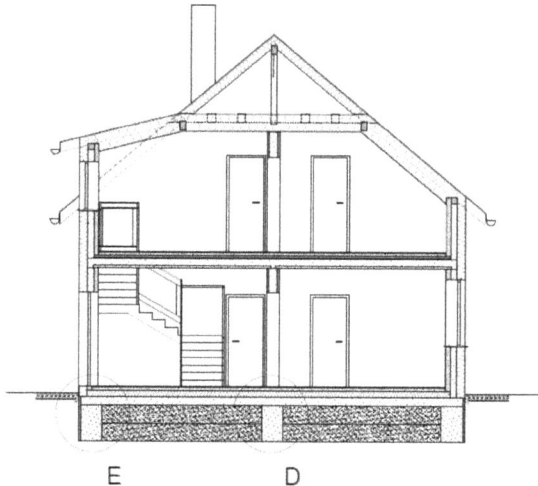

Figure 10. Cross section showing the location of details D and E.

Detail A (Figure 11) shows the corner joint of two perimeter panels. Detail B (Figure 12) shows the through joint of two perimeter panels. Detail C (Figure 13) shows a continuous joint of two perimeter panels with an internal load-bearing panel. Detail D (Figure 14) shows the connection of the inner load-bearing panel to the base plate. Detail E (Figure 15) shows the joint of the perimeter panel with the base plate.

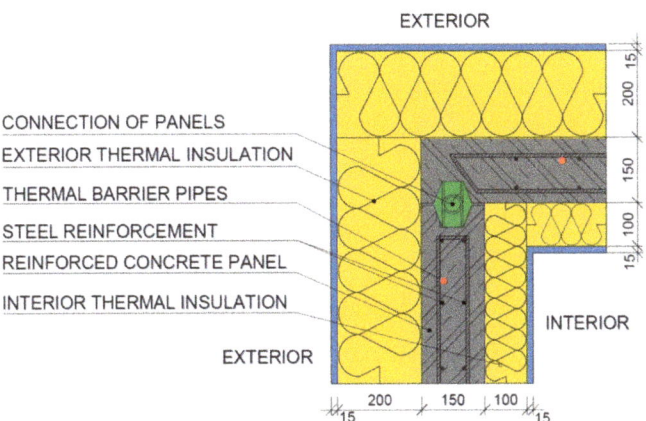

Figure 11. "Detail A", the corner joint of two perimeter panels. The colored points (red) in the picture are thermal barrier pipes.

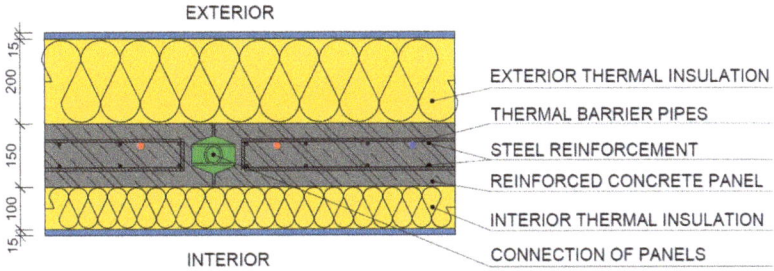

Figure 12. "Detail B", the through joint of two perimeter panels. The colored points (red and blue) in the picture are thermal barrier pipes.

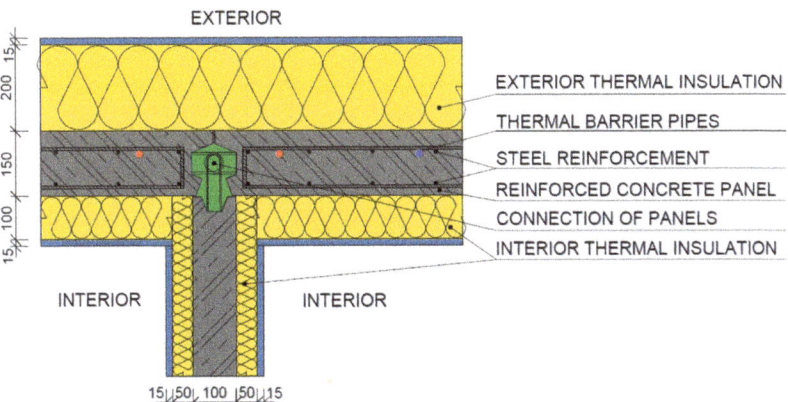

Figure 13. "Detail C", a through joint of two perimeter panels with an internal load-bearing panel. The colored points (red and blue) in the picture are thermal barrier pipes.

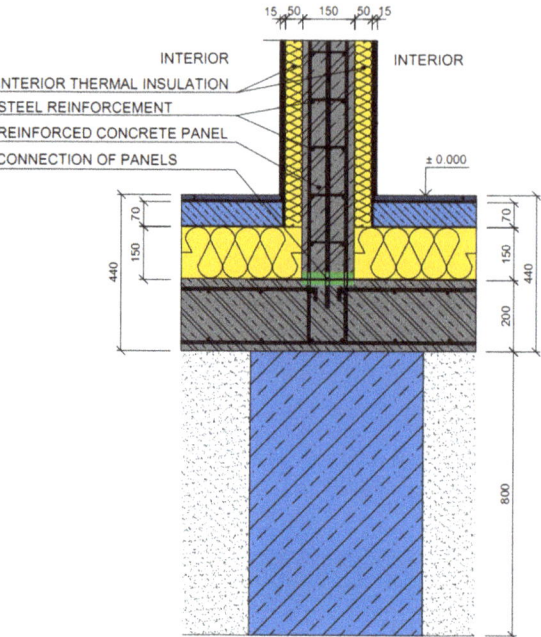

Figure 14. "Detail D", the connection of the inner load-bearing panel to the base plate.

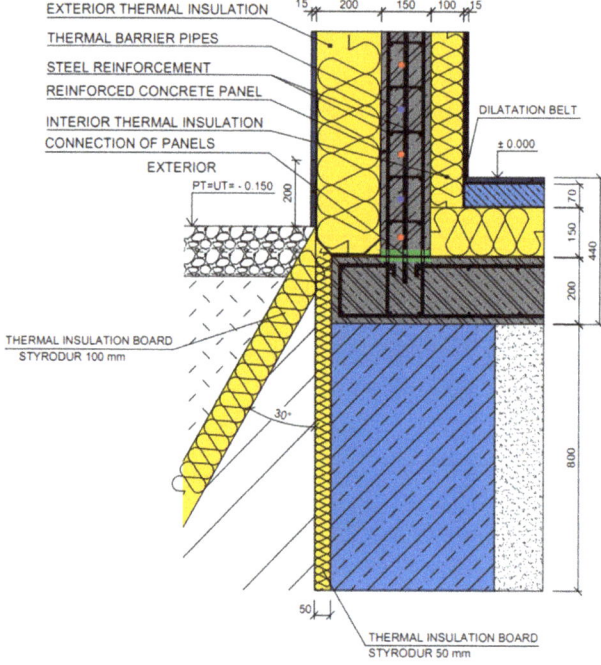

Figure 15. "Detail E", the joint of the perimeter panel with the base plate. The colored points (red and blue) in the picture are thermal barrier pipes.

4. Results and Discussion

In this section, we describe and analyze the most important results of our research:

- Induction and analog form of forming an innovative panel with integrated energy-active elements.
- Analysis of the energy potential of the retrofitted panel with a thermal barrier compared to the original panel.
- Synthesis of the knowledge obtained from the scientific analysis and transformation of the data into the design and implementation of the IDA I prefabricated house prototype.

4.1. Inductive and Analogous Form of the Formation of the Innovative Panel with Integrated Energy-Active Elements

As can be seen from the photos in Figures 2, 3, 16 and 17, the production of ISO-MAX panels is quite laborious, lengthy, and, on the construction site, requires gradual implementation—pouring concrete. It is problematic to achieve uniform placement of the tubes in the center of the load-bearing reinforced concrete wall and the compaction itself. It is also not possible to pour the lost formwork over the entire height of the wall, which causes cracks and nonconnection of the individual concrete layers along the height of the wall. We proposed removing these deficiencies by prefabricating the individual panels directly in the precast using vibratory tables and applying the finished panels to the building structure (Figure 18).

Figure 16. Construction with patented ISOMAX panels [2].

Figure 17. Construction with patented ISOMAX panels [2].

Figure 18. Construction with upgraded panels [1].

The application of the thermal barrier in innovative panels can be on the auxiliary reinforcement in the center of the panel structure (Figure 19, left), or on the outer reinforcement of the panel structure (Figure 19, right).

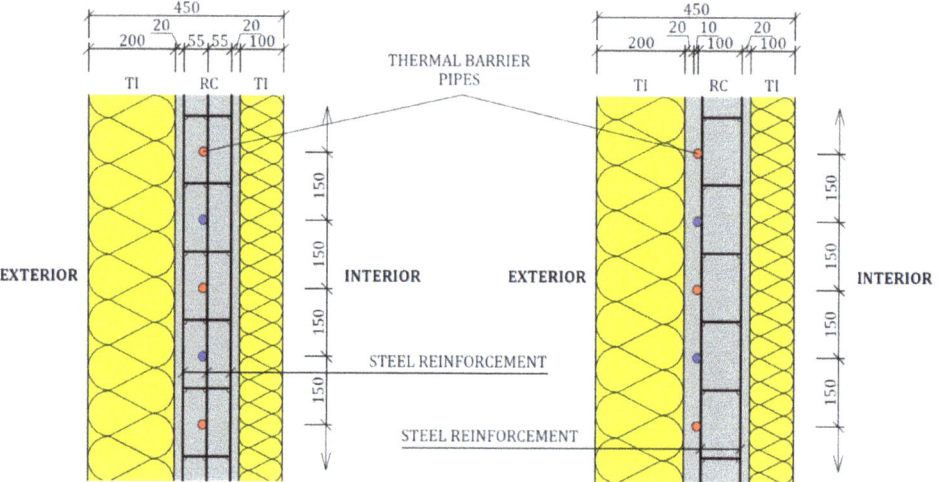

Figure 19. Principle of thermal barrier application in innovative panels (left—thermal barrier pipes are installed on the auxiliary reinforcement in the center of the panel structure; right—thermal barrier pipes are installed on the outer reinforcement of the panel structure). The colored points (red and blue) in the picture are thermal barrier pipes [1].

The prototype of the prefabricated house IDA I was additionally insulated (see Figures 1, 6 and 18). For the complex production of modernized panels already with insulation, a technological procedure has been designed. In the first step, the forms of the panels are created, the interior thermal insulation is inserted, and then the individual layers of reinforcement are created—the distances of which are precisely defined by the auxiliary demarcation elements between each other and the thermal insulation. The next step is to fix the thermal barrier pipes to the reinforcement (center layer or exterior reinforcement layer, Figure 19). Finally, the exterior thermal insulation is inserted, which is also delineated from the panel reinforcement. After these steps, the panel form is closed on all sides, leaving only the upper part of the form free, through which the concrete is poured. Before pouring the concrete, the forms are vertically erected and fixed on the vibrating table.

4.2. Analysis of the Energy Potential of the Retrofitted Panel with a Thermal Barrier Compared to the Original Panel

This technical solution of the building envelope, in addition to the function of a thermal barrier, fulfills the function of a large-capacity heat reservoir. In the mass of the load-bearing part of the walls or roofs made of reinforced concrete, heat/cold is accumulated, which significantly influences the passage of heat/cold through the building structure. When designing the innovative panel, we did not consider the possibility of using concrete with air to increase the thermal properties. In this panel construction, it is important that the load-bearing part is thermally well-conductive to create a uniform thermal layer = thermal barrier and simultaneously have the highest possible heat/cool accumulation capacity. For this reason, we consider the upgraded construction of the envelope panel with a greater thickness of thermal insulation, especially on the exterior side, to be justified and important from energy, economic, and environmental points of view.

The temperature in the thermal barrier of the thermal insulation panel of the envelope can be regulated as required in all four seasons. The role of building structures with integrated energy-active elements (in this case with a thermal barrier) is to actively control the heat transfer through the building envelope (i.e., to adjust the thermal resistance of the building envelope). Our research aimed to design an optimal structure in terms of energy efficiency, economic efficiency, and environmental friendliness.

Sections 3.2 and 3.3 described the compositions of the ISOMAX panel and the upgraded panel designs. Mathematical and physical models were developed and subsequently used to calculate the thermal resistance of the panels R ((m²·K)/W, the heat transfer coefficient U (W/(m²·K)) of the panels, the heat transfer coefficients of the external thermal insulation U_{TIe} (W/(m²·K)), and the temperature θ_j (°C) between the different layers of the panel structures for different thicknesses of the external thermal insulation through a parametric study. Figures 20–24 show a comparison of the most important physical parameters of the ISOMAX panel and the upgraded panel.

The graphs show that, for example, a mean temperature of θ_{TB} (°C) = +15 °C in the thermal barrier layer of this panel design during heating represents the equivalent thermal resistance $R_{equivalent}$ ((m²·K)/W) or the equivalent heat transfer coefficient $U_{equivalent}$ (W/(m²·K)) of the panel as would be achieved with a 500 mm thick exterior thermal insulation. By analogy, this can be applied to the cooling period, wherein a mean temperature of θ_{TB} (°C) = +27 °C in the thermal barrier layer for this panel design represents an external thermal insulation thickness of 500 mm.

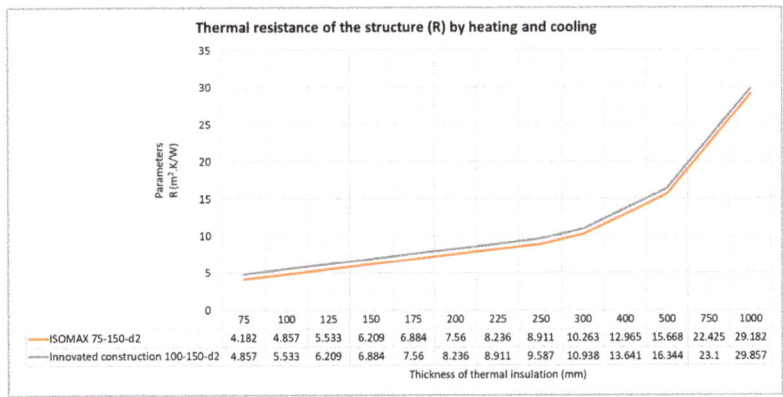

Figure 20. Dependence of the thermal resistance R of the ISOMAX panel and the innovative panel on the thickness of the exterior thermal insulation.

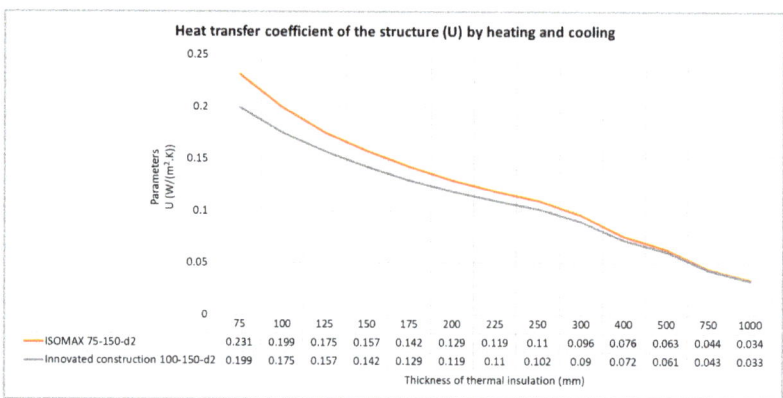

Figure 21. Dependence of the heat transfer coefficient U of the ISOMAX panel and the innovative panel on the thickness of the external thermal insulation.

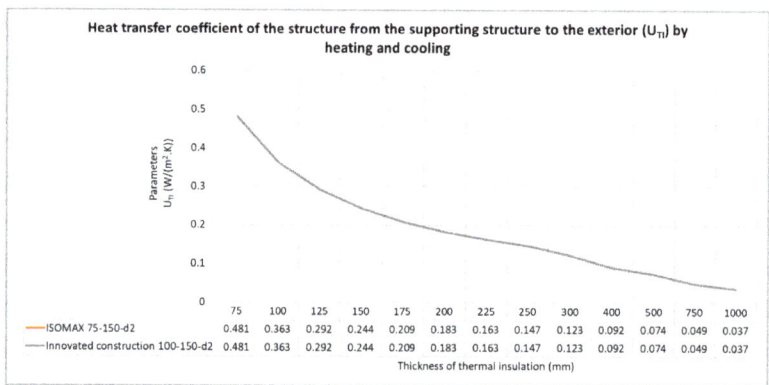

Figure 22. Dependence of the heat transfer coefficient U_{TI} of the exterior thermal insulation on the thickness.

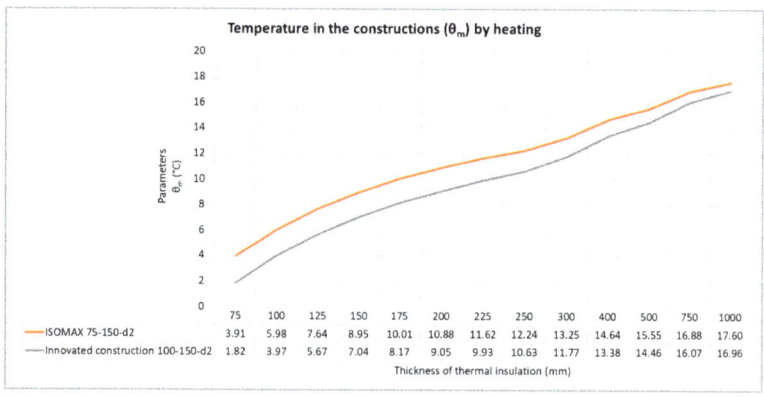

Figure 23. Dependence of the average temperature θ_m in the thermal barrier layer of the ISOMAX panel and the innovative panel in the heating period on the thickness of the external thermal insulation.

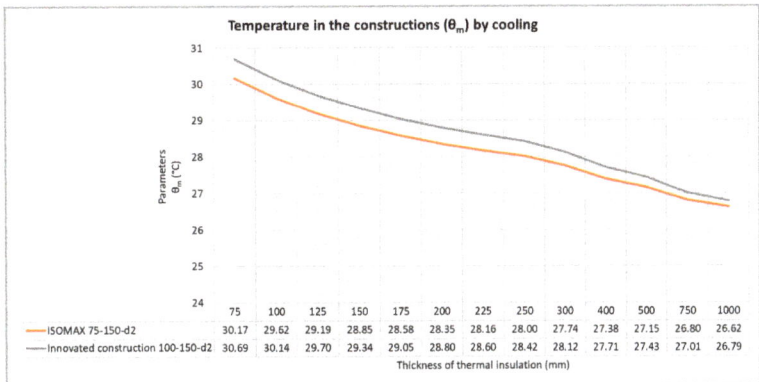

Figure 24. Dependence of the average temperature θ_m in the thermal barrier layer of the ISOMAX panel and the innovative panel during the cooling period on the thickness of the external thermal insulation.

Figures 25 and 26 show the isotherms characterizing the heat transfer through the ISOMAX panel structure and our proposed panel structure in the heating (winter) and cooling (summer) periods. The area between the isotherms of the two designs expresses the heat saving/loss for these alternatives. The area above the isotherms, when heated to a thermal barrier temperature equal to the interior temperature of 20 °C, indicates the potential for heat savings. Similarly, the area below the isotherms, when cooling to a thermal barrier temperature equal to the interior temperature of 26 °C, indicates the potential for cold savings.

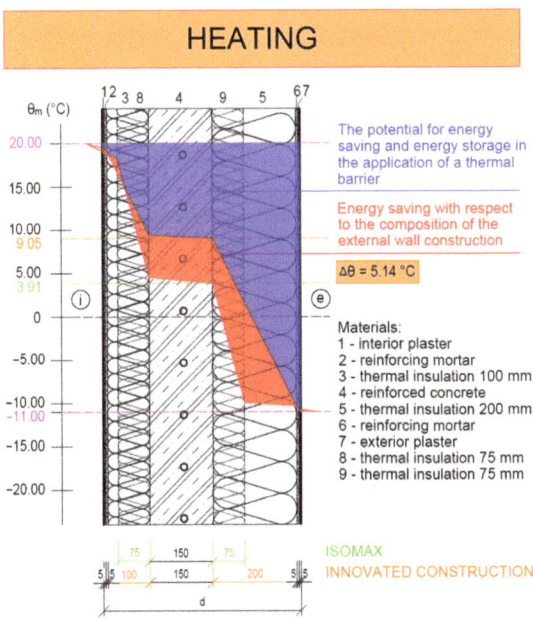

Figure 25. The course of isotherms characterizing the heat transfer through the ISOMAX panel construction and the upgraded panel construction in the heating season. q_i—heat flow towards the interior (W/m^2), q_e—heat flow towards the exterior (W/m^2), $\theta_i = 20$ °C—internal calculation temperature (°C), $\theta_e = -11$ °C—outdoor calculation temperature (°C), θ_{pi}—interior surface temperature (°C), θ_{pe}—exterior surface temperature (°C), $\Delta\theta$—temperature difference, i—interior, e—exterior.

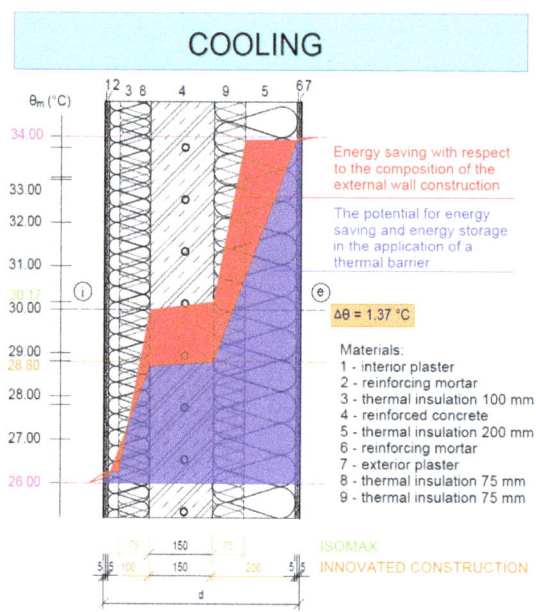

Figure 26. The course of isotherms characterizing the heat transfer through the ISOMAX panel design and the upgraded panel design in the summer period. q_i—heat flow towards the interior (W/m^2), q_e—heat flow towards the exterior (W/m^2), θ_i = 26 °C—internal calculation temperature (°C), θ_e = 34 °C—outdoor calculation temperature (°C), θ_{pi}—interior surface temperature (°C), θ_{pe}—exterior surface temperature (°C), $\Delta\theta$—temperature difference, i—interior, e—exterior.

4.3. Synthesis of the Knowledge Obtained from the Scientific Analysis and Transformation of the Data into the Design and Implementation of the IDA I Prefabricated House Prototype

In the following section, we describe the realization of the prototype of the prefabricated house IDA I, wherein we describe the construction part and the energy part of the building in more detail.

The prototype of the prefabricated house IDA I serves as an administrative building and is located in the Vrakuňa district of Bratislava and the site of the Paneláreň Vrakuňa, a.s. plant (Figure 27). The prefabricated house is a two-story building.

Figure 27. Location of the prototype of the prefabricated house IDA I within Bratislava, Slovak Republic. (https://www.google.com/maps (accessed on 20 June 2022)).

4.3.1. Structural System

From a structural point of view, a prefabricated longitudinal load-bearing system was applied, with load-bearing peripheral walls and one central wall formed by reinforced concrete panels (Figure 28). The roof is gable with a ridge parallel to the front façade

(Figure 29). The ground floor contains an entrance hall with a staircase, offices, and sanitary facilities. The attic contains a corridor, offices, and sanitary facilities.

Figure 28. Completion of the ground floor assembly (photo archive: Kalús, D.) [1].

Figure 29. View of the realization of the wooden truss (photo archive: Kalús, D.) [1].

The foundation strips were designed from 450–800 mm wide B-15 concrete, and the underlying concrete was designed from 150 mm thick B-15 concrete reinforced with welded KARI mesh.

The building walls are made of assembled prefabricated panels, which were subsequently insulated with a contact insulation system. The walls with a thermal barrier were designed with thermal insulation on both sides: 100 mm thermal insulation in the interior and 200 mm in the exterior.

The ceiling structures were reinforced concrete monolithic slabs cast into lost formwork. The total thickness of the ceiling slab is 250 mm.

The method of laying the roof covering was dry laying using clamps. Roof ventilation was ensured by breaking the waterproofing film at the ridge and using ventilation tiles [1].

Figure 30 shows a view of the insulated prototype of the IDA I prefabricated house before the final modifications of the external facade [1].

Figure 30. View of the implementation of the insulation of the perimeter panels (photo archive: Kalús, D.) [1].

4.3.2. Energy System

Figure 31 shows a simplified wiring diagram of the technical design of the energy systems of the IDA I prototype prefabricated house. The prototype of the panel house contains a solar energy absorber (i.e., an energetic solar roof) which was designed from 20×2 mm, or 16×2 mm PP pipes, with lengths of 100–120 m. This system is connected in the attic space to the distributor and collector.

The panel house also contains an underground heat storage tank, accumulating solar energy captured by the solar absorbers. The ground heat reservoir consists of three zones: two located under the base plate and the third zone located directly in the base plate. Excess heat from the heating water tank or directly from the fireplace with a hot water heat exchanger is also stored in the third zone located in the base plate. Individual zones are designed with 20×2 mm PP pipes with lengths of 120–200 m.

Low-temperature radiant heating is designed for heating. To eliminate thermal gains and losses, thermal barrier circuits were integrated in the perimeter construction. These systems are connected to the distributor and collector located in the space under the stairs. All other heat and cold sources and energy systems are also connected here. The building's control system cabinet, pumps, expansion vessels, and control valves are also located here.

An energy roof connects the individual circuits of the energy systems: an underground heat storage tank, a peak heat source—a fireplace with a hot water heat exchanger, and a heating water storage tank equipped with an electric heating insert. This is so that the supply of the necessary energy for heating is possible at any time and from any heat source. The circuits of the thermal barrier are also connected to liquid circuits stored in the ground outside the building, which are mainly used for passive cooling.

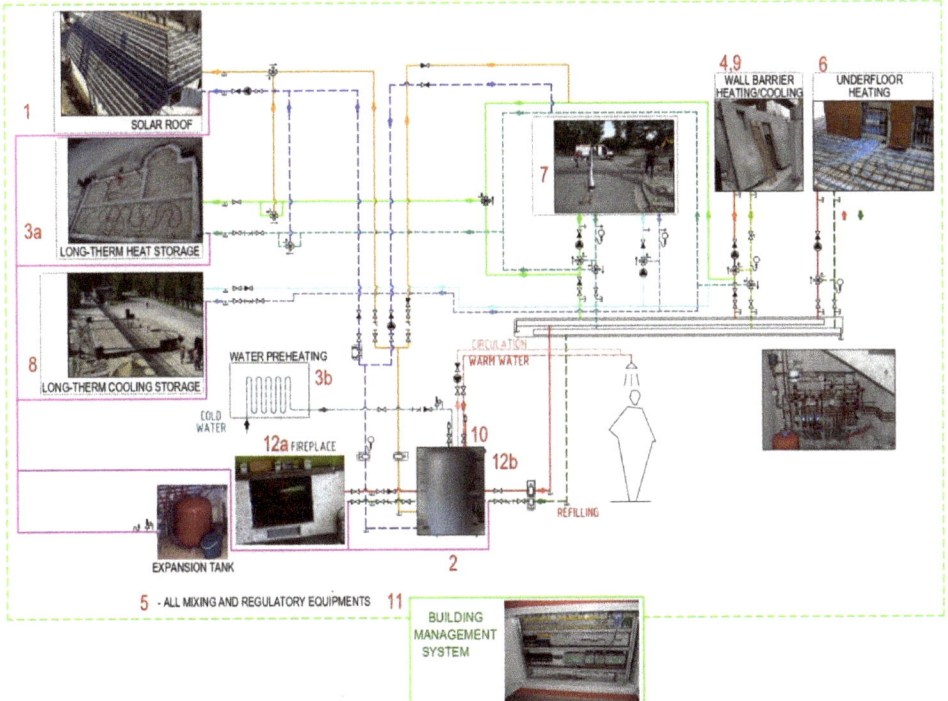

Figure 31. Simplified wiring diagram of the technical solution of the energy systems of the IDA I prefabricated house prototype [1]. 1—solar absorber (energy/solar roof), 2—short-term heat storage, 3a—long-term heat storage, 3b—long-term heat storage, 4—active thermal protection circuits (building structure with an internal heat source), 5—mixing and control equipment, 6—low-temperature heating circuits, 7—heat recovery ventilation equipment, 8—cooling circuits located in the ground outside the building, 9—high-temperature cooling circuits, 10—waste heat from the drainage system, 11—building control system, 12a—fireplace, 12b—peak heat source. The red and green arrows in the figure represent the direction of flow of the heat transfer fluid.

In the building, a pipe-in-pipe countercurrent recuperation exchanger (ISOMAX system) was designed, which was made of a special stainless-steel pipe with an antimicrobial surface of DN 180 for the inner pipe and DN 250 for the outer pipe.

The air supply to the rooms was solved using a plastic distribution under the base plate directly from the combined distributor and air collector on the first floor. On the second floor, the supply distributions are embedded in the thermal insulation of the floor. Air extraction on the first and second floors is handled through the pipe in the soffit.

An underground cooling circuit was designed for cooling the building—the pipes were made of PP 20 × 2 mm, and it was placed outside the building at a depth of 2 m below the ground level. The cooling system is connected through the distributor and the collector to the thermal barrier circuits in the external perimeter walls and also to the heat exchanger in the ventilation air cooling pipe.

A two-stage preparation of hot water was proposed. First, the water in the ground heat storage tank is preheated from a temperature of about 10 °C to about 25–30 °C. The hot water is then heated to the required temperature of 55 °C to 60 °C in a trivalent heating water tank with an integrated hot water tank (which uses solar energy), heating the water in a hot water exchanger in the fireplace or electricity.

After the completion of the assembly work, all necessary pressure and tightness tests were carried out. Both the heating systems (floor heating) and the thermal barrier were hydraulically regulated, the operating values of the control devices were set, and they were subjected to verification of functional characteristics.

5. Conclusions

Based on this study, we can draw the following conclusions:

- The analysis and synthesis of the knowledge from the production and implementation of ISOMAX panels have determined the shortcomings of these panels. We have designed and developed innovative panels that eliminate the lengthy and complicated production as well as on-site implementation. We eliminated the static problems associated with insufficient concrete compaction when the original panels were poured on-site by fabricating the panels on vibratory tables in the panel factory;
- The thermal barrier is one of the functions of building structures with integrated energy-active elements;
- From the review of the scientific literature, it is clear that this is a very progressive area of research. So far, most studies on active thermal protection are based on calculations, computer simulations, and experimental measurements. Few studies have focused on the economic and environmental aspects of the use of active thermal protection;
- For the analysis of both the ISOMAX panel and the upgraded panel, we developed mathematical–physical models and analyzed the energy potential of both panels based on a parametric study;
- The analysis shows that, for example, a mean temperature of θ_{TB} (°C) = +15 °C in the thermal barrier layer of this panel design during heating represents the equivalent thermal resistance $R_{equivalent}$ ((m^2·K)/W) or equivalent heat transfer coefficient $U_{equivalent}$ (W/(m^2·K)) of the panel—as would be achieved with a 500 mm thick exterior thermal insulation. By analogy, this can be applied to the cooling period, where a mean temperature of θ_{TB} (°C) = +27 °C in the thermal barrier layer for this panel design represents an external thermal insulation thickness of 500 mm;
- The energy analysis and design of the upgraded thermal barrier panel show an energy potential of the thermal barrier and heat/cold accumulation in the mass of the reinforced concrete load-bearing part of the panel. The potential was up to 2.6 times higher than that of the panel in the original ISOMAX design of the system;
- The results of the analysis of the innovative panel design with integrated energy-active elements show high potential for the use of RES and waste heat with the technology;
- In addition to a higher equivalent thermal resistance compared to the ISOMAX panel, our innovative building envelope panel has significantly lower requirements for the operation of the circulators, making the building's energy intensity lower, more economically efficient, and more environmentally friendly;
- The ISOMAX panels and the innovative panels with integrated energy-active elements only fulfil the energy functions of a thermal barrier and heat/cold storage. The design of building envelope panels (by application without external thermal insulation) offers additional energy functions, namely low-temperature radiant heating and high-temperature radiant cooling;
- Further variants of the self-supporting thermal insulation panels for systems with active heat transfer control are presented in the utility model SK 5729 Y1 [25];
- Among the most significant results and novelty of our research in this area can be considered the realization of the prototype of the prefabricated house IDA I.

The objectives of our further research are to:

1. Develop further design variants of thermal insulation envelope panels with integrated energy-active elements.
2. Develop a methodology for the installation of envelope panels with ATP.

3. Implement selected types of perimeter thermal insulation panels with integrated energy-active elements on a laboratory building.
4. Apply the proposed calculation methodology, selection, and assessment for selected combined building-energy systems using RES in buildings.
5. Conduct experimental measurements of selected types of building envelope thermal insulation panels with integrated energy-active elements using RES as part of a laboratory building object in different operating modes.
6. Measure usable energy of selected types of thermal insulation panels with integrated energy-active elements using RES in the application of active thermal protection in the functions of thermal barriers, cooling, and preparation of TV or heating water.
7. Measure the efficiency of selected types of thermal insulation panels with integrated energy-active elements using RES in the application of active thermal protection for the elimination of overheating of the envelope and the interior depending on the intensity of solar radiation, shading, and the outdoor temperature.
8. Develop software for designing, calculating, and assessing envelope thermal insulation panels with integrated RES-using active elements.
9. Develop a methodology for applying building envelope thermal insulation panels with integrated RES energy components in a building information modeling (BIM) model.
10. Ensure the automated transfer of the proposed database of envelope thermal insulation panels with integrated energy-active elements using RES to the BIM model.
11. Verify the proposed solution on a concrete building project created in the BIM model.

6. Patents

The novelty of the research described in this study lies in the innovation of the design of the envelope panel with a thermal barrier. Partial results of our research have been published in several scientific articles and are also part of three utility models (UM SK 5749 Y1, [26], UM SK 5729 Y1, [25], and UM SK 5725 Y1, [27]) and one European patent (EP 2 572 057 B1, [28]).

Author Contributions: Conceptualization, D.K. (Daniel Kalús), D.K. (Daniela Koudelková), M.K., V.M., and M.S.; methodology, D.K. (Daniel Kalús), D.K. (Daniela Koudelková), and M.K.; validation, D.K. (Daniel Kalús), D.K. (Daniela Koudelková), and M.K.; formal analysis, D.K. (Daniel Kalús); investigation, D.K. (Daniel Kalús), D.K. (Daniela Koudelková), and M.K.; resources, D.K. (Daniel Kalús), D.K. (Daniela Koudelková), M.K., and V.M.; data curation, D.K. (Daniel Kalús), D.K. (Daniela Koudelková), M.K., V.M., and M.S.; writing—original draft preparation, D.K. (Daniel Kalús), D.K. (Daniela Koudelková), M.K., V.M., and M.S.; writing—review and editing, D.K. (Daniel Kalús), D.K. (Daniela Koudelková), M.K., V.M., and M.S. All authors have read and agreed to the published version of the manuscript.

Funding: The publication of this work was financially supported by EHBconsulting s. r. o. At the same time, we express our sincere thanks to the private investor Ing. Tomáš Ircha, who significantly supported the research in the field of combined building and energy systems.

Institutional Review Board Statement: Not applicable.

Informed Consent Statement: Not applicable.

Data Availability Statement: Not applicable.

Acknowledgments: This research was supported by the Ministry of Education, Science, Research and Sport of the Slovak Republic (Grant VEGA 1/0304/21). This research was also supported by the Ministry of Education, Science, Research and Sport of the Slovak Republic (Grant KEGA 005STU-4/2021).

Conflicts of Interest: The authors declare no potential conflict of interest with respect to the research, authorship, and publication of this article.

References

1. Kalús, D. *The Contract for Work HZ 04–309–05—Design of a Passive House Using Solar and Geothermic Energy*; K–TZB SvF STU: Bratislava, Slovakia, 2006.
2. Available online: http://www.isomax-terrasol.eu/home.html (accessed on 20 June 2022).
3. Lehmann, B.; Dorer, V.; Koschenz, M. Application range of thermally activated building systems tabs. *Energy Build.* **2007**, *39*, 593–598. [CrossRef]
4. Gwerder, M.; Lehmann, B.; Tödtli, J.; Dorer, V.; Renggli, F. Control of thermally-activated building systems (TABS). *Appl. Energy* **2008**, *85*, 565–581. [CrossRef]
5. Rijksen, D.O.; Wisse, C.J.; Van Schijndel, A.W.M. Reducing peak requirements for cooling by using thermally activated building systems. *Energy Build.* **2010**, *42*, 298–304. [CrossRef]
6. Krzaczek, M.; Kowalczuk, Z. An effective control system for heating and cooling in residential buildings using Thermal Barrier. *Build. Environ.* **2010**.
7. Lehmann, B.; Dorer, V.; Gwerder, M.; Renggli, F.; Tödtli, J. Thermally activated building systems (TABS): Energy efficiency as a function of control strategy, hydronic circuit topology and (cold) generation system. *Appl. Energy* **2011**, *88*, 180–191. [CrossRef]
8. Stojanovic, B.; Janevski, J.; Mitkovic, P.; Stojanovic, M.; Ignjatovic, M. Thermally activated building systems in the context of increasing the energy efficiency of buildings. *Therm. Sci.* **2014**, *18*, 1011–1018. [CrossRef]
9. Jan, B.; Georgios, V. Thermally Activated Building System (TABS): Efficient cooling and heating of commercial buildings. In Proceedings of the Climamed, Juan-Les-Pins, France, 10 September 2015.
10. Yu, Y.; Niu, F.; Guo, H.-A.; Woradechjumroen, D. A thermo-activated wall for load reduction and supplementary cooling with free to low-cost thermal water. *Energy* **2016**, *99*, 250–265. [CrossRef]
11. Kisilewicz, T.; Fedorczak-Cisak, M.; Barkanyi, T. Active thermal insulation as an element limiting heat loss through external walls. *Energy Build.* **2019**, *205*, 109541. [CrossRef]
12. Figiel, E.; Leciej-Pirczewska, D. Outer wall with thermal barrier. Impact of the barrier on heat losses and CO_2 emissions. *Prz. Nauk. Inż. Kształt. Śr.* **2020**, *29*, 223–233. [CrossRef]
13. Kalús, D.; Gašparík, J.; Janík, P.; Kubica, M.; Šťastný, P. Innovative building technology implemented into facades with active thermal protection. *Sustainability* **2021**, *13*, 4438. [CrossRef]
14. Kalús, D.; Janík, P.; Koudelková, D.; Mučková, V.; Sokol, M. Contribution to research on ground heat storages as part of building energy systems using RES. *Energy Build.* **2022**, *267*, 112125. [CrossRef]
15. Maruyama, S.; Viskanta, R.; Aihara, T. Active thermal protection system against intense irradiation. *J. Thermophys. Heat Transf.* **1989**, *3*, 389–394. [CrossRef]
16. Olesen, B.W.; De Carli, M.; Scarpa's, M.; Koschenz, M. Dynamic Evaluation of the Cooling Capacity of Thermo-Active Building Systems. *ASHRAE Trans.* **2006**, *112*, 350–357.
17. Krecké, E.; Ulbrich, R.; Radlak, G. Connection of solar and near-surface geothermal energy in Isomax technology. In Proceedings of the CESB 07 PRAGUE Conference, Prague, Czech Republic, 24–26 September 2007; pp. 622–628.
18. Gwerder, M.; Tödtli, J.; Lehmann, B.; Dorer, V.; Güntensperger, W.; Renggli, F. Control of thermally activated building systems (TABS) in intermittent operation with pulse width modulation. *Appl. Energy* **2009**, *86*, 1606–1616. [CrossRef]
19. Xie, J.; Zhu, Q.; Xu, X. An active pipe-embedded building envelope for utilizing low-grade energy sources. *J. Cent. South Univ. Technol.* **2012**, *19*, 1663–1667. [CrossRef]
20. Doležel, M. Alternative Way of Thermal Protection by Thermal Barrier. *AMR* **2014**, *899*, 107–111. [CrossRef]
21. Ibrahim, M.; Wurtz, E.; Anger, J.; Ibrahim, O. Experimental and numerical study on a novel low temperature façade solar thermal collector to decrease the heating demands: A south-north pipe-embedded closed-water-loop system. *Sol. Energy* **2017**, *147*, 22–36. [CrossRef]
22. K-TZB SvF STU. The contract for work HZ 04-210-05. In *Assessment of Thermal Comfort State in an Experimental House*; K-TZB SvF STU: Bratislava, Slovakia, 2006.
23. STN 73 0540-2 + Z1 + Z2: Thermal Protection of Buildings. In *Thermal Performance of Buildings and Components. Part 2: Functional Requirements*; Úrad pre Normalizáciu, Metrológiu a Skúšobníctvo Slovenskej Republiky: Bratislava, Slovak Republic, 2019.
24. Krzaczek, M.; Kowalczuk, Z. Thermal Barrier as a technique of indirect heating and cooling for residential buildings. *Energy Build.* **2011**, *43*, 823–837. [CrossRef]
25. Available online: https://wbr.indprop.gov.sk/WebRegistre/Patent/Detail/5018-2010 (accessed on 20 June 2022).
26. Available online: https://wbr.indprop.gov.sk/WebRegistre/Patent/Detail/5014-2010 (accessed on 20 June 2022).
27. Available online: https://wbr.indprop.gov.sk/WebRegistre/Patent/Detail/5019-2010 (accessed on 20 June 2022).
28. Available online: https://patents.google.com/patent/WO2011146025A1/und (accessed on 20 June 2022).

Article

Low Fluorinated Oligoamides for Use as Wood Protective Coating

Yuqing Zhang [1,2,3], Laura Vespignani [1], Maria Grazia Balzano [1], Leonardo Bellandi [1], Mara Camaiti [4,*], Nadège Lubin-Germain [2,3] and Antonella Salvini [1,*]

[1] Department of Chemistry, University of Florence, 50019 Sesto Fiorentino, Italy; yqing.z@outlook.com (Y.Z.); laura.vespignani@unifi.it (L.V.); mariagraziabalzano@libero.it (M.G.B.); leonardo.bellandi@stud.unifi.it (L.B.)
[2] CY Cergy Paris Université, CNRS, BioCIS, 95000 Cergy Pontoise, France; nadege.lubin-germain@cyu.fr
[3] Université Paris-Saclay, CNRS, BioCIS, 92290 Châtenay-Malabry, France
[4] CNR-Institute of Geosciences and Earth Resources, 50121 Florence, Italy
* Correspondence: mara.camaiti@igg.cnr.it (M.C.); antonella.salvini@unifi.it (A.S.); Tel.: +39-0552757558 (M.C.); +39-0554573455 (A.S.)

Abstract: New highly hydrophobic fluorinated oligoamides were synthesized and studied as materials for the protection of non-varnishable wooden artifacts. The new oligoamides were designed to achieve the best performance (including high chemical affinity to the wood material) and the lowest environmental impact. In order to minimize the risk of bioaccumulation, short perfluoroalkyl side chains were reacted with oligoethylene L-tartaramide (ET), oligoethylene adipamide-L-tartaramide (ETA), oligoethylene succinamide-L-tartaramide (EST), oligoethylene succinamide (ES), and oligodiethylenetriamino-L-tartaramide (DT). Favorable reaction conditions were also adopted to obtain low molecular weight compounds characterized by non-film-forming properties and solubility or dispersibility in environmentally friendly organic solvents. Their behavior in terms of modification of the wood surface characteristics, such as wettability, moisture absorption, and color, was analyzed using a specific diagnostic protocol to rapidly obtain preliminary, but reliable, results for optimizing a future synthesis of new and tailored protectives. The influence of different monomer units on the reactivity, solubility, and hydrophobic properties of different oligoamides was compared showing ESF (contact angle 138.2°) and DF (132.2°) as the most effective products. The study of stability to photochemical degradation confirms ESF as promising protective agents for artefacts of historical and artistic interest in place of long-chain perfluoroalkyl substances (PFAS), products currently subject to restrictions on use.

Keywords: sustainable coating; hydrophobic coating; diagnostic protocol; wood conservation; wood finishing

Citation: Zhang, Y.; Vespignani, L.; Balzano, M.G.; Bellandi, L.; Camaiti, M.; Lubin-Germain, N.; Salvini, A. Low Fluorinated Oligoamides for Use as Wood Protective Coating. *Coatings* **2022**, *12*, 927. https://doi.org/10.3390/coatings12070927

Academic Editor: Giorgos Skordaris

Received: 27 May 2022
Accepted: 28 June 2022
Published: 30 June 2022

Publisher's Note: MDPI stays neutral with regard to jurisdictional claims in published maps and institutional affiliations.

Copyright: © 2022 by the authors. Licensee MDPI, Basel, Switzerland. This article is an open access article distributed under the terms and conditions of the Creative Commons Attribution (CC BY) license (https:// creativecommons.org/licenses/by/ 4.0/).

1. Introduction

The degradation of wooden objects is an important topic in the conservation field [1] and is influenced by various factors related to the characteristics of the material itself, such as the wood species, the site of origin, the section of the piece of wood, the defects, and the treatments applied, as well as the environmental conditions to which the wooden artefact was subjected in its history, i.e., temperature, pH, oxygen concentration, and relative humidity [2]. Water is one of the main causes of deterioration of wood. Water vapor can condense on exposed surfaces, carrying many impurities and making possible the action of liquid water even on surfaces protected from rain. Water promotes the acid hydrolysis of polysaccharides and solubilizes both the fragments thus formed and some the extractive components of the wood, causing the loss of material as well as a radical change in the mechanical properties of wood [3]. In addition, the presence of water allows the development of algae, fungi, lichens, and bacteria, giving rise to biological degradation [4]. A specific problem of wood is also due to the rapid equilibrium that it establishes with the external humidity. Excursions of temperature and relative humidity induce changes in the wood moisture content. As a consequence, moisture exchanges induce dimensional variations, i.e., any loss in moisture content leads to a decrease in volume (shrinkage), while any increase induces a volume increase (swelling). Moisture exchanges can cause cracks, reaching even permanent deformations and collapses [5]. Furthermore, the impact of environmental factors such as atmospheric pollutants and microorganism colonization, responsible for decohesion phenomena, requires hydrophobic surface treatment to halt or slow down degradation, in particular under outdoor conditions, but also for storage indoors in certain stressful conditions. Fundamental action in contrasting the degradation of wood is carried out by protectives and varnishes. Since ancient times, hydrophobic compounds (e.g., natural oils, waxes, creosote oil, acrylic polymers, silicon compounds) have been used for the protection of wood, to both limit moisture content exchanges and creating a physical barrier for insects and fungi [4,6,7]. More recently, with the increased interest in the fabrication of superhydrophobic and self-cleaning surfaces, superhydrophobic and superparamagnetic composite films have recently been obtained on wood by a soft lithography technique, using Fe_3O_4 nanoparticles and polydimethylsiloxane treated with fluorine silane [8], or using nanoscale copper compound particles [9] or with a layer-by-layer deposition of TiO_2 nanoparticles modified with 1H, 1H, 2H, 2H-perfluoroalkyltriethoxysilane [10]. In general, the use of nanotechnologies in the treatment of wood promotes deep penetration by significantly modifying the properties of the wood [11]. Furthermore, superhydrophobic surfaces with flame retardancy has been obtained with polydimethylsiloxane (PDMS) and stearic acid (STA)-modified kaolin (KL) particles [12].

However, when the application field concerns wooden artefacts of historic and artistic interest, two further considerations need to be evaluated. First, the long-term efficacy and durability of these compounds is unsatisfactory. Natural oils, waxes, other natural compounds, and acrylic polymers are known to have moderate hydrophobic effects and undergo photo-oxidative and/or hydrolytic reactions [13–16]. Silicon compounds, after an in situ sol-gel process [17], can react with the substrate [7,18] and (like other polymeric materials do) form a rigid structure, susceptible to cracking with inevitable loss of both protective efficacy and the aesthetic features of the surface. Secondly, the complexity of application, the aesthetic characteristics of the products, or their high ability to form irreversible, non-vapor-permeable and chromatically visible films make some of these treatments very often not applicable in the field of Cultural Heritage [19,20]. Indeed, as in the conservation of stone artworks, in some specific wooden artefacts, the original aesthetic appearance of the surface after protection must be unaltered.

Recently, partially fluorinated oligoamides, containing short-pendant perfluoropolyether (PFPE) chains, have been successfully synthesized and tested as protective agents for stone [21–23]. The simultaneous presence on the molecule of hydrophilic (-C=O-NH- units) and hydrophobic (perfluoropolyether chain) groups, their waxy consistency and solubility in hydro-alcoholic solvents (environmentally friendly solvents), their good photo-oxidative and chemical stability, as well as their ability to fabricate superhydrophobic surfaces, make these compounds good candidates as protective agents for wooden artifacts [21–23].

At the same time, hydroxylated and non-hydroxylated water-soluble oligoamides with a high affinity for polar materials have been synthesized by a polycondensation reaction between ethylene diamine or amino acids and diesters of dicarboxylic natural acids (e.g., L-tartaric acid, D(+)-glucaric acid and α,α-trehaluronic acid), and successfully tested as consolidation agents for waterlogged archaeological wood [24–26].

In this study, the high affinity of hydroxylated oligoamides for wooden materials and the high hydrophobic effect of short perfluorinated chains have been exploited to obtain new compatible and hydrophobic oligoamides for the protection of wooden artefacts.

Due to the unavailability of mono-functional low molecular weight PFPEs and their derivatives on the European market, the identification of new perfluorinated compounds has been necessary. An epoxy with a C6 perfluorinated chain (EC6F), soluble in 2-propanol and easily reactive with amino groups, was used for the functionalization of (hydroxylated) oligoamides containing terminal amino groups. Moreover, compared to PFPEs, the short perfluorinated chain of EC6F is considered to have a low risk of bioaccumulation [27] and therefore a lower environmental impact.

To obtain and study products with different structural characteristics, several partially fluorinated (hydroxylated) oligoamides have been synthesized by reaction of EC6F with the terminal amino groups of oligoamides obtained from different combinations of polyamines (i.e., ethylene diamine and diethylenetriamine) and esters of (hydroxylated) dicarboxylic acids (i.e., L-tartaric, succinic, adipic acid). Namely, fluorinated oligoethylene L-tartaramide (ETF), fluorinated oligoethylene adipamide-L-tartaramide (ETAF), fluorinated oligoethylene succinamide-L-tartaramide (ESTF), fluorinated oligoethylene succinamide (ESF), and fluorinated oligodiethylenetriamino-L-tartaramide (DTF) have been synthesized. The fluorinated diethylenetriamine (by reaction of EC6F with diethylenetriamine) (DF) has been also synthesized as a reference product.

To evaluate the performance of the new compounds, a specific diagnostic protocol was set up to quickly obtain preliminary, but reliable results. The diagnostic protocol includes contact angles, moisture absorption, and color measurements on treated wood samples, and photo-oxidative test on the neat synthesized compounds. The photo-oxidative stability is followed by FT-IR and NMR analysis, color measurements, and mass variations.

2. Materials and Methods

2.1. Materials

Ethylenediamine (99.5%), diethylenetriamine (99.9%), dimethyl L-tartrate (99%), diethyl succinate (99%), methanol (99.8%), anhydrous ethyl alcohol (99.5%), anhydrous ethyl ether (99.8%), and acetone (99.9%) were purchased from Sigma-Aldrich (St. Louis, MO, USA). Dimethyl adipate was kindly supplied by Radici Chimica S.p.A. (Bergamo, Italy). Triethylamine (97%) and 2-propanol (99.9%) was purchased from Carlo Erba, and 3-perfluorohexyl-1,2-epoxypropane (95%) (Hexafor IM-P6, EC6F) was kindly supplied by Maflon S.p.A. (Bergamo, Italy). All chemicals were used without further purification. The new products were characterized by Fourier-transform infrared spectroscopy (FT-IR) and nuclear magnetic resonance spectroscopy (^{1}H NMR, ^{13}C NMR, gCOSY and gHSQC NMR). Wooden samples (beech or oak prismatic specimens with a square base with dimensions of $5 \times 5 \times 2$ cm^3 for colorimetric and contact angle measurements or $2.5 \times 2.5 \times 1$ cm^3 for water vapor absorption test and contact angle measurement) were used to test the performance of the surface-wetting modification agents.

A 70:30 poly(ethylmethacrylate-co-methylacrylate) (Paraloid B72) was used as coating reference material.

2.2. Instruments

^{1}H NMR, ^{13}C NMR, gCOSY, and gHSQC spectra were recorded with a Varian Mercury Plus 400 (Palo Alto, CA, USA) spectrometer and a Varian INOVA (Palo Alto, CA, USA) spectrometer, both working at 399.921 MHz, on D$_2$O or CD$_3$OD solutions. All spectra are reported in ppm and referred to TMS as internal standard. ^{19}F NMR spectra were recorded with a Varian INOVA spectrometer operated at 376.5 MHz and shifts are reported in ppm respect to CFCl$_3$ at 0 ppm. Spectra elaboration was performed with the software Mestre-C (1996, Mestrelab Research S.L., Santiago de Compostela, Spain) 4.3.2.0. FT IR spectra were recorded with a Shimadzu FT-IR IR Affinity-1S model (Kyoto, Japan) and elaborated with the Lab Solution IR v. 2.16 (2017, Shimadzu, Kyoto, Japan) program. The spectra of solid samples were recorded as KBr pellets or as such without further manipulation in transmission mode using a diamond anvil cell (Specac, Slough, UK). The spectra were collected from 400 to 4000 cm^{-1} with a resolution of 2 cm^{-1} and 64 scans.

2.3. Synthesis

The syntheses of oligoethylene-L-tartaramide (ET) and its fluorinated derivative (ETF), oligodiethylenetriamino-L-tartaramide (DT) and its fluorinated derivative (DTF), and oligoethylene-adipamide-L-tartaramide (ETA) and its fluorinated derivative (ETAF) were performed according to the reported methods for the other oligoamides as described in the following paragraphs. Details of all of these syntheses, as well as the FT-IR and NMR (^{1}H, ^{13}C, ^{19}F) characterization of all synthesized compounds are given in the supplementary materials.

2.3.1. Fluorinated Diethylenetriamine (DF)

In a Sovirel® tube (SciLabware Limited, Stoke on Trent, UK), diethylenetriamine (D) (20.1 mg, 0.194 mmol), 3-perfluorohexyl-1,2-epoxypropane (EC6F) (291.8 mg, 0.776 mmol) and 4 mL of solvent (2-propanol) were added under nitrogen atmosphere. The reaction mixture was allowed to react at 70 °C for 48 h. A yellow solution was observed at the end of the reaction. After evaporating the reaction mixture, the residue was washed with water. A light orange oil (310 mg, 99% yield) was obtained.

^{1}H NMR (CD$_3$OD) δ: 2.07–2.83 (m, 2H, CH$_2$CH(OH)CH$_2$(CF$_2$)$_5$CF$_3$), (m, 8H, NHCH$_2$CH$_2$NHCH$_2$CH$_2$NH) and (m, 2H, NHCH$_2$CH(OH)), 4.11 and 4.19 (m, 1H, CH$_2$CH(OH)CH$_2$(CF$_2$)$_5$CF$_3$) ppm.

^{13}C NMR (CD$_3$OD) δ: 35.4 (CH$_2$CH(OH)CH$_2$(CF$_2$)$_5$CF$_3$), 55.0 (CH$_2$NH), 61.0 (CH$_2$CH(OH)CH$_2$(CF$_2$)$_5$CF$_3$), 63.1 (CH$_2$CH(OH)CH$_2$(CF$_2$)$_5$CF$_3$), 108.8, 111.7, 113.5, 115.9, 118.6, 121.0 (CF$_3$(CF$_2$)$_5$) ppm.

^{19}F NMR (CD$_3$OD) δ: −82.6 (CF$_3$), −123.0, −124.1, −124.8 (CF$_2$), −127.5 (CF$_2$CF$_3$) ppm.

Using the same synthetic procedure, different molar ratios of D/EC6F were employed to obtain mono- or di-substituted fluorinated primary amino groups (Table 1).

Table 1. Synthesis of fluorinated diethylenetriamine (DF).

Molar Ratio (D/EC6F) [2]	RNHF [1] (%)	RNF$_2$ [1] (%)	R$_2$NF [1] (%)	Yield (%)	Solubility (%) (Ethanol, 1%) [3]
1:1	100	0	0	76	100
1:2	72	28	0	97	100
1:4	28	71	1	99	100
1:6	10	88	2	100	100

[1] Percentage ratio between different amino groups: RNHF = mono-substituted fluorinated primary amino group; RNF$_2$ = di-substituted fluorinated primary amino group; R$_2$NF = fluorinated secondary group. [2] D = diethylenetriamine, EC6F = 3-perfluorohexyl-1,2-epoxypropane. [3] The solubility was determined as percentage of product dissolved for 1% concentration (w/w solute/solvent). Similar solubility was found in 2-propanol.

2.3.2. Oligoethylene-Succinamide-L-Tartaramide (EST) and Its Fluorinated Derivative (ESTF)

(a) EST—Diethyl succinate (S) (348.4 mg, 2 mmol), ethylenediamine (E) (240.4 mg, 4 mmol) and 1.5 mL of solvent (methanol) were added under nitrogen atmosphere to dimethyl L-tartrate (T) (356.3 mg, 2 mmol) in a Sovirel® tube. After stirring at 80 °C for 72 h, the formation of a light yellow solid was observed. In the work-up process, the reaction mixture was filtered on a Büchner Funnel, washed with ethyl ether, and then dried at a reduced pressure, giving 584.6 mg of product. A light yellow and water-soluble solid was recovered, with an average molecular weight of 887.92 g/mol, 100% yield, and a succinic/tartrate unit ratio of 1:1.

With the same procedure, but stirring at 80 °C for only 24 h instead of 72 h, a light yellow and water-soluble solid (EST^2) was obtained, with an average molecular weight of 633.86 g/mol, 72% yield, and a succinic/tartrate unit ratio of 1:2.

^{1}H NMR (D$_2$O) δ: 2.52 and 2.54 (m, 4H, COCH$_2$CH$_2$CO), 2.97 (m, 2H, CH$_2$NH$_2$), 3.2–3.5 (m, 2H, CH$_2$NHCO), 4.56 (m, 1H, CHOH) ppm.

^{13}C NMR (D$_2$O) δ: 31.0 (CH$_2$CONH), 38.5 (CH$_2$NHCO), 39.3 (m, 2H, CH$_2$NH$_2$), 72.3 (CH-OH), 174.0 and 175.0 (CONH) ppm.

(b) ESTF—In a Sovirel® tube, oligoamide EST2 (120 mg, 0.189 mmol), EC6F (284.3 mg, 0.756 mmol), and 4 mL of solvent (a mixture of 2-propanol/MilliQ water 3/1) were added under nitrogen atmosphere. After 48 h at 70 °C, the reaction mixture was evaporated to dryness under reduced pressure, and the solid residue was washed first with water and then with cold acetone (−20 °C) obtaining a yellow solid (281.8 mg, yield 70%).

^{1}H NMR (CD$_3$OD) δ: 2.46 (m, 4H, COCH$_2$**CH$_2$**CO), 2.10–2.80 (m, 4H, CH$_2$CH(OH)CH$_2$(CF$_2$)$_5$CF$_3$), 2.8 (m, 2H, CH$_2$NHCH$_2$CH(OH)CH$_2$(CF$_2$)$_5$CF$_3$), 3.0–3.5 (m, 2H, CH$_2$NHCO), 4.10 (m, 1H, CH$_2$CH(OH)CH$_2$(CF$_2$)$_5$CF$_3$), 4.47 (m,2H, CH(OH)CH(OH)) ppm.

^{13}C NMR (CD$_3$OD) δ: 30.7 (COCH$_2$), 35.5 (CH$_2$CH(OH)CH$_2$(CF$_2$)$_5$CF$_3$), 38.5 (CH$_2$NHCO), 54.7 (CH$_2$NH), 60.9 and 65.4 (CH$_2$CH(OH)CH$_2$(CF$_2$)$_5$CF$_3$), 61.9, 63.5, 63.5 (CH$_2$CH(OH)CH$_2$(CF$_2$)$_5$CF$_3$), 72.7 (COCH(OH)CH(OH)CO), 108.3, 111.5, 113.5, 115.7, 118.4, 121.0 (CF$_3$(CF$_2$)$_5$), 173.8 (CONH) ppm.

^{19}F NMR (CD$_3$OD) δ: −82.5 (CF$_3$), −122.9, −124.0, −124.6 (CF$_2$), −127.5 (CF$_2$CF$_3$) ppm.

2.3.3. Oligoethylenesuccinamide ES and Its Fluorinated Derivative ESF

(a) ES—In a Sovirel® tube, S (348.4 mg, 2 mmol), E (240.4 g, 4 mmol), and 1.5 mL of solvent (methanol) were added under nitrogen atmosphere. After stirring at 80 °C for 24 h and carrying out the same work-up used for EST, a white solid was obtained with an average molecular weight of 486 g/mol and a yield of 73%.

^{1}H NMR (D$_2$O) δ: 2.38, 2.51 (m, 4H, COCH$_2$CH$_2$CO), 2.71 (m, 2H, CH$_2$NH$_2$), 3.14, 3.31(m, 2H, CH$_2$NHCO) ppm.

^{13}C NMR (D$_2$O) δ: 31.1 (CH$_2$CONH), 38.6 and 41.3 (CH$_2$NHCO), 39.7 (m, 2H, CH$_2$NH$_2$), 175.0 (CONH) ppm.

(b) ESF—In a Sovirel® tube, oligoamide ES (120 mg, 0.247 mmol), EC6F (371.6 mg, 0.988 mmol) and 2 mL of solvent (2-propanol) were added under nitrogen atmosphere. After 48 h at 70 °C, a white dispersion was observed. The reaction mixture was evaporated at reduced pressure and the residue was washed with water to remove the unreacted oligoamide ES. A light yellow solid (438 mg) was obtained with a 89% yield.

^{1}H NMR (CD$_3$OD) δ: 2.46 (COCH$_2$CH$_2$CO), 2.05–2.77 (m, 4H, CH$_2$CH(OH)CH$_2$(CF$_2$)$_5$CF$_3$), 2.6–2.7 (m, 2H, CH$_2$NHCH$_2$CH(OH)CH$_2$(CF$_2$)$_5$CF$_3$), 3.26 (m, 2H, CH$_2$NHCO), 4.11 (m, 1H, CH$_2$CH(OH)CH$_2$(CF$_2$)$_5$CF$_3$) ppm.

^{13}C NMR (CD$_3$OD) δ: 30.7 (COCH$_2$), 34.6 (CH$_2$CH(OH)CH$_2$(CF$_2$)$_5$CF$_3$), 38.2 (CH$_2$NHCO), 52.6 and 54.1 (CH$_2$NH),61.1 and 61.2 (CH$_2$NHCH$_2$CH(OH)CH$_2$(CF$_2$)$_5$CF$_3$), 62.3 and 63.2 (CH$_2$CH(OH)CH$_2$(CF$_2$)$_5$CF$_3$), 108.4, 110.8, 112.9, 114.7, 118.9, 121.3 (CF$_3$(CF$_2$)$_5$), 173.6 (CONH) ppm.

^{19}F NMR (CD$_3$OD) δ: −82.5 (CF$_3$), −122.9, −124.1, −124.6, −124.9(CF$_2$), −127.6 (CF$_2$CF$_3$) ppm.

2.4. Diagnostic Protocol

The diagnostic protocol was designed to investigate the relationship between the chemical structure of the coating agents and their performance on wood samples. Various tests were selected to compare the changes in some wood properties obtained as a consequence of different treatments.

Beech and oak clear samples, oriented according to the anatomical directions, were prepared in the shape of parallelepipeds with a square base of $5 \times 5 \times 2$ cm^3 or $2.5 \times 2.5 \times 1$ cm^3. To evaluate the effect of wood extractives on the behavior of the tested products, samples as such and pre-extracted with ethanol (two extractions of 24 h at room temperature) were compared. In accordance with the known peculiarity of the wooden material, the samples presented an appreciable heterogeneity of appearance between the different faces of the same specimen, visible to the naked eye, and sometimes even between different portions of the same face. To reduce the error in the data evaluation, it was necessary to carry out the measurements by monitoring the behavior of the selected samples before and after treatment, identifying, when necessary, the precise positions on the samples and repeating more measurements even in different positions of the same sample.

Static water contact angle (CA), water vapor absorption, and chromatic changes were the measurements selected for evaluating the performance of the protective treatments.

The static CA of the neat and treated wood samples was measured using 5 µL of distilled water. Preliminary measurements were recorded through a video using an Osmo pocket (DJI, Shenzhen, China) camera, while definitive data were collected by a Ramé-Hart Model 190 CA (Succasunna, NJ, USA) Goniometer. To calculate the average value of five different drops on a same position of each sample, a specific position was selected and fixed through a mask, and the measurement was repeated after complete evaporation of the previous drop of water (approximately after 48 h).

The water vapor absorption test was performed introducing the dried samples in a climate-controlled room (temperature set at 20 °C) inside a closed container, where the relative humidity (RH%) was controlled by means of saturated salt solutions (UNI EN ISO 483—2006 [28]). The relative humidity was gradually increased, applying the following RH% values:

- Saturated solution of NH_4NO_3 (RH 65%);
- Saturated solution of KCl (RH 86%);
- Distilled water (RH 100%).

The samples were regularly weighed at time intervals ranging from 2 to 24 h. Only when the difference between two subsequent measurements was less than 0.5% were they moved on to the next humidity step. As a starting point for the absorption tests, according to standard methods [29,30], it would have been correct to use the wood in an anhydrous state after drying the specimens in an oven at 103 °C for 24 h. However, in this case, the high temperature would have modified the interaction between the oligoamides and the wood, inducing a migration of the product and altering the treatment. It was therefore decided to use the weight of the samples kept for 72 h in the desiccator at room temperature as an initial dry value. In this way, the weight obtained is not the anhydrous weight of the sample, but the weight in equilibrium with the moisture not retained by the calcium chloride present inside the desiccator.

The chromatic changes induced by the coatings were examined by colorimetric analysis performed before and after the application of the coating, following the standard method UNI-EN 15886–2010 [31]. Only one sample for each type of treatment was tested to select samples as similar as possible in initial chromatic behavior. In fact, the chromatic variation in the wood is also affected by the different shade of color present in different areas, even in the same initial piece of wood. Five measurements for each sample (before and after coating) were collected on the same spot, previously located by using a mask, exploiting a portable X-Rite SP60 (Grand Rapids, MIichigan, USA) spectrophotometer in specular component excluded mode. The results were analyzed and reported in the CIE-L*a*b* standard color system, and the color alterations (ΔE^*) were expressed as: $\Delta E^* = \sqrt{\Delta L^{*2} + \Delta a^{*2} + \Delta b^{*2}}$, where L* indicates lightness, and a* and b* are the color axes. ΔL^*, Δa^*, and Δb^* were calculated as $\Delta X = X_a - X_b$, where X_a and X_b are the L*, a*, or b* values after and before the application of the coating, respectively.

2.5. Treatment of Wood Samples

Based on the solubility test in ethanol or 2-propanol (see results and discussion), the products with the best solubility were selected for the coating treatments (i.e., ETAF, ESTF, ESF, DTF, DF), and 2-Propanol was selected as the solvent. Paraloid B72 was dissolved in acetone at the same concentration of the synthesized compounds.

Each product was tested on three wood samples (one $5 \times 5 \times 2$ cm^3 and two $2.5 \times 2.5 \times 1$ cm^3). Propanol solution (ETAF, ESTF, DTF, DF) or dispersion (ESF), 8 mg/mL, was freshly prepared and the volume of each formulation, as needed for each treatment, was measured using a graduated pipette (accuracy 0.01 mL). The treatment was performed by a Pasteur pipette using the classical method "wet on wet". A surface concentration of 30 g/m^2 was applied on one 5×5 cm^2 surface for the wettability tests and the colorimetric measurements, while all the surfaces of the sample were treated with an excess of product (40 g/m^2) for the water vapor absorption tests. After solvent evaporation in lab conditions for 3 days under a fume hood, all samples were put in a desiccator until constant weights were reached.

2.6. Photo-Oxidative Test

The stability of the synthetic partially fluorinated oligoamides and amine was estimated by means of accelerated aging tests exploiting UV radiations. Their behavior was followed via ^{1}H NMR, FT-IR spectroscopies and colorimetric measurements. The weight change was also recorded in order to track the change of the compounds after the UV exposure test.

Exposure to UV radiation was carried out using a Spectroline (Melville, NY, USA) lamp, Model ENF-260C/FE, with an emission in UV-A range at wavelength of 365 nm (tube of 6 W).

3. Results and Discussion

3.1. Non-Fluorinated Oligoamides Synthesis

In the first step, the synthesis of non-fluorinated oligoamides was performed by condensation of an ester and an amine group through step-growth polymerization. The monomers were selected to obtain products with the structural characteristics required for their use as wood protective agents. When it was possible, the monomers recoverable from biomass were also chosen.

Dimethyl L-tartrate was chosen as one of the monomers for the production of oligoamides because it has two hydroxyl groups in the molecule, which can make the final product affine to a polar material as wood. The dextrorotatory enantiomer of (R,R)-L-(+)-tartaric acid is a renewable resource widely distributed in nature (in many plants, particularly grapes, bananas, and tamarinds as well as in other fruits).

Diethyl succinate or diethyl adipate were selected to improve the solubility in alcohol and the hydrophobicity of the final fluorinated derivatives. Diethyl succinate derives from succinic acid, which is a naturally occurring four-carbon dicarboxylic acid, naturally formed by most living cells as an outcome of anaerobic digestion, and it is a coproduct of particular interest in biorefineries production. Adipic acid can be obtained in biorefinery processes by the fermentation of sugars to muconic acid followed by hydrogenation, direct fermentation, or by the anaerobic oxidation of sugar to glucaric acid and subsequent hydrodeoxygenation of glucaric acid.

Ethylenediamine and diethylenetriamine were selected as the amine blocks for the synthesis to compare the role of two different amines in the performance of the final products.

The selected amines can react with esters of dicarboxylic acid to form the –CONH– functional group, which make the final product affine to polar materials by interacting with their components through dipolar interaction or hydrogen bonding. Moreover, the amide group is more stable than other polar groups such as the ester ones.

By combining different amine blocks with different esters of diacids, it is possible to modulate the molecular weights and the physical characteristics of the products, such as hydrophilic or hydrophobic properties and solubility.

In the diethylenetriamine, the secondary –NH– group, generally not involved in the amidation reaction, can interact with polar material with additional hydrogen bonding, providing different physical characteristics and properties.

In summary, the choice of monomers was made in order to balance the presence of polar groups, to favor interactions with polar substrates and hydrophobic parts.

Synthetic procedures for the synthesis of oligoamides were designed to obtain products with low molecular weight and/or unwanted degradation products. In fact, in previous preliminary syntheses carried out in accordance with the industrial synthesis of common nylons (i.e., salts of hydroxylated monomers heated at high temperature), we obtained polymers with too-high molecular weights or a high percentage of byproducts. Therefore, esters of dicarboxylic acids were used for amides production in mild reaction conditions, avoiding salt formation between diacid and diamine, which requires higher temperatures for the condensation reaction. Starting from diesters, the low temperature (80 °C) avoids the alteration of the hydroxylated portions. In order to obtain oligoamides with terminal amine groups for successive reactions, the molar amount of amine was kept in slight excess with respect to the selected stoichiometric ratio diamine/diester (1:1 or 2:1). The syntheses of oligoamides were typically carried out in methanol as polar solvent, suitable for solubilizing all of the reagents.

The different reactivity of each monomer affects chain growth, resulting in different molecular weights, different monomer unit ratios, and different conversions for different oligoamide syntheses. Solid products were generally obtained for all of the oligoamides.

The reaction conditions used in the various syntheses are reported in Table 2 together with the yields, molecular weights, and the ratios between the monomer units. The reaction schemes for the synthesis of ES, EST, and ES are shown in Figure 1. As expected, different monomers influence the reactivity and, consequently, the yield. The co-products present in the reaction mixture are typically unreacted products, rather than byproducts. In fact, by prolonging the reaction time, the yield increases, as observed for EST where the yield respectively passes from 72% to 100% after 24 and 72 h (Table 2). In Figure 2, the oligoamide yield for different monomers is compared after 24 h and after long reaction times.

Table 2. Synthesis of non-fluorinated oligoamides performed at 80 °C with methanol as solvent.

Oligoamide	Reaction Time (h)	Initial Molar Ratio (Diacids/Diamine)	Units Ratio x:y:z [1]	Yield (%)	MW [2] (g/mol)
ET	60	1:1	1:1:0	64	700
ETA	24	1:1	3:4:1	62	750
DT	48	1:1	1:1:0	70	1734
EST	24	1:1	2:3:1	72	633.9
EST	72	1:1	1:2:1	100	887.9
ES	24	1:2	0:1:1	73	486

[1] x = tartaric unit, y = amine unit, z = succinic or adipic unit. [2] Molecular weight evaluated through the ^{1}H-NMR spectra as described below.

The results show that diethylenetriamine is more reactive than ethylenediamine, while the reactivity of the aliphatic diesters is in the order succinate > adipate. As the literature reports that diesters with hetero atom groups (including hydroxylated diesters) show enhanced reactivity in mild conditions compared with aliphatic diesters [32,33], the higher reactivity of ES (yield 73% after 24 h) in respect to ET (64% after 60 h) may be explained with the different diamine:diester ratio (2.1 instead of 1:1). However, when aliphatic and hydroxylated esters (i.e., tartrate) are contemporarily present in the reaction mixture, the tartrate shows a greater reactivity. This behavior is evidenced by comparing the higher yield of EST (73%) after 24 h (Figure 2a) with the lower yield of ET (64%) after 60 h. (Figure 2b). Moreover, in agreement with the reported higher reactivity of diesters with hetero atom groups, a higher tartaric:succinic ratio is observed at low reaction times (Table 2).

During the step-growth polymerization, the monomers initially form oligomers with low molecular weight, which grow gradually, but molecules of different lengths are synthesized and the final product, although at low average molecular weight, is composed of macromolecules with a different degree of polymerization. The ^{1}H NMR spectral integration method was previously applied to estimate the number of average molecular weights (M_n) [26,34]. Similarly, in this study, specific equations were applied for each oligoamide in agreement with its structure, selecting suitable signals. First of all, the characterization of new compounds with the attribution of all of the signals present in the ^{1}H and ^{13}C NMR spectra was also performed using, when necessary, 2D NMR spectroscopy (gCOSY, gHSQC). In particular, using ^{1}H and ^{13}C NMR spectroscopy, the presence of amino groups at the end of the polymeric chain was determined (signals respectively at 2.70–3.00 ppm and near 40 ppm). The absence in the ^{1}H NMR of a singlet signal in the 3.60–3.80 ppm range and a signal at 52 ppm in the ^{13}C NMR, both attributable to the –OCH$_3$ in the terminal ester group, is in agreement for all products with the presence of two terminal amino groups. As the CH$_2$-NH$_2$ signal is present in a quite clean region of the ^{1}H NMR spectrum, the corresponding integral was used as the reference value and set as 2 when one amino group was present at the end of chain, or 4 when both end groups were amino groups. Then, the integrals of the other signals were evaluated with respect to this value. Since the integral of the signal area is proportional to the number of protons, it is possible to calculate the number "x", "y", "z" of the repeating units of the oligoamide using the equations reported in Table 3 for each compound. As an example, the ^{1}H NMR spectrum of EST is shown in Figure 3, while the other NMR and FT-IR spectra of all synthesized products are shown in the Supplementary Materials (Figures S1–S36).

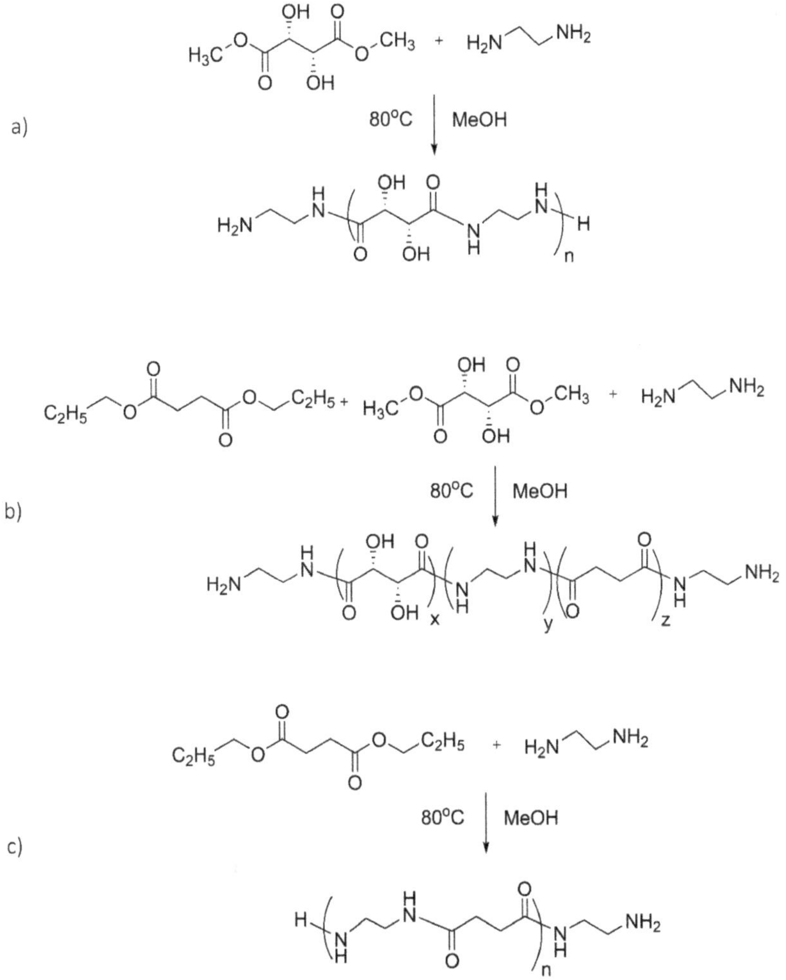

Figure 1. Schemes of the synthesis of oligoamides ET (**a**), EST (**b**) and ES (**c**). Scheme of synthesis for DT is the same as scheme (**a**) with diethylenetriamine (D) instead of ethylenediamine (E). Scheme of synthesis for ETA is the same as scheme b with diethyl adipate (A) instead of diethyl succinate (S).

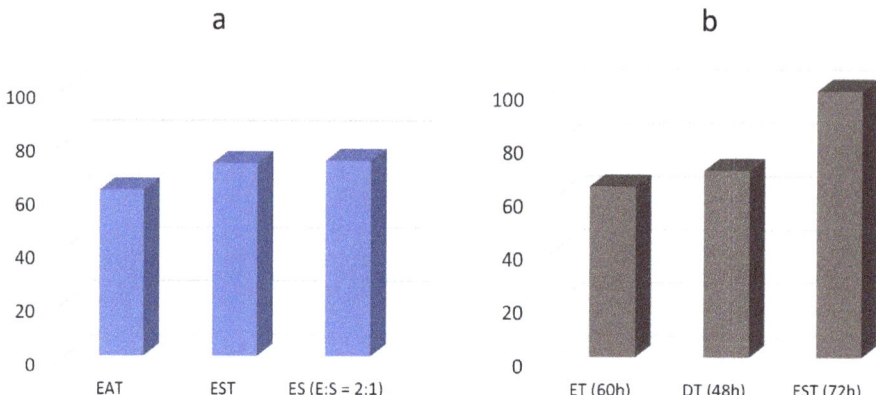

Figure 2. Oligoamide yield (%) for different monomers after different reaction times (**a**) 24 h; (**b**) ≥48 h at 80 °C.

Table 3. Synthesis of oligoamides: equations to calculate the number of repeating units of the oligoamide using the ^{1}H NMR integral values [1].

OA [2]	Tartaric Units (x)	Diamine Units (y + 1)	Diamine Units (y + 1)	Succinic Units (z′)	Adipic Units (z″)
ET	$I_{CHOH} = 2x$	$I_{CH2NHCO} = 4y + 4$	-	-	-
ETA	$I_{CHOH} = 2x$	$I_{CH2NHCO} = 4y + 4$	-	-	$I_{CH2CONH} = 4z$ $I_{CH2CH2CONH} = 4z$
DT	$I_{CHOH} = 2x$	$I_{CH2NHCO} = 4y + 4$	$I_{CH2NHCH2} = 4y + 8$	-	-
EST	$I_{CHOH} = 2x$	$I_{CH2NHCO} = 4y + 4$	-	$I_{CH2CONH} = 4z$	-
ES		$I_{CH2NHCO} = 4y + 4$	-	$I_{CH2CONH} = 4z$	-

[1] I = integral value of signal attributable to the labeled group, calculated with respect to the integral of the CH_2NH_2 signal set equal to 4 (two terminal amino groups). The x, y, and z values refer to the units within the chain in the presence of two terminal amino groups. The value y + 1 refers to the total number including two terminal amino groups. [2] OA = oligoamide.

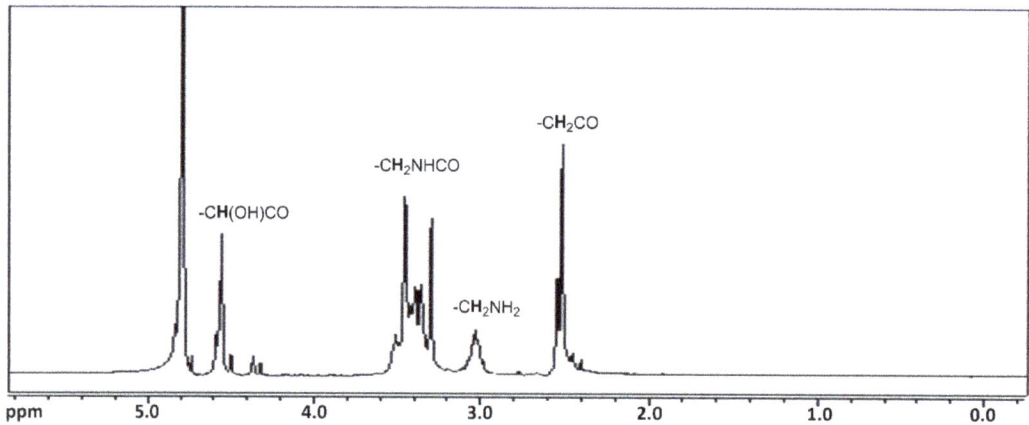

Figure 3. ^{1}H NMR spectrum in D_2O of oligoamide EST.

Based on the ^{1}H NMR characterization, the formation of the amide group is confirmed by the presence of signals between 3.2 and 3.5 ppm attributable to the CH_2 group in α position to this group: CH_2NHCO (2 methylene groups in the middle of the chain and the CH_2 in the terminal amino groups $CONHCH_2CH_2NH_2$). Similarly, in the ^{13}C NMR spectrum, the amide group formation is confirmed by the presence of signals at 174.0 and 175.0 ppm (CONH) and 38.5 ppm (CH_2NHCO).

The presence of different monomeric units is confirmed by the presence of characteristic signals. In particular, the signal at 4.56 ppm in the ^1H NMR spectrum is attributable to protons of tartrate units $COCH(OH)CH(OH)CO$, while a signal at 72.3 ppm is present in the ^{13}C NMR spectrum related to the CHOH. Signals between 2.30 and 2.55 ppm in the ^1H NMR spectrum are attributable to protons of 2 methylene groups in the α position of each of the 2 carbonyl groups CH_2CONH of the succinic or adipic units, while in the ^{13}C NMR spectrum a signal is present at 31.0–35.5 ppm (CH_2CONH). In the ^1H NMR, a signal between 2.73 and 2.77 attributable to protons of CH_2NHCH_2 is present when diethylenetriamine has been used as monomer, while in the ^{13}C NMR, the signals at 47.0 and 45.7 are attributable to CH_2NHCH_2. Finally, a signal at 1.61 ppm (24.7 ppm in the ^{13}C NMR) is attributable to the presence of adipic units (CH_2CH_2CONH).

3.2. Partially Fluorinated Oligoamides Synthesis

The synthesis of fluorinated derivatives was carried out by the nucleophilic ring-opening reaction between the terminal amino groups of oligoamides and a perfluorinated epoxy derivative (3-perfluorohexyl-1,2-epoxypropane, EC6F, Figure 4). This fluorinated compound was selected in order to use a reagent with good solubility in alcoholic and hydro-alcoholic solvents and with a low environmental impact.

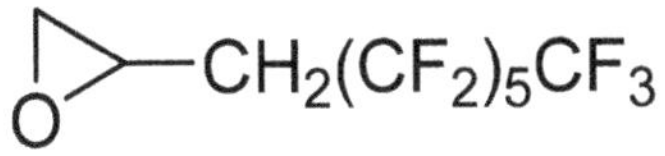

Figure 4. 3-perfluorohexyl-1,2-epoxypropane.

Preliminarily, the reactivity of the fluorinated epoxide was studied with hydroxyl nucleophiles such as the OH groups of sodium tartrate, selected as a reference for the reactivity of the tartaric unit. The reactivity with water was also studied for possible use of alcohol/water mixtures as solvent, if necessary, for the dissolution of oligoamides containing tartrate units. Regarding the evaluation of reactivity with the hydroxyl groups of the tartrate, no reaction was observed in the presence of EC6F with a molar ratio tartrate/EC6F = 2:1 after heating 48 h at 70 °C with 2-propanol. On the contrary, in the presence of water as a co-solvent, the formation of the diol by opening the epoxy ring prevailed. As confirmation, the formation of diol was also obtained by reacting EC6F with water. In the ^1H NMR spectrum (Figure S1), new signals are present at 4.06 ppm (CHOH) and between 3.65 and 3.45 ppm attributable to CH_2OH. Therefore, the presence of water as a co-solvent requires an excess of reagents for the competitive reaction of the epoxy group, which produces the corresponding diol and the subsequent purification of the product from the diol.

On the other hand, no conversion of the epoxide is observed after heating for 48 h at 70 °C in 2-propanol, as confirmed by the presence of the characteristic signals of EC6F at 3.30, 2.85, and 2.60 ppm (Figure S2) in the ^1H NMR spectrum of the solid residue recovered at the end of the reaction.

Similarly, to evaluate the reactivity of the epoxide with the amines, preliminary tests were carried out in 2-propanol using diethylenetriamine as a reagent with different molar ratios with respect to the epoxide (1:1, 1:2, 1:4, 1:6). In this way, it was possible to observe the formation of products (DF) with mono and di-substituted amino groups and to compare their behavior with respect to oligoamides.

In the ^1H NMR spectrum (Figure 5), the signals at 4.11 and 4.19 ppm are attributable to CH in the $CH_2CH(OH)CH_2(CF_2)_5CF_3$ respectively for the amino group with two fluorinated chains and for the mono-substituted one. In fact, the presence of different integral ratios between the two signals attributable to $CH_2CH(OH)CH_2(CF_2)_5CF_3$ agrees with the different molar ratio used between the two reagents (Table 1). This integral ratio, along with the molar ratio of the reagents, allowed the identification of the signals at 4.11 ppm for the presence of amino groups with two fluorinated chains, while the signal at 4.19 ppm can be assigned to the presence of amino groups with only one substituent. Finally, the low signal at 4.25 pm can be attributed to the reaction of the secondary amino group with the epoxide. The gCOSY spectrum (Figure S4) confirms the coupling between both signals at 4.11 or 4.19 ppm and signals between 2.07 and 2.83 ppm, which are attributable to $NHCH_2CH(OH)CH_2(CF_2)_5CF_3$ overlapped with the signals ascribable to CH_2NH.

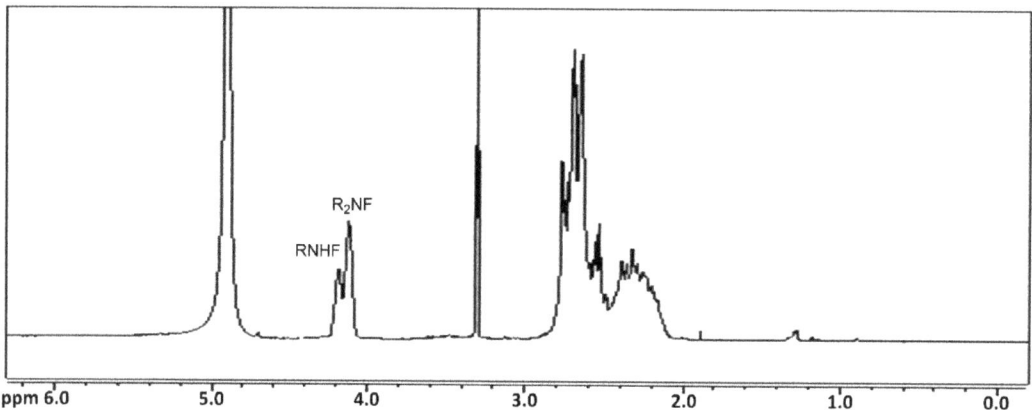

Figure 5. ^{1}H NMR spectrum in CD_3OD of oligoamide DF.

On the basis of the preliminary results, the conditions for the reactions between oligoamides and epoxide were selected considering that the starting oligoamides show different solubility in pure alcoholic solvent depending on the monomers present and on the molecular weight. The reaction schemes for the synthesis of ES, EST, and ES are shown in Figure 6. In particular, for oligoamides without tartaric units, 2-propanol was selected as the solvent, while oligoamides containing tartrate units are partially soluble in 2-propanol and soluble in water or a mixture of 2-propanol/water. Consequently, for the products containing tartrate units, the presence of undissolved oligoamide in the synthesis with 2-propanol can cause low conversion. On the contrary, the dissolution of both reactants can be achieved by working with a mixture of 2-propanol/water favoring the conversion. Then mixtures of water and 2-propanol in different ratios (1:1 or 1:3) were tested as a solvent to optimize the conversion, while different oligoamide/epoxide (OA/EC6F) molecular ratios between 1/1 and 1/12 were also studied to obtain different functionalization degrees. The most relevant results are shown in Table 4.

The best reaction conditions resulted in a molar ratio OA/EC6F of 1:4 and heating at 70 °C for 48 h. Using a 2-propanol/water mixture as a solvent in the ratio 3:1, higher conversion and degree of substitution values were obtained compared to the 1:1 ratio. Furthermore, a lower quantity of diol was obtained in the presence of a smaller quantity of water, allowing for better purification. As observed for the synthesis of non-fluorinated oligoamides, the co-products were typically unreacted reagents, except for the reaction performed with 2-propanol/water mixture as a solvent in which the fluorinated diol was also formed.

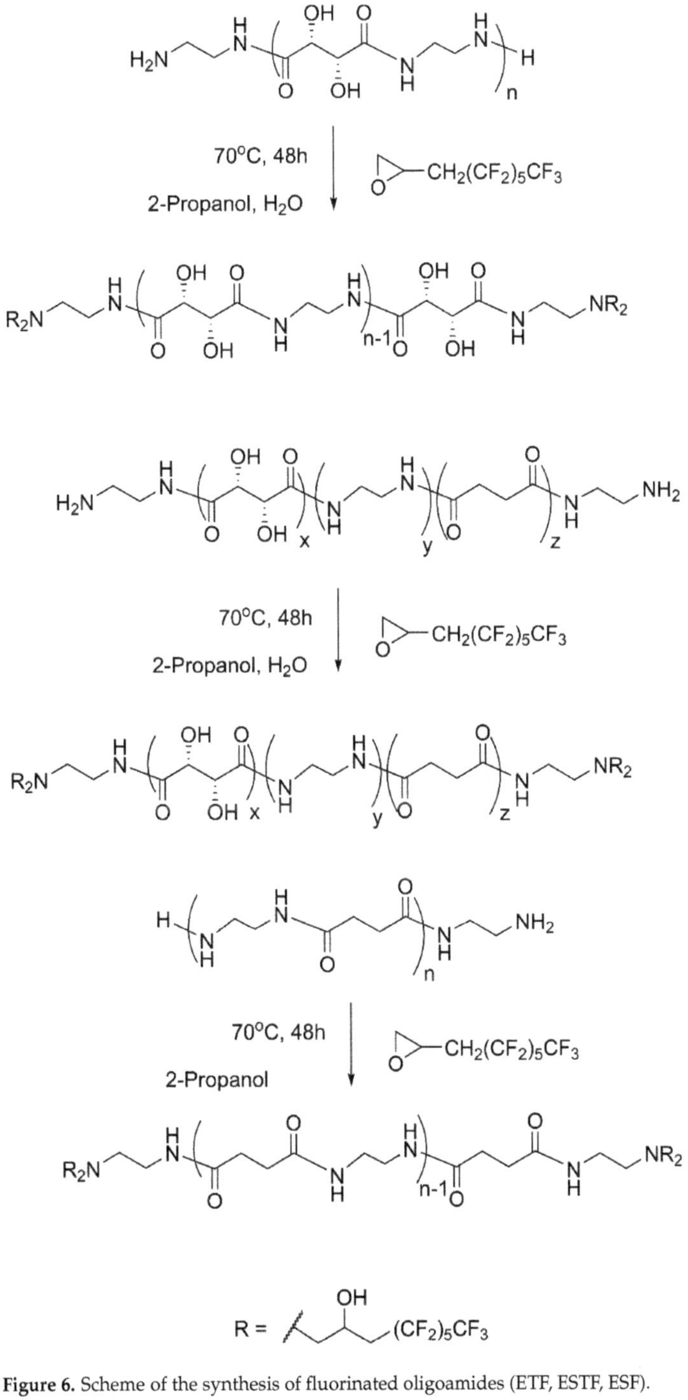

Figure 6. Scheme of the synthesis of fluorinated oligoamides (ETF, ESTF, ESF).

Table 4. Synthesis of fluorinated oligoamides [1].

OAF [2]	Solvent	Molar Ratio (OA:EC6F)	Yield (%)	DS_F [3]	Solubility (%) (Ethanol, 1%) [4]
ETF	2-propanol	1:2	trace	Not evaluable	-
	2-propanol/water (1:1)	1:2	30	0.6	-
	2-propanol/water (1:1)	1:4	48	2.5	57
DTF	2-propanol/water (1:1)	1:4	56	3.0	68
ETAF	2-propanol/water (1:1)	1:4	48	3.7	83
ESTF	2-propanol	1:4	trace	>3	-
	2-propanol/water (1:1)	1:4	61	3.6	65
	2-propanol/water (3:1)	1:4	70	3.9	65
ESF	2-propanol	1:4	89	~4.0	85

[1] Reaction conditions: T = 70 °C, 48 h. [2] OAF = fluorinated oligoamide. [3] DS_F: substitution degree with fluorinated group. [4] The solubility was determined as percentage of product dissolved for 1% concentration (w/w solute/solvent). Similar solubility was found in 2-propanol.

In the general work-up procedure, the solvent and the excess epoxide were removed by evaporation, and then water was used to wash the residue from the unreacted oligoamide. Finally, cold acetone (−20 °C) was used to wash the residue again in order to remove the diol formed as a byproduct of the reaction between the epoxy and water.

Oligoethylene-L-tartaramide was used as the first oligoamide in order to highlight the influence of the -OH groups of the hydroxylated diacid on the reactivity with EC6F and on the properties of the final product (i.e., solubility, color, physical state). Using 2-propanol as a solvent and an OA:EC6F molar ratio of 1:2, no reaction was observed due the low solubility of the starting oligoamide. Low conversion was also obtained using 2-propanol:water 1:1 as solvent and the same OA/EC6F molar ratio, while a yield of 48% with a degree of substitution of 2.5 was achieved by increasing the concentration of EC6F up to a OA/EC6F molar ratio of 1:4 (Table 4). Unfortunately, the fluorinated product shows a reduced solubility in all common solvents, and ethanol or methanol only dissolves 57% of the reaction solid residue. This behavior makes the product difficult to use for subsequent applications, and for this reason, ethylenediamine was replaced with diethylenetriamine to increase the polarity of the final product. As a result, a slightly higher conversion (56%) was observed compared with a better solubility in alcohol of the fluorinated product (68% instead of 57% of ETF) (Table 4), confirming that the reactivity with the epoxide is mainly controlled by the reduced solubility of the oligotartaramide. The partial solubility of the fluorinated products is attributable to the presence of oligomers with different molecular weight and also with different degrees of functionalization, which give rise to fractions of product with different solubility.

To increase the solubility of the starting oligoamide and fluorinated products in alcoholic solvents, products containing units of succinic or adipic acid together with tartaric acid were synthesized. The reactivity of the ETA or EST oligoamides was subsequently tested with the EC6F epoxide.

Due to the few secondary interactions between the aliphatic chains and the consequent better affinity for the solvent, greater solubility in alcohol is expected for the fluorinated products when using adipic or succinic units.

As reported in Table 4, the best results as conversion were obtained for ESTF. In this synthesis, the reaction was performed in the mixture 2-propanol:H$_2$O with a volume ratio of 1:1 or 1:3 for 48 at 70 °C and with a molar ratio OA:EC6F of 1:4. After the work-up procedure, the final product was obtained as a yellow solid with yields of 61% (in 2-propanol:water 1:1) or 70% (in 2-propanol:water 3:1) and was found to be soluble in 2-propanol or ethanol (8 mg/mL, 1%). However, the milky solution maintained for a long time (typically 1 day) at room temperature easily forms a fine solid. The spectroscopic characterization (^{1}H, gCOSY, ^{13}C, ^{19}F NMR) confirms the presence of fluorinated groups attached to the oligoamide with a degree of functionalization (DS) > 3.5. For this compound, DS was evaluated as the integral ratio between the signal at 4.10 (ascribing to 1H) and 4.47 ppm (ascribing to 2H for an average of 2.18 tartaric units on each oligoamide chain, value obtained by processing NMR data according to the equations reported in Table 3). In the ^{1}H NMR spectrum recorded in CD$_3$OD (Figure 7), the presence of the fluorine chain is confirmed by the signals between 2.10 ppm and 2.80 ppm attributable to CH$_2$ in CH$_2$NHCH$_2$CH(OH)CH$_2$(CF$_2$)$_5$CF$_3$, while the signal at 4.10 ppm is attributable to CH in the fragment CH$_2$CH(OH)CH$_2$(CF$_2$)$_5$CF$_3$, which is formed by the opening of the epoxy group during the reaction with the amino group. The signals ascribable to the hydroxylated oligoamide fragment are also present. The signals present in the ^{19}F NMR spectrum confirm the presence of fluorinated chains (−82.5 (CF$_3$), −122.9, −124.0, −124.6 (CH$_2$(CF$_2$)$_4$CF$_2$CF$_3$) and −127.5 (CH$_2$(CF$_2$)$_4$CF$_2$CF$_3$) ppm).

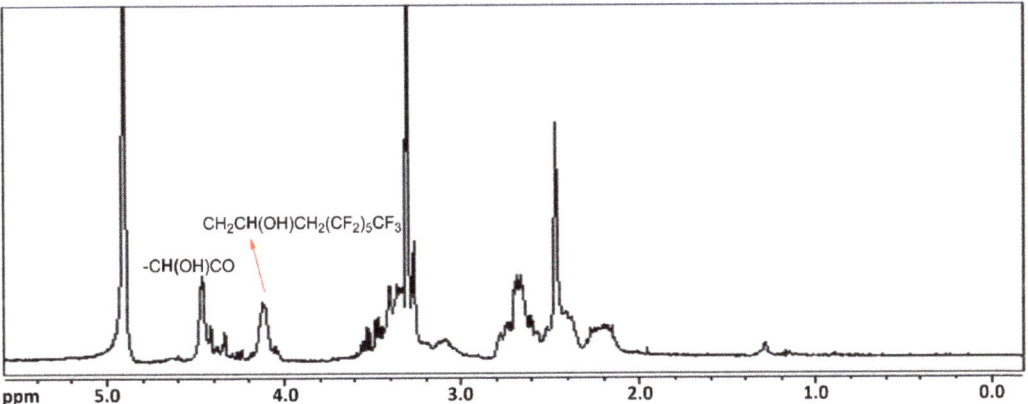

Figure 7. ^{1}H NMR spectrum in CD$_3$OD of oligoamide ESTF.

The presence of succinic units improved the solubility of the final product and it therefore seemed interesting to synthesize the fluorinated derivative of oligoethylenesuccinamide to evaluate the effect of structural variations from an applicative point of view.

The molar ratio between the reagents and the reaction conditions were the same as those used for the ESTF synthesis, but pure 2-propanol was used as a solvent. In fact, the solubility of the starting oligoamides in alcohol increases when the ratio of alkyl groups increases in its structures and when tartaric units are not present. The fluoro-oligoamide ESF was obtained as a white solid with an 89% yield.

The signals in the ^{1}H, ^{13}C, ^{19}F NMR, and gCOSY spectra confirm the presence of fluorinated groups bonded to oligoamide. However, the overlap between numerous signals does not allow an evaluation of the degree of functionalization for this compound, although also in this case, the presence of the signal at 4.11 ppm and the considerations reported on the characterization of the previous fluorinated products (ESTF and DF) allows one to hypothesize the presence of amino groups with two substituent groups and therefore a high degree of functionalization.

3.3. Treatments Evaluation

In this study, a specific protocol was developed to evaluate the efficacy of the synthesized products as hydrophobic agents and to evaluate the usefulness of their use on a highly hydrophilic and inhomogeneous material such as wood. To define the tests for the protocol, the need to obtain preliminary information to optimize the design of a suitable protective was also taken into consideration. However, based on the results obtained in this preliminary phase of the research, it will have to be optimized.

Diagnostic protocols for the evaluation of wood treatment generally concern the field of painted artefacts or preservative products, and are only partially regulated. However, this study aims to evaluate the behavior of materials capable of minimally altering the aesthetic appearance of the wood, but modifying its wettability and moisture content (mainly vapor absorbed to reach equilibrium with environmental humidity). Therefore, this specific diagnostic protocol is intended to be applied to those products that assume high importance in the conservation of materials of historical and artistic interest, such as the compounds developed in this work.

To the best of our knowledge, research in this area has not provided a very extensive previous background and few studies or patents cover this field of interest [8–12,35].

For sake of clarity, we remember that wood is a highly inhomogeneous material whose physical properties change anisotropically and its aesthetic appearance, even for the same wood species, is also affected by its different origin. The properties of wood are also influenced by manufacturing processes and environmental conditions. Consequently, the choice of the wood species, the type of cut and surface processing, and the presence or absence of concentrated extractives on the surface are the parameters that must firstly be taken into consideration for the selection criteria of the samples for the tests. The weight (moisture content variation) and other chemical–physical characteristics are finally influenced by the ambient humidity with which the wood is quickly balanced and a monitor of the behavior as a function of different RH values is also required.

Therefore, to reduce errors in the different measurement tests, and to understand the influence of some of the peculiarities of the wood, we decided to test two types of wood, beech and oak, with the same type of cut, two types of processing (planed and polished), and virgin and pre-extracted samples (i.e., wood extracted in ethanol for the removal of extractives).

On the basis of the data on the solubility in 2-propanol, the oligoamides ETAF, ESTF, ESF, and DF were selected for wood treatments.

The tested products, as expected and desired, did not show film-forming behavior, but they could vary the aesthetic appearance of the wood from a colorimetric point of view. The surface color variations, performed on the same position of each specimen before and after the treatment and expressed as ΔL^*, Δa^*, Δb^*, and ΔE, are reported in Table 5. Only for the samples treated with ESF was the chromatic variation below the detection limit of the human eye ($\Delta E = 3$). For the other samples, appreciable chromatic variations mainly influenced by the L^* parameter were observed (Table 5). However, it should be noted that the aesthetic appearance of the wood is not significantly altered compared to the typical and variable color of each species for all tested products, as can be seen in Figure 8, where the samples treated with ESF, DF, or ESTF are compared with the wood without a coating as a reference.

Table 5. Color variations of wood surfaces after coating.

Fluorinated Products	ΔL*	Δa*	Δb*	ΔE
ETF [1]	−5.26 ± 0.02	1.94 ± 0.02	4.23 ± 0.03	7.02
DTF [1]	−1.74 ± 0.62	1.83 ± 0.10	3.67 ± 0.15	4.46
ESTF [2]	7.88 ± 0.14	2.88 ± 0.04	3.77 ± 0.05	9.20
ESF [2]	2.28 ± 0.21	0.63 ± 0.13	1.67 ± 0.04	2.90
DF [2]	9.57 ± 0.63	4.25 ± 0.25	4.6 ± 0.1	11.46
DF [1]	−2.81 ± 0.54	1.29 ± 0.08	5.69 ± 0.14	6.48

[1] beech sample; [2] oak sample.

Figure 8. Oak samples after treatment with ESF, DF, or ESTF compared with the wood without coating as reference samples. (**a**) Virgin samples. (**b**) Samples pre-extracted in ethanol.

The hydrophobic properties of the synthesized compounds were evaluated by detecting the wettability change of the coated wood surface through contact angle (CA) measurements. However, due to the inhomogeneity of wood, as previously mentioned, a great variability of the CA values is expected, even on different areas of the same surface. For this reason, to estimate the degree of variability, preliminary measurements of CA were carried out on both untreated (virgin) and ethanol-extracted (for the removal of extractives) wood samples. As expected, the preliminary measurements obtained on different positions of the same surface of each sample showed a great variability. In fact, the contact angle of the uncoated wood, evaluated after one second from the deposition of the water drop, varied between 50° and 105°, whether the measurement was performed in different positions of the same sample or on different samples of the same wood species. After 5 s from the deposition of the water drop, the CA becomes generally non-evaluable. These results confirm the need to adopt the strategy of always measuring the contact angle in the same area (before and after the coating) to obtain reliable data on the hydrophobic effect of the protective agent. Moreover, to provide statistical data, at each step of the test (i.e., before coating, after coating, or after aging), it is necessary to repeat the CA measurement several times, in different moments and after the complete evaporation of the previous drop.

Furthermore, the wood extractives could influence the CA values. In fact, the extraction treatment with ethyl alcohol decreased the natural hydrophobicity of wood attributable to the presence of extractives on the surface. For a reference oak sample, the contact angle changed from 105.9° ± 1.2° to 84.5° ± 0.9°, after extraction in ethanol. As these compounds are soluble in alcohols, their dissolution and migration during the application of the alcoholic solution of the coating is expected. For this reason, some coatings were also applied to oak wood after extraction with alcohol. Similar to the extraction treatment, a decrease in the contact angle can be observed after abrasion of the surface (59.2° ± 0.1°), while the simultaneous abrasion and extraction with alcohol determines a contrasting effect of the two actions on wettability (69.3° ± 1.1°). In fact, the abrasion of the wood surface likely removes the extractives concentrated on the surface, while it can expose a new surface where the extractives were not completely removed during the extraction process. Furthermore, abrasion with varying roughness can influence the hydrophobic behavior of the surfaces. Finally, it is necessary to consider the natural variability of the wood when different samples are used as a reference. In all cases, after extraction with alcohol and/or abrasion, the values of the contact angles are considerably lower than those of raw wood and always less than 90°.

Based on these observations, the comparison of the performance between different products was carried out in the following conditions: (a) by using wood samples with or without the preliminary extraction with ethanol; (b) by evaluating the CA several times, over time, in a well-defined position of each sample after 1 and 5 s. In the end, a final evaluation of the CA was carried out after 8 min, although this measurement could be affected by the minimal evaporation of water. The results are reported in Table 6.

Table 6. Water contact angle (ϑ_c, °).

Fluorinated Products	Pretreatment	After 1 s	After 5 s	After 8 min
Beech (reference sample)	None	66.3 ± 1.2	n.e. [3]	n.e. [3]
	Ethanol extraction	59.9 ± 4.3	n.e. [3]	n.e. [3]
Oak (reference sample)	None	105.9 ± 1.2	75.4 ± 0.2	n.e. [3]
	Ethanol extraction	84.5 ± 0.9	n.e. [3]	n.e. [3]
ETF [1]	None	89.8 ± 2.6	64.1 ± 1.5	n.e. [3]
DTF [1]	None	78.2 ± 0.3	53.1 ± 1.8	n.e. [3]
ESTF [2]	None	82.4 ± 0.5	54.2 ± 0.2	n.e. [3]
	Ethanol extraction	76.4 ± 0.4	63.7 ± 0.6	n.e. [3]
ESF [2]	None	138.2 ± 1.6	132.6 ± 1.2	87.8 ± 0.5
	Ethanol extraction	132.4 ± 1.5	126.1 ± 1.4	90.0 ± 0.8
DF [1]	None	130.4 ± 1.8	114.9 ± 0.8	99.6 ± 1.5
DF [2]	None	132.2 ± 1.4	119.3 ± 0.2	104.7 ± 1.1
	Ethanol extraction	121.2 ± 0.4	112.4 ± 0.3	87.8 ± 0.3

[1] beech sample; [2] oak sample. [3] n.e.: not evaluable.

Treatment with fluorinated oligoamides containing tartaric unit show low CA values, always lower than 90° and generally lower than the initial value of the wood sample before treatment. This result highlights a negative effect of the presence of hydroxyl groups on the oligoamide chain, which can be attributed to a concomitant hydrophilic effect or to a greater ability to penetrate into the surface layers of the wood. In the latter case, the greater affinity with the polysaccharides constituting the wood instead of favoring the anchoring on the surface would favor its penetration and/or do not favor the distribution of fluorinated fragment towards the outside. The use of 2-propanol as a solvent for the deposition can also act on the surface of the wood favoring the migration of the extractives if not previously removed.

On the contrary, after treatment with ESF and DF, the wood samples show higher contact angle values, with a greater increase in the contact angle for the samples pre-treated with ethanol. The best hydrophobicity was founded in the sample without pre-treatment with ethanol and treated with ESF (138.2°). The presence of a non-functionalized NH group in the DF compound can favor anchoring to the support, but could reduce the hydrophobicity deriving from the presence of the fluorinated chain. However, the values obtained without pre-extraction in ethanol are around 132°. It is interesting to observe the comparison of these data with the contact angle observed in the presence of a commercial product such as Paraloid B72, a product widely used in the Cultural Heritage field. The application of equal amounts of product under the same conditions used for the fluorinated products shows a contact angle of (66 ± 4)° on wood as it is, and (63 ± 4)° on wood pre-extracted with ethanol. With this acrylic product, the color variation is relatively low (ΔE respectively 2.3 and 4.3), but the poor stability of the product reduces the interest in acrylic products.

After 5 s from the deposition of the water drop, a significant change in the contact angle is observed for samples with hydrophilic behavior ($\vartheta c < 90°$) compared to those with initial $\vartheta c > 90°$, in agreement with the hydrophobic properties and a reduced ease of water absorption of the wood specimens coated with ESF and DF. Finally, the contact angle was assessed after 8 min, but it was only possible for the treatments with ESF and DF.

The water vapor absorption test was performed using oak wood samples treated with ESTF, ESF, and DF on all surfaces with a suitable quantity of product to completely cover them (40 g/m^2). All of the samples were preliminarily extracted in ethanol, and after coating, they were kept in a desiccator for 72 h until a constant weight was reached. The weight variations observed by keeping the coated samples in the presence of progressively increasing humidity values (RH 65%, 86%, and 100%) were compared with those obtained in similar uncoated samples (one raw and one pre-extracted). For all humidity values, high water absorption is observed in the initial phase, which progressively decreases to zero in correspondence with the reached equilibrium, with a higher absorption rate on anhydrous samples (uncoated samples) (Figure 9). The mass of water absorbed per unit mass of the specimen is lower for the samples coated with the fluorinated products, in the order ESF < DF < ESTF, compared to the one observed in the uncoated wood, with or without pre-extraction in ethanol (Table 7 and Figure 10).

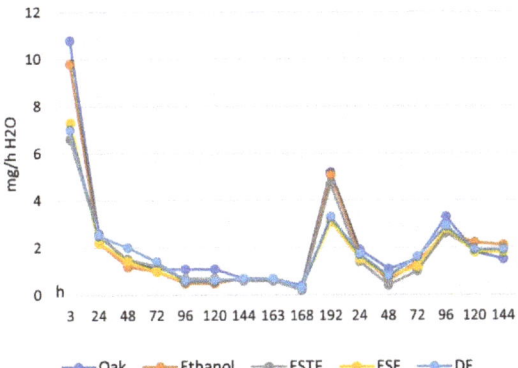

Figure 9. Variation of water vapor absorption over time at different RH values for oak sample, oak sample after ethanol extraction, and oak samples after coating with ESTF, ESF, or DF.

Table 7. Total mass of water absorbed at different RH values and mass per unit of weight of the sample (%).

Oak Samples	RH 65%		RH 86%		RH 100%	
	mH$_2$O (mg)	Δm/m(%) [1]	mH$_2$O (mg)	Δm/m(%) [1]	mH$_2$O (mg)	Δm/m(%) [1]
Without coating	290	4.7	461	7.3	769	11.7
After extraction in ethanol	277	4.5	438	6.9	736	11.1
ESTF coating [2]	262	4.3	411	6.5	678	10.4
ESF coating [2]	257	4.1	410	6.4	680	10.2
DF coating [2]	261	4.2	417	6.6	671	10.2

[1] Mass of water absorbed per unit mass of the specimen (%). [2] Oak pre-treated with ethanol.

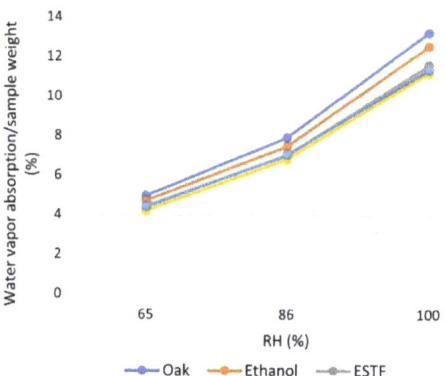

Figure 10. Water vapor absorption per unit of weight (%) over time at different RH values for oak sample, oak sample after ethanol extraction, and oak samples after coating with ESTF, ESF, or DF.

The increase in the surface wettability of wood (i.e., lower CA, Table 6) and the reduction in vapor absorption (i.e., lower mass of vapor absorbed, Table 7) for the specimens extracted in ethanol, compared to the raw wood, can be explained with the removal of saccharide components and other compounds. The extraction in ethanol of saccharide components of different complexity is confirmed both by ^{1}H-NMR (Figure S37), for the presence of the signals attributable to the saccharide skeleton at 3.40–4.00 ppm, and by FT-IR (Figure S38), for the signals at 3390 (O–H stretching), 1113, 1076, 1049 cm^{-1} (C–O stretching). Low-intensity signals at 2927 (C–H stretching) and 1732 cm^{-1} (C=O stretching), attributable to the presence of triglycerides and/or free fatty acids, were also present in the FT-IR spectrum. A different mobility of the various extractive components can be speculated, which can determine a lower concentration of hydrophobic components (e.g., triglycerides, free fatty acids) on the surface after extraction with ethanol, while the saccharide components are probably removed mainly from the wood mass, thus reducing its affinity towards water.

Finally, the stability of the synthetic partially fluorinated oligoamides and amine, deposed on glass slides, was estimated by means of accelerated aging tests exploiting UV radiations (wavelength of 365 nm). Their behavior was followed via ^{1}H NMR, FT-IR spectroscopy and colorimetric measurements. The weight change was also recorded, in order to monitor the decomposition or oxidation (mainly absorption of oxygen) reactions.

The colorimetric measurements were made before and after irradiation to evaluate the color variations due to the interaction between the partially perfluorinated compounds and the UV radiation. No appreciable weight changes were noted for any product. The results of the colorimetric measurements on UV irradiated glass slices are shown in the Table 8.

Table 8. Color changes after UV aging on glass slide after 254 h of irradiation.

Fluorinated Products	ΔL^*	Δa^*	Δb^*	ΔE
ESTF	−0.62 ± 0.23	−0.23 ± 0.02	0.85 ± 0.06	1.07
ESF	0.06 ± 0.19	−0.20 ± 0.02	−0.95 ± 0.08	0.97
DF	−1.27 ± 0.23	−2.27 ± 0.04	8.76 ± 0.14	9.14

DF, which contains diethylenetriamine units, shows the highest ΔE^* (9.14) after 254 h of irradiation among the three irradiation products tested, and is much higher than the threshold limit imperceptible to the human eye ($\Delta E^* = 3$). In particular, DF has a significant increase in Δb^*, which means the samples turned yellow under UV radiations. On the contrary, both ESF and ESTF, which contain ethylenediamine units, show ΔE^* and Δb^* values around 1.

The FT-IR and the ^{1}H NMR spectra of ESF after irradiation do not show changes, while the FT-IR of the irradiated ESTF displays slight shifts of the signals at about 3300 cm^{-1} (–OH and –NH stretching) and 1550 cm^{-1} (–NH bending, Amide II) compared to the non-irradiated ESTF. This modification may be justified with hydrogen bonding involving mainly –OH and –NH groups. ESTF did not show a significant yellowing after irradiation; however, in the ^{1}H NMR spectrum, the reduction of signal intensity to 4.11 ppm and of signals between 2.6 and 3.0 ppm were observed (Figure 11). These variations, as for the FT-IR results, are in agreement with a reduction in solubility due to hydrogen bonds between the OH and NH groups present in the compound.

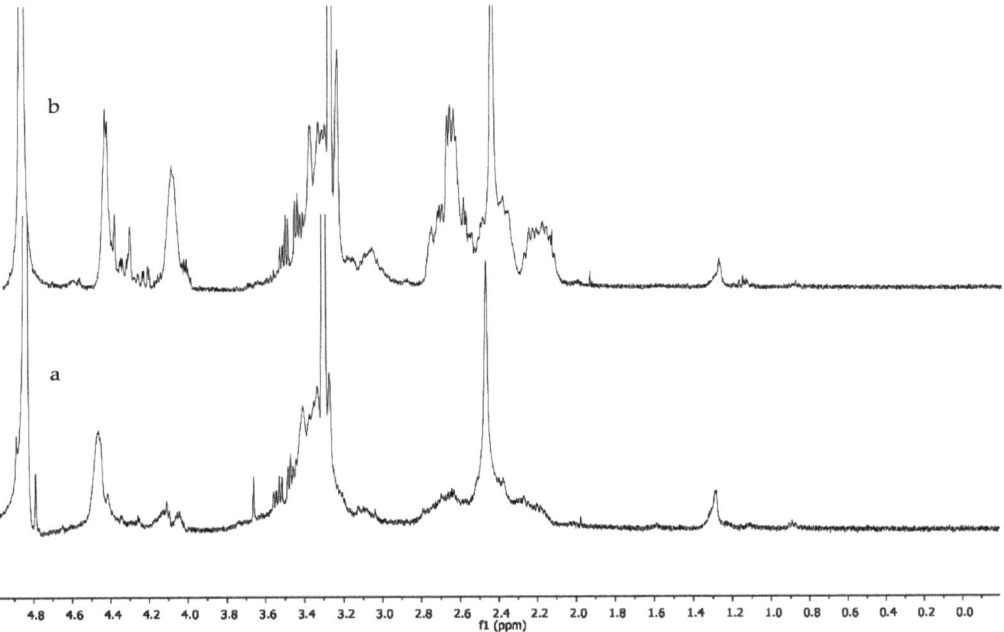

Figure 11. ^{1}H NMR spectra of ESTF after (**a**) and before (**b**) aging.

The FT-IR spectrum of irradiated DF (Figure S39) shows changes mainly in the region 1600–1700 cm^{-1}, probably attributable to a partial oxidation on the triamine units. In the ^{1}H NMR spectrum (Figure 12), the variation of the shape and relative intensity of the signals of CHNH in the 2–3 ppm zone can be observed. In particular, a new signal at 3.0 ppm appeared and the shape of the original CHNH area also changed.

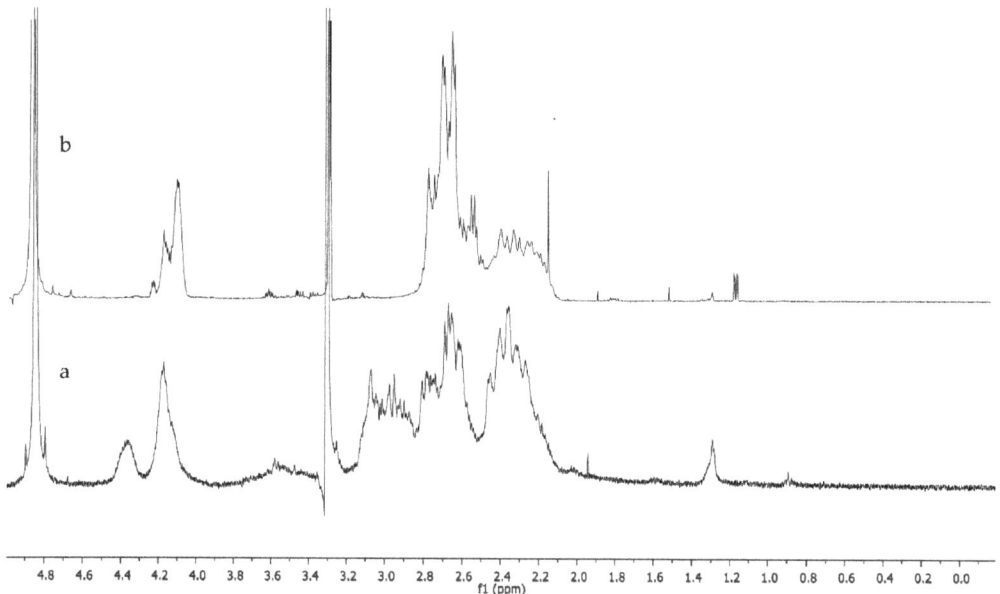

Figure 12. ^{1}H NMR spectra of DF after (**a**) and before (**b**) aging.

4. Conclusions

New fluorinated oligoamides and a diethylenetriamine derivative were successfully synthesized using 3-perfluorohexyl-1,2-epoxypropane. A shorter length of the fluorinated segment, compared to similar compounds previously studied as protective agents for stone, was used to minimize the risk of bioaccumulation with a consequent reduced impact on the environment. The influence of different monomer units, deriving from diacids and polyamines of different complexity, on reactivity, solubility, and hydrophobic properties was studied by comparing different oligoamides. The effectiveness of the new products as protective agents for wood was studied by developing a diagnostic protocol capable of rapidly providing information on some specific properties given to wood surfaces (i.e., hydrophobicity, water vapor uptake, color) in order to refine the design of new tailored materials.

All new synthesized compounds, soluble in 2-propanol, were tested as wood protective coatings, analyzing the variations in color and contact angle of the water drop. The best results were obtained with ESF, while DF, which also shows good hydrophobicity, provides a visible colorimetric variation, even if aesthetically acceptable for a wooden component.

On the contrary, the oligoamides containing the tartaric units show low contact angles. However, if the water drop is monitored on time, the value of the CAs decreased more slowly than that calculated on wood without treatment, which is in agreement with a reduced absorption capacity.

The absorption of water vapor in controlled humidity conditions in the presence of the protective coatings shows a reduction both in the speed of reaching equilibrium and in the mass of water absorbed with an efficiency that improves in the order ESTF < DF < ESF.

The comparison of the behavior of the wood with and without pre-extraction in ethanol highlighted the difficulty of applying a diagnostic protocol due to the heterogeneity of the samples. Reliable results were obtained by introducing some targeted changes to the protocol to favor the optimization of the measurements. The different type of wood (oak or beech) does not seem to influence the protective efficacy, which is, instead, influenced by the presence of extractives on the surface. However, the color variations are influenced not only by the chosen wood, but also by the area from which the samples were obtained from the same piece of wood. Aging by UV irradiation allowed the highlighting of greater stability for the ESF compound, which is confirmed as the best protective agent for wood. ESTF and DF showed lower photostability: indeed, ESTF reduced the solubility, while DF turned yellow.

Supplementary Materials: The following supporting information can be downloaded at: https://www.mdpi.com/article/10.3390/coatings12070927/s1, Synthesis of oligoethylene-L-tartaramide (ET) and its fluorinated derivative (ETF); Synthesis of oligodiethylenetriamino-L-tartaramide (DT) and its fluorinated derivative (DTF); Synthesis of oligoethylene-adipamide-L-tartaramide (ETA) and its fluorinated derivative (ETAF); Figure S1: ^{1}H NMR spectrum in CD_3OD of diol obtained from the reaction of EC6F with water; Figure S2: ^{1}H NMR spectrum in CD_3OD of EC6F; Figure S3: ^{1}H NMR spectrum in CD_3OD of DF; Figure S4: gCOSY NMR spectrum in CD_3OD of DF; Figure S5: ^{19}F NMR spectrum in CD_3OD of DF; Figure S6: ^{13}C NMR spectrum in CD_3OD of DF; Figure S7: FT IR spectrum of ET; Figure S8: ^{1}H NMR spectrum in D_2O of ET; Figure S9: ^{13}C NMR spectrum in D_2O of ET; Figure S10: ^{1}H NMR spectrum in CD_3OD of ETF (methanol-soluble fraction); Figure S11: ^{19}F NMR spectrum in CD_3OD of ETF; Figure S12: FT IR spectrum of DT; Figure S13: ^{1}H NMR spectrum in D_2O of DT; Figure S14: ^{13}C NMR spectrum in D_2O of DT; Figure S15: ^{1}H NMR spectrum in CD_3OD D of DTF; Figure S16: gCOSY NMR spectrum in CD_3OD of DTF; Figure S17: ^{19}F NMR spectrum in CD_3OD of DTF; Figure S18: FT IR spectrum of ETA; Figure S19: ^{1}H NMR spectrum in D_2O of ETA; Figure S20: ^{13}C NMR spectrum in D_2O of ETA; Figure S21: ^{1}H NMR spectrum in CD_3OD of ETAF (ethanol-soluble fraction); Figure S22: gCOSY NMR spectrum in CD_3OD of ETAF (ethanol-soluble fraction); Figure S23: ^{19}F NMR spectrum in CD_3OD of ETAF; Figure S24: ^{1}H NMR spectrum in D_2O of EST; Figure S25: ^{13}C NMR spectrum in D_2O of EST; Figure S26: ^{1}H NMR spectrum in CD_3OD of ESTF; Figure S27: gcosy NMR spectrum in CD_3OD of ESTF; Figure S28: ^{19}F NMR spectrum in CD_3OD of ESTF; Figure S29: ^{13}C NMR spectrum in CD_3OD of ESTF; Figure S30: ^{1}H NMR spectrum in D_2O of ES; Figure S31: ^{13}C NMR spectrum in D_2O of ES; Figure S32: ^{1}H NMR spectrum in CD_3OD of ESF; Figure S33: gCOSY NMR spectrum in CD_3OD of ESF; Figure S34: ^{19}F NMR spectrum in CD_3OD of ESF; Figure S35: ^{13}C NMR spectrum in CD_3OD of ESF; Figure S36: gHSQC NMR spectrum in CD_3OD of ESF; Figure S37: ^{1}H NMR spectrum in CD_3OD of the products extracted in ethanol from oak samples; Figure S38: FT-IR spectrum in CD_3OD of the products extracted in ethanol from oak samples; Figure S39: FT-IR spectra of DF before (black) and after (red) UV aging; Figure S40: ^{1}H NMR spectra of DF before (up) and after (down) aging; Figure S41: FT-IR spectra of ESTF before (black) and after (red) UV aging; Figure S42: ^{1}H NMR spectra of ESTF before (down) and after (up) aging; Figure S43: FT-IR spectra of ESF before (black) and after (red) UV aging; Figure S44: ^{1}H NMR spectra of ESF before (down) and after (up) aging.

Author Contributions: Conceptualization, A.S. and M.C.; methodology, A.S. and M.C.; formal analysis, Y.Z., L.B. and L.V.; investigation, Y.Z., L.B., L.V. and M.G.B.; resources, A.S., N.L.-G. and M.C.; data curation, A.S., M.C., Y.Z. and L.V.; writing—original draft preparation, A.S. and M.C.; writing—review and editing, A.S., L.V., Y.Z., N.L.-G. and M.C.; supervision, A.S., M.C. and N.L.-G.; project administration, A.S., M.C. and N.L.-G.; funding acquisition, A.S., M.C and N.L.-G. All authors have read and agreed to the published version of the manuscript.

Funding: This research received no external funding.

Institutional Review Board Statement: Not applicable.

Informed Consent Statement: Not applicable.

Data Availability Statement: Not applicable.

Acknowledgments: We thank MIUR-Italy ("Progetto Dipartimenti di Eccellenza 2018–2022" for the funds allocated to the Department of Chemistry "Ugo Schiff") and Laboratorio Congiunto VALORE and REGIONE TOSCANA (PAR FAS 2007–2013 projects).The authors thank Benedetto Pizzo for supplying wood samples.

Conflicts of Interest: The authors declare no conflict of interest.

References

1. Unger, A.; Schniewind, A.P.; Unger, W. *Conservation of Wood Artefacts*; Springer: Berlin/Heidelberg, Germany, 2001.
2. Hunt, D. Properties of wood in the conservation of historical wooden artifacts. *J. Cult. Herit.* **2012**, *13S*, S10–S15. [CrossRef]
3. Evans, P.D.; Banks, W.B. Degradation of wood surfaces by water: Weight losses and changes in ultrastructural and chemical composition. *Holz Roh- Werkstoff* **1990**, *48*, 159–163. [CrossRef]
4. Reinprecht, L. Biological Degradation of Wood. In *Wood Deterioration, Protection and Maintenance*; John Wiley & Sons: Hoboken, NJ, USA, 2016.
5. Kalita, K.; Boruah, P.K.; Sarma, U. Studies on change of strain developed in different wood samples due to change in relative humidity. *Sens. Bio-Sens. Res.* **2019**, *22*, 100264. [CrossRef]
6. Freeman, M.H.; Shupe, T.F.; Vlosky, R.P.; Barnes, H.M. Past, present and future of the wood preservation industry. *For. Prod. J.* **2003**, *53*, 8–15.
7. Mai, C.; Militz, H. Modification of wood with silicon compounds. Treatment systems based on organic silicon compounds—A review. *Wood Sci. Technol.* **2004**, *37*, 453–461. [CrossRef]
8. Chen, Y.; Wang, H.; Yao, Q.; Fan, B.; Wang, C.; Xiong, Y.; Jin, C.; Sun, Q. Biomimetic taro leaf-like films decorated on wood surfaces using soft lithography for superparamagnetic and superhydrophobic performance. *J. Mater. Sci.* **2017**, *52*, 7428–7438. [CrossRef]
9. *CN1883897B*; Wood Protective Agent. Jingjiang Guolin Wood Co., Ltd.: Jingjiang, China, 2010.
10. Lu, X.; Hu, Y. Layer-by-layer deposition of TiO_2 nanoparticles in the wood surface and its superhydrophobic performance. *Bioresources* **2016**, *11*, 4605–4620. [CrossRef]
11. Papadopoulos, A.N.; Taghiyari, H.R. Innovative Wood Surface Treatments Based on Nanotechnology. *Coatings* **2019**, *9*, 866. [CrossRef]
12. Wang, Z.; Shen, X.; Qian, T.; Wang, J.; Sun, Q.; Jin, C. Facile Fabrication of a PDMS@Stearic Acid-Kaolin Coating on Lignocellulose Composites with Superhydrophobicity and Flame Retardancy. *Materials* **2018**, *11*, 727. [CrossRef]
13. Charola, A.E. Water Repellents Treatments for building stones: A practical overview. *Apt. Bull.* **1995**, *26*, 10–17. [CrossRef]
14. Tabasso, L.M. Acrylic polymers for the conservation of stone: Advantages and drawbacks. *APT Bull.* **1995**, *26*, 17–21. [CrossRef]
15. Chiantore, O.; Lazzari, M. Photo-oxidative stability of paraloid acrylic protective polymers. *Polymer* **2001**, *42*, 17–27. [CrossRef]
16. Melo, M.J.; Bracci, S.; Camaiti, M.; Chiantore, O.; Piacenti, F. Photodegradation of acrylic resins used in the conservation of stone. *Polym. Degrad. Stab.* **1999**, *66*, 23–30. [CrossRef]
17. Charola, A.E.; Tucci, A.; Koestler, R.J. On the Reversibility of treatments with acrylic/silicone resin mixtures. *JAIC* **1986**, *25*, 83–92. [CrossRef]
18. Sèbe, G.; Brook, M.A. Hydrophobization of wood surfaces: Covalent grafting of silicone polymers. *Wood Sci. Technol.* **2001**, *35*, 269–282. [CrossRef]
19. Charola, E. Water Repellents and Other "Protective" Treatments: A Critical Review. In Proceedings of the 3rd International Conference on Surface Technology with Water Repellent Agents, Hannover, Germany, 25–26 September 2001; Aedificatio Publishers: Freiburg, Germany, 2001; pp. 3–20.
20. Evans, P.D.; Michell, A.J.; Schmalzl, K.J. Studies of the degradation and protection of wood surfaces. *Wood Sci. Technol.* **1992**, *26*, 151–163. [CrossRef]
21. Camaiti, M.; Brizi, L.; Bortolotti, V.; Papacchini, A.; Salvini, A.; Fantazzini, P. An environmental friendly fluorinated oligoamide for producing nonwetting coatings with high performance on porous surfaces. *ACS Appl. Mater. Interfaces* **2017**, *9*, 37279–37288. [CrossRef]
22. Cao, Y.; Salvini, A.; Camaiti, M. Facile design of "sticky" near superamphiphobic surfaces on highly porous substrate. *Mater. Des.* **2018**, *153*, 139–152. [CrossRef]
23. Camaiti, M.; Bortolotti, V.; Cao, Y.; Papacchini, A.; Salvini, A.; Brizi, L. High Efficiency Fluorinated Oligo(ethylenesuccinamide) Coating for Stone. *Coatings* **2021**, *11*, 452. [CrossRef]
24. Cipriani, G.; Salvini, A.; Fioravanti, M.; Di Giulio, G.; Malavolti, M. Synthesis of hydroxylated oligoamides for their use in wood conservation. *J. Appl. Polym. Sci.* **2013**, *127*, 420–431. [CrossRef]
25. Oliva, R.; Ortenzi, M.A.; Salvini, A.; Papacchini, A.; Giomi, D. One-pot oligoamides syntheses from L-lysine and L-tartaric acid. *RSC Adv.* **2017**, *7*, 12054–12062. [CrossRef]
26. Oliva, R.; Albanese, F.; Cipriani, G.; Ridi, F.; Giomi, D.; Malavolti, M.; Bernini, L.; Salvini, A. Water soluble trehalose-derived oligoamides. *J. Polym. Res.* **2014**, *21*, 496. [CrossRef]
27. Conder, J.M.; Hoke, R.A.; De Wolf, W.; Russell, M.H.; Buck, R.C. Are PFCAs bioaccumulative? A critical review and comparison with regulatory criteria and persistent lipophilic compounds. *Environ. Sci. Technol.* **2008**, *42*, 995–1003. [CrossRef]
28. *EN ISO 483-2005*; Plastics—Small Enclosures for Conditioning and Testing Using Aqueous Solutions to Maintain the Humidity at a Constant Value. ISO: Geneva, Switzerland, 2005.
29. *EN 13183-1:2002*; Moisture Content of a Piece of Sawn Timber—Part 1: Determination by oven dry method. European Committee for Standardization. CEN: Brussels, Belgium, 2002.
30. *ASTM D4442-20*; Standard Test Methods for Direct Moisture Content Measurement of Wood and Wood-Based Materials. ASTM International: West Conshohocken, PA, USA, 2020.
31. *UNI-EN 15886-2010*; Conservation of Cultural Heritage-Test Methods-Measurement of the Color of the Surface. UNI: Milan, Italy, 2010.

32. Ogata, N.; Sanui, K.; Ohtake, T.; Nakamura, H. Solution polycondansation of diesters and diamines having hetero atom groups in polar solvents. *Polym. J.* **1979**, *11*, 827–833. [CrossRef]
33. Hoagland, P.D. The formation of intermediate lactones during aminolysis of diethyl galactarate. *Carbohydr. Res.* **1981**, *98*, 203–208. [CrossRef]
34. Kiely, D.E.; Chen, L.; Lin, T.H. Hydroxylated nylons based on unprotected esterified D-glucaric acid by simple condensation reactions. *J. Am. Chem. Soc.* **1994**, *116*, 571–578. [CrossRef]
35. Szubert, K.; Dutkiewicz, A.; Dutkiewicz, M.; Maciejewski, H. Wood protective coatings based on fluorocarbosilane. *Cellulose* **2019**, *26*, 9853–9861. [CrossRef]

Article

PPy-Modified Prussian Blue Cathode Materials for Low-Cost and Cycling-Stable Aqueous Zinc-Based Hybrid Battery

Yushan Ruan [1], Lineng Chen [1,*], Lianmeng Cui [1] and Qinyou An [1,2,*]

[1] State Key Laboratory of Advanced Technology for Materials Synthesis and Processing, International School of Materials Science and Engineering, Wuhan University of Technology, Wuhan 430070, China; asura2015@whut.edu.cn (Y.R.); cuilianmeng0530@whut.edu.cn (L.C.)

[2] Foshan Xianhu Laboratory of the Advanced Energy Science and Technology Guangdong Laboratory, Xianhu Hydrogen Valley, Foshan 528200, China

* Correspondence: clnsmiles@whut.edu.cn (L.C.); anqinyou86@whut.edu.cn (Q.A.)

Abstract: Prussian blue analogs are promising cathode materials in aqueous ion batteries that have attracted increasing attention, but their low specific capacity and limited cycling stability remain to be further improved. Effective strategies to optimize the electrochemical performance of Prussian blue cathode materials are the aspects of electrolyte and structure modification. In this work, $Na_2MnFe(CN)_6$@PPy nanocubes were prepared by a simple co-precipitation method with PPy coating. Compared with the uncoated electrode material, the discharged capacity of the $Na_2MnFe(CN)_6$@PPy cathode material is raised from 25.2 to 55.0 mAh g^{-1} after 10 cycles in the Na-Zn hybrid electrolyte, while the capacity retention is improved from 63.5% to 86.5% after 150 cycles, indicating higher capacity and better stability. This work also investigates the electrochemical performances of $Na_2MnFe(CN)_6$@PPy cathode material in hybrid electrolyte of Li-Zn and K-Zn adjusted via different mixed ion solutions. The relevant results provide an innovative way to optimize advanced aqueous hybrid batteries from the perspective of cycling stability.

Keywords: Prussian blue analogues; cathode materials; aqueous zinc-ion batteries; hybrid ion electrolyte

Citation: Ruan, Y.; Chen, L.; Cui, L.; An, Q. PPy-Modified Prussian Blue Cathode Materials for Low-Cost and Cycling-Stable Aqueous Zinc-Based Hybrid Battery. *Coatings* **2022**, *12*, 779. https://doi.org/10.3390/coatings12060779

Academic Editor: Alexandru Enesca

Received: 10 May 2022
Accepted: 2 June 2022
Published: 5 June 2022

Publisher's Note: MDPI stays neutral with regard to jurisdictional claims in published maps and institutional affiliations.

Copyright: © 2022 by the authors. Licensee MDPI, Basel, Switzerland. This article is an open access article distributed under the terms and conditions of the Creative Commons Attribution (CC BY) license (https://creativecommons.org/licenses/by/4.0/).

1. Introduction

As a new type of secondary energy storage battery, aqueous zinc-ion battery (AZIB) has the advantages of environmental friendliness, high safety, non-toxicity, etc., which suggests rather broad application prospects. AZIBs can be assembled in air, which is convenient and efficient compared to lithium-ion batteries, which must be assembled in a glove box [1–3]; thus, it is considered to be a promising alternative to lithium-ion batteries, which have been incorporated widely in portable electronic devices. Zinc metal anode has low redox potential (−0.76 V vs. standard hydrogen electrode), high-quality specific capacity (820 mAh g^{-1}) [4,5], etc. Combined with high electrical conductivity (5.91 µΩ cm) and fast reaction kinetics of aqueous electrolyte, AZIBs exhibit advantages such as high specific capacity, high safety, low cost, and long cycle life, making them gradually become a hot research point in aqueous battery systems [6,7]. However, AZIBs still face great challenges, especially the cathode materials in AZIBs which show irresistible fading of capacity in long-cycle tests, and the study of ion storage mechanisms still remains to be further investigated. The current research on the cathode materials of AZIBs can be broadly classified as the following: vanadium-based cathode materials, manganese-based cathode materials, organics cathode materials, and Prussian blue analogue (PBA) cathode materials [8–13].

Compared with other cathode materials of rechargeable AZIBs, the biggest advantage of PBA material lies in its high working voltage platform, even up to 1.7 V [14,15]. However, PBA materials are prone to suffer complex phase transformation in the reaction process, and unstable structure will lead to collapse of the framework, leading to low specific capacity and poor cycling performance. For example, Svensso et al. found that CuHCF electrode material is prone to irregular structural damage during battery cycle. The specific capacities of CuFe(CN)$_6$ electrode materials is only 53 mAh g^{-1} at 60 mA g^{-1} [16]. Munseok et al. demonstrated nickel hexacyanoferrate K$_{0.86}$Ni[Fe(CN)$_6$]$_{0.954}$(H$_2$O)$_{0.766}$ as the cathode material which shows a rather low capacity of 55.6 mAh g^{-1} at 11.2 mA g^{-1}, and the coulombic efficiency exhibits a rapid decay in 20 cycles in aqueous zinc ion electrolyte [17], which is lower than vanadium and manganese cathode materials in AZIBs [4]. Therefore, it is of prime importance to improve the electrochemical performance of PBA cathode material; strategies of coating polymer materials on the surface of PBA have been reported [18,19], indicating that improvement in the electrochemical properties of PBA cathode materials can be achieved by coating or mixed methods. Conductive polymer materials have attracted attention due to their excellent electrical conductivity, for example, polypyrrole (PPy), polyamide (PI), polyethylene glycol (PEG), etc., which have been proved to be conductive skeletons of electrode materials [20]. The electronic conductivity of PPy is relatively high (up to 102 S cm^{-1}) among all conductive polymers because of the alternating conjugation of single and double bonds in the macromolecular structure, and the extra electrons in the double bond can move freely across the polymer chain. Furthermore, to eliminate the effect of the vacancies in the PBA material, researchers have used approaches of surface modification and composition optimization to reduce the lattice defects, which suggests that by coating with PPy the [Fe(CN)$_6$] vacancies in PBA material can be reduced and the electrochemical performance of the electrode is improved [21,22]. By coating with PPy, the damage to the structure can also be alleviated due to the mitigation of Mn^{2+} emissions of Mn^{II}-Fe^{III}-based PBA material, resulting in better stability of electrochemical performance [23]. Therefore, intensive research studies have been carried out to promote the electrochemical properties of PBA cathode materials by coating with PPy. Xue et al. proposed a polypyrrole-modified KHCF@PPy cathode material for potassium-ion batteries via an in situ polymerization coating method which shows better electronic conductivity and electrochemical properties [21]. Chen et al. synthesized K$_2$Mn[Fe(CN)$_6$] cathode material for aqueous zinc batteries which can achieve superior rate capability and prolonged cycle life due to the better electronic conductivity enhanced by coating with PPy [24]. Previous reports have been focusing on the effect of PPy in PBA materials in various types of ion batteries. However, the PBA cathode materials coated with PPy in aqueous hybrid batteries have not received much attention, which needs to be further investigated.

Herein, Na$_2$MnFe(CN)$_6$ (NMHFC) nanocube material was synthesized by simple co-precipitation method. Then, NMHFC sample was coated with PPy under ice bath condition and is referred to as NMHFC@PPy. The specific capacity and cyclic stability of the electrode material were improved by the coating of Ppy, of which the capacity retention after 150 cycles was enhanced in the Na-Zn hybrid battery. Electrochemical performances were further conducted via different types of electrolyte, which gives the proof of considerable promotion of cycle capacity in hybrid electrolyte. We also investigated the kinetic properties of NMHFC@PPy in the Na-Zn hybrid electrolyte using multiple scan rates of CV (Cyclic Voltammetry) tests, while the mechanism of zinc storage in the reaction was revealed by means of ex situ XRD (X-Ray Diffraction) and XPS (X-ray Photoelectron Spectroscopy). This work shares the insight into the research of enhancing the electrochemical performance of the Ppy-coated PBA cathode materials, which provides a new approach to the design of cathode materials in aqueous hybrid batteries.

2. Materials and Methods

2.1. Material Synthesis

Na$_2$MnFe(CN)$_6$ nanocubes were synthesized by a simple co-precipitation method. Typically, 2 mmol MnSO$_4$, 4 mmol K$_4$Fe(CN)$_6$, and 15 g NaCl were dissolved in 50 mL of water, respectively, denoted as solutions A, B, and C, respectively. Then, solution A and solution B were added dropwise to solution C simultaneously under magnetic stirring for 10 min. The white suspension solution obtained was maintained at room temperature for 12 h, and then the resulting precipitates were filtered and wash with deionized water and dried at 60 °C overnight. Finally, a dark gray powder was obtained and recorded as $Na_2MnFe(CN)_6$, marked as NMHFC.

Synthesis of $Na_2MnFe(CN)_6$@PPy was performed as follows: 0.0345 g NMHFC powder prepared above, 0.5 mL pyrrole, and 0.1 mmol $FeCl_3$ (applied as the initiator) were dispersed in 40 mL deionized water and stirred for 6 h at 0–5 °C to complete the reaction. The final composites obtained were washed three times with deionized water and ethanol, respectively, then dried in a vacuum oven at 60 °C for 6 h. The black powder obtained was noted as NMHFC@PPy.

2.2. Material Characterization

The crystal structure of the samples was characterized by X-ray diffraction (XRD, Bruker, MA, USA) with a D8 Advance X-ray diffractometer equipped with Cu Kα X-ray source. Field-emission scanning electron microscopy (SEM, JEOL-7100F microscope, Tokyo, Japan) was employed to analyze the morphology and elemental composition of the samples. Transmission electron microscopy (TEM, JEM-1400Plus, Tokyo, Japan) was conducted to adopt further studies on morphologies of the prepared materials. Raman spectra were obtained using a micro-Raman spectroscopy system (Raman, Renishaw INVIA, London, UK), while Fourier transform infrared spectroscopy (FTIR, Nicolet 6700, Thermo Fisher Scientific, Waltham, MA, USA) was performed to obtain the unique functional groups and molecular bonds in compound molecules. The surface elemental valence and molecular structure were analyzed by X-ray photoelectron spectroscopy (XPS, AXIS SUPRA, Kratos, Tokyo, Japan). STA449F3 (NETZSCH F3, Selb, Germany) was used to conduct the thermogravimetric analysis at a heating rate of 10 °C/min from room temperature to 800 °C under a N_2 atmosphere.

2.3. Electrochemical Measurements

The three different hybrid electrolytes were prepared by dissolving $LiCF_3SO_3$ and $Zn(CF_3SO_3)_2$; $NaCF_3SO_3$ and $Zn(CF_3SO_3)_2$; and KCF_3SO_3 and $Zn(CF_3SO_3)_2$ in distilled water, then marked as Li-Zn, Na-Zn, and K-Zn, respectively, and the concentration of all electrolytes was 1 mol L^{-1}. Electrochemical performance tests were carried out with coin batteries (CR2016) assembled in the air. The cathode consisted of active material, acetylene black, and binder (PTFE) in a weight ratio of 6:3:1. High-purity zinc foil with a thickness of 0.1 mm was directly used as anode after cutting and polishing. Glass fiber film (GF/A, Whatman) was applied as separator of the cathode and anode. Galvanostatic discharge–charge tests were conducted with the LAND battery testing system (CT2001A, Wuhan, China). Cyclic voltammetry (CV) tests were performed using a VMP3 multichannel electrochemical workstation (Bio-Logic, Grenoble, France).

3. Results and Discussion

The XRD test results of NMHFC and NMHFC@PPy nanocubes are shown in the Figure 1a. All diffraction peaks can be attributed to the cubic phase space group (FM-3m) with good crystallinity (JCPDS: 73-0687) [25,26]. The XRD patterns of coated NMHFC@PPy nanocubes are rather close to those of the uncoated ones. All of the diffraction peaks exhibit sharp shapes, while the intensity of the diffraction peaks was only weakened after coating with PPy. Compared with bare NMHFC, the PPy-modified nanocubes show no peak of other phases, since PPy on the surface is amorphous, revealing that the coating process of PPy does not change the crystal structure of the material [27]. With the intention of confirming the coating of PPy, FTIR tests of the samples before and after coating were performed (Figure 1b); according to previous literature, the characteristic peaks belonging to PPy appear at about 1559 and 1650 cm^{-1}, which can be attributed to the =C–H vibration of the pyrrole in the benzene ring and the C–N stretching vibration, respectively [20,28,29], indicating the appearance of specific vibrational IR peaks of PPy in the material, while the peaks at ~597, 1617, 2066, and 3418 cm^{-1} can be ascribed to the PBAs [30]. Therefore, NMHFC@PPy nanocubes have been successfully obtained by a chemical polymerization method. Figure S1 shows the Raman spectra of NMHFC@PPy ranging from 100 to 2000 cm^{-1}. Four peaks were observed around 110.1, 460.5, 519.5, and 582.6 cm^{-1}. These peaks reflect the bonding of the (C≡N)$^-$ ions with metal ions in different valence transition. Two peaks were observed at 110.1 cm^{-1} and 460.5 cm^{-1}, which can be attributed to the bonding vibrations of both Fe^{II}-CN-Mn^{II} and Fe^{III}-CN-Mn^{III}, respectively, since the cyanogen coordination with the lower valence Fe^{II} and Mn^{II} is higher than that with the higher valence Fe^{III} and Mn^{III} at lower wave positions [26]. From the TG profile (Figure S2), the mass loss before 200 °C is caused by the evaporation of interstitial water [31]. The water content in NMHFC@PPy is about 8.7%. Owing to the fact that the temperature of decomposition of PPy is close to the evaporation temperature of the coordinating water [32], the PPy content can only be roughly calculated to be 7.3 wt. % for NMHFC@PPy.

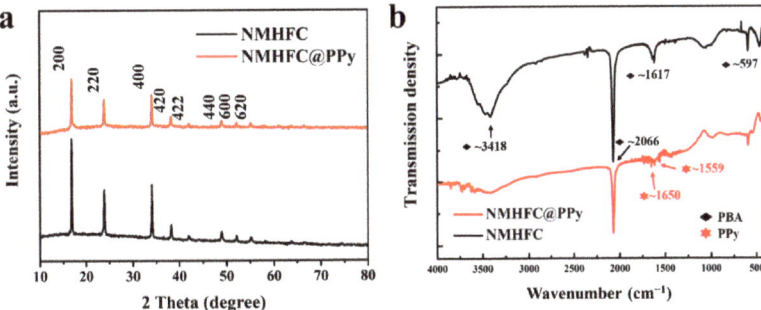

Figure 1. Characterization of as-synthesized NMHFC and NMHFC@PPy. (a) XRD patterns, (b) FTIR spectra.

The morphology of NMHFC and NMHFC@PPy nanocubes were characterized by SEM, of which the results are shown in Figure 2a,b. The edges of the nanocubes after coating are blurred. The presence of Na, Mn, and Fe in the material was confirmed by EDS-mapping analysis (Figures 2e and S3); the obvious increase in C and N was evident in the EDS-mapping pattern after coating, which, combined with the previous IR results, further indicates that PPy had been obtained on the NMHFC nanocubes. TEM characterizations of NMHFC and NMHFC@PPy nanocubes were shown in Figure 2c,d, which reveals that the samples are nano-cubic particles. After PPy coating, the morphology and size of NMHFC@PPy remain unchanged; the edges of the nanocubes turn into a rough and irregular state, indicating the existence of PPy.

In order to analyze the electrochemical performance of the as-synthesized PBA cathode material, in this work the CR2016 coin-type cells with different electrolytes and polished zinc as anode were assembled. Since the rechargeable aqueous hybrid ion batteries (RAHBs) system has been proved to be an efficient way to improve the electrochemical properties [33–38], we applied a Na-Zn hybrid battery in the following tests (1 mol L^{-1} NaCF$_3$SO$_3$ and Zn(CF$_3$SO$_3$)$_2$ mixed ion solution as the Na-Zn hybrid electrolyte). Figure 3a shows the charge and discharge curves of NMHFC@PPy in the first 10 cycles at 200 mA g^{-1}; similar discharging plateaus near 1.4 V were observed. The capacity of NMHFC@PPy reaches about 55.0 mAh g^{-1} after the first 10 cycles, while the uncoated NMHFC shows only about 25.2 mAh g^{-1} (Figure S5a). This result proves that the coating of PPy can significantly improve the electrochemical performance of the material, mainly on the grounds that the PPy coating on the material surface increases the electrical conductivity and improves the kinetics of ion intercalation in the structure, and the PPy on surface also protects the integrity of the electrode material during charging and discharging, thus improving the cycling performance of the electrode material. During the cycles, compared with the uncoated electrode material (Figure S5a), the charge and discharge curves of NMHFC@PPy do not show any obvious shape change, which proves that the electrochemical properties of this material are more stable after coating. In addition, the CV diagram shows that the PPy-coated material (Figure 3b) has a smaller change in reaction capacity from the first to the tenth cycle compared with the uncoated one (Figure S5b), which further indicates that the PPy-coated electrode material has a better capacity retention. Long-cycle performance tests were conducted at a current density of 200 mA g^{-1}. As shown in Figure 3d, the PPy-coated electrode materials exhibited higher cycling capacity and cycling stability. The discharged capacity of NMHFC@PPy is 50.1 mAh g^{-1} after 150 cycles with a retention of 86.5%, while the uncoated electrode material showed a significant decrease in the first 30 cycles, and the capacity was only 25.1 mAh g^{-1} after 150 cycles with a retention of 63.5%, indicating that the ionic conductivity of the material was enhanced after coating of PPy, thus improving the kinetics of ion deintercalation in the framework of cathode material and demonstrating a higher reaction of capacity. The XRD patterns of the electrode materials after 100 cycles in the Na-Zn hybrid electrolyte (Figure S6) show that the NMHFC@PPy cathode retains a better crystal structure after 100 cycles compared with the electrode material without coating with PPy, indicating better preservation of the structure improved by PPy. Furthermore, PPy exhibits higher tensile strength, good flexibility, and corrosion resistance, which can buffer the volume change of the cathode material during the charging and discharging process, resulting in the improved cycling stability of the material.

Figure 2. SEM and TEM images of NMHFC (**a,c**) and NMHFC@PPy (**b,d**) and (**e**) EDS-mapping elemental analyses of NMHFC@PPy.

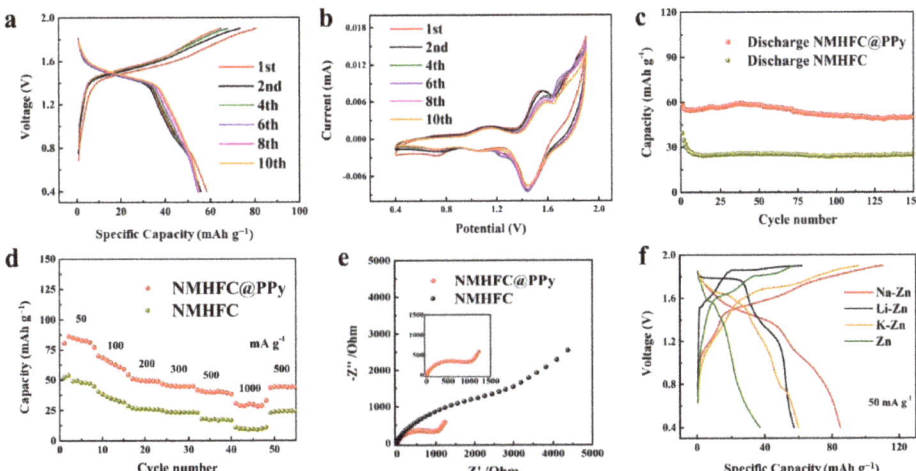

Figure 3. (a) Charge and discharge curves of NMHFC@PPy from first to tenth cycle at a current density of 200 mA g^{-1}; (b) CV curves of NMHFC@PPy at a scan rate of 0.1 mV s^{-1}; (c) long-cycle performance at 200 mA g^{-1} of NMHFC@PPy and NMHFC; (d) rate performance; (e) EIS tests of NMHFC@PPy and NMHFC; (f) charge and discharge curves in different electrolytes at a current density of 50 mA g^{-1}.

To further investigate the electrochemical properties in different electrolytes, this work continues with an in-depth analysis of electrochemical measurements. In the following work, different RAHBs will be taken into electrochemical tests, and their special advantages will be analyzed. Three more types of electrolyte include: $Zn(CF_3SO_3)_2$ at a concentration of 1 mol L^{-1}, mixed ionic solution of Li-Zn ($LiCF_3SO_3$ and $Zn(CF_3SO_3)_2$), and K-Zn (KCF_3SO_3 and $Zn(CF_3SO_3)_2$); all concentrations were specified as 1 mol L^{-1}. Figure 3f shows the charging and discharging curves of NMHFC@PPy cathode material in four different electrolytes of plain Zn, Li-Zn, Na-Zn, and K-Zn at a current density of 50 mA g^{-1} all in the third cycle to reduce the effect of polarization. The pure Zn electrolyte shows the lowest discharged capacity of 37.1 mAh g^{-1}, while Li-Zn, Na-Zn, and K-Zn hybrid electrolytes exhibit discharged capacities of 57.3, 84.7, and 59.6 mAh g^{-1}, respectively, indicating the highest discharged capacity in the Na-Zn hybrid electrolyte. Meanwhile, the Na-Zn hybrid electrolyte shows the longest platform of midpoint voltage. As shown in Figure S4, the cycling performance tests of the Li-Zn, Na-Zn, and K-Zn samples were conducted at a current density of 50 mA g^{-1}, which corresponded to the charge–discharge curves in Figure 3f. The discharge capacities of the Li-Zn and K-Zn hybrid electrolytes were both lower than 50 mAh g^{-1} after 100 cycles, with the coulombic efficiencies less than 80%, while the discharged capacity in the Na-Zn hybrid electrolyte reaches 99.3 mAh g^{-1} after 100 cycles with coulombic efficiency up to 93.3%, indicating higher capacity and stability. It is presumed that irreversible side reactions may appear during the charging process, which may lead to damage to the material structure resulting in electrochemical degradation. The capacity in K-Zn hybrid electrolyte is close to that in Li-Zn; both suffer from rapid capacity decays due to the fact that the framework of the material collapses in the continuous reaction. The Li-Zn hybrid electrolyte exhibits capacity decay after the first few cycles, relating to the solventization reaction of Zn^{2+} in the electrolyte, which could affect the pH of the solution and, thus, the cation–solvent interaction. It has been reported that the pH of Li-Zn hybrid electrolyte ranges from pH = 3 at 1 mol L^{-1}, which leads to hydrolysis side reactions, all the way to pH ≈ 7 at high concentrations of Li-Zn hybrid solution; hydrolysis could be effectively inhibited by neutral pH, implying that

Li-Zn hybrid electrolyte requires higher concentrations of the mixed solution to inhibit the occurrence of hydrolysis side reactions for better cycling stability [39].

In view of the electrochemical performance of NMHFC@PPy cathode materials in different hybrid electrolytes, it was found that NMHFC@PPy showed the best cycling efficiency and stability in the Na-Zn hybrid electrolyte among the four comparison samples. To further investigate the zinc storage performance by coating with PPy, Na-Zn mixed solution was used as the electrolyte for the following tests, with NMHFC@PPy as the cathode material.

In order to further investigate the kinetic properties of NMHFC@PPy cathode material, CV tests with different scan rates of the electrode reaction was conducted. In CV testing, it is possible to distinguish whether the cell behaves diffusively or pseudocapacitively during charging and discharging by mathematically analyzing the scan rate (v, mV s^{-1}) with the resulting peak current response (i, A). With the increase of scan rate, the peak current of redox increases gradually, and the peak potential difference between oxidation peak and reduction peak also rises. The oxidation peak shifts to high potential and reduction peak shifts to low potential, indicating that the electrochemical polarization increases with the increase of scan rate. The following formula exists between the i and the v, which is discussed [40]:

$$i = av^b \tag{1}$$

For the cathode electrode material, the electrochemical reaction process generally contains both battery properties and pseudocapacitance properties. Therefore, the current response at a specific voltage consists of two components: the diffusion contribution ($k_2 v^{1/2}$) and the capacitance contribution ($k_1 v$), which can be shown as:

$$i = k_1 v^{\frac{1}{2}} + k_2 v \tag{2}$$

where k_1 and k_2 are constants whose value magnitude can reflect the percentage of different property response contributions at a specific voltage. The current magnitudes at different sweep speeds and specific voltages can be obtained by multi-sweep speed cyclic voltammetry tests. Therefore, the corresponding k_1 and k_2 values can be gained by linearly fitting the current response at different sweep speeds using the above equations, and the diffusion contribution and capacitance contribution at specific sweep speeds can be obtained by integrating the k_1 and k_2 values obtained at each voltage.

Based on the theory discussed above, CV tests at different scan rates were performed for NMHFC@PPy cathode material (voltage ranging from 0.4 to 1.9 V) with scan rates of 0.1, 0.2, 0.4, 0.6, 0.8, 1.0, and 2.0 mV s^{-1}, and the results are shown in Figure 4, where it can be seen that as the scan speed increases, the corresponding redox peaks shift toward the anode and cathode, respectively. This can be attributed to the polarization effect at high currents, which will lead to poor cycling performance at high currents if the polarization is too large. In addition, it can be clearly observed that the end current changes significantly during the test from 0.4 to 1.9 V. The electrochemical analysis of the NMHFC@PPy electrode material was carried out using the above equation. Based on Figure 4c, we can find the proportions of pseudocapacitive and diffusion-controlled contributions from the total capacity at various scan rates of 0.1 to 2.0 mV s^{-1}. The contribution of the pseudocapacitance process (red shaded region) is 58.04% at a scan rate of 0.8 mV s^{-1} (Figure 4d), due to good rate performance of the NMHFC@PPy material.

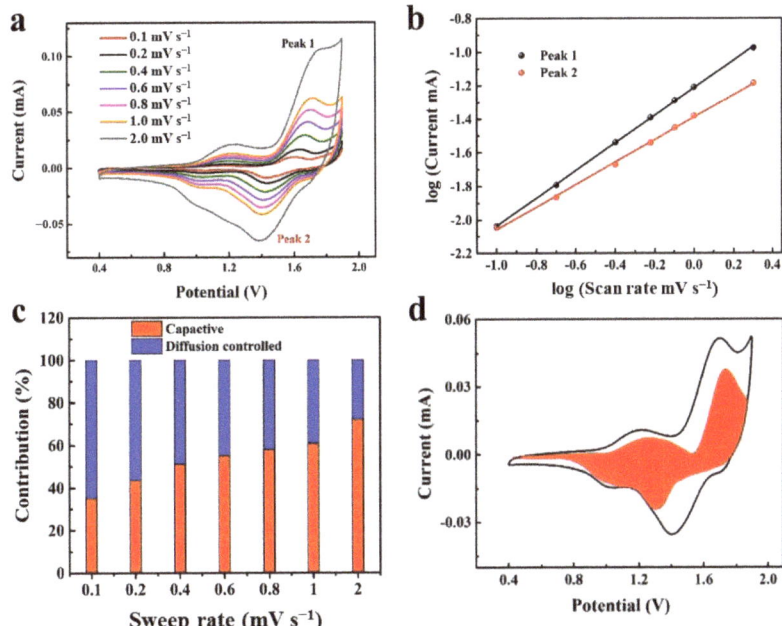

Figure 4. (**a**) CV curves at different scan rates; (**b**) logarithmic relationship between peak current and scan rate; (**c**) the diffusion and capacitance contributions at different a scan rates; (**d**) the contribution proportion of pseudocapacitance at 0.8 mV s^{-1}.

To investigate the changes of the crystal structure during the charging and discharging process of the aqueous hybrid Na-Zn battery, we conducted ex situ XRD tests of NMHFC@PPy during the first cycle in Figure 5a. As can be seen from the ex situ XRD patterns in the Na-Zn hybrid electrolyte, the diffraction peaks during the charge and discharge process indicates that no phase transition occurred during the electrochemical redox process, and the results show similar patterns of the initial and fully charged state, which indicates good reversibility of NMHFC@PPy electrode material. Moreover, the peak at 18.1° belongs to PTFE, which acts as the binder in the electrode material, while the peaks at 16.8°, 34.0°, and 38.2° are attributed to the NMHFC@PPy cathode; they all exhibited weakened intensities when discharged to 0.4 V, indicating the intercalation of Zn^{2+} according to previous reports [41]. Besides, the peak at 23.8° shows a slight shift to a larger angle when the electrode was fully discharged and charged, which gives further proof of the intercalation and deintercalation of Zn^{2+} [21,24]. Figure S7 shows the ex situ XRD and XPS patterns of NMHFC electrode material. As can be seen from the ex situ XRD results, the cathode without coating with PPy exhibits alike peaks compared with the coated one, while the peaks at fully discharged state maintained worse crystal structure when discharged to 0.4 V in comparison to NMHFC@PPy, suggesting that the coating of PPy can improve the reversibility of the electrode material during the cycle [21]. In addition, the ex situ Fe 2p XPS results of NMHFC demonstrate that the changes of peaks are relatively small (0.3 eV), indicating that the volume change during the charge–discharge process can be reduced via PPy coating [24]. To further characterize the situation of the material before and after the charging and discharging process, EDS-mapping analysis was performed on the elemental distribution of the material before and after discharge. As shown in Figure S8, the results show that when the electrode material is discharged to 0.4 V, after the Zn^{2+} intercalation occurs in the corresponding charge–discharge curve, the presence of Zn element can be clearly detected in the elemental spectrum. At the same time, the morphology of the material does not change significantly after discharge, and the material still maintains the morphology of nanocubes, indicating that the morphology of the clad material is well preserved after the discharge, which demonstrates the excellent stability of the material. The ex situ XPS Zn 2p spectra of NMHFC@PPy electrode material in different states were also collected in Figure 5b (initial, full discharge at 0.4 V, and full charge at 1.9 V states). There exists no signal for Zn in the XPS spectrum of the initial NMHFC@PPy electrode. When discharged to 0.4 V, an obvious Zn $2p_{3/2}$–$2p_{1/2}$ spin–orbit doublet was detected, indicating the successful intercalation of Zn^{2+} into the structure of the cathode material, while the signal of Zn exhibits a weakened state when the cathode is fully charged to 1.9 V. The XPS results indicate the successful insertion/extraction of Zn^{2+} ions into/from the NMHFC@PPy electrode during discharge/charge processes.

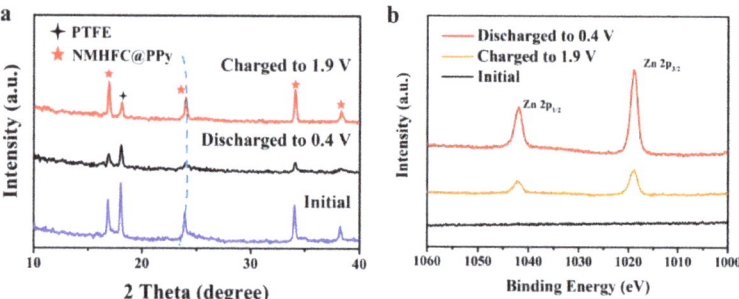

Figure 5. (a) Ex situ XRD and (b) ex situ XPS patterns of NMHFC@PPy cathode material during the first cycle in different states.

To explore the chemical composition of NMHFC@PPy cathode material, XPS spectra of NMHFC@PPy in different states are shown in Figure S9; the elements of C, N, O, and Fe are found in the initial state of NMHFC@PPy. XPS spectra of Fe 2p and C 1s in different states are shown in Figure 6. After coating with PPy, the initial Fe $2p_{3/2}$ XPS peak (Figures 6a and S10a) of NMHFC@PPy cathode material can be fitted by the peak related to Fe^{II} (705.4 eV) and the peak attributed to Fe^{III} (708.8 eV) [42], and the C 1s XPS peak (Figure 6d) can be fitted by the peaks related to C–C/C=C (281.4 eV) and C–N (289.9 eV) [41,43]. When the electrode is fully discharged to 0.4 V (Figure S10b), the peak changes attributed to Fe 2p (0.6 eV) are larger than that of NMHFC (0.3 eV, Figure S7b), indicating that the coating of PPy can reduce the volume change in the charge and discharge process, while the strength of the peak fitted to Fe^{III} increases. The C 1s shows a new peak of C–H in 285.5 eV (Figure 6c) in fully discharged state. However, after the cathode material has been fully charged in the Na-Zn hybrid electrolyte, the peak of Fe^{II} shows a slight change (705.5 eV) compared with the initial one (Figure S10c), while C 1s is almost the same as the initial one (Figure 6b), and Fe 2p shows weak Fe^{III} peaks in initial and charged states, mainly due to the oxidation of Fe in the synthetic process. This indicates that the intercalated Zn^{2+} in the PPy-coated cathode material mainly interacts with the redox reactions of Fe ions of both Fe^{II} and Fe^{III} when the electrode material is fully discharged, and C–H bonds of PPy also contribute to the ionic conductivity of the cathode material during the discharge process, which is similar to the previous reports [30,44].

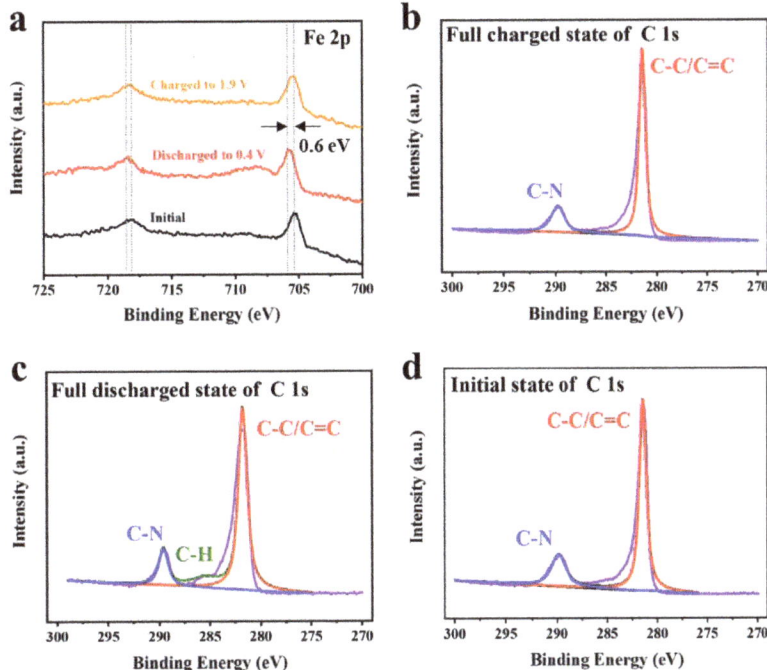

Figure 6. XPS spectra of NMHFC@PPy in different states; (**a**) Fe 2p and (**b**–**d**) C 1s.

4. Conclusions

In this paper, NMHFC nanocubes were synthesized by a simple co-precipitation method and modified by PPy coating to obtain NMHFC@PPy nanocubes. Compared with bare NMHFC, the as-prepared NMHFC@PPy cathode material exhibited an enhanced cycling stability and rate performance due to the better electronic conductivity provided by PPy. The capacity retention was improved from 63.5% to 86.5% by coating with PPy after 150 cycles in a Na-Zn hybrid battery, while the rate performance of NMHFC@PPy cathode material in high current densities is also superior to that of the uncoated one. These advances provide new research views for development of modified PBA cathode materials for cycling-stable aqueous hybrid batteries in the future.

Supplementary Materials: The following supporting information can be downloaded at: https://www.mdpi.com/article/10.3390/coatings12060779/s1, Figure S1: Raman spectra of NMHFC@PPy; Figure S2: TG curve of NMHFC@PPy; Figure S3: EDS-mapping elemental analysis of NMHFC; Figure S4: Cycling performance curves of NMHFC@PPy in Li-Zn, Na-Zn, and K-Zn mixed ionic electrolytes at current density of 50 mA g^{-1}; Figure S5: Charge and discharge curves of NMHFC from first to tenth cycle at a current density of 200 mA g^{-1}, CV curves of NMHFC at a scan rate of 0.1 mV s^{-1}; Figure S6: XRD patterns of NMHFC@PPy and NMHFC electrode material after 100 cycles; Figure S7: (a) Ex situ XRD and (b) ex situ Fe 2p XPS patterns of NMHFC cathode material during the first cycle in different states; Figure S8: SEM and EDS-mapping analyses of NMHFC@PPy electrode material before discharge and after discharge to 0.4 V; Figure S9: XPS spectra of NMHFC@PPy in different states; Figure S10: Fe 2p XPS spectra of NMHFC@PPy in different states.

Author Contributions: Conceptualization, methodology, Q.A., L.C. (Lineng Chen) and L.C. (Lianmeng Cui); data curation, writing—original draft preparation, Y.R.; writing—review and editing, Y.R. and L.C. (Lineng Chen); project administration, funding acquisition, Q.A. All authors have read and agreed to the published version of the manuscript.

Funding: This work was supported by the National Natural Science Foundation of China (51972259, 52172231), Foshan Xianhu Laboratory of the Advanced Energy Science and Technology Guangdong Laboratory (XHT2020-003), the National Key Research and Development Program of China (2017YFE0127600), and the Fundamental Research Funds for the Central Universities (WUT: 2021III024GX and WUT: 2021III001GL).

Institutional Review Board Statement: Not applicable.

Informed Consent Statement: Not applicable.

Data Availability Statement: Data sharing is not applicable to this work.

Conflicts of Interest: The authors declare no conflict of interest.

References

1. Bo, X.; Qian, D.; Wang, Z.; Ying, S.M. Recent progress in cathode materials research for advanced lithium ion batteries. *J. Cheminform.* **2012**, *73*, 51–65.
2. Fergus, J.W. Recent developments in cathode materials for lithium ion batteries. *J. Power Sources* **2010**, *195*, 939–954. [CrossRef]
3. Aurbach, D.; Gofer, Y.; Lu, Z.; Schechter, A.; Chusid, O.; Gizbar, H.; Cohen, Y.; Ashkenazi, V.; Moshkovich, M.; Turgeman, R. A short review on the comparison between Li battery systems and rechargeable magnesium battery technology. *J. Power Sources* **2001**, *97*, 28–32. [CrossRef]
4. Song, M.; Tan, H.; Chao, D.; Fan, H.J. Recent Advances in Zn-Ion Batteries. *Adv. Funct. Mater.* **2018**, *28*, 1802564. [CrossRef]
5. Zeng, X.; Hao, J.; Wang, Z.; Mao, J.; Guo, Z. Recent progress and perspectives on aqueous Zn-based rechargeable batteries with mild aqueous electrolytes. *Energy Storage Mater.* **2019**, *20*, 410–437. [CrossRef]
6. Ming, J.; Guo, J.; Xia, C.; Wang, W.; Alshareef, H.N. Zinc-ion batteries: Materials, mechanisms, and applications. *Mater. Sci. Eng. R Rep.* **2018**, *135*, 58–84. [CrossRef]
7. Guduru, R.; Icaza, J. A Brief Review on Multivalent Intercalation Batteries with Aqueous Electrolytes. *Nanomaterials* **2016**, *6*, 41. [CrossRef] [PubMed]
8. Wei, S.; Fei, W.; Hou, S.; Yang, C.; Wang, C. Zn/MnO$_2$ Battery Chemistry with H$^+$ and Zn^{2+} Coinsertion. *J. Am. Chem. Soc.* **2017**, *139*, 9775–9778.
9. Shoji, T.; Hishinuma, M.; Yamamoto, T. Zinc-manganese dioxide galvanic cell using zinc sulphate as electrolyte. Rechargeability of the cell. *J. Appl. Electrochem.* **1988**, *18*, 521–526. [CrossRef]
10. Yan, M.; He, P.; Chen, Y.; Wang, S.; Wei, Q.; Zhao, K.; Xu, X.; An, Q.; Shuang, Y.; Shao, Y. Water-Lubricated Intercalation in V$_2$O$_5$·nH$_2$O for High-Capacity and High-Rate Aqueous Rechargeable Zinc Batteries. *Adv. Mater.* **2017**, *30*, 1703725. [CrossRef] [PubMed]
11. Liang, Y.; Yan, J.; Gheytani, S.; Lee, K.Y.; Liu, P.; Facchetti, A.; Yao, Y. Universal quinone electrodes for long cycle life aqueous rechargeable batteries. *Nat. Mater.* **2017**, *16*, 841–848. [CrossRef] [PubMed]
12. Chen, L.; Ruan, Y.; Zhang, G.; Wei, Q.; Jiang, Y.; Xiong, T.; He, P.; Yang, W.; Yan, M.; An, Q. An Ultrastable and High-Performance Zn/VO$_2$ Battery Based on a Reversible Single-Phase Reaction. *Chem. Mater.* **2019**, *31*, 699–706. [CrossRef]
13. Li, Z.; Young, D.; Xiang, K.; Carterw, C.; Chiang, Y.M. Towards high power high energy aqueous sodium-ion batteries: The NaTi$_2$(PO$_4$)$_3$/Na$_{0.44}$MnO$_2$ System. *Adv. Energy Mater.* **2012**, *3*, 290–294. [CrossRef]
14. Paolella, A.; Faure, C.; Timochevskii, V.; Marras, S.; Bertoni, G.; Abdelbast, G.; Vijh, A.; Armand, M.; Zaghib, K. A review on hexacyanoferrate-based materials for energy storage and smart windows: Challenges and perspectives. *J. Mater. Chem. A* **2017**, *5*, 18919–18932. [CrossRef]
15. Du, G.; Pang, H. Recent advancements in Prussian blue analogues: Preparation and application in batteries. *Energy Storage Mater.* **2021**, *36*, 387–408. [CrossRef]
16. Trocoli, R.; La Mantia, F. An aqueous zinc-ion battery based on copper hexacyanoferrate. *Chem. Sustain. Energy Mater.* **2015**, *8*, 481–485. [CrossRef]
17. Chae, M.S.; Heo, J.W.; Kwak, H.H.; Lee, H.; Hong, S.T. Organic electrolyte-based rechargeable zinc-ion batteries using potassium nickel hexacyanoferrate as a cathode material. *J. Power Sources* **2017**, *337*, 204–211. [CrossRef]
18. Wang, H.; Xu, E.; Yu, S.; Li, D.; Quan, J.; Li, X.; Li, W.; Yang, J. RGO-Anchored Manganese Hexacyanoferrate with Low Interstitial H$_2$O for Superior Sodium-Ion Batteries. *Am. Chem. Soc. Appl. Mater. Interfaces* **2018**, *10*, 34222–34229. [CrossRef]
19. Zhu, W.; Wang, W.; Xue, W.; Kong, K.; Zhao, R. Achieving a Zn-ion battery-capacitor hybrid energy storage device with a cycle life of more than 12,000 cycles. *Compos. Part B Eng.* **2021**, *207*, 108555. [CrossRef]
20. Cao, J.; Hu, G.; Peng, Z.; Du, K.; Cao, Y. Polypyrrole-coated LiCoO$_2$ nanocomposite with enhanced electrochemical properties at high voltage for lithium-ion batteries. *J. Power Sources* **2015**, *281*, 49–55. [CrossRef]
21. Xue, Q.; Li, L.; Huang, Y.; Huang, R.; Wu, F.; Chen, R. Polypyrrole-Modified Prussian Blue Cathode Material for Potassium Ion Batteries via In Situ Polymerization Coating. *Am. Chem. Soc. Appl. Mater. Interfaces* **2019**, *11*, 22339–22345. [CrossRef]
22. Yin, J.; Sheng, Y.; Bao, W.; Yong, L.; Wen, S. Prussian Blue@C Composite as an Ultrahigh-Rate and Long-Life Sodium-Ion Battery Cathode. *Adv. Funct. Mater.* **2016**, *26*, 5315–5321.

23. Hayashi, H. Cs sorption of Mn–Fe based Prussian blue analogs with periodic precipitation banding in agarose gel. *Phys. Chem. Chem. Phys.* **2022**, *24*, 9374–9383. [CrossRef]
24. Chen, M.; Li, X.; Yan, Y.; Yang, Y.; Xu, Q.; Liu, H.; Xia, Y. Polypyrrole-Coated $K_2Mn[Fe(CN)_6]$ Stabilizing Its Interfaces and Inhibiting Irreversible Phase Transition during the Zinc Storage Process in Aqueous Batteries. *Am. Chem. Soc. Appl. Mater. Interfaces* **2022**, *14*, 1092–1101. [CrossRef]
25. Wang, L.; Lu, Y.; Liu, J.; Xu, M.; Cheng, J.; Zhang, D.; Goodenough, J.B. A superior low-cost cathode for a Na-ion battery. *Angew. Chem. Int. Ed.* **2013**, *52*, 1964–1967. [CrossRef] [PubMed]
26. Hu, F.; Li, L.; Jiang, X. Hierarchical Octahedral $Na_2MnFe(CN)_6$ and $Na_2MnFe(CN)_6$@PPy as Cathode Materials for Sodium-Ion Batteries. *Chin. J. Chem.* **2017**, *35*, 415–419. [CrossRef]
27. Wu, C.; Fang, X.; Guo, X.; Mao, Y.; Ma, J.; Zhao, C.; Wang, Z.; Chen, L. Surface modification of $Li_{1.2}Mn_{0.54}Co_{0.13}Ni_{0.13}O_2$ with conducting polypyrrole. *J. Power Sources* **2013**, *231*, 44–49. [CrossRef]
28. Xiong, X.; Wang, Z.; Li, X.; Ding, D.; Huang, B. Surface modification of $LiNi_{0.8}Co_{0.1}Mn_{0.1}O_2$ with conducting polypyrrole. *J. Solid State Electrochem.* **2014**, *18*, 2619–2624. [CrossRef]
29. Xiong, P.; Zeng, G.; Zeng, L.; Wei, M. Prussian blue analogues $Mn[Fe(CN)_6]_{0.6667} \cdot nH_2O$ cubes as an anode material for lithium-ion batteries. *Dalton Trans.* **2015**, *44*, 16746–16751. [CrossRef]
30. Yang, Q.; Mo, F.; Liu, Z.; Ma, L.; Li, X.; Fang, D.; Chen, S.; Zhang, S.; Zhi, C. Activating C-Coordinated Iron of Iron Hexacyanoferrate for Zn Hybrid-Ion Batteries with 10,000-Cycle Lifespan and Superior Rate Capability. *Adv. Mater.* **2019**, *31*, 1901521. [CrossRef]
31. Sun, Y.; Liu, C.; Xie, J.; Zhuang, D.; Zheng, W.; Zhao, X. Potassium manganese hexacyanoferrate/graphene as a high-performance cathode for potassium-ion batteries. *New J. Chem.* **2019**, *43*, 11618–11625. [CrossRef]
32. Li, W.; Chou, S.; Wang, J.; Wang, J.; Gu, Q.; Liu, H.; Dou, S. Multifunctional conducing polymer coated $Na_{1+x}MnFe(CN)_6$ cathode for sodium-ion batteries with superior performance via a facile and one-step chemistry approach. *Nano Energy* **2015**, *13*, 200–207. [CrossRef]
33. Lu, K.; Song, B.; Zhang, J.; Ma, H. A rechargeable Na-Zn hybrid aqueous battery fabricated with nickel hexacyanoferrate and nanostructured zinc. *J. Power Sources* **2016**, *321*, 257–263. [CrossRef]
34. Xi, C.; Lulu, W.; Ji-Tao, C.; Junrong, Y. A Low-Cost Mg^{2+}/Na^+ Hybrid Aqueous Battery. *J. Mater. Chem. A* **2018**, *6*, 15762–15770.
35. Tao, G.; Han, F.; Zhu, Y.; Suo, L.; Chao, L.; Kang, X.; Wang, C. Hybrid Mg^{2+}/Li^+ Battery with Long Cycle Life and High Rate Capability. *Adv. Energy Mater.* **2015**, *5*, 1401507.
36. Yu, F.; Zhang, S.; Fang, C.; Liu, Y.; He, S.; Xia, J.; Yang, J.; Zhang, N. Electrochemical characterization of P2-type layered $Na_{2/3}Ni_{1/4}Mn_{3/4}O_2$ cathode in aqueous hybrid sodium/lithium ion electrolyte. *Electrolytes* **2017**, *43*, 9960–9967. [CrossRef]
37. Huang, M.; Meng, J.S.; Huang, Z.J.; Wang, X.P.; Mai, L.Q. Ultrafast cation insertion-selected zinc hexacyanoferrate for 1.9 V K-Zn hybrid aqueous batteries. *J. Mater. Chem. A* **2020**, *8*, 6631–6637. [CrossRef]
38. Jing, Y.; Jing, W.; Hao, L.; Bakenov, Z.; Gosselink, D.; Chen, P. Rechargeable hybrid aqueous batteries. *J. Power Sources* **2012**, *216*, 222–226.
39. Fei, W.; Borodin, O.; Tao, G.; Fan, X.; Wang, C. Highly reversible zinc metal anode for aqueous batteries. *Nat. Mater.* **2018**, *17*, 543–549.
40. Mao, Y.; Chen, Y.; Qin, J.; Shi, C.; Liu, E.; Zhao, N. Capacitance controlled, hierarchical porous 3D ultra-thin carbon networks reinforced prussian blue for high performance Na-ion battery cathode. *Nano Energy* **2019**, *58*, 192–201. [CrossRef]
41. Li, Z.; Huang, Y.; Zhang, J.; Jin, S.; Zhang, S.; Zhou, H. One-step synthesis of MnO_x/PPy nanocomposite as a high-performance cathode for a rechargeable zinc-ion battery and insight into its energy storage mechanism. *Nanoscale* **2020**, *12*, 4150–4158. [CrossRef]
42. Lu, K.; Song, B.; Zhang, Y.X.; Ma, H.Y.; Zhang, J.T. Encapsulation of zinc hexacyanoferrate nanocubes with manganese oxide nanosheets for high-performance rechargeable zinc ion batteries. *J. Mater. Chem. A* **2017**, *5*, 23628–23633. [CrossRef]
43. Liao, X.; Pan, C.; Pan, Y.; Yin, C. Synthesis of three-dimensional β-MnO_2/PPy composite for high-performance cathode in zinc-ion batteries. *J. Alloys Compd.* **2021**, *888*, 161619. [CrossRef]
44. Xia, W.A.; Yc, A.; Hua, S.B.; Wh, A.; Ying, Z.A.; Yl, A.; Gh, A.; Yx, C. Codoping triiodide anion in polypyrrole cathode: An effective route to increase the capacity of zinc-ion battery. *J. Electroanal. Chem.* **2022**, *912*, 116232.

Article

Effects of Additive and Roasting Processes on Nitrogen Removal from Aluminum Dross

Shuai-Shuai Lv [1], Yu Zhang [1], Hong-Jun Ni [1,2,*], Xing-Xing Wang [1], Wei-Yang Wu [1] and Chun-Yu Lu [1]

[1] School of Mechanical Engineering, Nantong University, Nantong 226019, China; lvshuaishuai@ntu.edu.cn (S.-S.L.); 18501470690@163.com (Y.Z.); wangxx@ntu.edu.cn (X.-X.W.); wuweiyang.woy@gmail.com (W.-Y.W.); 2109310022@stmail.ntu.edu.cn (C.-Y.L.)

[2] Jiangsu Province Engineering Research Center of Aluminum Dross Solid Waste Harmless Treatment and Resource Utilization, Nantong 226019, China

* Correspondence: ni.hj@ntu.edu.cn

Abstract: Taking high nitrogen aluminum dross as the research object, the effects of the additives sodium carbonate and cryolite and the roasting process on the denitrification effect of aluminum dross were studied. The principle of additive denitrification was studied by XRD, SEM and TG-DSC. The experimental results show that sodium carbonate and cryolite can quickly reduce the content of aluminum nitride in aluminum dross. The optimum denitrification process parameters were also obtained simultaneously. When the mass ratio of cryolite to aluminum dross was 0.4, the roasting temperature was 900 °C, and the roasting time was 3 h, the denitrification degree could reach 96.19%. When the mass ratio of sodium carbonate to aluminum dross was 0.8, the roasting temperature was 1000 °C, and the roasting time was 4 h, the denitrification degree could reach 91.32%. This study provides a reference for the non-harmful treatment of high nitrogen aluminum dross.

Keywords: high nitrogen aluminum dross; additive; nitrogen removal; process parameter

Citation: Lv, S.-S.; Zhang, Y.; Ni, H.-J.; Wang, X.-X.; Wu, W.-Y.; Lu, C.-Y. Effects of Additive and Roasting Processes on Nitrogen Removal from Aluminum Dross. *Coatings* **2022**, *12*, 730. https://doi.org/10.3390/coatings12060730

Academic Editor: Günter Motz

Received: 19 April 2022
Accepted: 24 May 2022
Published: 25 May 2022

Publisher's Note: MDPI stays neutral with regard to jurisdictional claims in published maps and institutional affiliations.

Copyright: © 2022 by the authors. Licensee MDPI, Basel, Switzerland. This article is an open access article distributed under the terms and conditions of the Creative Commons Attribution (CC BY) license (https://creativecommons.org/licenses/by/4.0/).

1. Introduction

Aluminum dross is a kind of waste produced in the process of electrolytic aluminum furnace front casting and aluminum alloy casting. According to the National Hazardous Waste List (2021 Edition) [1], aluminum dross is recognized as hazardous waste. According to the content of nitride and fluoride, aluminum dross is divided into high-nitrogen aluminum dross and high-fluoride aluminum dross. Between these, high-nitrogen aluminum dross is mainly produced in the secondary aluminum process, and high fluoride aluminum dross is mainly produced in the aluminum electrolysis process [2–4]. At present, the treatment method for aluminum dross is still mainly stacking and landfill, but when aluminum dross that has not undergone non-harmful treatment is landfilled or stacked, inorganic salts dominated by NaCl and KCl will enter the soil, resulting in the aggravation of land salinization, and fluoride ions will penetrate into the soil and groundwater. When exceeding the discharge standard in the national standard, it is easy to cause high-fluoride water and high-fluoride soil and to induce immune diseases [5–8]. Therefore, a low-cost and high-efficiency non-harmful aluminum dross treatment method is a top priority. Many scholars have studied the non-harmful treatment of aluminum dross. Based on the response surface method, Wang [9] carried out research on the influence of cryolite (Na_3AlF_6), roasting temperature and soaking time on the denitrification degree of secondary aluminum dross. Li [10] achieved nitrogen removal and fluoride retention with the addition of calcium salt ($CaCl_2$) and the roasting of aluminum dross at a high temperature (1300 °C). With sodium carbonate (Na_2CO_3) and cryolite as additives, this study attempted to explore the effect of the additive ratio, roasting temperature and roasting time on the nitrogen removal effects of high-nitrogen aluminum dross, as well as the phase transformation law of high-nitrogen aluminum dross under different roasting conditions.

The aluminum dross obtained after non-harmful treatment can be used in many fields. Taking the treatment results of this paper as an example, the experimental samples treated with cryolite could be returned to the electrolytic cell for reuse and $NaAlO_2$ could be obtained from the samples treated with sodium carbonate by solid–liquid separation and other purification methods.

2. Materials and Methods

2.1. Raw Material, Reagent and Instrument

Raw material: high-nitrogen aluminum dross obtained by recovering part of the aluminum from the primary aluminum dross (Jiangsu Haiguang Metal Co., Ltd., Suqian, China).

Reagents: sodium carbonate (analytical purity, Jiangsu Qiangsheng Functional Chemical Co., Ltd., Suzhou, China), cryolite (analytical purity, Shanghai Maclean Biochemical Technology Co., Ltd., Shanghai, China), sodium hydroxide (analytical purity, Xilong Chemical Co., Ltd., Shanghai, China), hydrochloric acid (analytical purity), Shanghai Lingfeng Chemical Reagent Co., Ltd., Shanghai, China), methyl red (indicator, Shanghai Yuanye Biotechnology Co., Ltd., Shanghai, China), methylene blue (indicator, COOLABER SCIENCE&TECHNOLOGY, Beijing, China), and boric acid (chemical purity, Shanghai Myreel Chemical Technology Co., Ltd., Shanghai, China).

Instruments: programmable box furnace (SXL-1230, Hangzhou Zhuo Chi Instrument Co., Ltd., Hangzhou, China), universal electric furnace (DK-98-II, Xinghua Chushui Electric Heating Appliance Factory, Jiangsu Province, Xinghua, China), ultra-pure water system (EPED-10TH, Nanjing Yipu Technology Development Co., Ltd., Nanjing, China), X-ray fluorescence spectrometer (ZSX PRIMUS III+, Japan Science Co., Ltd., Osaka, Japan), X-ray diffractometer (D8 Advance, Bruker (Beijing) Technology Co., Ltd., Beijing, China), field emission scanning electron microscope system (ZEISS sigma HD, Carl Zeiss (Shanghai) Management Co., Ltd., Shanghai, China), and synchronous thermal analyzer (STA 449 F3, NETZSCH Instrument Manufacturing Co., Ltd., Selb, Germany).

Technical parameters of the relevant test instruments:

(1) XRD: the maximum output power was 3 KW, the voltage was 20–60 kV, the current was 2–60 mA, and the radius of the goniometer was 185 mm. MDI jade 9 was used for the XRD analysis. The XRD database was in PDF 2009 format.
(2) SEM-EDS: the acceleration voltage was 10 kV, the magnification was 100–200 k, and the secondary electron resolution was 1 nm.
(3) TG-DSC: the heating rate was 10 °C/min, the weighing resolution was 0.1 µg, and the vacuum degree was 10^{-4} mbar.

2.2. Experimental Method

First, a total of 50 g aluminum dross and additives were placed in the corundum crucible according to the additive ratio. The mixture was mixed in the corundum crucible evenly with a stirring rod. Then, the corundum crucible was placed in a programmable box furnace and fired at a certain temperature (50–1000 °C). Subsequently, the corundum crucible was taken out and cooled at room temperature. Finally, the content of aluminum nitride in the aluminum dross was measured, and the denitrification degree was calculated.

The denitrification degree test method adopted the distillation separation neutralization titration method [11,12]. The treated aluminum dross was ground, 2 g weighed, and 150 mL NaOH solution poured with a mass fraction of 20 wt.% into the conical flask, and the cork of the conical flask quickly tightened. The conical flask was heated on the asbestos net of the universal electric furnace, the mixed solution heated in the conical flask to boiling point and kept boiling for 2 h, and the ammonia distilled from the conical flask absorbed with 200 mL and 40 g/L boric acid solution. After distillation, the solution was titrated with 0.05 mol/L dilute hydrochloric acid solution, with standard methyl red–methylene blue as the indicator. The end point of titration was the sudden change of the solution from blue to purple red. Alumina was used as the blank control group. The formula for calculating the content of aluminum nitride in aluminum dross was as shown in Formula (1). The calculation formula of the denitrification degree was as in Formula (2) [13].

$$W_1 = \frac{2.05C(V_2 - V_1)}{bM} \times 100\%,\quad(1)$$

$$X_b = \frac{W_2 - W_1}{W_2} \times 100\%,\quad(2)$$

where W_1 is the aluminum nitride content in the sample (%); V_1 is the volume of dilute hydrochloric acid consumed by the test sample (mL); V_2 is the volume of dilute hydrochloric acid consumed by the blank control group (mL); C is the concentration of dilute hydrochloric acid (mol/L); M is the mass of the test sample taken during measurement (g); b is the total mass of aluminum dross added to the test sample under the current additive ratio (g); X_b is the denitrification degree (%); and W_2 is the aluminum nitride content in the original aluminum dross (%).

3. Results and Discussion

3.1. Nature of the Aluminum Dross

As shown in Figure 1, the XRD pattern of the high-nitrogen aluminum dross mainly contained Al_2O_3, AlN, NaCl, and other substances. Figure 2 shows the TG-DSC analysis result for the high-nitrogen aluminum dross. According to the TG curve in Figure 2, during the roasting process, the mass of the high-nitrogen aluminum dross showed a decreasing trend. The total weight loss of the sample was 6.03%, which was mainly due to the evaporation of adsorbed water on the sample surface or the decomposition of other surface materials and impurities. There was no obvious endothermic peak or exothermic peak in the DSC curve. Figure 3 shows the micromorphology and energy spectrum analysis of the high-nitrogen aluminum dross. The high-nitrogen aluminum dross was mainly composed of a variety of irregular and adherent small particles. According to the corresponding EDS analysis results, it could be seen that in the high-nitrogen aluminum dross, AlN did not exist alone, but was mixed with alumina—part of it was even wrapped by alumina, which made it difficult to remove AlN efficiently.

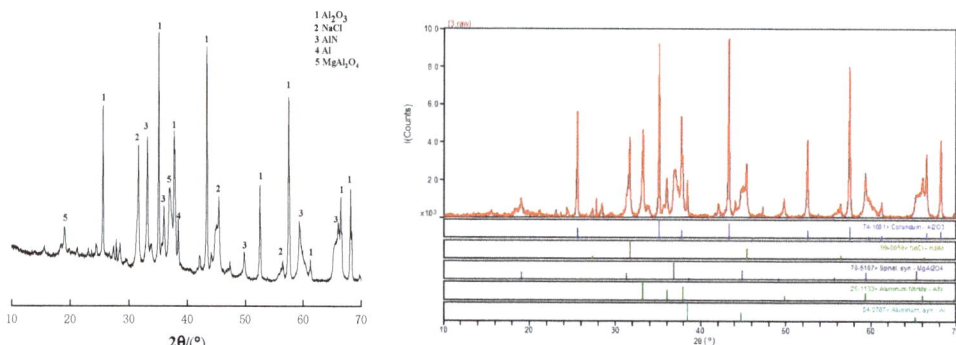

Figure 1. XRD pattern of the raw material.

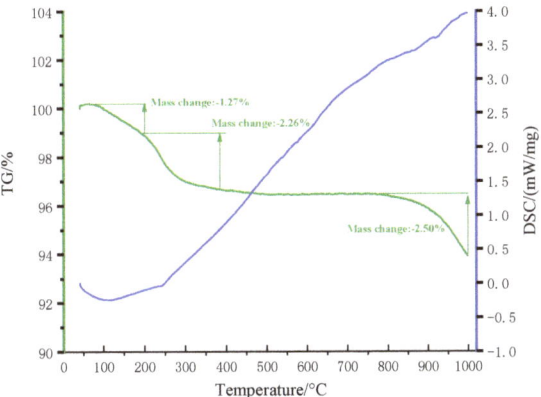

Figure 2. TG-DSC of the raw material.

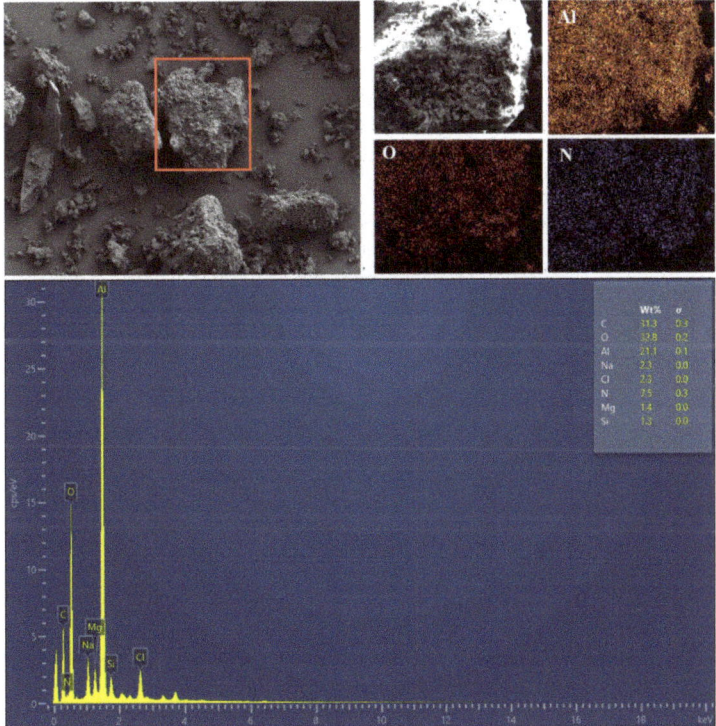

Figure 3. SEM-EDS of the raw material.

3.2. The Effect of Cryolite and Sodium Carbonate on the Nitrogen Removal of High-Nitrogen Aluminum Dross by Roasting

Cryolite is a commonly used flux for industrial aluminum production that destroys the oxide film wrapped on the surface of the AlN by melting alumina [14]—thereby exposing the AlN to the air more and improving the efficiency of the nitrogen removal. In contrast, the properties of sodium carbonate are relatively stable. However, a small part of sodium carbonate will decompose to produce oxygen during the roasting process. [15] As shown in Formulas (3) and (4), the oxygen content in the sample was increased, which is conducive to the oxidation of AlN. Above 500 °C, sodium carbonate will react with alumina, as shown in Formulas (5) and (6). The generated sodium aluminate will dissolve in sodium carbonate, which can destroy the AlN surface oxide film and increase the denitrification degree. The relevant experimental parameters are shown in Table 1 [16,17].

$$Na_2CO_3 = Na_2O + CO_2, \quad (3)$$

$$2Na_2O = 4Na + O_2, \quad (4)$$

$$Al_2O_3 + Na_2CO_3 = 2NaAlO_2 + CO_2 \quad (5)$$

$$4AlN + 3O_2 = 2Al_2O_3 + 2N_2 \quad (6)$$

Table 1. Relevant experimental parameters.

Additive	Additive Ratio (g/g)	Roasting Temperature (°C)	Roasting Time (h)
Cryolite	Different additive ratio	900	4
	0.4:1	Different temperature	4
	0.4:1	900	Different time
Sodium carbonate	Different additive ratio	900	4
	0.8:1	Different temperature	4
	0.8:1	1000	Different time

3.2.1. The Effect of the Additive Ratio on the Nitrogen Removal of High-Nitrogen Aluminum Dross by Roasting

At 900 °C, samples prepared with different additive ratios were placed in an electric resistance furnace and roasted for 4 h to study the effect of the addition of cryolite and sodium carbonate on the denitrification degree of the high-nitrogen aluminum dross. The experimental results are shown in Figure 4.

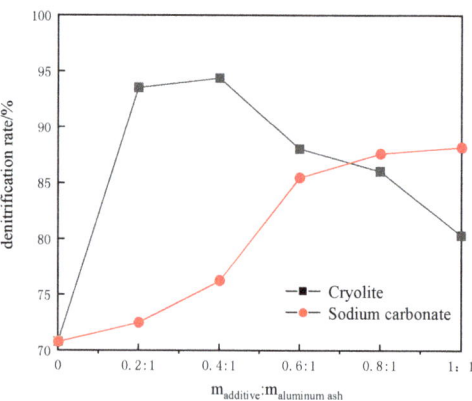

Figure 4. Effects of different additives on the nitrogen removal rate of high-nitrogen aluminum dross.

It can be seen from Figure 4 that the incorporation of cryolite (cr) and sodium carbonate (sc) will significantly affect the denitrification degree of high-nitrogen aluminum dross (ad). When the amount of cryolite is increased, the denitrification degree will increase first and then decrease. Since excessive cryolite will float on the surface of the sample, this will reduce the contact area between the sample and the air and then cause agglomeration and hardening. Therefore, excessive cryolite will cause a significant decrease in denitrification degree. When the incorporation of sodium carbonate increased, the denitrification degree showed an upward trend. When $m_{sc}:m_{ad} = 0.4:1$, the denitrification degree increased sharply. When $m_{sc}:m_{ad}$ = 0.8:1, the denitrification degree basically tended to the maximum. This is mainly because sodium carbonate can react with alumina to destroy the dense oxide film, thereby increasing the denitrification degree. In addition, during the roasting process, sodium carbonate can decompose to produce oxygen, causing the oxygen concentration in the sample to increase. XRD tests are conducted on samples with different additive ratios, so as to study the phase transformation law between different additive ratios. The XRD pattern is shown in Figure 5.

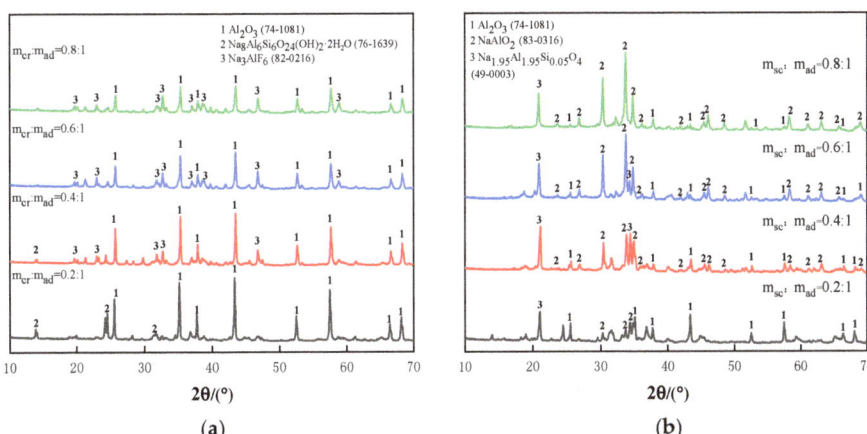

Figure 5. XRD patterns of different ingredient ratios. (**a**) XRD pattern of sample with cryolite; (**b**) XRD pattern of sample with sodium carbonate.

It can be seen from Figure 5 that as the additive ratio increased, the diffraction intensity of the alumina gradually decreased. For samples with cryolite added, the products after roasting denitrification were mainly Al_2O_3, Na_3AlF_6, and $Na_8Al_6Si_6O_{24}(OH)_2 \cdot 2H_2O$. When $m_{cr}:m_{ad}$ = 0.4:1, the diffraction peak of the cryolite appeared in the XRD spectrum. Moreover, with the increase of the proportioning ratio, the number and intensity of cryolite diffraction peaks increased, while the diffraction peak intensity of the Al_2O_3 did not change significantly at this time. Therefore, it can be considered that when the proportioning ratio exceeds 0.4, the promotional effect of adding cryolite on aluminum dross denitrification is less than that of the inhibition effect of sintering on aluminum dross denitrification. With the increase of cryolite addition, the diffraction peak of the sodalite ($Na_8Al_6Si_6O_{24}(OH)_2 \cdot 2H_2O$) decreased slowly until it disappeared; this was mainly because the Al/Si ratio in the ideal cell composition of sodalite is one. Therefore, when the amount of cryolite increased, the Al/Si was farther and farther away from the ideal cell composition, which made it difficult to form sodalite. Based on Figure 4, it can be concluded that $m_{cr}:m_{ad}$ = 0.4:1 is the optimal additive ratio, with the denitrification degree of 94.39%.

For the sample with sodium carbonate added, the products after roasting denitrification were mainly Al_2O_3, $NaAlO_2$, and $Na_{1.95}Al_{1.95}Si_{0.05}O_4$. $Na_{1.95}Al_{1.95}Si_{0.05}O_4$, known as zeolite, is a compound with a cellular structure. With the increase of sodium carbonate addition, the diffraction intensity of alumina decreased, and the diffraction intensity of the sodium meta-aluminate increased. Therefore, it can be proved that the chemical reaction involved in Formula (5) did occur in the roasting process, which transformed the dense alumina into sodium aluminate and placed the oxygen more fully in contact with the AlN. Based on Figure 4, it can be concluded that $m_{sc}:m_{ad}$ = 0.8:1 is the optimal additive ratio, with the denitrification degree of 87.63%.

3.2.2. The Effect of Roasting Temperature on the Nitrogen Removal of High-Nitrogen Aluminum Dross by Roasting

Under different temperature conditions, the experimental samples with the additive ratios of $m_{cr}:m_{ad} = 0.4:1$ and $m_{sc}:m_{ad} = 0.8:1$ were placed in electric resistance furnace and roasted for 4 h to explore the effects of different additives on the denitrification degree of high-nitrogen aluminum dross under different temperature conditions. The experimental results are shown in Figure 6.

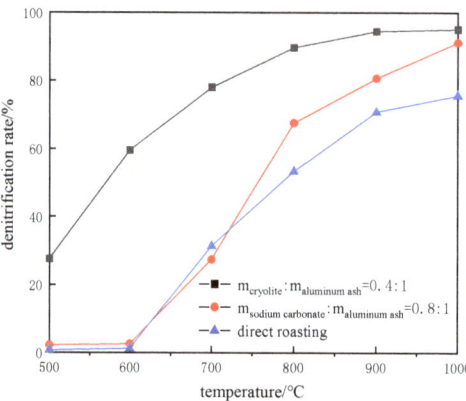

Figure 6. Effects of different temperatures on the nitrogen removal rate of high-nitrogen aluminum dross.

It can be seen from Figure 6 that the denitrification degree showed an upward trend as the temperature increased. For samples with cryolite, as the temperature increased, the denitrification degree gradually increased. At around 900 °C, the denitrification degree reached 94.41%. When the temperature increased to 1000 °C, the denitrification degree was 95.02%. The two did not change much. Therefore, the optimal roasting temperature for nitrogen removal with cryolite was 900 °C. The denitrification degree of direct roasting at the same temperature was only 70.79%. For the sample with sodium carbonate added, the effect was not obvious at the low temperature stage. As the temperature increased above 600 °C, the denitrification degree increased sharply. This was mainly because the alumina reacted with sodium carbonate when the temperature rose. At this time, more AlN was exposed to the air atmosphere, resulting in an increase in the denitrification degree. At 1000 °C, the denitrification degree reached the maximum of 91.15%. Therefore, the optimal roasting temperature for nitrogen removal with sodium carbonate added was 1000 °C. The denitrification degree of direct roasting at the same temperature was only 75.05%. The samples treated with different roasting temperatures were tested by XRD to study the phase transformation law under different roasting temperature conditions, as shown in Figure 7.

For the sample with cryolite added, there was mainly Al_2O_3, Na_3AlF_6, NaCl, and $Na_8Al_6Si_6O_{24}(OH)_2·2H_2O$ in the sample during roasting. With increases in temperature, the diffraction intensity of the Al_2O_3 did not change significantly. Therefore, it can be known that the principle of denitrification via the addition of cryolite does not directly destroy the oxide film wrapped on the surface of AlN, but the oxygen atom in the oxygen can replace the fluoride atom in the cryolite during roasting to produce a gas similar to O_xF_{2x}. Since O_xF_{2x} is more oxidizing than oxygen, AlN can be oxidized at a lower temperature to form N_yF_x and Al_2O_3, and the reaction between N_yF_x and O_2 can form O_xF_{2x} again, so that the reaction can continue.

For the sample with sodium carbonate added, Al_2O_3, $NaAlO_2$, and $Na_{1.95}Al_{1.95}Si_{0.05}O_4$ were mainly present in the sample during roasting. With increases in temperature, the diffraction peak intensity of the Al_2O_3 decreased—some even peaks disappeared, while the diffraction peak intensity of the NaAlO$_2$ increased. Therefore, the addition of sodium carbonate to remove AlN in high-nitrogen aluminum dross was mainly achieved by converting Al$_2$O$_3$ into NaAlO$_2$ and destroying the dense oxide film wrapped on the surface of the AlN. In addition, Na$_{1.95}$Al$_{1.95}$Si$_{0.05}$O$_4$ will be produced during roasting, as shown in Formula (7). CO$_2$ will be produced during the formation of Na$_{1.95}$Al$_{1.95}$Si$_{0.05}$O$_4$, which makes the surface of the material loose and porous, with many micropores inside. Therefore, more air will enter the sample, which improves the denitrification degree [18].

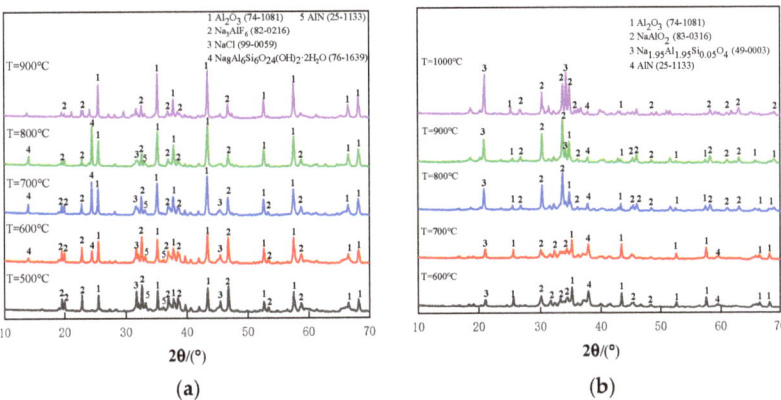

Figure 7. XRD patterns of different temperatures. (a) XRD pattern of the sample with cryolite; (b) XRD pattern of the sample with sodium carbonate.

$$1.95 NaCO_3 + 1.95\, Al_2O_3 + SiO_2 = 2Na_{1.95}Al_{1.95}Si_{0.05}O_4 + 1.95 CO_2, \qquad (7)$$

3.2.3. The Effect of Roasting Time on the Nitrogen Removal of High-Nitrogen Aluminum Dross by Roasting

The experimental samples with the additive ratios of $m_{cr}:m_{ad} = 0.4:1$ and $m_{sc}:m_{ad} = 0.8:1$ were roasted in an electric resistance furnace at 900 °C and 1000 °C for different times to study the effect of roasting time on the denitrification degree of high-nitrogen aluminum dross. The experimental results are shown in Figure 8.

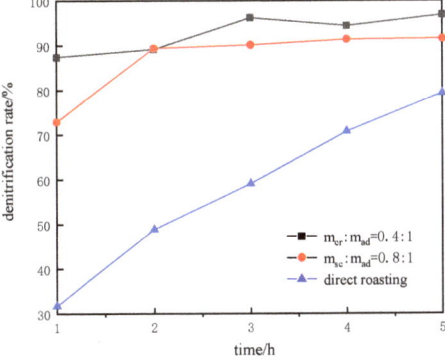

Figure 8. Effect of different times on the nitrogen removal rate of high-nitrogen aluminum dross.

It can be seen from Figure 8 that as the roasting time increased, the denitrification degree of the high-nitrogen aluminum dross showed an upward trend. For samples with cryolite added, the denitrification degree increased significantly after 2 h of calcination, then remained basically unchanged, and then reached the maximum value of 96.19% at 3 h. Therefore, the optimal roasting time for nitrogen removal with cryolite was 3 h. Under the same roasting temperature and roasting time, the denitrification degree of the direct roasting was only 59.12%. For the sample with sodium carbonate added, the denitrification degree was basically maintained at about 90% after 4 h of calcination. After that, increasing the roasting time had no significant effect on the denitrification degree. Therefore, the optimal roasting time for nitrogen removal with sodium carbonate added was 4 h, with a denitrification degree of 91.32%. Under these conditions, the denitrification degree of direct roasting was only 70.79%.

3.3. TG-DSC Analysis

The experimental samples with the additive ratios of $m_{cr}:m_{ad} = 0.4:1$ and $m_{sc}:m_{ad} = 0.8:1$ were analyzed by TG-DSC to study the law of weight change, as well as the endothermic and exothermic changes of the sample during calcination from room temperature to 1000 °C. The experimental results are shown in Figure 9.

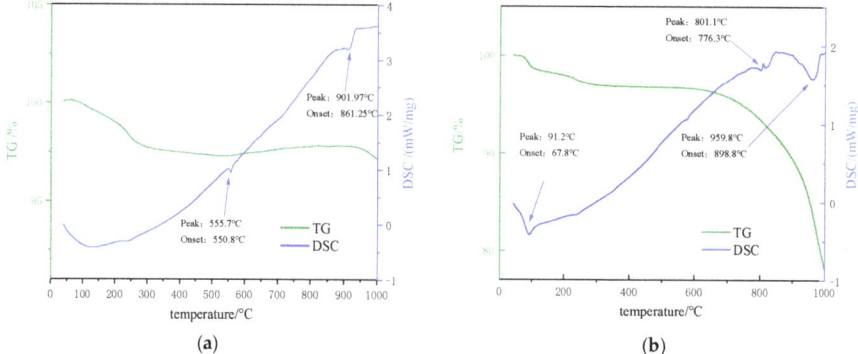

Figure 9. TG-DSC curve. (**a**) with cryolite added; (**b**) with sodium carbonate added.

It can be seen from Figure 9a that at about 100 °C, the TG curve dropped significantly, and the endothermic peak occurred in the DSC curve. This was mainly due to the evaporation of free water absorbed on the sample surface. At 550.8 °C, the TG curve did not change significantly, and a small endothermic peak appeared in the DSC curve. This was mainly due to the endothermic melting of a small amount of metallic aluminum in the sample. At 901.97 °C, an obvious endothermic peak appeared in the DSC curve, and the TG curve had no obvious change. This was mainly because cryolite melts Al_2O_3 in this temperature range. When the temperature increased to 1000 °C, there were no obvious endothermic peaks and exothermic peaks in the DSC curve, while the TG curve showed a downward trend. This was mainly because a small number of substances in the aluminum dross volatilized. It can be seen from Figure 9b that at about 100 °C, the endothermic peak appeared in the DSC curve, and the mass in the TG curve began to decrease. This was mainly due to the evaporation of adsorbed water on the sample surface. At about 800 °C, the DSC curve showed an endothermic peak, and the TG curve showed an obvious downward trend. This was mainly due to the reaction between the alumina and sodium carbonate. At about 960 °C, the DSC curve had an obvious endothermic peak, and the quality of the TG curve was reduced. This was mainly due to the decomposition of part of the sodium carbonate.

3.4. Morphology of the Samples before and after Roasting

Samples with different additive ratios and roasting process conditions were analyzed by micromorphology. The SEM picture is shown in Figure 10. Figure 10a shows the micromorphology of the high-nitrogen aluminum dross before roasting. It can be seen that there were a lot of irregularly shaped particles agglomerated together. The larger ones were alumina particles, and the rest were AlN and some small impurities. Figure 10b shows the sample micromorphology when $m_{sc}:m_{ad} = 0.8:1$, the roasting temperature was 1000 °C, and the roasting time was 4 h. It can be seen that the micro shape of the sample was relatively regular, that there were pores on the sample surface, and that there were relatively few heterogeneous grains. Figure 10c shows the sample micromorphology when $m_{cr}:m_{ad} = 0.4:1$, the roasting temperature was 900 °C, and the roasting time was 3 h. It can be seen that the sample particle shape was mostly flakes that crossed each other. There were still a few impure phases on the surface of the flaky particles, and they were looser compared to Figure 10b,c, which was more conducive to the contact between AlN and oxygen and the increase in denitrification degree.

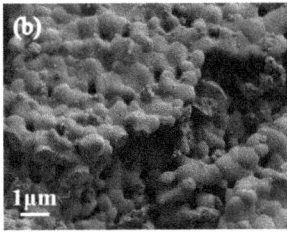

Figure 10. SEM images of different additive ratios and roasting processes. (**a**) Micromorphology of the high-nitrogen aluminum dross; (**b**) with sodium carbonate added; (**c**) with cryolite added.

The samples after the different roasting processes were compared and analyzed by micromorphology. According to the relevant experimental results, the micromorphology of the high-nitrogen aluminum dross after roasting became more regular, and the specific surface area became larger, which made the AlN more fully in contact with the air and improved the denitrification degree. Therefore, the poor effect of the direct roasting nitrogen removal was mainly due to the insufficient contact of AlN with the air, due to being wrapped by other substances.

4. Conclusions

In view of a single non-harmful treatment method for high-nitrogen aluminum dross and the difficulties in the efficient removal of AlN, the roasting denitrification degree was accelerated by adding cryolite and sodium carbonate, and the optimum process parameters were studied. The conclusions are as follows:

(1) The optimal process of nitrogen removal for adding cryolite: for $m_{cr}:m_{ad} = 0.4:1$, a roasting temperature of 900 °C, and a roasting time of 3 h, the denitrification degree was 96.19%. The optimal process of nitrogen removal for adding sodium carbonate: for $m_{sc}:m_{ad} = 0.8:1$, a roasting temperature of 1000 °C, and a roasting time of 4 h, the denitrification degree was 91.32%.

(2) The main reason for adding cryolite for denitrification is that in the roasting process, the micro morphology of the sample became flakey with a layered structure, which effectively increased the contact area between the sample and air, so as to improve the denitrification degree.

(3) The main reason for adding sodium carbonate to improve the denitrification degree of aluminum dross is that sodium carbonate and Al_2O_3 can produce stress at about 800 °C, which destroys the oxide film wrapped on the surface of AlN—thus increasing the contact area between the N and the air and improving the denitrification degree.

(4) The poor effect of the direct roasting nitrogen removal was mainly due to insufficient contact with the air caused by AlN being wrapped by alumina and other impurities. After roasting, the microscopic shape of the sample was more regular and the specific surface area became larger, which made the contact between the AlN and air more sufficient and improved the denitrification degree.

Author Contributions: Conceptualization, H.-J.N. and X.-X.W.; methodology, Y.Z.; validation, Y.Z.; formal analysis, H.-J.N. and S.-S.L.; investigation, Y.Z. and X.-X.W.; resources, H.-J.N. and S.-S.L.; writing—original draft preparation, Y.Z.; writing—review and editing, H.-J.N. and S.-S.L.; visualization, C.-Y.L. and W.-Y.W.; supervision, H.-J.N. All authors have read and agreed to the published version of the manuscript.

Funding: This research was supported by the Priority Academic Program Development of Jiangsu Higher Education Institutions, grant number PAPD.

Institutional Review Board Statement: Not applicable.

Informed Consent Statement: Not applicable.

Data Availability Statement: The authors confirm that the data supporting the findings of this study are available within the article.

Conflicts of Interest: The authors declare no conflict of interest.

References

1. Ministry of Ecology and Environment of the People's Republic of China. *National List of Hazardous Wastes (2021 Edition), Order No. 15 of the Ministry of Ecology and Environment*; Ministry of Ecology and Environment of the People's Republic of China: Beijing, China, 2021.
2. Sarker, S.; Alam, Z.; Qadir, R.; Gafur, M.; Moniruzzaman, M. Extraction and characterization of alumina nanopowders from aluminum dross by acid dissolution process. *Int. J. Miner. Metall. Mater.* **2015**, *22*, 429–436. [CrossRef]
3. Lv, S.S.; Zhang, J.Q.; Ni, H.J.; Wang, X.X.; Zhu, Y.; Gu, T. Study on Preparation of Aluminum Ash Coating Based on Plasma Spray. *Appl. Sci.* **2019**, *9*, 4980. [CrossRef]
4. Ni, H.J.; Zhang, J.Q.; Lv, S.S.; Gu, T.; Wang, X.X. Performance Optimization of Original Aluminum Ash Coating. *Coatings* **2020**, *10*, 831. [CrossRef]
5. Tsakiridis, P.; Ouatadakis, P.; Agatzini, L. Aluminium recovery during black dross hydrothermal treatment. *J. Environ. Chem. Eng.* **2013**, *1*, 23–32. [CrossRef]
6. Mahinroosta, M.; Allahverdi, A. Enhanced alumina recovery from secondary aluminum dross for high purity nanostructured γ-alumina powder production: Kinetic study. *J. Environ. Manag.* **2018**, *212*, 278–291. [CrossRef] [PubMed]
7. Mirian, C.S.; Raphael, H. Effect of disposal of aluminum recycling waste in soil and water bodies. *Environ. Earth Sci.* **2016**, *75*, 1–10.
8. Bruckard, W.J.; Woodcock, J.T. Recovery of valuable materials from aluminium salt cakes. *Int. J. Miner. Process.* **2009**, *93*, 1–5. [CrossRef]
9. Wang, J.H.; Zhong, Y.Q.; Tong, Y.; Xu, X.L.; Lin, G.Y. Removal of AlN from secondary aluminum dross by pyrometallurgical treatment. *J. Cent. South Univ.* **2021**, *28*, 386–397. [CrossRef]
10. Li, Y.; Peng, L.; Wang, H.; Qing, Z.; Qu, Y.; Li, Y.; Li, C.; Wang, Y. Experimental study on denitrification and fluorine fixation of secondary aluminum dross by high temperature roasting. *Conserv. Util. Miner. Resour.* **2020**, *40*, 133–140. (In Chinese)
11. Cong, L.L.; Wang, F.; Jin, Y.; Yuan, Y.J. Determination of aluminum nitride in aluminum dross by Kjeldahl method. *Chem. Anal. Meterage* **2020**, *29*, 63–66. (In Chinese)
12. Huang, Y.X.; Sun, W.Z. *Common Chemical Element Analysis Methods*; Chemical Industry Press: Beijing, China, 2008; pp. 217–220. (In Chinese)
13. Lv, S.S.; Ni, W.; Ni, H.J.; Wang, X.X. Study on hydrolysis of aluminum dross based on orthogonal experiment and nonlinear regression analysis. *Nonferr. Met. Eng.* **2019**, *9*, 52–56. (In Chinese)
14. Bazhin, V.Y.; Boikov, A.V.; Sman, A.V.; Ivanov, P.V. Optoelectronic method for monitoring the state of the cryolite melt in aluminum electrolyzers. *Russ. J. Non-Ferr. Met.* **2015**, *56*, 6–9. [CrossRef]
15. Jong, W.; Lee, H. Thermal and carbothermic decomposition of Na_2CO_3 and Li_2CO_3. *Metall. Mater. Trans. B* **2001**, *32*, 17–24.
16. Guo, H.; Wang, J.; Zhang, X.; Zheng, F.; Li, P. Study on the extraction of aluminum from aluminum dross using alkali roasting and subsequent synthesis of mesoporous γ-alumina. *Metall. Mater. Trans. B* **2018**, *49*, 2906–2916. [CrossRef]
17. Tang, L. *Study on Conversion of Aluminum Nitride in Aluminum Dross during Roasting and Hydrolysis*; Northeastern University: Boston, MA, USA, 2015. (In Chinese)
18. Li, Y.; Chen, X.; Liu, B. Experimental Study on Denitrification of Black Aluminum Dross. *JOM* **2021**, *73*, 2635–2642. [CrossRef]

Article

PLLA/Graphene Nanocomposites Membranes with Improved Biocompatibility and Mechanical Properties

Yaoting He [1], Jiafei Yan [2], Xuzhao He [1], Wenjian Weng [1] and Kui Cheng [1,*]

[1] School of Materials Science and Engineering, State Key Laboratory of Silicon Materials, Center of Rehabilitation Biomedical Materials, Cyrus Tang Center for Sensor Materials and Applications, Zhejiang University, Hangzhou 310027, China; 22026060@zju.edu.cn (Y.H.); hf@zju.edu.cn (X.H.); wengwj@zju.edu.cn (W.W.)

[2] The Affiliated Sir Run Run Shaw Hospital of Medical College, Zhejiang University, Hangzhou 310016, China; yan_jiafei@163.com

* Correspondence: chengkui@zju.edu.cn; Tel.: +86-0571-8795-3945

Abstract: In this work, nanocomposite membranes based on graphene and polylactide were evaluated for mechanical properties and biocompatibility. Single-layer graphene (SLG), graphene nanosheets (GNS), and poly L-lactic acid (PLLA) were prepared through layer-by-layer deposition and homogeneous mixing. The results revealed that PLLA/SLG nanocomposites and PLLA/GNS nanocomposites could show enhanced mechanical properties and biocompatibility. The addition of a tiny amount of SLG significantly improved Young's modulus and tensile strength of the PLLA matrix by 15.9% and 32.8% respectively, while the addition of the same mass ratio of GNS boosted the elongation at break of the PLLA matrix by 79.7%. These results were ascribed to the crystallinity and interfacial interaction differences resulting from graphene incorporation. Also, improved biocompatibility was observed with graphene incorporation. Such nanocomposites membranes showed a lot of potential as environment-friendly and biomedical materials.

Keywords: single-layer graphene; mechanical properties; thermal stability; biocompatibility

Citation: He, Y.; Yan, J.; He, X.; Weng, W.; Cheng, K. PLLA/Graphene Nanocomposites Membranes with Improved Biocompatibility and Mechanical Properties. *Coatings* **2022**, *12*, 718. https://doi.org/10.3390/coatings12060718

Academic Editor: Giorgos Skordaris

Received: 29 April 2022
Accepted: 19 May 2022
Published: 24 May 2022

Publisher's Note: MDPI stays neutral with regard to jurisdictional claims in published maps and institutional affiliations.

Copyright: © 2022 by the authors. Licensee MDPI, Basel, Switzerland. This article is an open access article distributed under the terms and conditions of the Creative Commons Attribution (CC BY) license (https://creativecommons.org/licenses/by/4.0/).

1. Introduction

Biodegradable polymers attract more and more attention in many biomedical and environmental areas. As a typical biodegradable polymer, polylactic acid (PLA) is a polyester material with superior optical, physical, mechanical, and barrier properties, PLA has been widely used in the production of environment-friendly food packaging as freestanding membranes or coatings [1–4] and has great potential as a sustainable alternative to petrochemical-derived polymers [5]. In addition, PLA has also played a great role in advancing the development of biomedical materials due to its superior biocompatibility and degradability. Currently, sutures, stents, orthopedic implants, and drug delivery systems produced from polylactic acid and other biodegradable polymers have mature clinical applications [6–9].

However, rather poor tensile strength and weak impact toughness owing to the amorphous or semi-crystalline structure of PLA limited its application. Normal technical methods are found to be difficult to increase these crystalline characteristics [10,11]. Hence, modification of PLA becomes one of the research focuses. The mechanical characteristics and thermal stability of PLA have been improved using a variety of techniques. One of the most efficient ways to enhance the crystallization characteristics of PLA is to use nanofillers as nucleating agents to lower the nucleation activation energy [12–17]. Therefore, a variety of high-performance nano-reinforcing agents such as nanoclays, nanofibers, carbon nanotubes, etc. were developed [18–22].

Graphene is a nanomaterial with one-atom-thick carbon atoms arranged in a two-dimensional honeycomb structure. Since the first separation of graphene by Navoselov in 2004 [23], graphene has sparked the extensive interest of scientists in various fields due to its unique physical and chemical properties, including ultra-high specific surface area, exceptional electrical, thermal, and mechanical properties [24–26]. At the same time, these properties make graphene an ideal nano-enhancer, and some composite materials with higher comprehensive properties have been obtained [27–29]. It is noteworthy that the reinforcing impact of nanofillers like graphene in polymers is heavily reliant on their efficient interfacial interaction with the polymer matrix [30,31]. In general, the larger the contact area of the nanofiller with the polymer matrix, the better the dispersion and the performance of composites. Considering the application requirements of biomedical materials, the good biocompatibility of nanofillers is essential. The response of living cells to the graphene-filled polymer nanomaterials depends greatly on the properties of graphene such as layer number, lateral size, purity, dose, surface chemistry, and hydrophilicity. Therefore, the selection of nanofillers as reinforcing agents usually requires consideration of several factors including the surface area, type, structure, and biocompatibility of the filler [32].

Presently, numerous experiments and theoretical studies have been conducted on the PLA composites of graphene and modified graphene [33–35]. Nonetheless, research on single-layer graphene (SLG) and graphene nanosheets (GNS)-filled PLA composite materials is still in its early stages. This is primarily because the compounding of PLA with SLG based on completely preserving the properties of SLG is still challenging. Despite this, as a material that can enhance the ability of human osteogenic differentiation [36], the performance and prospective biomedical applications of SLG and GNS in PLA composites cannot be underestimated.

In this article, the impacts and properties of SLG and GNS in poly L-lactic acid (PLLA) nanocomposites are compared. The effects of the different interactions between SLG/GNS and PLLA were elucidated by a direct comparison of various properties of the two composites using the same preparation methods and process parameters. The tensile properties and fracture toughness of SLG and GNS-filled PLLA nanocomposites were investigated. The biocompatibility of the nanocomposites was also studied.

2. Materials and Experimental Procedure

2.1. Preparation of PLLA/SLG Films

The PLLA/SLG films were prepared following the method described in the literature [37]. Briefly, a single-layer Gr (Xianfeng Nanomaterials, Nanjing, China) grown on a Cu substrate by CVD was spin-coated with 20–30 mg/mL PLLA solution (Huanuo Biomaterials, Changchun, China, the solvent was dichloride methane, molecular weight 200,000). The spin-coated polymer film was dried at room temperature for 2 h to completely evaporate the solvent. Then, the copper substrate was etched away with an aqueous ammonium persulfate solution (Aladdin Chemical Reagent, 0.1 M, Wokai Biotechnology Co., Ltd., Shanghai, China) for less than 8 h. After the copper sheet was completely etched, slowly rinse the PLLA/SLG film with deionized water more than 4 times.

2.2. Fabrication of PLLA/SLG and PLLA/GNS Composites

The composites were prepared by the solution casting method. Using a film applicator to flatly and evenly cover the PLLA surface of the PLLA/SLG film with the 50 mg/mL PLLA solution (the solvent was 1,4-dioxane), cut the obtained PLLA/SLG (PSG) film into desired shapes after drying for 24 h. The mass ratio of PLA to monolayer graphene ($M_{PLLA}/M_{Graphene}$) in PSG composite films was calculated. Then, graphene nanosheets (GNS, Shanghai Pro-Graphene, Shanghai, China) containing the same $M_{PLLA}/M_{Graphene}$ were dispersed in 50 mg/mL PLLA solution to obtain PLLA/GNS solution. The PLLA/GNS solution was uniformly spread into a film with the same thickness parameters using the film applicator and then dried in an oven at 37 °C for 24 h to obtain composite films of PLLA/GNS (PGN).

2.3. Analysis and Characterization

Raman spectra were taken on samples to characterize the chemical components and structure of SLG and GNS using Micro-Raman (Renishaw plc, inVia-Reflex, NT-MDT Co., Moscow, Russia) with a DXR laser operating at 532 nm. X-ray diffraction (XRD) measurement was conducted by an X' pert PRO (PANalytical B.V., Almelo, The Netherlands) X-ray diffractometer at room temperature. The diffracted intensity of Cu Kα radiation (k = 1.54178 Å) was recorded with a scanning speed of 5°/min from 10° to 80°. The surface morphology of composites was observed with scanning electron microscopy (SEM, S-4800, SU-70 Hitachhi, Tokyo, Japan).

The mechanical properties of composites were tested by using a tensile machine (Zwick GmbH&Co.KG, Z005, Dongguan, China) at a crosshead speed of 5 mm/min at room temperature. The samples were cut into rectangular shapes of 5 mm wide and 20 mm long. The reported tensile strength and elongation at break were the average values of at least four samples. Crystallization evolution was observed using an E600POL polarizing microscope (POM, NIKON, Japan). To imitate the crystallization behavior during the fabrication process of composites, the melted samples were cooled to room temperature at 80 °C/min, and the crystallization morphology evolution during the cooling process was observed. The thermal stability of the samples was checked using an STA analyzer (TGA/DSC3+, Mettler Toledo, Switzerland) from the temperature range of 50–500 °C with a heating rate of 10 °C/min under the nitrogen atmosphere. Differential scanning calorimetry (DSC) measurement was performed on a TA-DSC calorimeter (PE, Mettler Toledo, Zurich, Switzerland) under a nitrogen flow of 50 mL/min. To eliminate the thermal history of the nanocomposite samples, the samples were heated from 20 °C to 200 °C at a heating rate of 10 °C/min under a nitrogen atmosphere for 5 min. Then, the melt was cooled to 20 °C at a rate of 10 °C/min and reheated to 200 °C. The first cooling and second heating scan curves were recorded.

Rat bone marrow mesenchymal stem cells (BMSCs) were used as model cells. Cell Counting Kit (CCK-8) was used to evaluate the cell adhesion and proliferation capacity of neat PLLA, PLLA/SLG, and PLLA/GNS. After BMSCs (20,000 cells/cm^2) had been seeded on the surface of the nanocomposites and cultured for 1 day and 3 days, the solution of CCK-8 (Dojingdo Laboratories, Kumamoto, Japan) with a concentration ratio of 1:10 to the culture medium was added. Then, after the cells were incubated in an atmosphere of 5% CO_2 and 37 °C for 2 h, the optical density (OD) of the solution was measured at 450 nm with a microplate reader (Multiskan MK3, Thermo Fisher Scientific, Shanghai, China). The contact angle test was used to evaluate the hydrophilicity and hydrophobicity of material surfaces (OCA 20, Dataphysics, Stuttgart, Germany). Contact angles reported are the average of at least three measurements per sample.

3. Results and Discussion

3.1. Composition, Structure and Morphology of the Nanocomposite Films

The compositions of neat PLLA and representative PLLA/SLG (PSG) and PLLA/GNS (PGN) composites at the same mass ratio were determined by confocal laser Raman spectroscopy (Figure 1). Different from pure PLLA, exhibiting characteristic bands at 2940 cm^{-1} identified as C-H stretching modes of -CH_3 groups (Figure 1a,b), new bands appeared in the Raman spectra of PSG and PGN, with wavelengths around 1580 cm^{-1} and 2700 cm^{-1}, corresponding to the G and 2D peaks of graphene (Figure 1a,b). Combined with the I_G/I_{2D} value of each composite material around 1:1 (Figure 1a,b), it can be seen that the graphene in the composite material is complete and uniformly dispersed. Fully confirmed the effective combination of SLG and PLLA, and GNS dispersed in the PLLA matrix. In addition, the appearance of graphene D peaks was not observed, indicating that the structures of the obtained SLG and GNS were complete and defect-free.

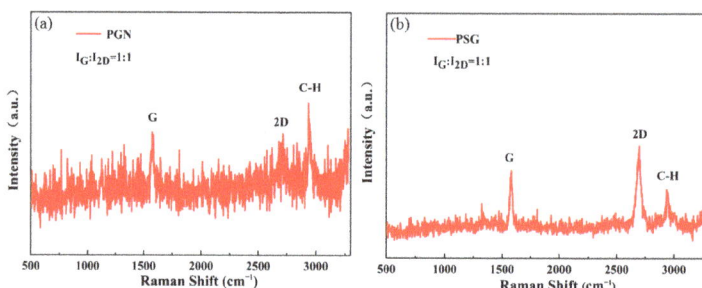

Figure 1. Raman spectra of (**a**) PSG and (**b**) PGN nanocomposites with the same $M_{PLLA}/M_{Graphene}$.

To characterize the morphology and dispersion of SLG and GNS in the composites, SEM images of PLLA, PSG, and PGN composite films were studied (Figure 2). Because of the various types and adding procedures of graphene, the surface of each film was unique. The surface of the composite film added with SLG was comparable to that surface of neat PLLA, with dense and fine dendrites distributed (Figure 2a,b). The addition of GNS will make the dendrites on the surface of the film connected to a certain extent, resulting in a thicker crystalline surface (Figure 2c). This might originate from the interfacial interaction between GNS and PLLA. It was also consistent with the subsequent crystallization test. It is worth mentioning, however, that because of the limited compatibility of GNS with the PLLA matrix, there were some GNS aggregation on the surface indicated by the red arrows (Figure 2c).

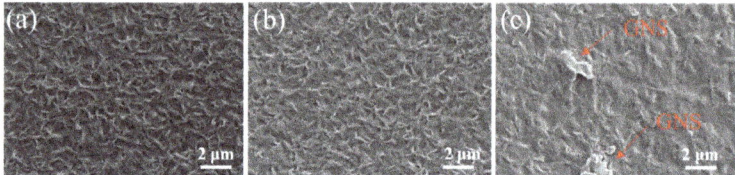

Figure 2. SEM images of surface morphologies of (**a**) neat PLLA (**b**) PSG and (**c**) PGN composites films.

3.2. Crystallization Behavior of the Nanocomposites

The crystallinity of a polymer plays a crucial role in its performance. As a semi-crystalline polymer, it is very meaningful to study the crystallization of PLLA and the effect of SLG and GNS as fillers on its crystallization behavior. The XRD patterns of nanocomposite films including pure PLLA, PSG, and PGN are shown in Figure 3. In general, crystals with two different structures, α, and β, can be detected in PLLA. The diffraction patterns of pure PLLA, PSG, and PGN in Figure 3 all have the strongest diffraction peaks at 2θ = 16.5°, corresponding to the (110) and (200) reflections of PLLA, which indicate that PLLA in nanocomposites is in the order of orthorhombic α-crystal form exists [38]. Moreover, less intense diffraction peaks were observed at 18.7° and 21.7°, corresponding to the (203) and (210) reflections of PPLA, respectively. The creation of a few mesomorphic structures in the PLLA due to an excessive cooling rate is primarily responsible for the appearance of these peaks [39,40]. All XRD patterns revealed the same diffraction peaks, which means that the incorporation of SLG and GNS did not change the crystal structure of PLLA in the nanocomposites. Interestingly, the diffraction peak intensity of PLLA was significantly increased in PSG and PGN nanocomposites, indicating an improvement in crystallinity and an increase in crystal order [41]. In addition, as shown in Figure 3, in the XRD patterns of PSG and PGN, the characteristic diffraction peak at 26.5° attributed to graphite was not detected, indicating that the dispersion of graphene in the PLLA matrix was good and close to the monolayer level, which was consistent with the results of Raman spectroscopy.

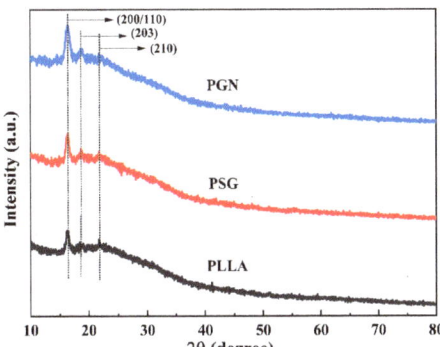

Figure 3. XRD patterns of neat PLLA, PSG and PGN nanocomposite films.

Since PLLA is a semicrystalline polymer, the mechanical properties are influenced by the degree of the crystallinity to some extent [42,43]. DSC was used to investigate the crystallization behavior of PLLA in the nanocomposite. Figure 4 showed the DSC heating and cooling curves of the neat PLLA, PSG and PGN nanocomposites at a scan rate of 10 °C/min, and the results of typical thermal properties, including T_g (glass transition temperature), T_c (cold crystallization temperature), T_m (melting point temperature), ΔH_c (the cold crystallization enthalpy and χ_c (the crystallinity) were listed in Table 1. The degree of crystallization of the films was calculated by the following Equation (1),

$$\chi_c = \left(\frac{\Delta H_c}{\Delta H_m^0}\right) \times 100\% \tag{1}$$

where χ_c is the percent crystallinity, ΔH_c was the enthalpy of the cold crystallization of nanocomposites, and ΔH_m^0 was the enthalpy of the melting with a value of 93.0 J/g, which was reported to be the melting enthalpy for 100% crystalline PLLA [44]. With the existence of SLG and GNS, the exothermic peaks become more noticeable and ΔH_c was higher. Furthermore, at the same mass ratio of graphene filling, PSG nanocomposites have distinct and sharper exothermic peaks than PGN nanocomposites. This is because graphene provides PLLA with heterogeneous nucleation sites, which improves the crystallization ability of PLLA, and the nucleation speed of SLG is faster than that of GNS. Both T_c and χ_c of the nanocomposites were significantly increased after the addition of SLG and GNS, indicating that the presence of SLG and GNS favored the non-isothermal cold crystallization behavior of PLLA due to the efficient nucleating effect of graphene for PLLA crystallization. For polymers, the high molecular mobility at high temperatures is very beneficial for crystal nucleation. In the presence of graphene, the exothermic amplitude of PLLA crystals increases and the crystallization exotherm sharpens, meaning that both crystallization nucleation and crystal growth of the PLLA matrix are enhanced. Ajala et al. [40] likewise came up with similar findings. The increased T_g values of PSG and PGN nanocomposites can be attributed to the effective linkage of PLLA to graphene that restricts the segmental motion of PLLA chains, and this increased stability also indicates a higher interfacial bond strength between PLLA and graphene [45]. Further, a rise in T_m in nanocomposites suggested that SLG and GNS are beneficial to the perfection of crystallites and prevent the formation of more defective PLLA crystallites.

Table 1. Calorimetric data derived from the DSC cooling curves and heating curves for the PLLA, PSG and PGN nanocomposites.

Samples	T_g (°C)	T_c (°C)	T_m (°C)	ΔH_c (J/g)	χ_c (%)
PLLA	60.4	116.9	176.5	33.4	35.9
PSG	62.1	132.2	177.6	38.0	40.9
PGN	65.3	122.8	178.6	39.6	42.6

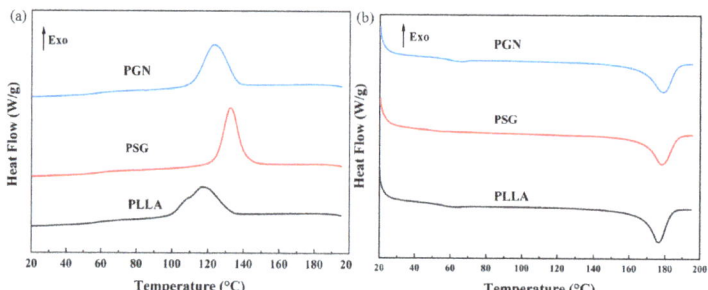

Figure 4. DSC curves of (**a**) the cooling curves and (**b**) the heating curves of PLLA, PSG and PGN nanocomposites with different filler contents.

To further specifically explore the nucleation effect of graphene on PLLA crystallization, polarizing microscopy (POM) was used to study the isothermal crystallization morphologies of neat PLLA and its nanocomposites at around their crystallization temperatures (120, 127 and 136 °C, respectively). Figure 5 showed the spherulitic morphology of neat PLLA, PSG, and PGN at different crystallization times. Only a few nuclei appear in neat PLLA whereas PSG and PGN show more nuclei after 1 min (Figure 5a,d,g). The morphological differences for PLLA, PSG, and PGN are more obvious after 5 min (Figure 5c,f,i). For neat PLLA, the number of spherulites is small, and their size is relatively big because the spherulites had large space to grow before impinging on each other (Figure 5c). With the addition of graphene, the number of PLLA spherulites increases significantly, and consequently, their size is dramatically reduced (Figure 5f,i). Spherulitic morphology studies here clearly suggest the strong heterogeneous nucleation effect of graphene, which is in agreement with DSC findings.

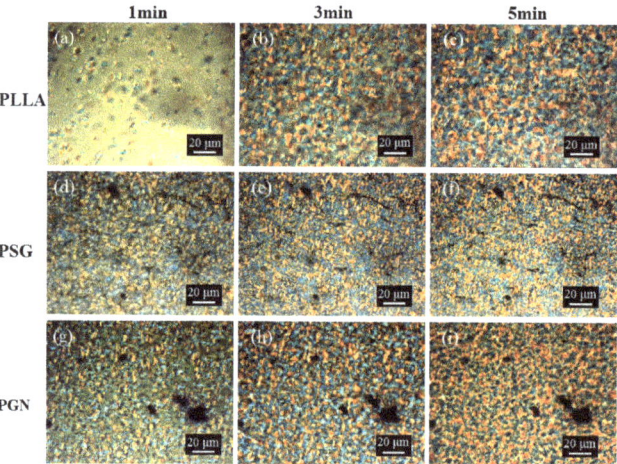

Figure 5. Polarized optical micrographs for 1 min, 3 min and 5 min of (**a**–**c**) neat PLLA isothermally crystallized at 120 °C (**d**–**f**) PSG nanocomposite isothermally crystallized at 136 °C (**g**–**i**) and PGN nanocomposite isothermally crystallized at 127 °C.

3.3. Mechanical, Thermal Properties and Biocompatibility

Considering the remarkable mechanical properties of graphene, such as high mechanical strength and high modulus, it is predicted that the mechanical properties of PLLA nanocomposites will be enhanced by SLG and GNS. The mechanical properties of pure PLLA, PSG, and PGN nanocomposite films were evaluated using tensile tests, and typical stress-strain curves are shown in Figure 6a. The variations of tensile strength, Young's modulus, and elongation at break are plotted in Figure 6b–d. As seen in Figure 6a, all nanocomposite films show a typical characteristic of ductile fracture behavior with a well-defined yield point, including the neat PLLA films. While as a brittle material, the reason why neat PLLA exhibit a typical characteristic of ductile fracture behavior may be due to the incomplete discharge of the dissolving agent. Nevertheless, some significant changes can be observed in the mechanical properties of polymer by graphene. The tensile strength and Young's modulus of PSG nanocomposite films were significantly increased as compared to that of neat PLLA film (30.47 MPa and 1.13 GPa), and reached a maximum of 40.45 MPa and 1.31 GPa (Figure 6a,b), indicating that the mechanical strength of PSG nanocomposite films is enhanced. Different from the previous study [46], the toughness of PSG was also improved compared with that of neat PLLA film, and the elongation at break increased from 114.07% to 123.95%. The reason for this simultaneous improvement of mechanical strength and toughness is mainly attributed to the good interfacial interaction between SLG and PLLA matrix. Meanwhile, GNS greatly improves the toughness of PGN nanocomposite films based on sacrificing some mechanical strength, and the elongation at break increases to 204.97%, an increase of 79.7%. (Figure 6d). Compared with neat PLLA, SLG can play the role of strengthening and toughening at the same time, while GNS can greatly improve stretch ability with low content. Therefore, choosing different kinds of graphene as reinforcing fillers has different effects, which is consistent with many previous reports of nanofiller-reinforced PLLA composites [47,48]. This result may be attributed to the different interfacial structures between nanofillers and polymer matrix.

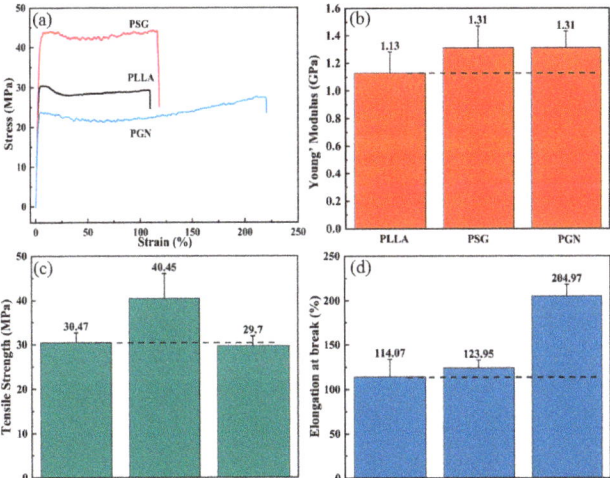

Figure 6. (**a**) Representative stress-strain curves for PLLA, PSG, PGN nanocomposite films; the variations of (**b**) Young's modulus, (**c**) tensile strength, (**d**) elongation at break as a function of PLLA number of layers.

In fact, the improvement in tensile strength and tensile modulus depends on factors such as the shape and dispersion of nanofillers in the matrix, nanofillers interfacial adhesion, and strength with the matrix material. However, the shape and dispersion of graphene play an important role in the enhancement of the tensile properties of PLLA. In PSG, the SLG is composited with the polylactic acid matrix in the mode of whole-sheet coverage, and the compatibility between SLG-PLLA acid and PLLA matrix is stronger, so the effective contact area between the nanofiller and the matrix is larger, and the adhesion energy is higher. When the composite nanomaterials are subjected to stress stretching, SLG can limit the expansion of cracks, avoid the formation of larger cracks, and improve the tensile strength of the composites. In addition, SLG also restricts the mobility of polymer chains under load by blocking the molecular chains. The increased crosslinking rate in the PLLA matrix and the restricted movement of the PLLA macromolecular chains ultimately improved the tensile modulus of the nanocomposite PSG. From the shape and dispersibility of nanofillers, GNS are complexed in the matrix in the form of small clusters. Due to the strong van der Waals forces between GNS, GNS are more likely to accumulate in the matrix and have poorer dispersion. The interfacial interaction force between GNS and PLLA is weaker, which also leads to a slight decrease in the strength of PGN. Therefore, the good interfacial interaction between SLG and PLLA can effectively transfer the load to SLG, enhancing the tensile strength and tensile modulus of PLLA/graphene nanocomposites.

On the other hand, since PLLA is semi-crystalline, including the interfacial interaction between the nanofillers and the polymer matrix, the crystal structure, and morphology of the polymer matrix are also key factors affecting the mechanical properties of nanocomposites. The results of XRD (Figure 3), DSC (Figure 4), and POM (Figure 5) confirmed that the addition of SLG and GNS both improved the crystallinity of PLLA, and benefited from the excellent heterogeneous nucleating ability of graphene on PLLA crystallization. The nanofiller-induced interfacial crystallization layer plays a key role in the non-covalent stress transfer between graphene and PLLA. As mentioned earlier, the larger contact area between SLG and PLLA matrix results in the formation of a larger interfacial crystalline layer, which can lead to a significant increase in mechanical strength. The addition of SLG and GNS not only improved the crystallinity of the nanocomposites, but also tended to induce the PLLA matrix to form spherulites with smaller diameters. It is very important for the improvement of toughness. In general, unreinforced amorphous polymers undergo plastic deformation through unhindered shear band propagation. The fine-grained PLLA crystals in PSG and PGN provide resistance to the propagation of shear bands, thus, the plasticity of the nanocomposites is greatly enhanced. Especially in the PGN nanocomposites with the highest crystallinity and the smallest spherulite size, this toughness enhancement is extremely obvious.

Aliphatic polyesters such as PLLA in particular are easy to hydrolyze and thermally degrade to monomers and oligomers, improving the thermal stability of PLLA an important part as regards processing. It was found that the incorporation of SLG and GNS not only improved the mechanical properties of PLLA, but also enhanced its thermal stability. Figure 7 and Table 2 show the TGA and corresponding differential thermogravimetric analysis (DTA) results of the neat PLLA, PSG, and PGN composite films. In PSG and PGN nanocomposites, the decomposition onset temperature (T_{onset}), the maximum decomposition temperature (T_{max}), and the decomposition temperature at 50% weight loss (T_{50}) of the composite films shifted to higher temperatures. Among them, the decomposition onset temperature and maximum decomposition temperature of PSG were observed to vary from 267 °C to 271 °C and from 348 °C to 349 °C, respectively. The improved thermal stability is mainly due to the presence of SLG and GNS as reinforcing fillers, they can delay oxygen permeation and volatile degradation products escape and coke formation by the so-called "tortuous path" effect of graphene [49]. Moreover, it can be seen from the DTA curves that the DTA peaks of PGN and PSG are higher in temperature than that of neat PLLA. This change can be ascribed that when PLLA is heated, free radicals are generated which cause the degradation of the PLLA matrix. The increase in the crystallinity of PSG and PGN makes the degradation rate slower, and the degradation requires more energy, so the degradation temperature is also higher [50]. Therefore, good interfacial interaction and high crystallinity also contribute to the thermal stability of PLLA nanocomposites.

Table 2. Thermal data derived from the TGA curves and DTA curves of neat PLLA, PSG and PGN nanocomposites.

Samples	T_{onset} (°C)	T_{max} (°C)	T_{50} (°C)
PLLA	267	354	348
PSG	271	355	349
PGN	259	357	350

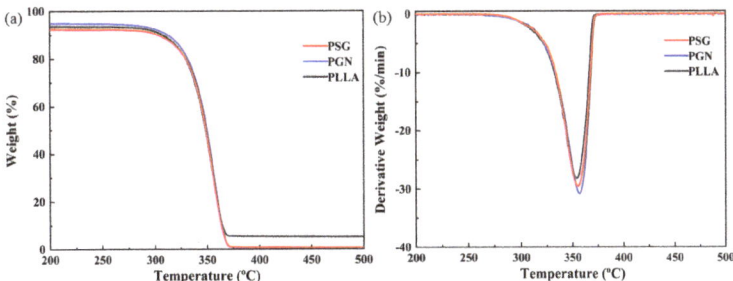

Figure 7. (a) TGA and (b) DTA curves for neat PLLA, PSG and PGN nanocomposite.

From the results obtained above, the fact can be known that SLG and GNS can be used to improve the mechanical properties and thermal stability of PLLA, but the biocompatibility of the composites determines whether PLLA/graphene nanocomposites can be used for the preparation of biomedical materials. The biocompatibility of the nanocomposites was evaluated by the CCK-8 test (Figure 8). All samples were 0.5×0.5 cm^2 in size. Rat bone marrow mesenchymal stem cells (BMSCs) were inoculated and cultured for 1 day and 3 days. As shown in Figure 8, the OD values of PSG and PGN increased slightly after 1 day of culture. But after 3 days, it was found that the OD values of the PLLA/SLG and PLLA/GNS groups were significantly higher than those of the neat PLLA group, indicating that the cells on the surface of the nanocomposites are more than that of neat PLLA. This may benefit from the improvement of SLG and GNS on the surface of the nanocomposites, and the hydrophilic surface improves the cell proliferation ability. From the contact angle test results, the contact of a pure PLLA film was 94.3° (±1.7°). However, after adding a layer of single-layer graphene, the contact angle of the PSG surface was reduced to 88.3° (±1.4°), and the hydrophilicity of the surface was further improved. Meanwhile, there was a small-decreased contact angle of PGN (91.2° ± 0.7°), and all the results revealed the surface of the nanocomposite materials is hydrophilic. All these tests proved that the existence of SLG and GNS provided a good adhesion environment for cells, which is conducive to cell proliferation and has good cell activity.

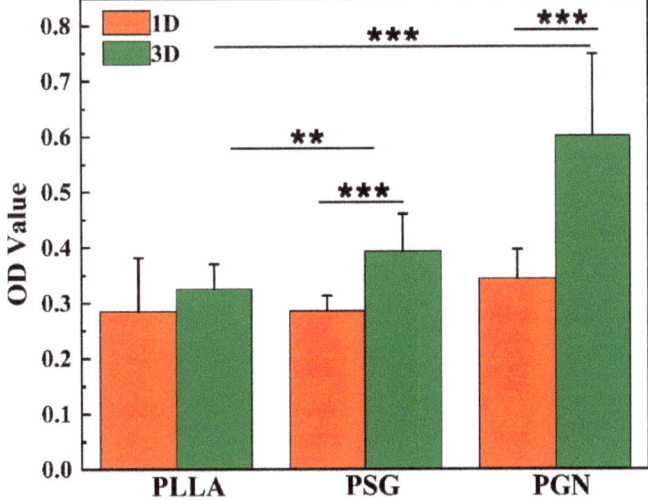

Figure 8. Cell viability measured with CCK-8 assay on PLLA, PSG and PGN nanocomposites monolayer film at different culture times (** $p < 0.01$, *** $p < 0.001$).

4. Conclusions

In this paper, two kinds of graphene, single-layer graphene (SLG) and graphene nanosheets (GNS), and poly L-lactic acid (PLLA) nanocomposite membranes were successfully prepared. The effects of SLG and GNS on the mechanical properties, crystallization ability, thermal stability, and cell viability of PLLA nanocomposites were investigated.

SLG and GNS act as nucleating agents to further enhance the crystallinity of PLLA. The good interaction between SLG and PLLA matrix makes PSG have high mechanical strength. Moreover, SLG and GNS greatly improve the toughness of PGN by changing the shape and crystallinity of PLLA crystal. The tensile strength and fracture toughness of PSG were increased by 32.8% and 8.7% respectively, and the toughness of PGN was increased by 79.7%.

In addition, CCK-8 results indicated that PSG and PGN were non-cytotoxic, and the nanocomposite membranes showed good cell survival and proliferation ability. These PSG and PGN nanocomposites with higher mechanical properties, thermal stability, and cell viability exhibit broad application prospects as biomedical materials.

Author Contributions: Y.H.: conceptualization, methodology, writing—original draft preparation; J.Y.: writing—review and editing and funding acquisition; K.C.: conceptualization, methodology, writing—review and editing, supervision, project administration; X.H.: formal analysis and investigation; W.W.: writing—review and editing. All authors have read and agreed to the published version of the manuscript.

Funding: This research was supported by the Natural Science Foundation of Zhejiang Province of China (Grant no. LGF20H030009).

Institutional Review Board Statement: Not applicable.

Informed Consent Statement: Not applicable.

Data Availability Statement: Not applicable.

Conflicts of Interest: The authors declare no conflict of interest.

References

1. Nampoothiri, K.M.; Nair, N.R.; John, R.P. An overview of the recent developments in polylactide (PLLA) research. *Bioresour. Technol.* **2010**, *101*, 8493–8501. [CrossRef] [PubMed]
2. Saeidlou, S.; Huneault, M.A.; Li, H.B.; Park, C.B. Poly(lactic acid) crystallization. *Prog. Polym. Sci.* **2012**, *37*, 1657–1677. [CrossRef]
3. Savaris, M.; Santos, V.D.; Brandalise, R.N. Influence of different sterilization processes on the properties of commercial poly(lactic acid). *Mater. Sci. Eng. C* **2016**, *69*, 661–667. [CrossRef] [PubMed]
4. Yang, F.; Murugan, R.; Ramakrishna, S.; Wang, X.; Ma, Y.X.; Wang, S. Fabrication of nanostructured porous PLLA scaffold intended for nerve tissue engineering. *Biomaterials* **2005**, *25*, 1891–1900. [CrossRef] [PubMed]
5. John, R.P.; Nampoothiri, K.M.; Pandey, A. Solid-state fermentation for L-lactic acid production from agro wastes using Lactobacillus delbrueckii. *Proc. Biochem.* **2006**, *41*, 759–763. [CrossRef]
6. Singhvi, M.S.; Zinjarde, S.S.; Gokhale, D.V. Polylactic acid: Synthesis and biomedical applications. *J. Appl. Microbiol.* **2019**, *127*, 1612–1626. [CrossRef]
7. Guerra, A.J.; Cano, P.; Rabionet, M.; Puig, T.; Ciurana, J. 3D-Printed PCL/PLLA Composite Stents: Towards a New Solution to Cardiovascular Problems. *Materials* **2018**, *11*, 1679. [CrossRef]
8. Zhang, B.Q.; Wang, L.; Song, P.; Pei, X.; Sun, H.; Wu, L.A.; Zhou, C.C.; Wang, K.F.; Fan, Y.J.; Zhang, X.D. 3D printed bone tissue regenerative PLLA/HA scaffolds with comprehensive performance optimizations. *Mater. Des.* **2021**, *201*, 109490. [CrossRef]
9. Liu, S.Q.; Yu, J.J.; Li, H.M.; Wang, K.W.; Wu, G.H.; Wang, B.W.; Liu, M.F.; Zhang, Y.; Wang, P.; Zhang, J. Controllable Drug Release Behavior of Polylactic Acid (PLLA) Surgical Suture Coating with Ciprofloxacin (CPFX)-Polycaprolactone (PCL)/Polyglycolide (PGA). *Polymers* **2020**, *12*, 288. [CrossRef]
10. Rasal, R.M.; Janorkar, A.V.; Hirt, D.E. Poly(lactic acid) modifications. *Prog. Polym. Sci.* **2010**, *35*, 338–356. [CrossRef]
11. Lim, L.T.; Auras, R.; Rubino, M. Processing technologies for poly(lactic acid). *Prog. Polym. Sci.* **2008**, *33*, 820–852. [CrossRef]
12. Bharadwaz, A.; Jayasuriya, A.C. Recent trends in the application of widely used natural and synthetic polymer nanocomposites in bone tissue regeneration. *Mater. Sci. Eng. C-Mater.* **2020**, *110*, 110698. [CrossRef]
13. Terzopoulou, Z.; Klonos, P.A.; Kyritsis, A.; Tziolas, A.; Avgeropoulos, A.; Papageorgiou, G.Z.; Bikiaris, D.N. Interfacial interactions, crystallization and molecular mobility in nanocomposites of Poly(lactic acid) filled with new hybrid inclusions based on graphene oxide and silica nanoparticles. *Polymer* **2019**, *166*, 1–12. [CrossRef]

14. Raquez, J.M.; Murena, Y.; Goffin, A.L.; Habibi, Y.; Ruelle, B.; DeBuyl, F.; Dubois, P. Surface-modification of cellulose nanowhiskers and their use as nanoreinforcers into polylactide: A sustainably-integrated approach. *Compos. Sci. Technol.* **2012**, *72*, 544–549. [CrossRef]
15. Li, Y.C.; Liao, C.Z.; Tjong, S.C. Synthetic Biodegradable Aliphatic Polyester Nanocomposites Reinforced with Nanohydroxyapatite and/or Graphene Oxide for Bone Tissue Engineering Applications. *Nanomaterials* **2019**, *9*, 590. [CrossRef]
16. Bai, T.T.; Zhu, B.; Liu, H.; Wang, Y.M.; Song, G.; Liu, C.T.; Shen, C.Y. Biodegradable poly(lactic acid) nanocomposites reinforced and toughened by carbon nanotubes/clay hybrids. *Int. J. Biol. Macromol.* **2019**, *151*, 628–634. [CrossRef] [PubMed]
17. Xia, S.; Liu, X.B.; Wang, J.F.; Kan, Z.; Chen, H.; Fu, W.X.; Li, Z.B. Role of poly(ethylene glycol) grafted silica nanoparticle shape in toughened PLLA-matrix nanocomposites. *Compos. B Eng.* **2019**, *168*, 398–405. [CrossRef]
18. Mayekar, P.C.; Castro-Aguirre, E.; Auras, R.; Selke, S.; Narayan, R. Effect of Nano-Clay and Surfactant on the Biodegradation of Poly(Lactic Acid) Films. *Polymers* **2020**, *12*, 311. [CrossRef]
19. Liu, X.Z.; He, X.; Jin, D.W.; Wu, S.T.; Wang, H.S.; Yin, M.; Aldalbahi, A.; El-Newehy, M.; Mo, X.M.; Wu, J.L. A biodegradable multifunctional nanofibrous membrane for periodontal tissue regeneration. *Acta Biomater.* **2020**, *108*, 207–222. [CrossRef]
20. Wu, S.H.; Zhou, R.; Zhou, F.; Streubel, P.N.; Chen, S.J.; Duan, B. Electrospun thymosin Beta 4 loaded PLGA/PLLA nanofiber/microfiber hybrid yarns for tendon tissue engineering application. *Mater. Sci. Eng. C-Mater.* **2020**, *106*, 110268. [CrossRef]
21. Tan, Y.J.; Li, J.; Tang, X.H.; Yue, T.N.; Wang, M. Effect of phase morphology and distribution of multi-walled carbon nanotubes on microwave shielding of poly(L-lactide)/poly(epsilon-caprolactone) composites. *Compos. Part A-Appl. Sci. Manuf.* **2020**, *137*, 106008. [CrossRef]
22. Yang, H.T.; Shi, B.B.; Xue, Y.J.; Ma, Z.W.; Liu, L.N.; Liu, L.; Yu, Y.M.; Zhang, Z.Y.; Annamalai, P.K.; Song, P.A. Molecularly Engineered Lignin-Derived Additives Enable Fire-Retardant, UV-Shielding, and Mechanically Strong Polylactide Biocomposites. *Biomacromolecules* **2021**, *22*, 1432–1444. [CrossRef] [PubMed]
23. Novoselov, K.S.; Geim, A.K.; Morozov, S.V.; Jiang, D.; Zhang, Y.; Dubonos, S.V.; Grigorieva, I.V.; Firsov, A.A. Electric field effect in atomically thin carbon films. *Science* **2004**, *306*, 666–669. [CrossRef]
24. Korkmaz, S.; Kariper, I.A. Graphene and graphene oxide based aerogels: Synthesis, characteristics and supercapacitor applications. *J. Energy Storage* **2020**, *27*, 101038. [CrossRef]
25. Afroj, S.; Tan, S.R.; Abdelkader, A.M.; Novoselov, K.S.; Karim, N. Highly Conductive, Scalable, and Machine Washable Graphene-Based E-Textiles for Multifunctional Wearable Electronic Applications. *Adv. Funct. Mater.* **2020**, *30*, 2000293. [CrossRef]
26. Cao, K.; Feng, S.Z.; Han, Y.; Gao Ly, T.H.; Xu, Z.P.; Lu, Y. Elastic straining of free-standing monolayer graphene. *Nat. Commun.* **2020**, *11*, 284. [CrossRef] [PubMed]
27. Bie, C.B.; Yu, H.G.; Cheng, B.; Ho, W.; Fan, J.J.; Yu, J.G. Design, Fabrication, and Mechanism of Nitrogen-Doped Graphene-Based Photocatalyst. *Adv. Mater.* **2021**, *33*, 2003521. [CrossRef]
28. Dai, W.; Lv, L.; Ma, T.F.; Wang, X.Z.; Ying, J.F.; Yan, Q.W.; Tan, X.; Gao, J.Y.; Xue, C.; Yu, J.H. Multiscale Structural Modulation of Anisotropic Graphene Framework for Polymer Composites Achieving Highly Efficient Thermal Energy Management. *Adv. Sci.* **2021**, *8*, 2003734. [CrossRef]
29. Raslam, A.; del Burgo, L.S.; Ciriza, J.; Pedraz, J.L. Graphene oxide and reduced graphene oxide-based scaffolds in regenerative medicine. *Int. J. Pharmaceut.* **2020**, *580*, 119226. [CrossRef]
30. Liang, J.J.; Huang, Y.; Zhang, L.; Wang, Y.; Ma, Y.F.; Guo, T.Y.; Chen, Y.S. Molecular level dispersion of graphene into poly(vinyl alcohol) and effective reinforcement of their nanocomposites. *Adv. Funct. Mater.* **2009**, *19*, 2297–2302. [CrossRef]
31. Cheng, H.K.F.; Sahoo, N.G.; Tan, Y.P.; Pan, Y.Z.; Bao, H.Q.; Li, L.; Chan, S.H.; Zhao, J.H. Poly(vinyl alcohol) nanocomposites filled with poly(vinyl alcohol)-grafted graphene oxide. *ACS Appl. Mater. Interfaces* **2012**, *4*, 2387–2394. [CrossRef] [PubMed]
32. Zakaria, M.R.; Kudus, M.H.A.; Akil, H.M.; Thirmizir, M.Z.M.; Malik, M.F.I.A.; Othman, M.B.H.; Ullah, F.; Javed, F. Comparative Study of Single-Layer Graphene and Single-Walled Carbon Nanotube-Filled Epoxy Nanocomposites Based on Mechanical and Thermal Properties. *Polym. Compos.* **2019**, *40*, 1840–1849. [CrossRef]
33. Caminero, M.A.; Chacon, J.M.; Garcia-PLLAza, E.; Nunez, P.J.; Reverte, J.M.; Becar, J.P. Additive Manufacturing of PLLA-Based Composites Using Fused Filament Fabrication: Effect of Graphene NanoPLLAtelet Reinforcement on Mechanical Properties, Dimensional Accuracy and Texture. *Polymers* **2019**, *11*, 799. [CrossRef] [PubMed]
34. Yang, L.; Zhen, W.J. Poly(lactic acid)/p-phenylenediamine functionalized graphene oxidized nanocomposites: Preparation, rheological behavior and biodegradability. *Eur. Polym. J.* **2019**, *121*, 109341. [CrossRef]
35. Mao, N.D.; Jeong, H.; Nguyen, T.K.N.; Nguyen, T.M.L.; Do, T.V.V.; Thuc, C.N.H.; Perre, P.; Ko, S.C.; Kim, H.G.; Tran, D.T. Polyethylene glycol functionalized graphene oxide and its influences on properties of Poly(lactic acid) biohybrid materials. *Compos. B Eng.* **2019**, *161*, 651–658. [CrossRef]
36. Liu, Y.S.; Chen, T.; Du, F.; Gu, M.; Zhang, P.; Zhang, X.; Liu, J.Z.; Lv, L.W.; Xiong, C.Y.; Zhou, Y.S. Single-Layer Graphene Enhances the Osteogenic Differentiation of Human Mesenchymal Stem Cells In Vitro and In Vivo. *J. Biomed. Nanotechnol.* **2016**, *12*, 1270–1284. [CrossRef]
37. Long, X.J.; Wang, X.Z.; Yao, L.L.; Lin, S.Y.; Zhang, J.M.; Weng, W.J.; Cheng, K.; Wang, H.M.; Lin, J. Graphene/Si-Promoted Osteogenic Differentiation of BMSCs through Light Illumination. *ACS Appl. Mater. Interfaces* **2019**, *11*, 43857–43864. [CrossRef]
38. Hoogsteen, W.; Postema, A.R.; Pennings, A.J.; Ten Brinke, G.; Zugenmaier, P. Crystal structure, conformation and morphology of solution spun poly(L-lactide) fibers. *Macromolecules* **1990**, *23*, 634–642. [CrossRef]

39. Stoclet, G.; Seguela, R.; Lefebvre, J.M.; Rochas, C. New insights on the strain-induced mesophase of poly(D, L-lactide):in situ WAXS and DSC study of the thermos-mechanical stability. *Macromolecules* **2010**, *43*, 7228–7237. [CrossRef]
40. Bao, C.L.; Song, L.; Xing, W.Y.; Yuan, B.H.; Wilkie, C.A.; Huang, J.L.; Guo, Y.Q.; Hu, Y. Preparation of graphene by pressurized oxidation and multiplex reduction and its polymer nanocomposites by masterbatch-based melt blending. *J. Mater. Chem.* **2012**, *22*, 6088–6096. [CrossRef]
41. Harris, A.M.; Lee, E.C. Improving mechanical performance of injection molded PLLA by controlling crystallinity. *J. Appl. Polym. Sci.* **2008**, *107*, 2246–2255. [CrossRef]
42. Yin, H.Y.; Wei, X.F.; Bao, R.Y.; Dong, Q.X.; Liu, Z.Y.; Yang, W.; Xie, B.H.; Yang, M.B. Enhancing thermomechanical properties and heat distortion resistance of poly(l-lactide) with high crystallinity under high cooling rate. *ACS Sustain. Chem. Eng.* **2015**, *3*, 654–661. [CrossRef]
43. Kalb, B.; Pennings, A.J. General crystallization behavior of poly(L-lactic acid). *Polymer* **1980**, *21*, 607–612. [CrossRef]
44. Ajala, O.; Werther, C.; Nikaeen, P.; Singh, R.P.; Depan, D. Influence of graphene nanoscrolls on the crystallization behavior and nano-mechanical properties of polylactic acid. *Polym. Adv. Technol.* **2019**, *30*, 1825–1835. [CrossRef]
45. Li, W.X.; Xu, Z.W.; Chen, L.; Shan, M.J.; Tian, X.; Yang, C.Y.; Lv, H.M.; Qian, X.M. A facile method to produce graphene oxide-g-poly(L-lactic acid) as a promising reinforcement for PLLA nanocomposites. *Chem. Eng. J.* **2014**, *237*, 291–299. [CrossRef]
46. Tsuji, H.; Aratani, T.; Takikawa, H. Physical properties, crystallization, and thermal/hydrolytic degradation of poly(L-lactide)/nano/micro-diamond composites. *Macromol. Mater. Eng.* **2013**, *298*, 1149–1159. [CrossRef]
47. Kim, I.H.; Jeong, Y.G. Polylactide/exfoliated graphite nanocomposites with enhanced thermal stability, mechanical modulus, and electrical conductivity. *J. Polym. Sci. Part B* **2010**, *48*, 850–858. [CrossRef]
48. Duan, J.K.; Shao, S.X.; Li, Y.; Wang, L.F.; Jiang, P.K.; Liu, B.P. Polylactide/graphite nanosheets/MWCNTs nanocomposites with enhanced mechanical, thermal and electrical properties. *Iran. Polym. J.* **2012**, *21*, 109–120. [CrossRef]
49. McLauchlin, A.R.; Thomas, N.L. Preparation and thermal characterization of poly(lactic acid) nanocomposites prepared from organoclays based on an amphoteric surfactant. *Polym. Degrad. Stabil.* **2009**, *94*, 868–872. [CrossRef]
50. Ramanathan, T.; Abdala, A.A.; Stankovich, S.; Dikin, D.A.; Herrera-Alonso, M.; Piner, R.D.; Adamson, D.H.; Schniepp, H.C.; Chen, X.; Ruoff, R.S. Functionalized graphene sheets for polymer nanocomposites. *Nat. Nanotechnol.* **2008**, *3*, 327–331. [CrossRef]

Article

Solid-Solution-Based Metal Coating Enables Highly Reversible, Dendrite-Free Aluminum Anode

Bo Hu [1], Kang Han [1], Chunhua Han [1,*], Lishan Geng [1], Ming Li [1], Ping Hu [1,*] and Xuanpeng Wang [2,3,*]

[1] State Key Laboratory of Advanced Technology for Materials Synthesis and Processing, Wuhan University of Technology, Wuhan 430070, China; 13311621655@163.com (B.H.); hankang@whut.edu.cn (K.H.); gls523@whut.edu.cn (L.G.); m18334708396@163.com (M.L.)
[2] Hainan Institute, Wuhan University of Technology, Sanya 572025, China
[3] Department of Physical Science & Technology, School of Science, Wuhan University of Technology, Wuhan 430070, China
* Correspondence: hch5927@whut.edu.cn (C.H.); huping316@whut.edu.cn (P.H.); wxp122525691@whut.edu.cn (X.W.)

Abstract: Aluminum-ion batteries have attracted great interest in the grid-scale energy storage field due to their good safety, low cost and the high abundance of Al. However, Al anodes suffer from severe dendrite growth, especially at high deposition rates. Here, we report a simple strategy for constructing a highly reversible, dendrite-free, Al-based anode through directly introducing a solid-solution-based metal coating to a Zn foil substrate. Compared with Cu foil substrates and bare Al, a Zn foil substrate shows a lower nucleation barrier of Al deposition due to the intrinsic, definite solubility between Al and Zn. During Al deposition, a thin, solid-solution alloy phase is first formed on the surface of the Zn foil substrate and then guides the parallel growth of flake-like Al on Zn substrate. The well-designed, Zn-coated Al (Zn@Al) anode can effectively inhibit dendrite growth and alleviate the corrosion of the Al anode. The fabricated Zn@Al–graphite battery exhibits a high specific capacity of 80 mAh·g^{-1} and an ultra-long lifespan over 10,000 cycles at a high current density of 20 A·g^{-1} in low-cost molten salt electrolyte. This work opens a new avenue for the development of stable Al anodes and can provide insights for other metal anode protection.

Keywords: aluminum-ion battery; metal coating; solid-solution alloy; dendrite-free anode

Citation: Hu, B.; Han, K.; Han, C.; Geng, L.; Li, M.; Hu, P.; Wang, X. Solid-Solution-Based Metal Coating Enables Highly Reversible, Dendrite-Free Aluminum Anode. *Coatings* **2022**, *12*, 661. https://doi.org/10.3390/coatings12050661

Academic Editor: Giorgos Skordaris

Received: 10 April 2022
Accepted: 9 May 2022
Published: 12 May 2022

Publisher's Note: MDPI stays neutral with regard to jurisdictional claims in published maps and institutional affiliations.

Copyright: © 2022 by the authors. Licensee MDPI, Basel, Switzerland. This article is an open access article distributed under the terms and conditions of the Creative Commons Attribution (CC BY) license (https://creativecommons.org/licenses/by/4.0/).

1. Introduction

With the continuous emission of greenhouse gases and rapid fossil energy consumption, the development and utilization of new-type, clean energy, such as wind energy, solar energy, wave energy, etc., is important [1,2]. However, due to the uneven distribution of new-type, clean energy in time and space, an efficient approach is required to convert them into reliable chemical energy through rechargeable batteries [3–5]. At present, lithium-ion batteries have practical applications in energy storage systems [6,7], but their high cost, limited cycling life and safety problems restrict their further development and application [8]. Therefore, the development of a low-cost, long-life and high-safety energy storage battery is important. Among energy storage batteries, the aluminum-ion battery is one of the promising alternatives [9–11].

However, aluminum-ion batteries also face some problems, especially in seeking a suitable cathode/anode and electrolyte. Aqueous electrolytes are safer and environmentally friendly, but the side reactions caused by water decomposition and the severe corrosion/passivation of Al anodes remain huge challenges [12,13]. Most aluminum-ion batteries research focuses on developing cathode materials with high capacity and wide operating voltage windows. The study of anode materials has also gradually become a concern. It has been reported that the metallic properties of Al are likely to induce dendrite formation during reversible plating and stripping of Al, and these dendrites may pierce the separator leading to battery failure [14]. In order to solve the above issues, the alloying of Al with other metals is a promising strategy [15]. Some pieces of literatures reported that Al-based alloy anodes can stabilize the aluminum/electrolyte interface to alleviate volume changes during Al plating/striping [16–18]. Wang et al. reported an aluminum-ion battery with an Al–Mg alloy anode. The aluminum-ion batteries delivered a long lifespan which was longer than that with a pure Al anode [19]. Additionally, other Al-based intermetallic compounds were also proposed to stabilize the Al/electrolyte interface, such as Al–Sn [20] and Al–Mg–Sn [21]. However, most of the current research focuses on room-temperature aluminum-ion batteries, and there is still no simple and effective way to suppress the violent growth of Al dendrites under extreme or high-temperature conditions.

Herein, we demonstrate that metallic Al can be controllably deposited into the Zn and Cu foils due to its definite, intrinsic solubility in Al. Moreover, it is found that the nucleation overpotential of Al ions (Al^{3+}) on zinc is the lowest. Based on this finding, we construct a general method for inhibiting Al dendrite growth by the in situ electrochemical deposition of Al on Zn foils to form a Zn@Al anode. The well-designed, Zn-coated Al (Zn@Al) anode can easily guide the homogeneous nucleation and uniform growth of Al. The Zn@Al–graphite battery delivers a high capacity of 80 mAh·g^{-1} and a high discharge voltage of 1.8 V. This work opens up a new technical route for the practical application of aluminum-ion batteries.

2. Materials and Methods

2.1. Material Selection

All chemicals and reagents were of analytical grade and were used without further purification. Anhydrous aluminum chloride ($AlCl_3$, 99.99%) and anhydrous sodium chloride (NaCl, 99%) and anhydrous potassium chloride (KCl, 99%) were purchased from Aladdin and Macklin. $AlCl_3$ was weighed with NaCl and KCl in a molar ratio of 5:2:1 and heated at 150 °C for 12 h under an inert gas atmosphere. When the three components were completely melted into a liquid, they were quickly poured into the agate mortar and ground to a fine powder in the glove box. Finally, the finely ground powder was sieved to remove large particles.

2.2. Preparation of Cathode and Anode Material

The graphite cathode was prepared by mixing graphite (2000 mush, 99%, purchased from Macklin) and polytetrafluoroethylene (PTFE) in a mass ratio of 9:1. The resulting mixture of dry powder was dispersed with isopropyl alcohol and thoroughly ground until the isopropyl alcohol was completely volatilized. This process was repeated three times until the slurry formed a film. Finally, the obtained film was rolled to obtain a fully compacted electrode. The final electrodes were dried in a vacuum oven at 60 °C for 12 h. The thickness of the graphite cathode was 50 μm, and the mass loading of a single electrode was about 1.5 mg. The Zn-coated Al (Zn@Al) anode was fabricated by in situ electrochemical deposition during cycling. Specifically, a thin, solid-solution alloy phase preferentially deposited on the surface of the Zn substrate and then guided the parallel growth of flake-like Al on the Zn substrate to form a Zn@Al anode.

2.3. Electrochemical Measurements

The symmetrical Al || Al cells, the Al || Zn cells and the Al || Cu cells were assembled in a special battery case with an Ar atmosphere in the glove box using Al foil (100 μm) as an anode and a glass fiber filter (Whatman, grade GF/D and GF/A) as the separator. All electrochemical tests were performed in a 110 °C oven. For Al–graphite and Zn@Al–graphite cells, Al foil and Zn foil were set as anodes, respectively, and graphite was set as the cathode. Moreover, the above electrochemical tests were carried out through the Neware CT4008-5V 50 mA-164 multichannel testing system. The above Al–graphite full cell was electrochemically tested at various currents from 1 A·g^{-1} to 100 A·g^{-1}. The CV and electrochemical window were tested by an Autolab PGSTAT 302N and CHI 600e electrochemical workstation. The scanning rates of CV curves were 1, 2, 4, 6, 8 and 20 mV/s, and the potential range was 0.6–2.3 V (vs. Al^{3+}/Al).

2.4. Characterizations

XRD characterization was measured using a D8 Discover X-ray diffractometer (Karlsruhe, Germany) with a non-monochromatic Cu Kα X-ray source (λ = 1.5406 Å). Scanning electron microscopy (SEM, JEOL, Japan) and energy dispersive spectrometer (EDS, JEOL, Japan) images were collected using a JEOL-7100F microscope at an acceleration voltage of 15 kV. X-ray photoelectron spectroscopy (XPS) was carried out using an ESCALAB 250Xi instrument (Thermo Fisher Scientific, Waltham, MA, USA).

3. Results

3.1. Dendrite Growth and Al Deposition on Various Substrates

In order to study the nucleation curve of Al^{3+} on different metal surfaces, Cu and Zn foils were selected as substrates for the electrochemical deposition of Al^{3+} (Figure 1a). The original metal foils were ground with sandpaper to remove the surface oxide layer (Figure S1). The battery was constructed with a pure Zn foil, and the Al was deposited to form Zn@Al in the setup. As shown in Figure 1b–d, Zn foils exhibited a lower overpotential (2.4 mV) than Cu and Al foils (12 and 6.6 mV, respectively), suggesting that Al^{3+} is easier to deposit on the Zn surface. As demonstrated in Figure 1e, there was uniform size and distribution of the intensively deposited Al on the Zn substrates in the shapes of spheres and ovals in the initial stage. When Cu foil was used as a substrate, a greater deposition of Al was observed under the same deposition conditions (Figure 1f). On the contrary, when Al foil was used as a substrate, non-uniform Al deposition was observed (Figure 1g). The main reason is that the solubility of the Zn is higher than the Cu in Al. Even after a 0.75 mA cm^{-2} deposition for 1 h, the flat and compact Al deposition on Zn foil was observed without Al dendrites (Figure 1h). At the same time, the deposition of Al on the surface of Cu foil was uneven, and a small number of Al dendrites was produced (Figure 1i). The deposition on the surface of the Al foil was more uneven than on the Cu foil, and a large number of Al dendrites were grown (Figure 1j). There were significant differences in Al nucleation behaviors between Zn metal and Cu metal because of their different solubility [22].

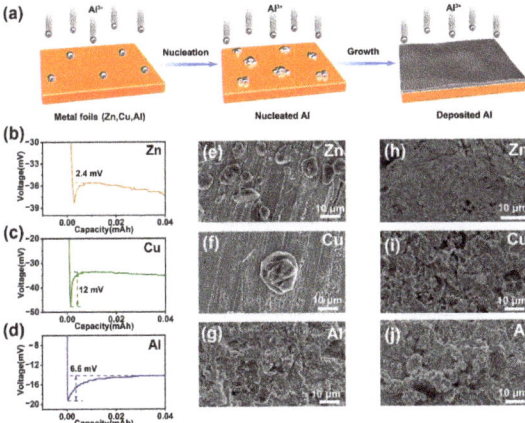

Figure 1. (a) Schematic diagram of nucleation and growth of Al deposited on different metal substrates. The nucleation curve of Al deposited on (b) metal Zn, (c) metal Cu and (d) metal Al. Scanning electron microscopy (SEM) images of deposited Al on metal foils, including (e) Zn, (f) Cu and (g) Al after depositing 0.1 mAh·cm^{-2} of Al. SEM images of deposited Al on (h) Zn, (i) Cu and (j) Al after depositing 2 mAh·cm^{-2} of Al.

3.2. Electrochemical Performance in Symmetrical Battery

Compared to traditional ionic liquid, molten salt eutectic electrolyte has the advantages of fast diffusion kinetics, high conductivity and low cost [23]. NaCl–AlCl$_3$–KCl was used as a molten salt eutectic electrolyte. To explore the stability of different metal anodes, symmetrical Al | | Al cells with pure Al and asymmetric Zn | | Al cells with pure Zn were fabricated. The electrochemical behaviors of Al plating and stripping on Zn and pure Al substrates were studied by comparing the voltage distribution in symmetrical cells. It is worth mentioning that a solid-solution Zn@Al alloy layer was formed during the initial stage of Al deposition [24]. As shown in Figure 2a, the battery using pure Al only ran for a few cycles and suffered from large voltage polarization (>40 mV), and the voltage dropped sharply in about 120 h. The main reason may be the uneven coating and peeling of Al on the surface of the pure Al, causing a branch crystal to penetrate the battery separator, resulting in a short circuit (Figure 2b). By contrast, the Zn@Al | | Al symmetrical battery exhibited a smooth and stable voltage curve under the same test conditions, achieving a long cycling life (300 h) without a short circuit. As can be seen from the enlarged figure, the voltage polarization was <30 mV (Figure 2c). Under a higher current density (5 mA·cm^{-2}) and cycling capacity (5 mA·cm^{-2}), the cell using pure Al substrate ran for only 60 h, and a dramatic voltage change occurred; thus, the cell failed. The polarization voltage increased from 0.1 V to 0.6 V in the above processes (Figure 2d,e). As shown in Figure 2f, the results show that the cell ran steadily for 260 h, and the polarization potential was only 0.1 V, even at a high current density of 5 mA cm^{-2}. After 150 h, during symmetrical battery cycling, loose structures with uneven Al chaotic clusters were observed on the pure Al substrate (Figure 2g). The deposition of Al was non-uniform, and the Al metals were stacked together to form dendritic crystals (Figure 2h). As shown in Figure 2i, the non-uniform electric field distribution on the Al matrix led to the non-uniform nucleation and deposition of Al. With the increase in cycle time, the non-uniform deposition of Al further developed into an Al dendrite (the thickness of the dendrite was 14.8 μm) by a self-diffusion mechanism. On the contrary, the Al coating on the Zn substrate was flat and uniform after cycling, and the coating thickness was about 8 μm (Figure 2j–l). After a certain period of deposition, the coexistence of Zn and Al elements in the deposition layer showed the formation of Zn@Al alloy (Figure S2), and no dendrites were produced [25].

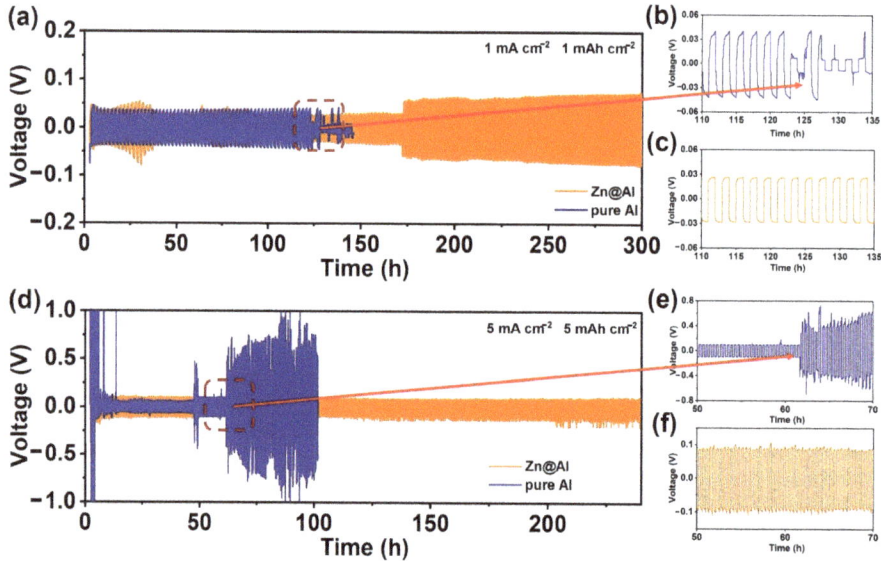

Figure 2. Cont.

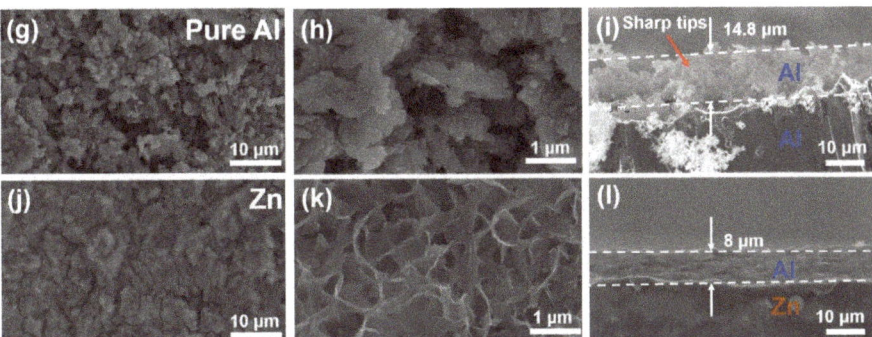

Figure 2. (**a–c**) Electrochemical performances of symmetric batteries and corresponding partial enlargement based on Zn and Al substrates at 1 mA·cm^{-2}/1 mAh·cm^{-2}. (**d–f**) Voltage profiles for symmetric cells and corresponding partial enlargement based on Zn and Al substrates at 5 mA·cm^{-2}/5 mAh·cm^{-2}. (**g–i**) SEM images of Al plating on pure Al foil with a capacity of 1 mAh·cm^{-2} though the Al | | Al cells. (**j–l**) SEM images of Al plating on Zn foil with a capacity of 1 mAh·cm^{-2} though the Al | | Zn cells.

3.3. Electrochemical Performance in Al Full Cell

In order to further prove the stable performance of Zn@Al anodes, the Al full cell was fabricated by commercial graphite cathode. The basic energy storage mechanism is shown in Figure 3a. During the charging processes, AlCl$_4^-$ ions were reversibly embedded in the graphite. Moreover, during the discharging processes, Al^{3+} deposited on the surface of the Zn@Al anode [26]. The CV curves-based Zn@Al and pure Al anode are compared in Figure 3b. The polarization potential of Zn@Al was lower than the pure Al anode, which was attributed to the rapid reaction kinetics of the solid-solution alloying of Zn with Al (Figure S3). The specific capacity of the Zn@Al–graphite full cell reached 78 mAh·g^{-1} when testing at the current density of 2 A·g^{-1}. The capacity retention was as high as 99.0% after 160 cycles, with a high Coulombic efficiency of 98% (Figure 3c). Different metal foils were used as the anode electrodes of the full cell. As expected, the Zn@Al–graphite full cell exhibited a better cycling performance than the Al–graphite full cell (pure Al anode) at a current density of 1 A·g^{-1} (Figures S4 and S5a). In contrast, the Al–graphite full cell showed obvious attenuation and increasing polarization after 200 cycles (Figure S5b). As shown in Figure 3d, the Zn@Al–graphite full cell had a flat charging platform of 1.8 V, and the discharge platform appeared at around 1.35 V. The charge termination voltage of the Zn@Al–graphite and Al–graphite full cell was 2.3 V, and the voltage of the Cu@Al–graphite full cell was 1.2 V (Figure S6a,b). The advantages of such a full cell were further evidenced by the attractive rate performance, realizing a high capacity of 63 mAh·g^{-1} of the Zn@Al–graphite full cell at a high current density of 50 A·g^{-1}. When returned to 1 A·g^{-1}, the discharge capacity could be completely recovered (80 mAh·g^{-1}) (Figure 3e). The excellent rate capability can be ascribed to the high ionic conductivity of the molten salt system and the uniform coating and stripping of Al^{3+} on the surface of the Zn anode. On the contrary, the rate performance of the Al–graphite full cell was poor (Figure S7a). In order to reveal the reason for its poor rate performance, ex situ SEM tests were performed for the Al anode after the Al–graphite full cell cycle, and the dendrites were also obvious observed on the Al substrate surface (Figure S7b). As shown in Figure 3f, with the increasing current density, the impulse discharge platform of the full cell gradually disappeared [27]. To further demonstrate the advantages of Zn@Al anode, the full cell showed high capacity retention of ~81% after 10,000 cycles at a high current density of 20 A·g^{-1}. However, the Al–graphite full cell experienced fast capacity decay and only worked for 2000 cycles (Figure 3g). The main reason for the poor performance may be attributed to the fact that the surface of the Al deposition on the pure Al anode was uneven. Al metal stacked together to form dendrite and pierced the battery separator, resulting in a short circuit. In addition, the Cu@Al–graphite full cell only worked for 1000 cycles and exhibited a poor rate performance (Figure S8).

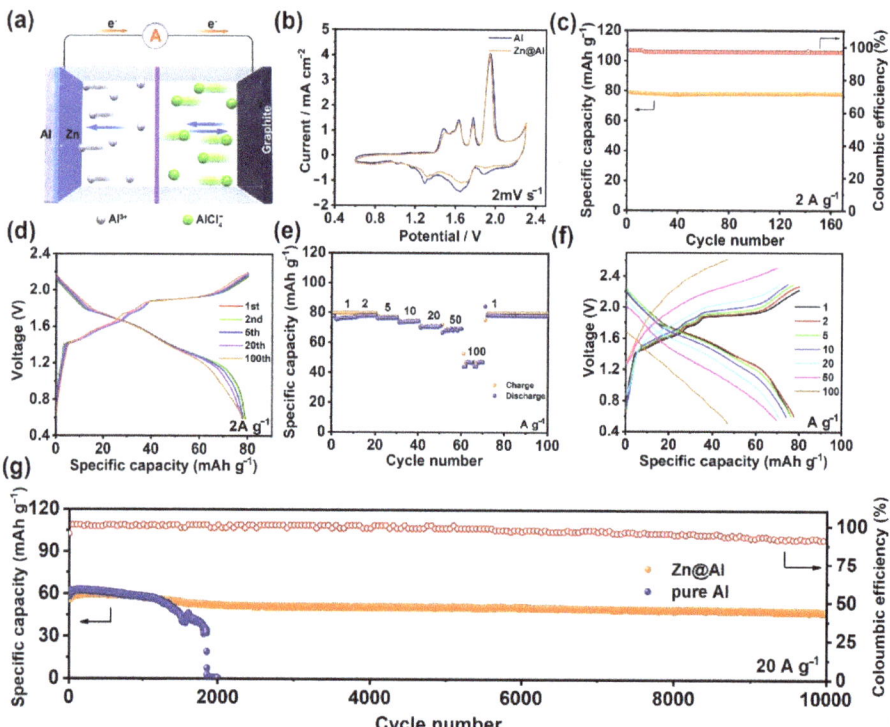

Figure 3. Electrochemical performances of Zn@Al–graphite and Al–graphite cells: (**a**) Working diagram of full battery. (**b**) CV curves in full battery at a scan rate of 2 mV·s^{-1}. (**c**) Cycling stability (**d**) and charge/discharge curves of Zn@Al–graphite full battery at 2 A·g^{-1}. (**e**) Rate performance and (**f**) charge/discharge curves of Zn@Al–graphite cell at various currents from 1 A·g^{-1} to 100 A·g^{-1}. (**g**) Cycling stability in Zn@Al–graphite and Al–graphite cells at 20 A·g^{-1}.

3.4. Investigation of Dendrite Growth Mechanism

As shown in Figure 4a, the schematic diagram of the coating and stripping of pure Al foil and Zn foil show that the main ions and molecules in the molten salt electrolyte were AlCl$_4^-$ and AlCl$_7^-$. Furthermore, in the early stage of deposition, the electrolyte corroded the surface of the Al matrix and caused pulverization [28]. It resulted in an uneven electric field distribution, producing Cl$_2$ and some by-products (Figure S9). When Al^{3+} were diffused through the electrolyte and then reduced on the surface of the anode, an irregular Al deposited layer was produced. As a result, only a part of the coated Al was stripped into the electrolyte [29,30]. Therefore, the thickness of the bare Al was thinned, and a lot of unstripped and heterogeneous Al accumulated many times to form Al dendrites during the repeated deposition and stripping processes [31,32]. On the contrary, on the surface of Zn, because of the high solubility of Zn in Al, Zn atoms dissolved into Al before producing a pure Al phase during the deposition process, forming a solid-solution surface coating layer which could be used as a buffer layer, effectively eliminating the nucleation barrier. The atomic energy of the Al nucleated and grew evenly on the Zn matrix, forming a flat deposition layer without the formation of Al dendrite.

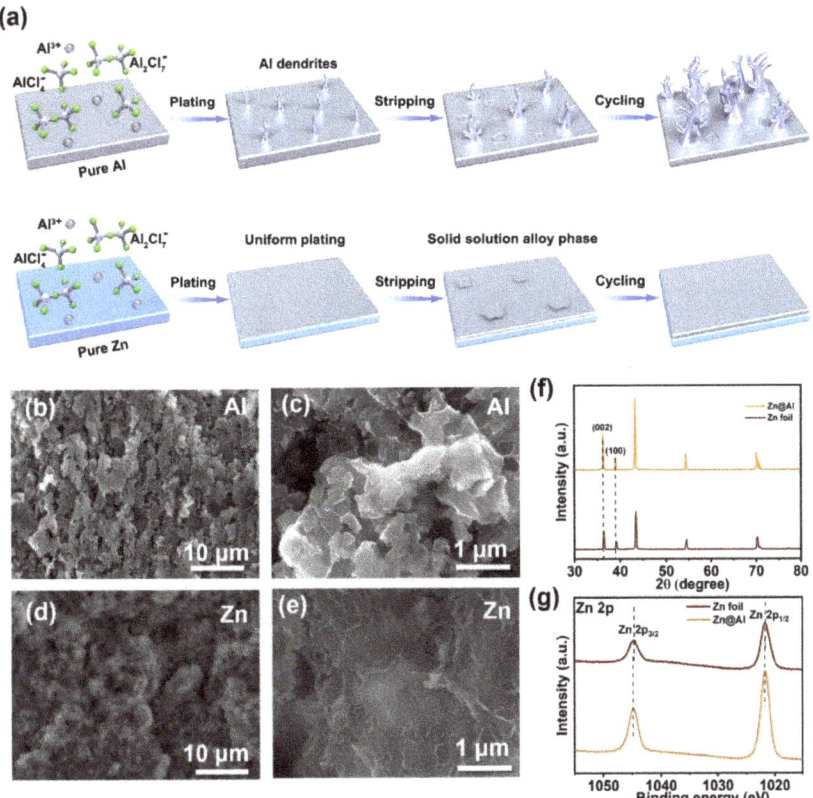

Figure 4. (**a**) Schematic illustration of Al plating/stripping in Al substrate and Zn substrate. (**b,c**) SEM images of Al plating on Al foil with a capacity of 20 A·g^{-1} in the Al–graphite cells. (**d,e**) SEM images of Al plating on Zn foil with a capacity of 20 A·g^{-1} in the Zn@Al–graphite cells. (**f**) XRD patterns of Zn foil and Zn@Al. (**g**) Zn 2p XPS spectra of Zn foil and Zn@Al.

SEM characterized the surface morphology of the anode after plating and stripping. After 2000 cycles, massive dendrites formed on the surface of the pure Al anode (Figure 4b,c). The deposition of Al on the surface of the Cu anode was also heterogeneous due to the low solid solubility (Figure S10). In contrast, the surface of the Zn anode showed a flat and compact Al coating, and no Al dendrite was observed (Figure 4d,e). In order to further determine the surface structure of the Zn anode, the XRD results showed only the diffraction peak of Zn, without Al or Al$_2$O$_3$ phase, indicating that Zn was completely dissolved in Al and formed an alloy (Figure 4f) [33–35]. Compared with pure Zn foil, the peak value of Zn@Al shifted to a small angle due to the alloying effect. Through the distinct change in the relative peak intensity, I (002)/I (100), the alloying of the Al and Zn substrates was further proved [35]. As shown in Figure 4g, the presence of the metal state of Al and Zn was explicitly verified, indicating a Zn@Al alloy. The XPS results showed that the Zn 2p peaks of Zn@Al were slightly different from the Zn foil, and the electronic state of Zn in the Zn@Al alloy changed due to the Al formation [36–38].

4. Conclusions

In conclusion, a new Zn@Al–graphite full cell configuration using a Zn@Al anode combined with commercial graphite as the cathode in a low-cost molten salt electrolyte was developed. During the Al deposition process, a thin, solid-solution alloy phase was first formed on the surface of the Zn and then guided the parallel growth of flake-like Al on the Zn substrate. The well-designed Zn@Al anode achieved uniform Al plating/stripping and effectively alleviated the corrosion and powdering of the negative electrode and inhibited the growth of Al dendrite. The fabricated Zn@Al–graphite full cell exhibited a high specific capacity of 80 mAh·g^{-1} and an ultra-long lifespan over 10,000 cycles at a high current density of 20 A·g^{-1}. This Zn@Al anode is expected to replace pure Al as an ideal anode electrode for aluminum-ion batteries. This work opens a new technical route for the practical application of aluminum-ion batteries.

Supplementary Materials: The following supporting information can be downloaded at: https://www.mdpi.com/article/10.3390/coatings12050661/s1, Figure S1: SEM images of (a) pure Al foils and (b) pure Zn foils; Figure S2: SEM image and EDS energy spectrum of Zn@Al alloy; Figure S3: CV curves of Zn@Al–graphite full cell at various scan rate from1 to 20 mV·s^{-1}; Figure S4: Cycling stability of different full cells at 1 A·g^{-1}; Figure S5: (a) Cycling stability the Zn@Al–graphite full cell at 1 A·g^{-1}. (b) Cycling stability in the Al–graphite full cell 1 A·g^{-1}; Figure S6: (a) The charge/discharge curves of graphite–Al full cells at 1 A·g^{-1}. (b) The charge/discharge curves Cu@Al–graphite full cells at 1 A·g^{-1}; Figure S7: (a) Rate performance of Al-graphite full cell at various current form1 A·g^{-1} to 100 A·g^{-1}. (b) SEM image of Al plating on Al foil after various currents from 1 A·g^{-1} to 100 A·g^{-1} though the Al–graphite full cell; Figure S8: Cycling stability of Cu@Al–graphite full cells at 20 A·g^{-1}; Figure S9: SEM images of aluminum anode at the initial stage of full cell operation; Figure S10: SEM images of Al plating on Zn foil with a capacity of 20 A·g^{-1} through the Cu@Al–graphite full cells.

Author Contributions: B.H. and K.H. contributed equally to this work. Conceptualization, methodology, X.W., M.L., C.H., L.G. and P.H.; data curation, writing—original draft preparation, writing—review and editing, B.H. and K.H.; project administration, funding acquisition, X.W. and P.H. All authors have read and agreed to the published version of the manuscript.

Funding: This work was supported by the the Hainan Provincial Joint Project of Sanya Yazhou Bay Science and Technology City (520LH055), the Sanya Science and Education Innovation Park of Wuhan University of Technology (2021KF0019, 2020KF0019),National Natural Science Foundation of China (51872218, 52172233, 21905218), the Key Research and Development Program of Hubei Province (2021BAA070), and the Fundamental Research Funds for the Central Universities (WUT: 2021CG014).

Institutional Review Board Statement: Not applicable.

Informed Consent Statement: Not applicable.

Data Availability Statement: Data sharing is not applicable to this work.

Conflicts of Interest: The authors declare no conflict of interest.

References

1. Yang, H.; Yin, L.; Liang, J.; Sun, Z.; Wang, Y.; Li, H.; He, K.; Ma, L.; Peng, Z.; Qiu, S.; et al. An Aluminum-Sulfur Battery with a Fast Kinetic Response. *Angew. Chem. Int. Ed. Engl.* **2018**, *57*, 1898–1902. [CrossRef] [PubMed]
2. Luo, W.; Ren, J.; Feng, W.; Chen, X.; Yan, Y.; Zahir, N. Engineering Nanostructured Antimony-Based Anode Materials for Sodium Ion Batteries. *Coatings* **2021**, *11*, 1233. [CrossRef]
3. Liu, H.; Zhao, G.; Li, H.; Tang, S.; Xiu, D.; Wang, J.; Yu, H.; Cheng, K.; Huang, Y.; Zhou, J. The Discharge Performance of Mg-3In-xCa Alloy Anodes for Mg–Air Batteries. *Coatings* **2022**, *12*, 428. [CrossRef]
4. Wang, J.; Chen, J.H.; Chen, Z.C.; Wu, Z.Y.; Zhong, X.N.; Ke, J.P. The LiTFSI/COFs Fiber as Separator Coating with Bifunction of Inhibition of Lithium Dendrite and Shuttle Effect for Li-SeS_2 Battery. *Coatings* **2022**, *12*, 289. [CrossRef]
5. Dunlap, R.A. Renewable Energy: Volume 3: Electrical, Magnetic, and Chemical Energy Storage Methods. Synthesis Lectures on Energy and the Environment. *Sci. Technol. Soc.* **2020**, *3*, i-99.
6. Goodenough, J.B.; Park, K.S. The Li-Ion Rechargeable Battery: A Perspective. *J. Am. Chem. Soc.* **2013**, *135*, 1167–1176. [CrossRef]
7. Dunn, B.; Kamath, H.; Tarascon, J.M. Electrical Energy Storage for the Grid: A Battery of Choices. *Science* **2011**, *334*, 928–935. [CrossRef]
8. Shen, L.; Du, X.; Ma, M.; Wang, S.; Huang, S.; Xiong, L. Progress and Trends in Nonaqueous Rechargeable Aluminum Batteries. *Adv. Sustain.* **2022**, *6*, 2100418. [CrossRef]
9. Jiang, M.; Fu, C.; Meng, P.; Ren, J.; Wang, J.; Bu, J.; Dong, A.; Zhang, J.; Xiao, W.; Sun, B.D. Challenges and Strategies of Low-Cost Aluminum Anodes for High-Performance Al-Based Batteries. *Adv. Mater.* **2022**, *34*, 2102026. [CrossRef]

10. Cai, Y.; Chua, R.; Srinivasan, M. Anode Materials for Rechargeable Aqueous Al-Ion Batteries: Progress and Prospects. *ChemNanoMat* **2022**, *8*, e202100507. [CrossRef]
11. Zaromb, S. The Use and Behavior of Aluminum Anodes in Alkaline Primary Batteries. *J. Electrochem. Soc.* **1962**, *109*, 1125. [CrossRef]
12. Abedin, S.; Endres, F. Electrochemical Behaviour of Al, Al—In and Al–Ga–In Alloys in Chloride Solutions Containing Zinc Ions. *J. Appl. Electrochem.* **2004**, *34*, 1071–1080. [CrossRef]
13. Arik, H. Design, Effect of Mechanical Alloying Process on Mechanical Properties of α-Si_3N_4 Reinforced Aluminum-Based Composite Materials. *Mater. Des.* **2008**, *29*, 1856–1861. [CrossRef]
14. Dillon, R.L. Observations on the Mechanisms and Kinetics of Aqueous Aluminum Corrosion (Part 2—Kinetics of Aqueous Aluminum Corrosion). *Corrosion* **1959**, *15*, 29–32. [CrossRef]
15. Lin, D.; Liu, Y.; Cui, Y. Reviving the Lithium Metal Anode for High-Energy Batteries. *Nat. Nanotechnol.* **2017**, *12*, 194–206. [CrossRef]
16. Tang, Y.; Lu, L.; Roesky, H.W.; Wang, L.; Huang, B.Y. The Effect of Zinc on the Aluminum Anode of the Aluminum–Air Battery. *J. Power Sources* **2004**, *138*, 313–318. [CrossRef]
17. Yu, Y.; Chen, M.; Wang, S.; Hill, C.; Joshi, P.; Kuruganti, T.; Hu, A. Laser Sintering of Printed Anodes for Al-Air Batteries. *J. Electrochem. Soc.* **2018**, *165*, A584. [CrossRef]
18. Ryu, J.; Jang, H.; Park, J.; Yoo, Y.; Park, M.; Cho, J. Seed-Mediated Atomic-Scale Reconstruction of Silver Manganate Nanoplates for Oxygen Reduction towards High-Energy Aluminum-Air Flow Batteries. *Nat. Commun.* **2018**, *9*, 3715. [CrossRef]
19. Wang, C.; Li, J.; Jiao, H.; Tu, J.; Jiao, S.Q. The Electrochemical Behavior of an Aluminum Alloy Anode for Rechargeable Al-Ion Batteries using an $AlCl_3$–Urea Liquid Electrolyte. *RSC Adv.* **2017**, *7*, 32288–32293. [CrossRef]
20. El Shayeb, H.A.; Abd El Wahab, F.M.; El Abedin, S.Z. Electrochemical Behaviour of Al, Al–Sn, Al–Zn and Al–Zn–Sn Alloys in Chloride Solutions Containing Stannous Ions. *Corros. Sci.* **2001**, *43*, 655–669. [CrossRef]
21. Ma, J.; Wen, J.; Ren, F.; Wang, G.; Xiong, Y. Electrochemical Performance of Al−Mg−Sn Based Alloys as Anode for Al-Air battery. *J. Electrochem. Soc.* **2016**, *163*, A1759. [CrossRef]
22. Davies, I.G.; Dennis, J.M.; Hellawell, A. The Nucleation of Aluminum Grains in Alloys of Aluminum with Titanium and Boron. *Metall. Trans.* **1970**, *1*, 275–280. [CrossRef]
23. Tu, J.; Wang, J.; Zhu, H.; Jiao, S. The Molten Chlorides for Aluminum-Graphite Rechargeable Batteries. *J. Alloys Compd.* **2020**, *821*, 153285. [CrossRef]
24. Yan, C.; Lv, C.; Wang, L.; Cui, W.; Zhang, L.; Dinh, K.N.; Tan, H.; Wu, C.; Wu, T.; Ren, Y.; et al. Architecting a Stable High-Energy Aqueous Al-Ion Battery. *J. Am. Chem. Soc.* **2020**, *142*, 15295–15304. [CrossRef]
25. Zhao, Q.; Zheng, J.; Deng, Y.; Archer, L. Regulating the Growth of Aluminum Electrodeposits: Towards Anode-Free Al Batteries. *J. Mater. Chem. A* **2020**, *8*, 23231–23238. [CrossRef]
26. Zhou, Q.; Zheng, Y.; Wang, D.; Lian, Y.; Ban, C.; Zhao, J.; Zhang, H. Cathode Materials in Non-Aqueous Aluminum-Ion Batteries: Progress and Challenges. *Ceram. Int.* **2020**, *46*, 26454–26465. [CrossRef]
27. Cai, T.; Zhao, L.; Hu, H.; Li, T.; Li, X.; Guo, S.; Li, Y.; Xue, Q.; Xing, W.; Yan, Z.; et al. Stable $CoSe_2$/Carbon Nanodice@reduced Graphene Oxide Composites for High-Performance Rechargeable Aluminum-Ion Batteries. *Energy Environ. Sci.* **2018**, *11*, 2341–2347. [CrossRef]
28. Liu, X.; Jiao, H.; Wang, M.; Song, W.-L.; Xue, J.; Jiao, S. Current Progresses and Future Prospects on Aluminium–Air Batteries. *Int. Mater. Rev.* **2021**, *11*, 1–31. [CrossRef]
29. Tong, X.; Zhang, F.; Ji, B. Carbon-Coated Porous Aluminum Foil Anode for High-Rate, Long-Term Cycling Stability, and High Energy Density Dual-Ion Batteries. *Adv. Mater.* **2016**, *28*, 9979–9985. [CrossRef]
30. Tong, X.; Zhang, F.; Chen, G. Core–Shell Aluminum@Carbon Nanospheres for Dual-Ion Batteries with Excellent Cycling Performance under High Rates. *Adv. Energy Mater.* **2018**, *8*, 1701967. [CrossRef]
31. Kim, D.J.; Yoo, D.J.; Otley, M.T. Rechargeable Aluminium Organic Batteries. *Nat. Energy* **2019**, *4*, 51–59. [CrossRef]
32. Pradhan, D.; Reddy, R.G. Dendrite-Free Aluminum Electrodeposition from $AlCl_3$-1-ethyl-3-Methyl-Imidazolium Chloride Ionic Liquid Electrolytes. *Metall. Mater. Trans. B* **2012**, *43*, 519–531. [CrossRef]
33. Xu, Y.C.; Cui, X.Q.; Wei, S.T.; Zhang, Q.H.; Gu, L.; Meng, F.Q.; Fan, J.C.; Zheng, W.T. Highly Active Zigzag-Like Pt-Zn Alloy Nanowires with High-Index Facets for Alcohol Electrooxidation. *Nano Res.* **2019**, *12*, 1173–1179. [CrossRef]
34. Gong, M.X.; Deng, Z.P.; Xiao, D.D.; Han, L.L.; Zhao, T.H.; Lu, Y.; Shen, T.; Liu, X.P.; Lin, R.Q.; Huang, T.; et al. One-Nanometer-Thick Pt_3Ni Bimetallic Alloy Nanowires Advanced Oxygen Reduction Reaction: Integrating Multiple Advantages into One Catalyst. *ACS Catal.* **2019**, *9*, 4488–4494. [CrossRef]
35. Yang, C.; Wu, Z.W.; Zhang, G.H.; Sheng, H.P.; Tian, J.; Duan, Z.L.; Sohn, H.; Kropf, A.J.; Wu, T.P.; Krause, T.R.; et al. Promotion of Pd Nanoparticles by Fe and Formation of a Pd_3Fe Intermetallic Alloy for Propane Dehydrogenation. *Catal. Today* **2019**, *323*, 123–128. [CrossRef]
36. Li, X.H.; Cao, X.X.; Xu, L.; Liu, L.X.; Wang, Y.; Meng, C.M.; Wang, Z.G. High Dielectric Constant in Al-Doped ZnO Ceramics using High-Pressure Treated Powders. *Alloys Compd.* **2016**, *657*, 90–94. [CrossRef]

37. Feliu, S., Jr.; Barranco, V. XPS Study of the Surface Chemistry of Conventional Hot-Dip Galvanised Pure Zn, Galvanneal and Zn-Al Alloy Coatings on Steel. *Acta Mater.* **2003**, *51*, 5413–5424. [CrossRef]
38. Bonova, L.; Zahoranová, A.; Kováčik, D.; Zahoran, M.; Mičušík, M.; Černák, M. Atmospheric Pressure Plasma Treatment of Flat Aluminum Surface. *Appl. Surf. Sci.* **2015**, *331*, 79–86. [CrossRef]

Article

Effect of EDM and Femtosecond-Laser Groove-Texture Collision Frequency on Tribological Properties of 0Cr17Ni7Al Stainless Steel

Liguang Yang [1,2], Wensuo Ma [1,*], Fei Gao [2,3] and Shiping Xi [2]

1. School of Mechatronics Engineering, Henan University of Science and Technology, Luoyang 471023, China; bmzgedu@126.com
2. Luoyang Bearing Research Institute Co., Ltd., Luoyang 471039, China; 13698807332@126.com (F.G.); yejinxsp@163.com (S.X.)
3. School of Mechanical Engineering, Tsinghua University, Beijing 100084, China
* Correspondence: mawensuo@haust.edu.cn

Abstract: Electric spark and femtosecond-laser surface texture are very effective in antifriction systems, but there are few applications and studies in dry friction. In this study, a groove texture was prepared on the surface of 0Cr17Ni7Al stainless steel via electric spark and femtosecond laser, respectively. The tribological properties of the two groove textures under different collision frequencies with the groove were studied under the condition of dry friction. The results show that the friction coefficient of the groove texture prepared by EDM and femtosecond laser is lower than that of the untextured surface. However, this does not mean that every groove-texture design will reduce wear rate. In addition, the groove texture seems to produce different tribological properties under different preparation methods. It is found that in the friction process of the same load, time and linear velocity, different collision frequencies will affect the friction and wear properties of the surface.

Keywords: collision frequency; electric spark; femtosecond laser; groove texture; tribology

1. Introduction

As we all know, the friction and wear between the contact surfaces of mechanical parts directly affect the reliability of the entire equipment or system. Friction and wear directly cause huge economic losses every year. With the improvement of tribological characteristics of mechanical systems, the surface-morphology design of mechanical support has become a relatively weak link in the tribological design of the system. In recent years, based on the lubrication and wear-reduction mechanism under different working conditions [1–3], bionic surface texture has been proven to effectively improve the lubrication performance of the friction pair surface and reduce friction and wear. It then becomes an effective lubrication and wear-reduction method [4–6]. It breaks the traditional view that the smoother the surface, the smaller the friction coefficient will be.

In addition, with the continuous development of the research on the influence of texture on friction and wear properties, there are more and more texture-processing methods, such as machining technology [7], laser-processing technology [8], chemical-etching technology [9], EDM technology [10], shot-peening technology [11], ultrasonic-machining technology [12], electrochemical-machining technology [13,14], etc. It meets the processing needs of different materials, different morphologies, complex textures and surface textures of various friction pairs, as well as promotes the research of texture lubrication and wear reduction. These method has made scholars' research on bionics gradually become a hot topic. Surface texture has been widely extended to the machinery industry, such as bearings [15,16], seals [17], cutters [18,19], drill bits [20], cylinder liner piston rings [21], etc. Texture has become an effective way to achieve high efficiency, miniaturization and high reliability of mechanical equipment [22].

Recent years, surface texture has become an effective means to improve the tribological properties of friction pairs. Scholars continue to refine and deepen the research on surface texture, and have conducted a lot of research in sliding direction [23], lubricating oil [24,25], additives [26], fabric [27], hydrophobicity [28], increasing wear [29], corrosion resistance [30] and drag reduction [31]. However, in practice, no matter how the surface texture is prepared or what texture shape is used, friction and collision with the edge of texture will generally occur between friction pairs in the process of sliding friction. This kind of friction and collision is often ignored in previous studies. There are few studies on the tribological characteristics of surface texture considering the collision frequency in the friction process.

In this paper, the groove texture of 0Cr17Ni7Al material was prepared by EDM and femtosecond-laser processing. The friction and wear experiments of groove texture were carried out by MFT-5000 friction and wear tester. Under the experimental conditions of the same load, time, linear velocity and different friction radius, combined with the collision frequency of the sliding-friction process, the surface morphology, friction and wear state of the samples were analyzed by three-dimensional white-light interferometer, based on the scanning electron microscope (SEM) and energy-dispersive spectrometer (EDS). Surface tribological properties of groove texture, untexture and two groove-texture preparation methods are compared, in which the friction and wear mechanism is discussed and the tribological properties of the groove-texture surface are studied.

2. Materials and Methods

2.1. Machining Electric Spark Plate

Electric-discharge machining (EDM) uses the electric corrosion phenomenon of positive and negative electrodes during pulse electric discharge to remove excess metal to meet the machining requirements of surface texture [32]. Micro-EDM technology can be used to reverse copy the array of microgroups with a small machining gap. At the same time, because there is no mechanical cutting force between the electrode and the workpiece, it can be added on rigid workpieces such as thin-wall and elastic parts to achieve high machining accuracy. The electrodes used in this experiment were machined; that is, 30 annular arrays with a size of 13.05 mm × 0.7 mm × 0.61 mm processed on the electrode's raised cuboid (as shown in Figure 1). Then, the prepared electrode was used to process the groove texture. The sample of surface texture of electric spark was provided by Luoyang Bearing Research Institute Co., Ltd. the specific parameters of electric-spark equipment are given in Table 1.

Figure 1. Electrode for preparing texture.

Table 1. Specific parameters of electric-spark equipment.

Parameters	Value	Unit
Productivity	500 (max)	mm^3/min
Processing current	50A (max)	nm
Machining dimension	600 × 400	mm
Total power	9	KVA
Electrode loss	0.1 (max)	%
Surface roughness	0.1	μm

2.2. Processing Femtosecond-Laser Plate

Femtosecond-laser processing (FS) irradiates the laser beam onto the surface of the workpiece. The laser beam interacts with the material. The material-removal mechanism is the vaporization of the matrix material, with almost no heat-affected zone [33], to realize the processes of cutting, welding, surface treatment, drilling and micromachining of materials (including metal and nonmetal). Femtosecond-laser surface-texture samples were provided by Shenzhen transcend laser Intelligent Equipment Co., Ltd. (Shenzhen, China). Table 2 shows the specific parameters of femtosecond-laser equipment.

Table 2. Specific parameters of femtosecond-laser equipment.

Parameters	Value	Unit
Pulse frequency	1–2000	KHz
Laser wavelength	1030	nm
Cutting format	700 × 600	mm
Cutting efficiency	800–7000	mm/s
Laser power	20 (max)	W
Comprehensive accuracy	±30	μm

According to the above two preparation methods, surface grooves with width of 800 μm, depth of 150 μm, length of 12 mm and 30 grooves were processed, as shown in Figures 2 and 3, respectively.

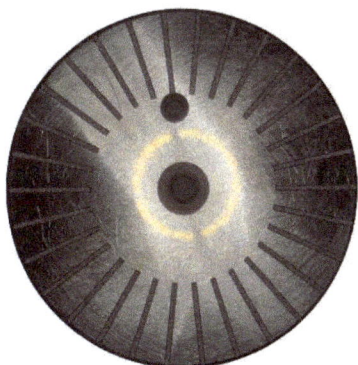

Figure 2. Electric-spark texture.

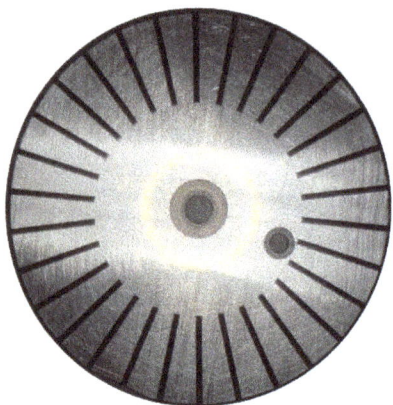

Figure 3. Femtosecond texture.

2.3. Friction and Wear Test Material

The friction characteristics of micromechanical groove texture on 0Cr17Ni7Al surface were experimentally studied using friction and wear testbed (MFT-5000). The testing machine is shown in Figure 4, with the relevant parameters of the test piece shown in Table 3.

Figure 4. MFT-5000 Friction and wear test bench.

Table 3. Parameters of texture samples.

Test Piece	Geometric Dimension	Hardness	Material	Surface Roughness
Upper test ball	Φ9.525 mm	64 HRC	9Cr18	0.014 μm
Lower test plate	Φ50.8 mm × 6.35 mm	42 HRC	0Cr17Ni7Al	0.05 μm

The relative position of the friction pair in the friction and wear test is shown in Figure 4. The experimental parameters are shown in Table 4.

Table 4. Experimental conditions for rotation test.

Specimen Name	Test Radius (mm)	Rotation Speed (r/min)	Load (N)	Time (min)
Electric-spark texture	15	60	10	20
	18	50	10	20
	22.5	40	10	20
Femtosecond texture	15	60	10	20
	18	50	10	20
	22.5	40	10	20
Untexture	15	60	10	20
	18	50	10	20
	22.5	40	10	20

Three different test pieces were prepared and compared during the experiment: smooth surface, electric-spark texture and femtosecond-laser texture. Before the test of each sample, acetone was used for further ultrasonic cleaning for 5 min to wipe the surface of the test piece. After the test, the wear morphology was detected by white-light interferometer, scanning electron microscope and energy spectrum. The experimental situation was analyzed combined with the change of friction coefficient.

2.4. Friction and Wear Calculation

The specific experimental parameters are listed in Table 4. The white-light interferometer is used to detect the cross-sectional profile of the wear mark; the wear area is calculated by integrating the cross-sectional profile; and the wear volume is obtained by multiplying the friction step and wear area:

$$W = \frac{V}{F \times S}$$

where W is the wear rate, 10^{-4} mm^3/N·mm; V is the wear volume, mm^3; F is the normal load, N; S is the running distance, mm. In this study, the error is reduced by the average value of the wear rate of three parallel tests.

3. Results and Discussion

3.1. Friction Coefficient and Wear Rate

It can be seen from Figure 5 that the textured surface can quickly enter the stable wear stage under the same friction conditions. Most of them enter the stable wear stage at 200 s. The groove texture mostly enters the stable wear stage at 200 s. After reaching the stable wear stage, the groove texture can be stabilized in a small variation range. It does not change greatly with the increase in friction time. The friction-coefficient curve of the electric-spark texture decreases with the increase in friction radius. When the radii are 18 mm and 22.5 mm, the friction coefficient is relatively stable in the stable wear stage. The friction coefficient of the femtosecond-laser texture first decreases and then increases with the increase in friction radius. The friction coefficient is relatively stable at the radius of 15 mm and 18 mm. It can be seen that the selection of different processing methods and processing principles needs further consideration when using surface texture to reduce friction. With the progress of friction, the untextured surface is difficult or takes a longer time to enter the stable wear. Regardless of the friction radius, the friction increases with the friction time. At least in the friction process of 1200 s, the friction coefficient increases gradually, and the friction is further intensified.

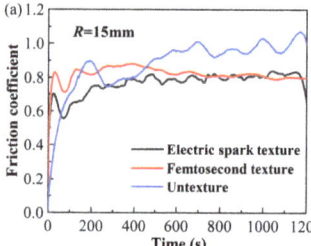 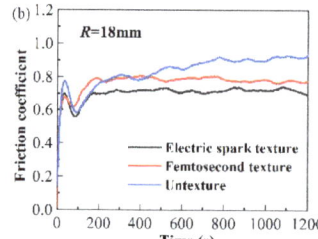 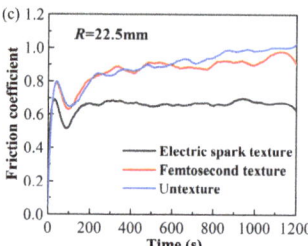

Figure 5. Friction-coefficient curves of different samples: (**a**) rotation radius 15 mm; (**b**) rotation radius 18 mm; (**c**) rotation radius 22.5 mm.

The reason why the friction coefficient of the groove texture is lower than that of the untextured surface may be due to the following three aspects: Firstly, this may be caused by the wear debris on the surface of the groove texture being taken away with friction and captured and stored by the groove texture during the friction process, which reduces the further aggravation of wear particles; Secondly, the ball collides with the groove continuously to result in friction-pair vibration, which makes the wear debris roll and reduces the friction; Thirdly, the existence of texture grooves can reduce the contact area and contact time of the friction pair in the friction process, reduce the increase of friction heat, achieve the effect of heat dissipation, and slow down the formation of oxidative wear and fretting wear. Under the same test conditions, the friction coefficient of the untextured surface increases gradually with the progress of the friction experiment. The main reason is that the wear debris generated in the friction process cannot be discharged in time and also forms three-body wear on the surface of the friction pair, which increases the heat generated by friction and aggravates the wear. Under the same preparation method, the variation of texture friction coefficient with different friction radius may be related to the collision frequency of texture in the friction process. Under different preparation methods, the different variation of texture friction coefficient with friction radius may be related to the influence of different preparation methods on the performance parameters of surface texture.

It can be seen from Figure 6 that the impact frequency has different effects on the groove texture of different preparation methods. The average friction coefficient of the spark texture decreases with the decrease in collision frequency, while the femtosecond texture decreases first and then increases. However, the law of untextured surface is not obvious. It can be seen from Tables 5 and 6 that when the friction radius is 22.5 mm and the collision frequency is 24,000 times, the average friction coefficient of the EDM texture is the minimum $C = 0.6755$. Under the conditions of 15, 18 and 22.5 mm rotation radius, the friction coefficient of the surface modified by electric spark and femtosecond groove texture is less than that of the untextured surface, indicating that the surface texture has good antifriction performance.

As shown in Table 6, under the same sliding linear speed, time and load conditions, the frequency of collision between the upper specimen and the surface groove of the lower specimen is different in the process of friction.

Table 5. Average friction coefficient and wear rate of different samples.

Specimen Name	Rotation Radius (mm)	Average Friction Coefficient	Wear Rate (10^{-4} mm^3/N·mm)
Electric-spark texture	15	0.7686	5.926
	18	0.7149	4.266
	22.5	0.6755	3.662

Table 5. Cont.

Specimen Name	Rotation Radius (mm)	Average Friction Coefficient	Wear Rate (10^{-4} mm^3/N·mm)
Femtosecond texture	15	0.7883	6.82
	18	0.7386	5.031
	22.5	0.8177	3.73
Untexture	15	0.8949	5.219
	18	0.8696	6.352
	22.5	0.8672	5.14

For the wear rate in Figure 7, with the increase in friction radius and the decrease in collision frequency, the electric spark texture and femtosecond texture show a downward trend. When the friction radius is 22.5 mm and the number of collisions is 24,000, the texture wear rate of EDM is the lowest $W = 3.662 \times 10^{-4}$ mm^3/N·mm. The wear rate of femtosecond texture reaches the maximum when the radius is 15 mm and the collision frequency is 36,000 times $W = 6.820 \times 10^{-4}$ mm^3/N·mm.

It can be seen that not all textures of friction radius are more wear-resistant than untextured surfaces. The regularity of the untextured surface is not strong, and it is difficult to predict and judge. The use of friction texture in engineering can not only reduce the wear effect, but also lead to the unreasonable use of friction texture in practice.

Table 6. Collision parameters with different radius of rotation.

Rotation Radius (mm)	Time (min)	Speed (r/min)	Number of Grooves per Turn	Total Number of Friction Turns	Total Times of Friction and Collision
15	20	60	30	1200	36,000
18	20	50	30	1000	30,000
22.5	20	40	30	800	24,000

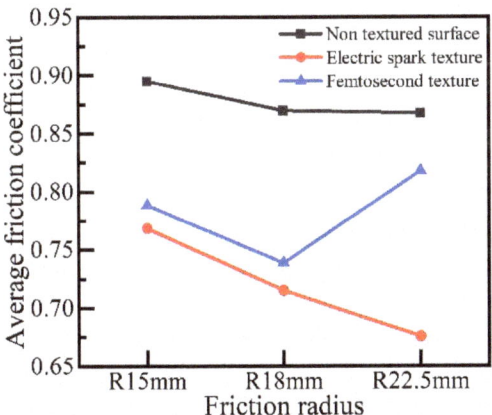

Figure 6. Average friction coefficient of different samples.

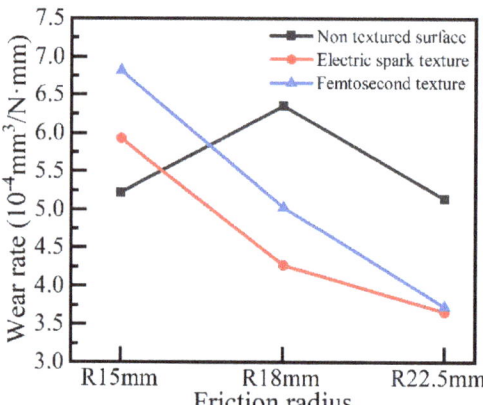

Figure 7. Wear rate of different samples.

3.2. Worn-Morphology Analysis

After the friction experiment, the flaked convex part formed on the surface of the friction pair is called the hard-phase peak [34]. The hard phase peak may be the migration of wear debris under the combined action of load and sliding speed during the friction process. Heat is generated rapidly during friction, and it is cooled rapidly during friction pair collision. Under this repeated action, it is formed by gluing and "cold welding", which is similar to cold-work hardening. According to the position of hard-phase peak, it can be divided into wear-mark hard-phase peak and wear-mark-edge hard-phase peak.

The hard-phase peak at the edge of the wear mark helps to form a supporting effect on the surface of the friction pair, which could protect the surface from further wear. The hard-phase peak of the wear mark will further aggravate the wear groove effect and even play an important role in the evolution from abrasive wear to adhesive wear.

It can be seen from the three-dimensional wear-trace morphology and wear-trace depth curve in Figure 8 that the hard-phase peak of the untextured surface is basically the hard-phase peak of wear trace under different friction radius. The textured surface is mostly the hard-phase peak at the edge of the wear mark, which is mainly due to the fact that the wear debris on the untextured surface cannot be discharged during the friction process. With the migration of friction, the cemented hard-phase peak is formed at the edge of the wear mark and the wear mark. Most of the wear debris is located in the middle of the wear mark, which is caused by the hard-phase peak of the wear mark being greater than the edge hard-phase peak. During the friction process of the textured surface, the wear debris in the middle of the wear mark, especially the large volume of wear debris, falls into the groove, which greatly reduces the wear debris at the wear mark. The wear debris at the edge of the wear mark failed to migrate far with the friction and was not well-captured by the texture. It remained at the edge of the wear mark and formed the edge hard-phase peak.

The shape of the wear-mark depth curve is concave. The smoother the curve is, the more fully the friction pair is run in. The longer the time to enter the stable stage, combined with the compound effect of the edge hard-phase peak, will help to achieve the optimal antifriction effect. However, the concave part of the wear-mark depth curve is disordered and irregular. The hard-phase peak of the wear mark is convex. The edge hard-phase peak is connected with the hard-phase peak of the wear mark. All these phenomena indicate that the friction is intensifying. It showed that when there are both wear-mark hard-phase peaks and edge hard-phase peaks, the reverse effect of the wear-mark hard-phase peak is much greater than the positive effect of the edge hard-phase peak.

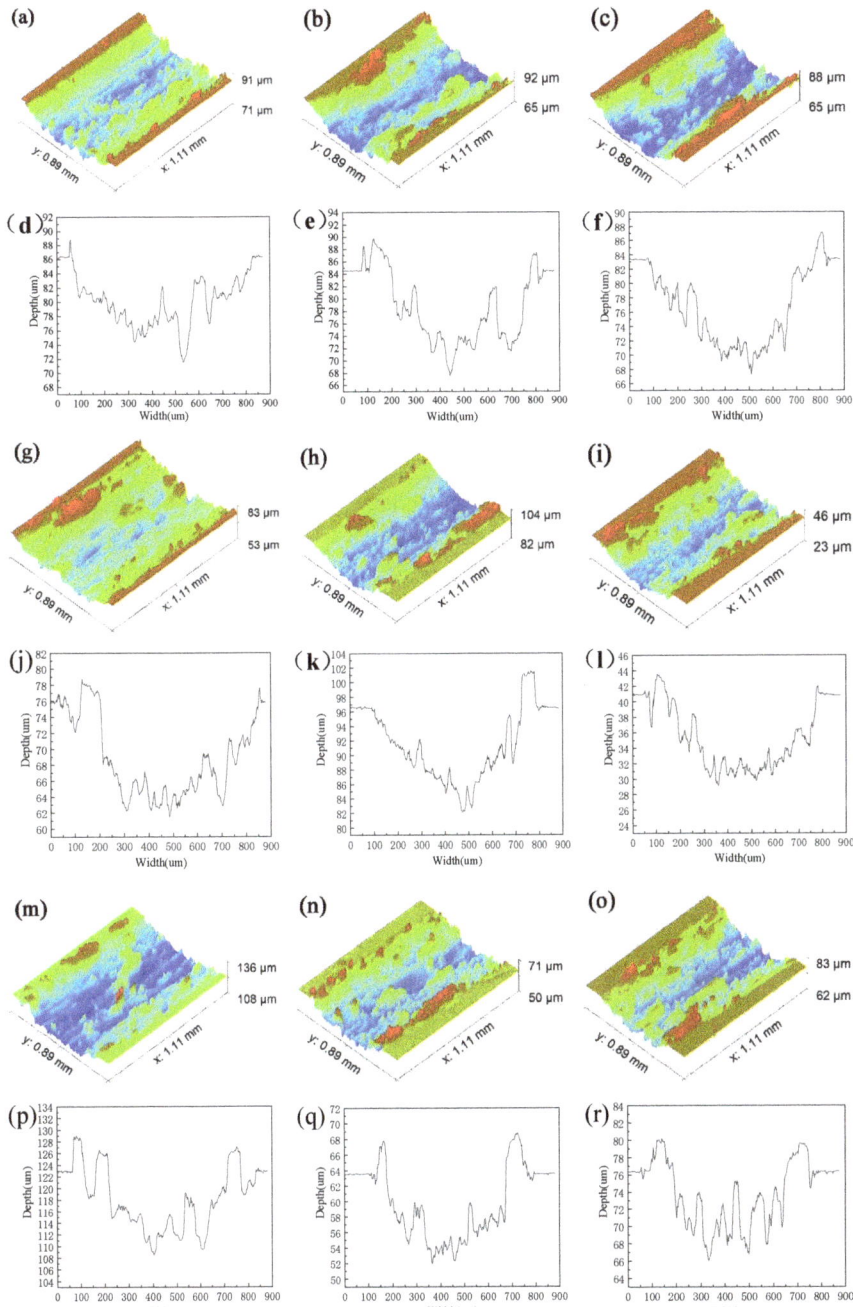

Figure 8. Morphology and depth of wear marks of different samples: (**a,d**) untexture rotation radius 15 mm; (**g,j**) untextured rotation radius 18 mm; (**m,p**) untextured rotation radius 22.5 mm; (**b,e**) electric-spark texture rotation radius 15 mm; (**h,k**) electric-spark texture rotation radius 18 mm; (**n,q**) electric-spark texture rotation radius 22.5 mm; (**c,f**) femtosecond texture rotation radius 15 mm; (**i,l**) femtosecond texture rotation radius 18 mm; (**o,r**) femtosecond texture rotation radius 22.5 mm.

3.3. Scanning Electron Microscope Analysis of the Worn Surface

As shown in Figure 9a–f, a large number of surface spalling has occurred on the untextured surface under any rotation radius. Moreover, there is less wear debris on the surface. However, there are large glued flakes, which are mainly concentrated in the wear-scar area. This is because at the initial stage of wear, abrasive wear is the main form. The wear debris generated by abrasive wear may accumulate and glue together due to the concentrated heat generated at the moment of friction, which will form the hard-phase peak of the wear mark. As mentioned earlier, the hard-phase peak of the wear mark exacerbates further wear. This is also the main reason why the friction coefficient of untextured samples increases gradually and it is difficult to enter stable wear. Under the same test parameters, the wear of the untextured surface is basically the same, which is not affected by the rotation radius.

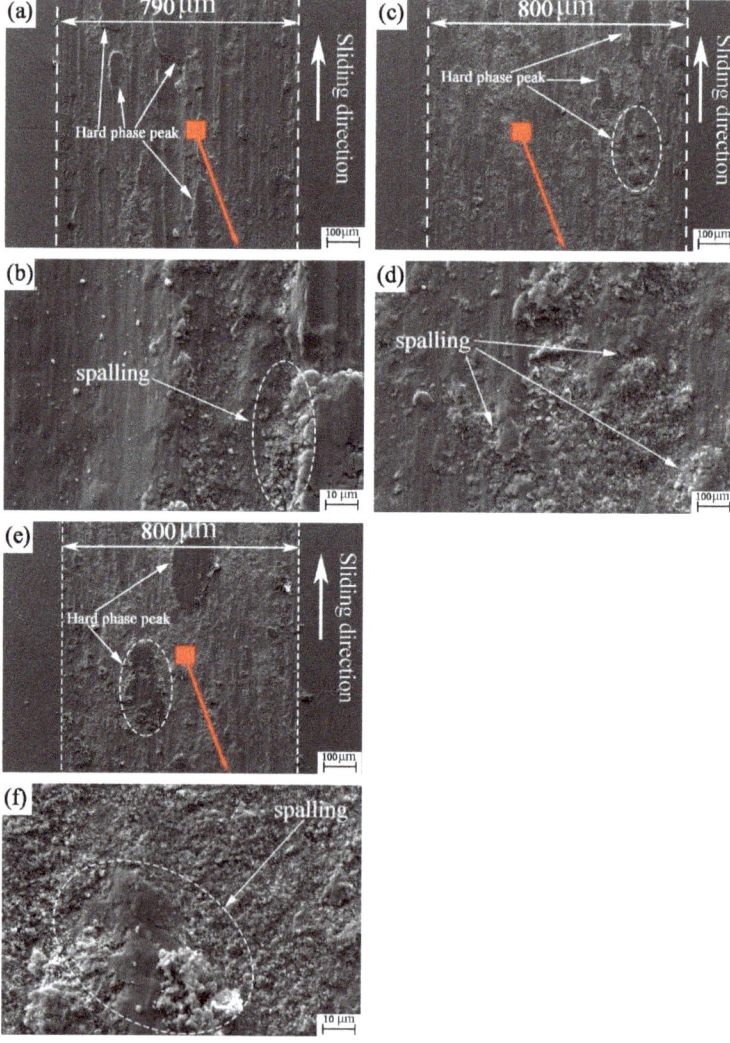

Figure 9. SEM images of the untextured worn plates: (**a**,**b**) rotation radius 15 mm; (**c**,**d**) rotation radius 18 mm; and (**e**,**f**) rotation radius 22.5 mm.

According to the SEM of the ball surface from Figure 10a–i, it can be seen that the wear mark of the ball is round. There are some large particle bulges on the wear-mark surface of the ball, and the wear particles accumulate at the edge of the wear mark along the sliding direction. Moreover, there are several obvious plows in Figure 10e. After the end of friction, these raised abrasive particles and plows can still exist, indicating that these raised abrasive particles have great hardness. It can be inferred that during the friction process, the bulges and plows accumulated by abrasive particles will cut the untextured surface like a sharp edge. Combined with Figure 9, it can be found that the wear marks on both the concave and convex surface of the friction pair are relatively consistent, which further confirms our inference.

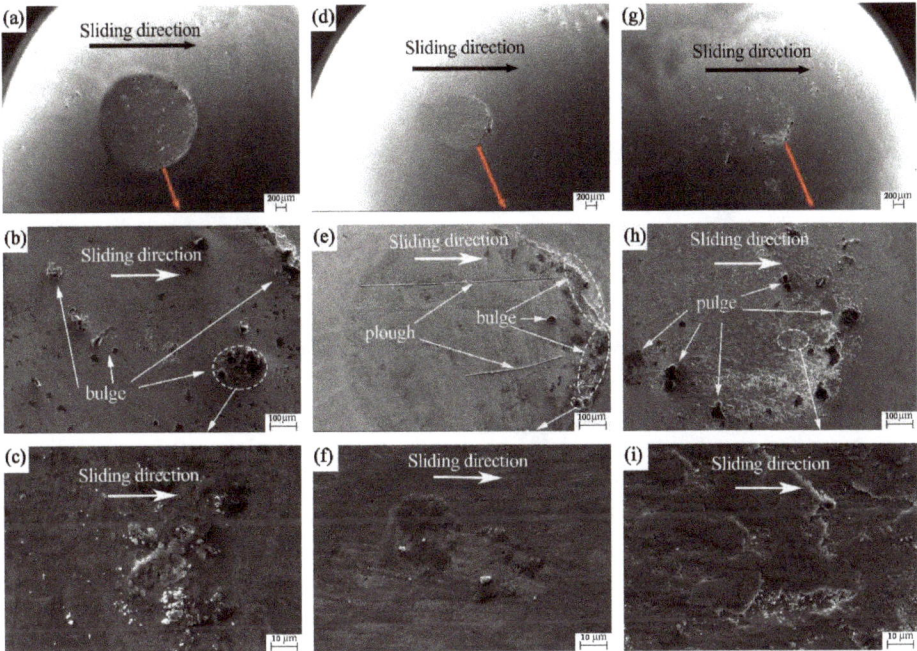

Figure 10. SEM images of the untextured worn balls: (**a–c**) rotation radius 15 mm; (**d–f**) rotation radius 18 mm; and (**g–i**) rotation radius 22.5 mm.

As can be seen from Figure 11a,b, there are many fine-wear debris on the surface of the electric-spark texture. However, no flake cement is found and no hard-phase peak is formed. It shows that the EDM texture is in the stage of abrasive wear. It can be seen from Figure 11c,d that the surface of the EDM texture is still mainly in the form of abrasive wear, with traces of adhesive wear and a few hard-phase peaks of wear marks, but the size is very small and has no obvious impact on wear. As shown in Figure 11e,f, there are both wear debris and flake cement on the surface of the EDM texture. This shows that the surface is in the common development stage of abrasive wear and adhesive wear. The hard-phase peak and edge hard-phase peak of the wear mark are formed at the same time, but the size and number of the edge hard-phase peaks are obviously greater. As mentioned earlier, when two kinds of hard-phase peaks exist at the same time, the positive effect of the edge hard-phase peak is much greater than the reverse effect of the wear-mark hard-phase peak. Thus, the total fusion effect is to reduce friction. It can be judged from Table 6 that with the increase in rotation radius, the collision frequency between ball and groove gradually decreases. The lower the wear rate of the EDM-textured surface, the more obvious the antifriction effect is.

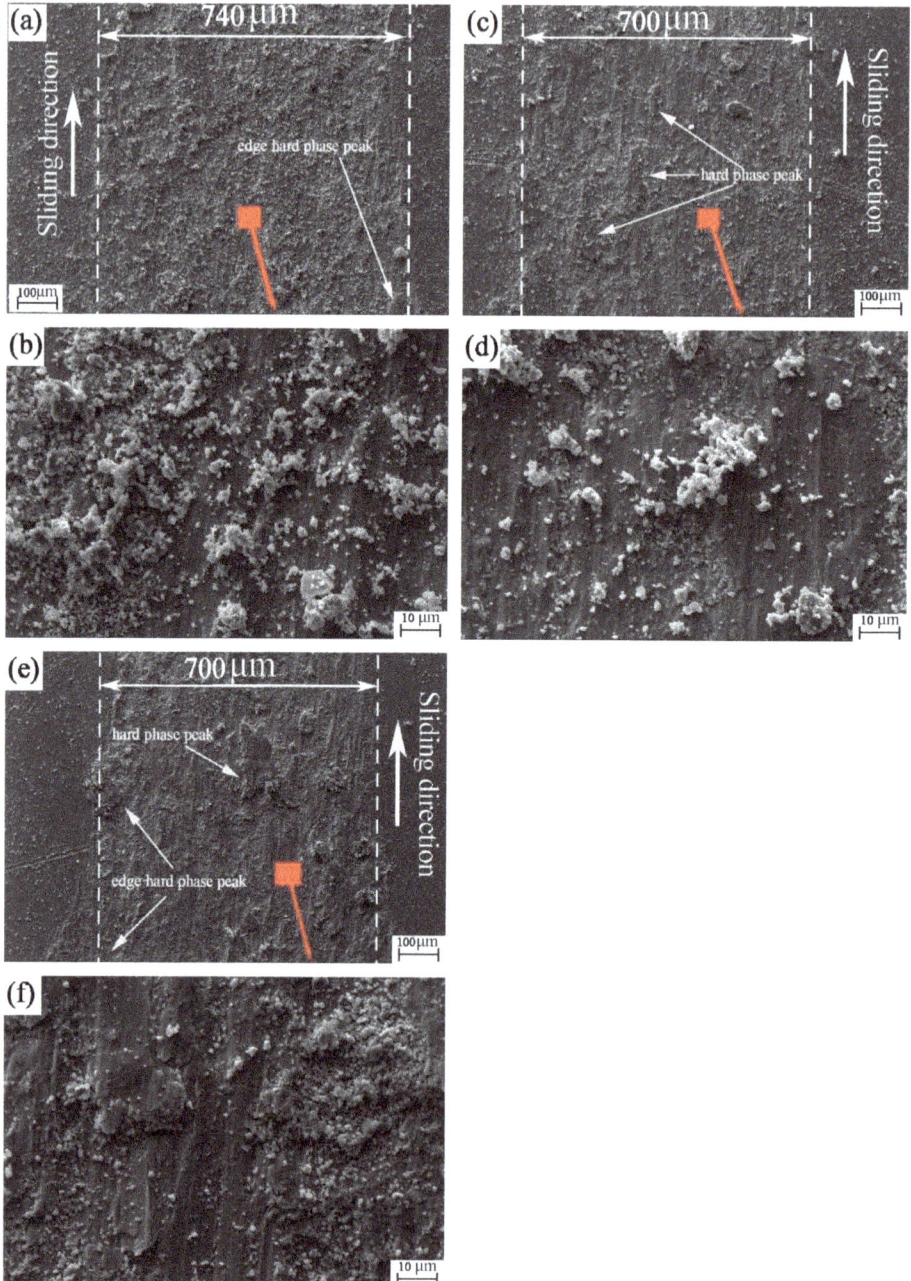

Figure 11. SEM images of the electric-spark-textured worn plates: (**a**,**b**) Rotation radius 15 mm; (**c**,**d**) Rotation radius 18 mm; and (**e**,**f**) Rotation radius 22.5 mm.

According to the SEM of the ball surface from Figure 12a–i, it can be seen that the wear mark of the ball is quadrilateral. The main reason is that during the wear process, the ball collides with the groove and the edge of the groove is equivalent to a sharp edge, which cuts the wear mark of the ball into a straight line and then becomes a quadrilateral. Figure 12a,d,g shows that the wear marks appear like waves along the sliding direction one by one. This may be because the ball will encounter 30 collisions with the groove texture for each revolution during the friction process. Under the combined action of pressure, speed in the moving direction and friction heat, the wear debris is impact-glued on the wear-mark surface of the ball. In the friction process, such layered waves will achieve the same effect as windsurfing on the sea, which is conducive to reducing the resistance on the surface of the friction pair and alleviating the increasing wear.

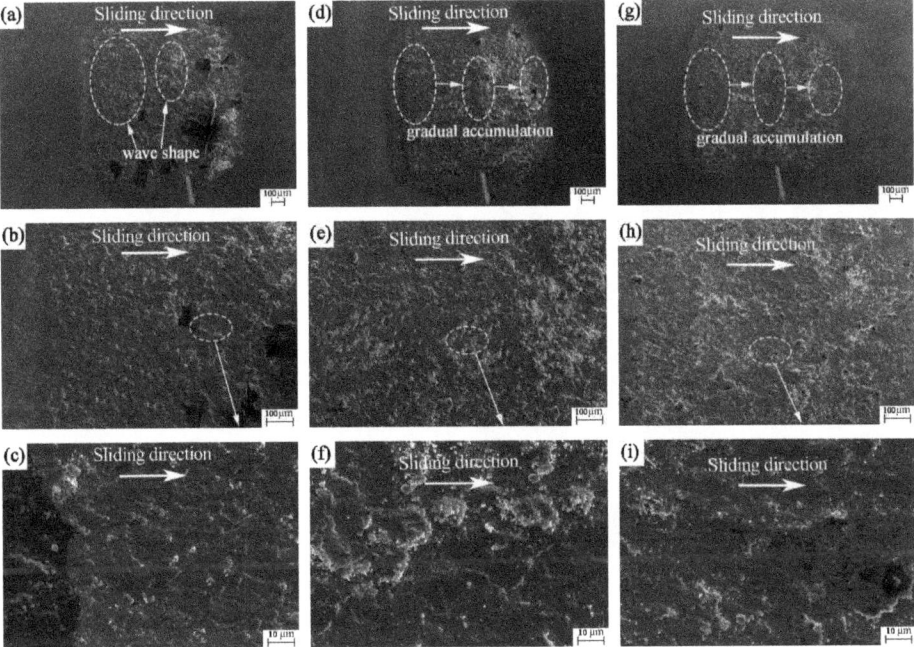

Figure 12. SEM images of the electric-spark-textured worn balls: (**a–c**) rotation radius 15 mm; (**d–f**) rotation radius 18 mm; and (**g–i**) rotation radius 22.5 mm.

As can be seen from Figure 13a–f, there are both wear debris and flake cement on the surface of the femtosecond texture, indicating that the surface of the femtosecond texture is in the common development stage of abrasive wear and adhesive wear. The lamellar glues (hard-phase peaks) are distributed from the edge of the wear mark to the wear-mark area. It can be seen from Figure 13a,b that the size and number of wear-mark hard-phase peaks on the femtosecond texture surface are larger than those on the edge. The two hard-phase peaks are separated from each other, with a long distance and no interaction. At this time, the hard-phase peak of the wear mark plays a decisive role and the wear is intensified. From Figure 13c,d, it can be seen that the hard-phase peak of the wear mark is close to the edge hard-phase peak, which affects each other and roughly offsets the effect. As shown in Figure 13e,f, two kinds of hard-phase peaks have formed a corridor of hard-phase peaks, which interact with each other. The edge hard-phase peak has greater advantages. At this time, the antifriction effect is the main effect.

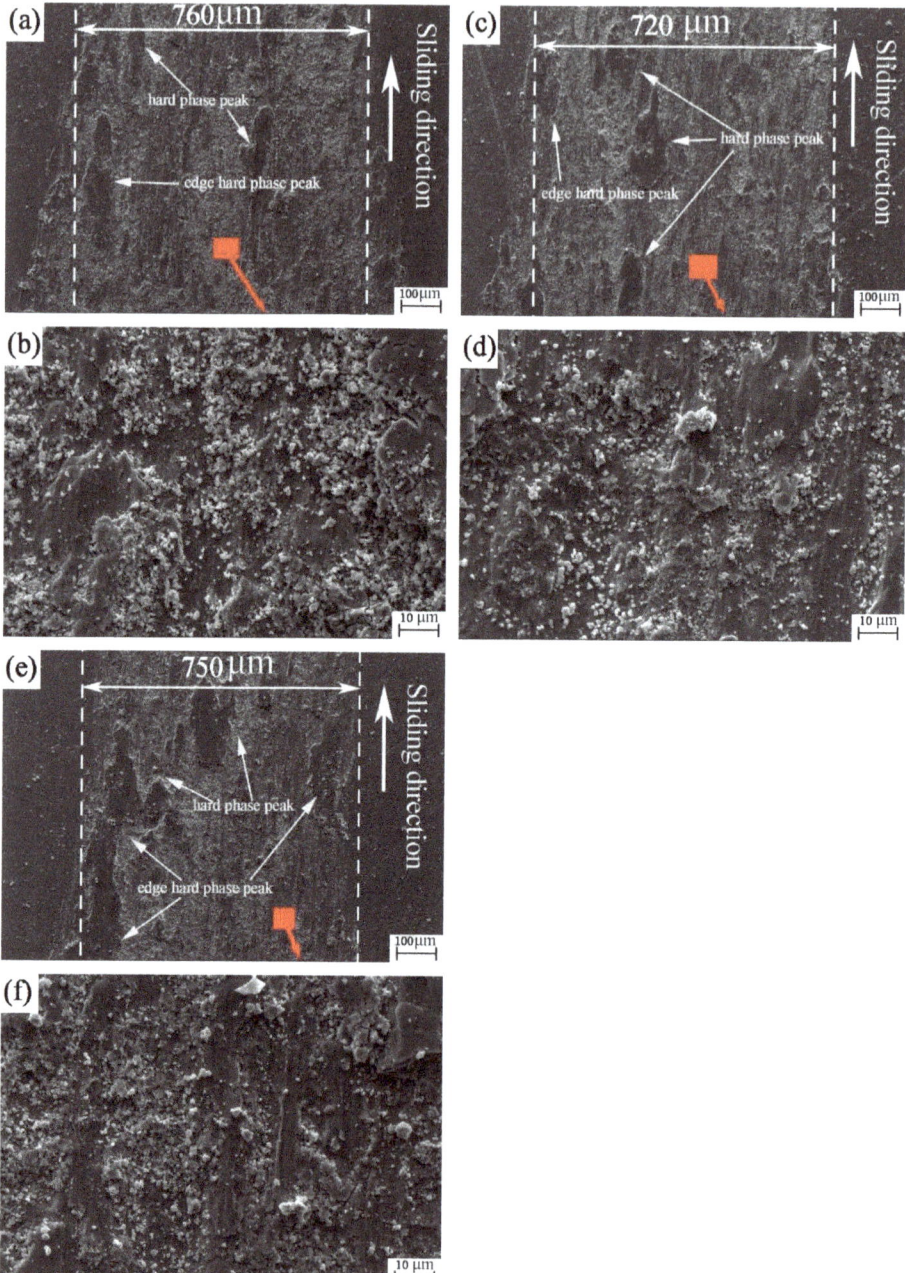

Figure 13. SEM images of the femtosecond-textured worn plates: (**a**,**b**) Rotation radius 15 mm; (**c**,**d**) Rotation radius 18 mm; and (**e**,**f**) Rotation radius 22.5 mm.

According to the SEM of the ball surface from Figure 14a–i, it can be seen that the wear mark of the ball is quadrilateral. The main reason is that during the wear process, the ball collides with the groove and the edge of the groove is equivalent to a sharp edge, which cuts the wear mark of the ball into a straight line and then becomes a quadrilateral. Figure 14a,d,g shows that the wear marks appear like waves along the sliding direction one by one. This may be because during the friction process, the ball will encounter 30 collisions with the groove texture for each revolution. As mentioned earlier, under the joint action of pressure, speed in the direction of motion and friction heat, the wear debris is impact-glued on the wear-mark surface of the ball. In the friction process, such layered waves will achieve the same effect as windsurfing on the sea, which is conducive to reducing resistance between friction pairs and alleviating the increasing wear. It can be found in Figure 14d that there are abrasive bulges in the ball-milling marks. In the process of friction, it will have a certain impact on the femtosecond texture. However, this bulge is flaked and has little impact.

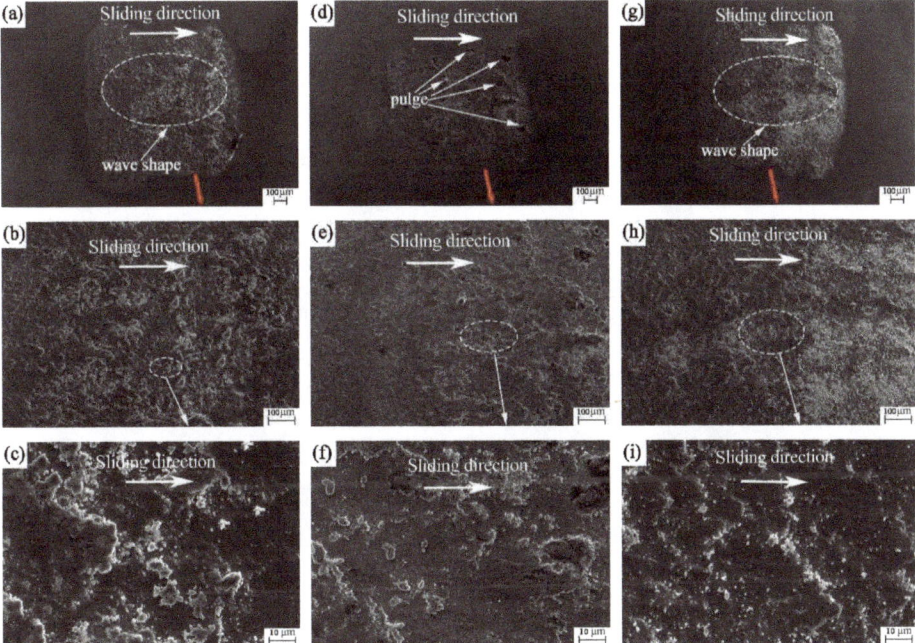

Figure 14. SEM images of the femtosecond-textured worn balls: (**a–c**) rotation radius 15 mm; (**d–f**) rotation radius 18 mm; and (**g–i**) rotation radius 22.5 mm.

From Figure 15a–l, it can be found that the bottom of the groove with electric-spark texture presents the wavy corrosion layer left after electrode discharge. The remelting layer left after femtosecond-laser processing is granular. The groove hardness of both EDM texture and femtosecond texture is higher than that of the original material. The grooves all play good roles in capturing and storing wear debris. As shown in Figure 15a,c,e,g,i,k, there are obvious marks of collision and cutting between the ball and the groove at the edge of the groove, which further confirms our judgment that the quadrilateral wear mark of the groove-texture ball is due to the ball being cut by the edge of the groove.

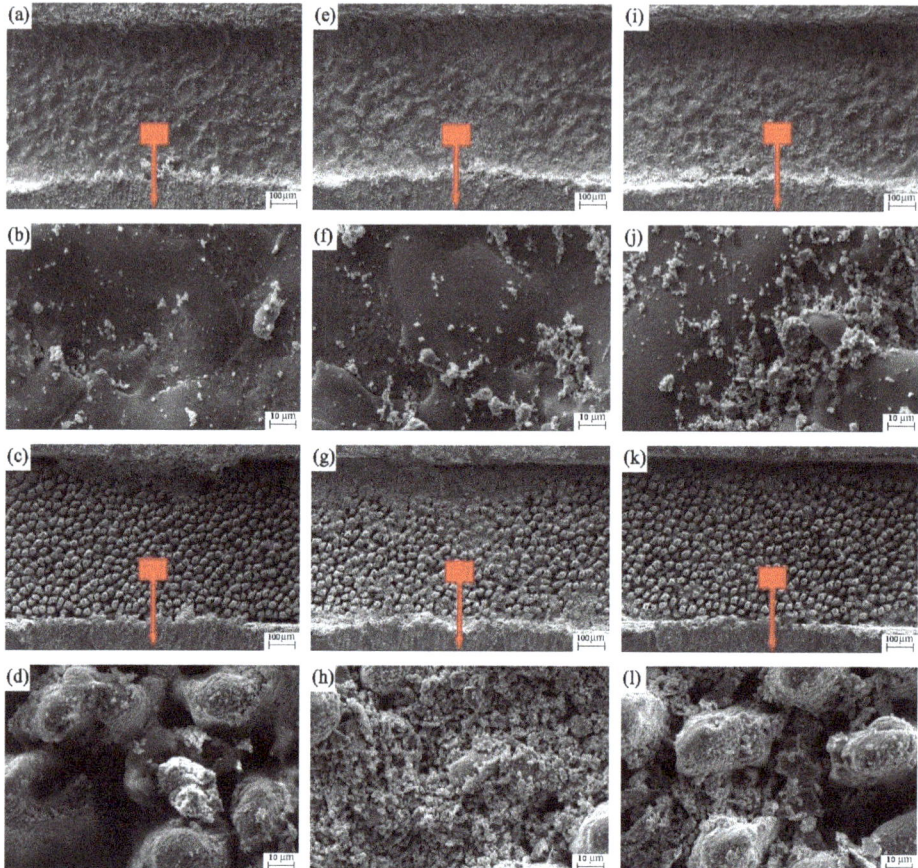

Figure 15. SEM images of the groove-textured worn plates: (**a,b**) electric-spark texture rotation radius 15 mm; (**e,f**) electric-spark texture rotation radius 18 mm; (**i,j**) electric-spark texture rotation radius 22.5 mm; (**c,d**) femtosecond texture rotation radius 15 mm; (**g,h**) femtosecond texture rotation radius 18 mm; (**k,l**) femtosecond texture rotation radius 22.5 mm.

3.4. Energy-Spectrum Analysis of Worn Surface

It can be seen from Table 7 that the chemical composition content of the upper and lower pairs of grinding parts is very similar. Only the content Ni and Al of 0Cr17Ni7Al is higher than that of 9Cr18, so it can be used as a key element. From the energy spectrum of all groups of experiments, the content of the energy-spectrum elements of each surface-modification method under different friction radii is almost the same. Thus, the energy spectrum with radius 15 mm is taken for analysis.

Table 7. Chemical composition and mass fraction of friction pair.

Specimen Name	Material	C	Si	Mn	P	S	Ni	Cr	Al
Plate	0Cr17Ni7Al	0.09	1.0	1.0	0.04	0.03	6.5~7.75	16~18	0.75~1.5
Ball	9Cr18	0.9~1.0	0.8	0.8	0.04	0.03	0.06	17~19	-

According to the energy spectrum in Figure 16a–f, it can be seen that O element is produced in all energy spectra, which proves that there is also oxidation-corrosion wear in the process of friction besides abrasive wear and adhesive wear that are caused by the action of friction heat and air. It can be seen that there are many kinds of wear forms in the process of this experiment. Comparing the energy spectra of the untextured plate and ball in Figure 16a,b, it is not easy to judge whether the material of the ball has transferred to the disk, but it can be seen from the energy spectrum of the ball that the material of the plate has not transferred to the ball. In Figure 16c,e, no elements of 9Cr18 material were found on the disk surface. It is hard to judge whether the material of the ball has transferred to the plate. However, in Figure 16d,f, in the energy spectrum of the ball, we can see that there is Al element, and the mass fraction of Ni element increases significantly. It can be boldly inferred that the material of the plate with spark texture and femtosecond texture is transferred to the ball. It further confirms our judgment that the wear debris produced by friction is under the joint action of load, sliding friction speed, collision, vibration and friction heat. One part is captured by the groove; another is glued and transferred to the surface of the ball specimen to form corrugated wear marks, which could protect the lower specimen and reduce friction.

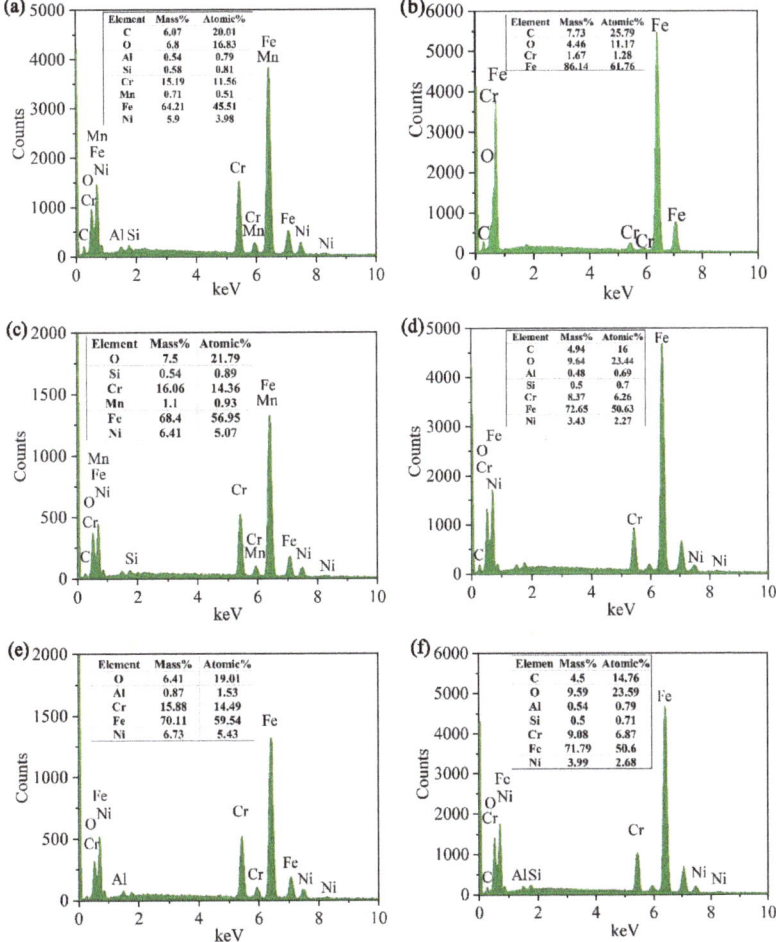

Figure 16. Comparison of EDS spectra of samples used at the rotation radius of 15 mm in the experiment: (**a**) untextured plate; (**b**) untextured ball; (**c**) electric-spark-textured plate; (**d**) electric-spark-textured ball; (**e**) femtosecond-textured plate; (**f**) femtosecond-textured ball.

4. Conclusions

This paper mainly studies the influence mechanism of different groove-collision frequencies on the friction and wear properties of EDM groove texture and femtosecond groove texture on the surface of 0Cr17Ni7Al stainless steel. Some of the main results can be summarized as follows:

(1) The groove texture can be prepared by electric spark and femtosecond laser. The largest friction coefficient of the electric-spark texture and femtosecond texture in the three rotating radii of 15, 18 and 22.5 mm is the friction coefficient of the femtosecond texture at the rotating radius of 22.5 mm $C = 0.8177$. The untextured surface has the lowest friction coefficient $C = 0.8672$. It can be seen that the friction coefficient of both the spark texture and femtosecond texture is significantly lower than that of the untextured surface. However, for the wear rate, not all groove textures are lower than the untextured surface. The wear rate of the femtosecond texture at the rotation radius of 15 mm reached $W = 6.820 \times 10^{-4}$ mm^3/N·mm which is higher than that under all rotation radii without texture. In addition, the friction coefficient of EDM texture at the rotation radius of 22.5 mm is the lowest $C = 0.6755$. Meanwhile, the wear rate is also the lowest $W = 3.662 \times 10^{-4}$ mm^3/N·mm. The best effect of friction reduction and wear resistance is achieved. Therefore, the choice of preparation method of groove texture is often very important.

(2) In the process of friction, the collision between the ball and groove texture plays a great role. The wear debris changes from sliding friction to rolling friction in order to reduce friction. On one hand, it is conducive to the capture and storage of wear particles by groove texture. On the other hand, the wear debris is moved to the edge of the wear mark, which is more conducive to the formation of the edge hard-phase peak while reducing the formation of the hard-phase peak of the wear mark. The edge hard-phase peak forms a support on the surface of the friction pair to protect the wear-mark area and reduce the wear degree.

(3) The influence of collision frequency on spark texture and femtosecond texture are different. In the experiment, with the increase in rotation radius, the collision frequency decreases from 36,000 to 30,000 to 24,000. The electric-spark texture also shows more and more excellent friction reduction and wear resistance with the decrease in collision frequency. The wear resistance of the femtosecond texture improves with the decrease in collision frequency. However, the friction-reducing performance of the femtosecond texture decreases at the beginning and then increases. It can be seen that for the groove texture, reasonable collision frequency performs better, and not more collision frequency.

(4) It can be seen from the oxygen-element content of the ball in the energy spectrum that there is also oxidative-corrosion wear in the friction process, except abrasive wear and adhesive wear. Al element is added to the energy spectrum of the spherical surface with spark texture and femtosecond texture. The mass content of Ni reached 3.99% and 3.43%, respectively. This is significantly higher than the mass content of the original Ni element on the surface of the ball by 0.06%. The ball collided with the groove texture to transfer and glue wear debris to the surface of the ball-mill mark through the joint action of pressure, rotating speed and friction heat. In this way, a wavy antifriction area was formed, which plays the role of reducing resistance and antifriction.

Author Contributions: Conceptualization, L.Y. and W.M.; methodology, L.Y. and W.M.; software, L.Y.; validation, L.Y., W.M. and F.G.; formal analysis, L.Y. and F.G.; investigation, L.Y. and S.X.; resources, L.Y.; data curation, L.Y. and S.X.; writing—original draft preparation, L.Y. and W.M.; writing—review and editing, L.Y. and W.M.; visualization, L.Y. and S.X.; supervision, L.Y. and S.X.; project administration, L.Y. and S.X.; funding acquisition, L.Y. All authors have read and agreed to the published version of the manuscript.

Funding: This research was supported by the National Key R&D Program of China (Grant No. 2018YFB2000100).

Institutional Review Board Statement: Not applicable.

Informed Consent Statement: Not applicable.

Data Availability Statement: Not applicable.

Conflicts of Interest: The authors declare no conflict of interest.

References

1. Pettersson, U.; Jacobson, S. Textured surfaces in sliding boundary lubricated contacts—Mechanisms, possibilities and limitations. *Tribol.-Mater. Surf. Interfaces* **2013**, *1*, 181–189. [CrossRef]
2. Kim, B.; Chae, Y.H.; Choi, H.S. Effects of surface texturing on the frictional behavior of cast iron surfaces. *Tribol. Int.* **2014**, *70*, 128–135. [CrossRef]
3. Shen, X.H.; Tao, G.C. Tribological behaviors of two micro textured surfaces generated by vibrating milling under boundary lubricated sliding. *Int. J. Adv. Manuf. Technol.* **2015**, *79*, 1995–2002. [CrossRef]
4. Han, J.; Fang, L.; Sun, J.; Ge, S. Hydrodynamic lubrication of microdimple textured surface using three-dimensional CFD. *Tribol. Trans.* **2010**, *53*, 860–870. [CrossRef]
5. Lu, X.; Khonsari, M.M. An experimental investigation of dimple effect on the stribeck curve of journal bearings. *Tribol. Lett.* **2007**, *27*, 169–176. [CrossRef]
6. Wang, X.; Liu, W.; Zhou, F.; Zhu, D. Preliminary investigation of the effect of dimple size on friction in line contacts. *Tribol. Int.* **2009**, *42*, 1118–1123. [CrossRef]
7. Ping, G.; Ehmann, K.F. An analysis of the surface generation mechanics of the elliptical vibration texturing process. *Int. J. Mach. Tools Manuf.* **2013**, *64*, 85–95. [CrossRef]
8. Tanvir Ahmmed, K.M.; Grambow, C.; Kietzig, A. Fabrication of micro/nano structures on metals by femtosecond laser micromachining. *Micromachines* **2014**, *5*, 1219–1253. [CrossRef]
9. Parreira, J.G.; Gallo, C.A.; Costa, H.L. New advances on maskless electrochemical texturing (MECT) for tribological purposes. *Surf. Coat. Technol.* **2012**, *212*, 1–13. [CrossRef]
10. Kumar, S.; Singh, R.; Singh, T.P.; Sethi, B.L. Surface modification by electrical discharge machining: A review. *J. Mater. Process. Technol.* **2009**, *209*, 3675–3687. [CrossRef]
11. Uehara, Y.; Wakuda, M.; Yamauchi, Y.; Kanzaki, S.; Sakaguchi, S. Tribological properties of dimpled silicon nitride under oil lubrication. *J. Eur. Ceram. Soc.* **2004**, *24*, 369–373. [CrossRef]
12. Fang, S.; Zhao, H.; Zhang, Q. The application status and development trends of ultrasonic machining technology. *J. Mech. Eng.* **2017**, *53*, 22–32. [CrossRef]
13. Natsu, W.; Ikeda, T.; Kunieda, M. Generating complicated surface with electrolyte jet machining. *Precis. Eng.* **2007**, *31*, 33–39. [CrossRef]
14. Kern, P.; Veh, J.; Michler, J. New developments in through-mask electrochemical micromachining of titanium. *J. Micromech. Microeng.* **2007**, *17*, 1168. [CrossRef]
15. Brizmer, V.; Kligerman, Y.; Etsion, I. A laser surface textured parallel thrust bearing. *Tribol. Trans.* **2003**, *46*, 397–403. [CrossRef]
16. Liu, X.; Dong, L.; Wang, S.; Li, G.; Liu, K. Influence of micro-cavity textured surface on tribological property of grease lubricated spherical bearing. *Tribology* **2014**, *34*, 387–392. [CrossRef]
17. Dong, H.; Liu, K.; Wang, W.; Liu, X. Laser textured shaft surfaces on the pumping action and frictional properties of lip seals. *Tribology* **2012**, *32*, 126–132. [CrossRef]
18. Enomoto, T.; Sugihara, T.; Yukinaga, S.; Hirose, K.; Satake, U. Highly wear-resistant cutting tools with textured surfaces in steel cutting. *CIRP Ann.-Manuf. Technol.* **2012**, *61*, 571–574. [CrossRef]
19. Xi, P.; Cong, Q.; Teng, F.; Guo, H. Design and experiment of bionics pit shape grinding roller for improving wear resistance and crushability. *Trans. Chin. Soc. Agric. Eng.* **2018**, *34*, 55–61. [CrossRef]
20. Ling, T.D.; Liu, P.; Xiong, S.; Grzina, D.; Cao, J.; Wang, Q.J.; Xia, Z.C.; Talwar, R. Surface texturing of drill bits for adhesion reduction and tool life enhancement. *Tribol. Lett.* **2013**, *52*, 113–122. [CrossRef]
21. Yin, B.; Lu, Z.; Liu, S.; Fu, Y.; Wang, Y. Theoretical and experimental research on lubrication performance of laser surface texturing cylinder liner. *J. Mech. Eng.* **2012**, *48*, 91–96. [CrossRef]
22. Bruzzone, A.A.G.; Costa, H.L.; Lonardo, P.M.; Lucca, D.A. Advances in engineered surfaces for functional performance. *CIRP Ann.-Manuf. Technol.* **2008**, *57*, 750–769. [CrossRef]
23. Saeidi, F.; Meylan, B.; Hoffmann, P.; Wasmer, K. Effect of surface texturing on cast iron reciprocating against steel under starved lubrication conditions: A parametric study. *Wear* **2016**, *348–349*, 17–26. [CrossRef]
24. Galda, L.; Sep, J.; Olszewski, A.; Zochowski, T. Experimental investigation into surface texture effect on journal bearings performance. *Tribol. Int.* **2019**, *136*, 372–384. [CrossRef]
25. Liu, C.; Guo, F.; Wong, P.; Li, X. Laser pattern-induced unidirectional lubricant flow for lubrication track replenishment. *Friction* **2021**, *10*, 1234–1244. [CrossRef]
26. Peng, J.; Shen, M.; Cai, Z. Nano diesel soot particles reduce wear and friction performance using an oil additive on a laser textured surface. *Coatings* **2018**, *8*, 89. [CrossRef]
27. Qi, X.; Wang, H.; Dong, Y.; Fan, B.; Zhang, W.; Zhang, Y.; Ma, J.; Zhou, Y. Experimental analysis of the effects of laser surface texturing on tribological properties of PTFE/Kevlar fabric composite weave structures. *Tribol. Int.* **2019**, *135*, 104–111. [CrossRef]

28. Volpe, A.; Covella, S.; Gaudiuso, C.; Ancona, A. Improving the laser texture strategy to get superhydrophobic aluminum alloy surfaces. *Coatings* **2021**, *11*, 369. [CrossRef]
29. Pranav, C.; Do, M.T.; Tsai, Y.C. Analysis of high-friction surface texture with respect to friction and wear. *Coatings* **2021**, *11*, 758. [CrossRef]
30. Wang, Z.; Song, J.; Wang, T.; Wang, H.; Wang, Q. Laser texturing for superwetting titanium alloy and investigation of its erosion resistance. *Coatings* **2021**, *11*, 1547. [CrossRef]
31. Bai, Q.; Bai, J.; Meng, X.; Ji, C.; Liang, Y. Drag reduction characteristics and flow field analysis of textured surface. *Friction* **2016**, *4*, 11. [CrossRef]
32. Pham, D.T.; Dimov, S.S.; Bigot, S.; Ivanov, A.; Popov, K. Micro-EDM—Recent developments and research issues. *J. Mater. Process. Technol.* **2004**, *149*, 50–57. [CrossRef]
33. Chichkov, B.N.; Momma, C.; Nolte, S.; Alvensleben, F.V.; Tünermann, A. Femtosecond, picosecond and nanosecond laser ablation of solids. *Appl. Phys. A* **1996**, *63*, 109–115. [CrossRef]
34. Yang, L.; Ma, W.; Gao, F.; Li, J.; Deng, M.; Liu, Z.; Ma, L.; Meng, H. Study on Tribological Properties of groove texture in surface micromachining. *Tool Technol.* **2021**, *55*, 73–76. [CrossRef]

Article

Self-Cleaning Coatings for the Protection of Cementitious Materials: The Effect of Carbon Dot Content on the Enhancement of Catalytic Activity of TiO₂

Charis Gryparis, Themis Krasoudaki and Pagona-Noni Maravelaki *

Laboratory of Materials of Cultural Heritage and Modern Building, School of Architecture, Technical University of Crete, 73100 Chania, Greece; cgryparis@isc.tuc.gr (C.G.); themis.krasoudaki@gmail.com (T.K.)
* Correspondence: pmaravelaki@isc.tuc.gr

Abstract: The urgent demand for pollution protection of monuments and buildings forced the interest towards specific preservation methods, such as the application of photocatalytic coatings with self-cleaning and protective activity. TiO₂ photocatalysts without and with a variety of carbon dots loading (TC0, TC25–75) were synthesized via a green, simple, low cost and large-scale hydrothermal method using citric acid, hydroxylamine and titanium isopropoxide (TTIP) and resulted in uniform anatase phase structures. In photocatalysis experiments, TC25 and TC50 composites with 1:3 and 1:1 mass ratio of C-dots solution to TTIP, respectively, showed the best degradation efficiency for methyl orange (MO) under UV-A light, simulated solar light and sunlight compared to TiO₂, commercial Au/TiO₂ (TAu) and catalysts with higher C-dot loading (TC62.5 and TC75). Treatment of cement mortars with a mixture of photocatalyst and a consolidant (FX-C) provided self-cleaning activity under UV-A and visible light. This study produced a variety of new, durable, heavy metal-free C-dots/TiO₂ photocatalysts that operate well under outdoor weather conditions, evidencing the C-dot dosage-dependent performance. For the building protection against pollution, nanostructured photocatalytic films were proposed with consolidation and self-cleaning ability under solar irradiation, deriving from combined protective silica-based agents and TiO₂ photocatalysts free or with low C-dot content.

Keywords: self-cleaning; TiO₂/Cdots; protective films; cultural heritage conservation; photocatalysts; sunlight irradiation

1. Introduction

Air pollution, the fourth greatest fatal health risk so far, is accounting for serious danger to human health, causing millions of premature deaths annually [1]. There is a demanding need to combat climate change as far as the global population grows; thus, highly efficient technologies can serve as the key to improving air quality and protecting the environment [2,3]. Moreover, air pollution and climate change cause physical and aesthetic damage to buildings and cultural heritage constructions, accelerating the decay process [4]. For the protective treatment of several types of building substrates, advanced coatings with self-cleaning ability have been developed throughout the years [5]. The main advantage of photocatalytic air and wastewater purification is that chemicals or external energy input are not requisite, except for ambient light or sunlight, which is not costly. Despite the strong effort for the development of various novel active materials, titanium dioxide (TiO₂) is by far the most promising and widely employed technology in photocatalysis among them [6].

TiO₂ was introduced in the field of photocatalysis in 1972 [7], and since then, it has become the most investigated compound for air purification due to its excellent physico-chemical stability and oxidation properties, as well as cost-effectiveness, high availability and eco-friendliness [8]. Photoinduced processes begin with the absorption of electromagnetic radiation (UV-light) by TiO_2, which cause an electron to be promoted from the valence band (VB) to the conduction band (CB), leaving a positively charged hole in the VB [9]:

$$TiO_2 + h\nu \rightarrow e^-_{CB} + h^+_{VB} \qquad (1)$$

introducing the energy gap (Eg), one of the most useful characteristics in semiconductor field (Equation (1)). Afterward, electron and hole separately may take part in reduction and oxidation reactions, respectively, with species such as H_2O, O_2 and OH^- adsorbed on the TiO_2 surface. These species can be oxidized to OH radical or reduced to $O_2\bullet$, which can react with the pollutant, leading to decomposition of the latter. Since the ideal photocatalyst offers a combination of low Eg and efficient charge separation, TiO_2, when solely used, has limited performance in outdoor conditions. TiO_2 appears in three crystalline phases: rutile and anatase are both tetragonal, and brookite is orthorhombic [10]. Although anatase has a larger bandgap (3.2 eV) compared to rutile and brookite (3.0 eV), its photocatalytic activity is apparently superior to that of rutile, while synthesis of pure brookite is a demanding challenge. Most research is focused on anatase due to the high surface adsorption capacity and the indirect bandgap, which leads to a much longer lifetime of photogenerated holes and electrons compared to the larger grain size and direct bandgap of rutile [11]. Even in the form of anatase, the bandgap of TiO_2 can be approached by UV light irradiation ($\lambda \leq 380$ nm, 5% of sunlight), but the low photoenergy of visible light is not sufficient for the excitation of electrons. Thus, in order to increase the photocatalytic activity and take advantage of solar light, modified TiO_2 photocatalysts, such as TiO_2 loaded with Carbon Dots (TiO_2/C-dots, TC), is necessary.

C-dots, byproducts of carbon nanotubes purification accidentally isolated in 2004 [12], have attracted intense attention in photocatalytic applications due to their low toxicity, good photostability, small size and low synthetic cost even on a large scale [13]. Moreover, their strong and tunable photoluminescence enables their application in biomedicine, optoelectronic devices and biosensing [14]. A typical C-dot is a small carbon nanoparticle functionalized on its surface by organic molecules, typically acids or amines, using various chemical reactions. This functionalization plays an important role in the characteristic properties of C-dots due to the defects of the organic groups on the surface of each C-dot, which can cause efficient charge separation, leading to radiative recombination of holes and electrons with subsequent fluorescence emission [15].

There are several reports regarding the synthesis of C-dots, or nitrogen-doped C-dots, and TiO_2 nanocomposites with applications in the photocatalytic degradation of organic dyes [16–25]. The synthetic process of these methodologies required either prolonged heating time for C-dots preparation [16,17] or calcination at 400–500 °C [18,19] and even at 700 °C [20] in the TiO_2/C-dots' step of the synthesis. In another study, a one-step solvothermal route was proposed, showing remarkable photocatalytic degradation of methylene blue (MB) under UV light [21]. However, for the successful degradation of methyl orange (MO) and Rhodamine B (RB) under UV, calcination of catalyst at 400 °C was inevitable. Other studies used specialized equipment, such as high-pressure Teflon-lined sealed autoclave containers [22,23] or a combination with spin coating method [24] for the controlled synthesis of C-dots, making large-scale production prohibitive. Finally, in all cases, the cycling performance of the synthesized photocatalysts exhibited moderate degradation ($\leq 70\%$ after six cycles) [23,25].

Recently, we developed an efficient, simple and low-cost method for the synthesis of pure TiO_2 and TiO_2/C-dots and studied comparatively into the photodegradation of MO under UV-A and visible light, with promising results [26]. MO was used as a model compound because it is a common, highly toxic dye that exhibits good photostability upon light irradiation under different conditions [27] and resistance to complete biodegradation. In this study, using a similar methodology, five TiO_2-based compounds were synthesized containing different C-dot loading, namely TC0, TC25, TC50, TC62.5 and TC75, and analyzed by several techniques, such as X-Ray diffraction (XRD), scanning and transmission electron microscopies (SEM, TEM) and UV-Vis/near-IR diffuse reflectance. These TiO_2/C-dots nanocomposites were used as model photocatalysts to investigate the effect of the C-dot content in photocatalytic degradation of MO under UV-A and visible light, but also direct sunlight radiation, showing remarkable results compared to pure TiO_2 (TC0), as well as commercially available Au/TiO_2 (TAu) and this is the first innovative result of this study. The most suitable photocatalysts were found to be TC25 and TC50, evidencing that this range of C-dot loading is superior for the decomposition of organic pollutants, another interesting finding that correlates Cdots amount and efficiency. Another innovation of this study, apart from the rapid and excellent photocatalytic activity of these catalysts under solar light irradiation, is the almost eternal nature, as they can be recycled and reused up to ten times without losing their catalytic ability (~90% degradation after 3 h of irradiation into the 10th cycle, >99% after 6 h).

Furthermore, based on the tetraethoxysilane (TEOS) effectiveness as a strengthening agent for Portland cement mortars [28], we synthesized an advanced hydrophobic consolidant, designated as FX-C, using a sol–gel process that combines TEOS, PDMS and nano-calcium oxalate (nano-CaOx) for the preservation of cultural heritage [29,30]. It was found that silicon in TEOS/PDMS copolymer can form strong Si–O bonds with hydroxyl groups of the material for preservation [31]. It is well-established that among other useful properties of TiO_2-based catalysts [32–34], coating application of nano-TiO_2 or $SiO_2@TiO_2$ [35–38] in mortars provides restoration and self-cleaning ability in building materials [39–41], while TEOS-PDMS-TiO_2 hybrid nanomaterials in limestone and marbles enhance these properties [42,43]. Our previous work that studied the self-cleaning ability of the composite coating FX-C with TiO_2/C-dots in concrete, limestones and lime mortars showed a moderate photocatalytic activity of this protective agent under visible light irradiation [26]. The fourth innovative result of this work refers to the accomplishment of a self-cleaning coating of FX-C and TC0–TC25 activated under visible light irradiation, ready to be applied onto cementitious mortars.

2. Materials and Methods

2.1. Materials

The citric acid (monohydrate, $\geq$99%) and hydroxylamine hydrochloride (99%) were employed from Sigma-Aldrich (St. Louis, MI, USA) as C-dots precursors and used as received. Titanium (IV) isopropoxide (TTIP, $\geq$97%) and hydrochloric acid (~37%) were purchased from Sigma-Aldrich (St Louis, MI, USA) and used as received for the synthesis of TC0–75 powders. Methyl Orange (for microscopy, Hist.) was obtained from Sigma Aldrich (St Louis, MI, USA) and utilized as the model compound for the degradation of organic pollutants. Gold 1% on titanium dioxide extrudates was employed from Strem Chemicals (Newburyport, MA, USA) and used as received. Commercially available solvents were used as received without further purification.

2.2. Analytical Techniques

High-resolution TEM micrographs of C-dots and TC0–75 were obtained using a JEM-2100 transmission electron microscope and a JEM-100C microscope (JEOL Ltd., Tokyo, Japan), operating at an accelerating voltage of 200 kV and 80 kV, respectively. The samples for TEM were prepared by drop-casting 8 µL of the diluted aqueous solution of each sample onto Formvar/carbon–copper-coated grids supplied by Agar Scientific (Essex, UK) and then left to dry at room temperature for 2 h. A transmission electron microscope was also used for statistical size-distribution histograms by counting and measuring the diameter size distribution via ImageJ software. SEM images were recorded on a JSM 6390LV scanning electron microscope (JEOL Ltd., Tokyo, Japan) operated at 20 kV electron voltage. TC nanoparticles were dried on a two-sided carbon tape and sputter-coated with ca. 100 Å of gold (Au), using an SCD 050 sputter coater (BAL-TEC, Los Angeles, CA, USA). UV-Vis spectra of samples were obtained on a Cary 1E UV-Vis spectrophotometer (Varian, Palo Alto, CA, USA), in the wavelength range between 200 and 700 nm, using a quartz cuvette with 10 mm path length and 3.5 mL chamber volume. UV-Vis/near-IR diffuse reflectance spectra were recorded on a UV-2600 spectrophotometer (Shimadzu, Kyoto, Japan) using $BaSO_4$ as the 100% reflectance reference. Powder X-ray diffraction data (XRD) of the TC0–75 catalysts were obtained from a D8 Advance X-ray diffractometer (Bruker, Billerica, MA, USA) equipped with a Lynx Eye strip silicon detector (Bruker, Billerica, MA, USA). Measurements were applied at room temperature, using Cu Kα radiation at 35 mA and 35 kV. Fourier Transform Infrared spectroscopy (FT-IR) in powdered samples was performed with an iS50 FT-IR spectrometer (Thermo Fischer Scientific, Waltham, MA, USA) in the spectral range between 4000 and 400 cm^{-1}. For Attenuated Total Reflection (ATR) measurements, a built-in, all-reflective diamond was used as the internal reflection element. The measurement of the color change via photocatalytic degradation was performed with a CM-2600d spectrophotometer (Konica Minolta, Tokyo, Japan) with a D65 illuminant at 8-degree viewing, in the wavelength range between 360 and 740 nm. A 36 W UV EBN001W LED curing lamp (340–380 nm, maximum emission at 365 nm, Esperanza sp. j. Poterek, Ożarów Mazowiecki, Poland) provided the UV-A irradiation source, while a laboratory constructed stimulated solar box equipped with two L15W/840 tubular fluorescent lamps (400–630 nm, maximum emission at 540 nm, OSRAM, Munich, Germany) was used as the solar simulator. A GM3120 (BENETECH, Shenzhen, China) electromagnetic radiation tester was used to measure light irradiance.

2.3. Synthesis of Carbon Dots

Regarding the C-dots preparation, the procedure was slightly modified from an already published process [44]. A solution of citric acid (1.65 g) and hydroxylamine hydrochloride (1.65 g) in deionized water (16.5 mL) was heated at 300 °C for 12 min. The resulting dark brown, sticky solid was dissolved in water, sonicated via ultrasonic agitation for 10 min to separate the aggregates and filtered through a 7–12 μm Whatman filter paper to remove undissolved byproducts. A final 2.5% w/v C-dots solution was collected as an opaque liquid (Figure 1).

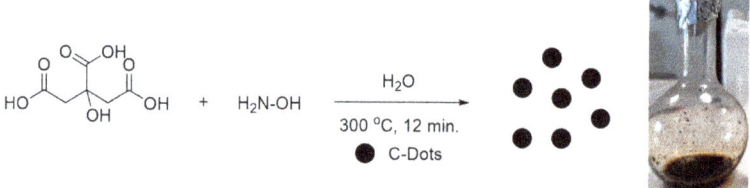

Figure 1. Hydrothermal synthesis of C-Dots.

2.4. Synthesis of TiO_2 (TC0) and TC25–75 Photocatalysts

For the preparation of TiO_2 with different loading of C-dots, an already published method was followed [26], employing slight modifications (Figure 2). Briefly, a 2.5% w/v aqueous C-dots solution, specifically 1.5 g for TC25 up to 13.5 g for TC75, was dissolved in a 1:1 mixture of ethanol and deionized water (50 mL). For this solution, a solution of TTIP (4.5 g) in ethanol (20 mL) was added dropwise, and the pH of the resulting mixture was adjusted to 2.5 by the addition of concentrated HCl (~5 drops) to facilitate the hydrolysis of TTIP. The mixture was allowed to stir at ambient temperature overnight, followed by heating at 80 °C for 8 h to facilitate titania polymerization and C-dots incorporation. Afterward, the mixture was cooled to ambient temperature, centrifuged for 10 min at 3000 rpm, and the wet solid was dried at 80 °C and ground using a mortar. Finally, the catalyst was calcinated at 200 °C for 2 h to produce the crystalline material as a yellow (TC25) to brown solid (TC50–75) in 85%–98% yield.

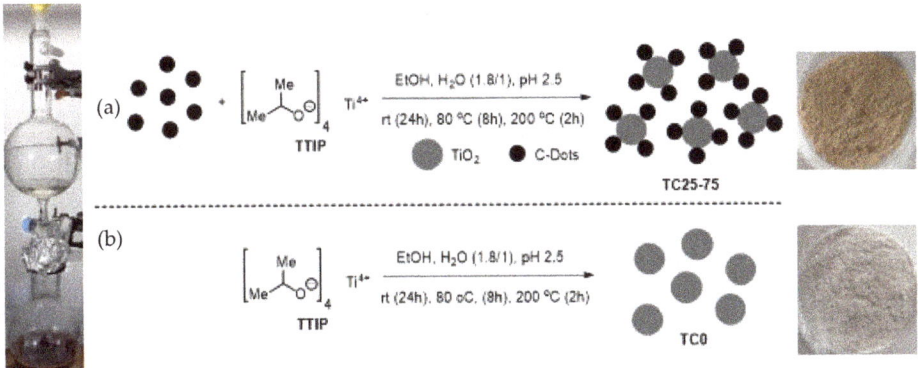

Figure 2. Synthesis of (**a**) TiO$_2$ enriched with C-dots (TC25–75) and (**b**) TiO$_2$ (TC0).

The same process, in the absence of the aqueous C-dots solution, was applied for the synthesis of TiO$_2$ as a control substance (Figure 2). It is worth mentioning that a crucial factor in this experiment is the adjustment of pH at 2.5 since lower pH leads to a reduction in the isolated yield of the catalyst. For example, at pH 1.5, the isolated yield of TC50 drops to 46%. Furthermore, in the process of synthesis of the control substance TC0, acidification up to pH 1.5 leads to a mixture of brookite and anatase phases of TiO$_2$, whereas at pH 2.5, only the anatase phase is present, as indicated in the XRD analysis.

2.5. Incorporation of Photocatalysts into the Protective Agent

Consolidant FX-C was prepared in the laboratory by an already published method [29]. Each solid catalyst (4% w/w) was added to the sol FX-C, and the mixture was sonicated via ultrasonic agitation for 10 min to ensure homogeneity (Figure S1).

2.6. Application of the Protective/Photocatalytic Film onto Cementitious Mortars

The performance of the photocatalytic films in the conservation of cement-based materials was evaluated using cement mortars prepared in the laboratory, with a binder/sand ratio of 1:3, based on local 0–3 mm of grain size standard sand of carbonaceous nature. Further information on the mix design of the treated mortars can be found in Table S1. The synthesized photocatalytic films were brushed three times on the cement mortar substrates. Control experiments were also performed by measuring the untreated cement mortar and by brushing the cementitious material with FX-C in the absence of a photocatalyst.

2.7. Photodegradation of MO Using Photocatalysts and Their Reusability

The activity of a series of TC nanocomposites on the photodegradation of MO was initially evaluated under UV-A light, using a 36 W LED curing lamp (maximum emission at 365 nm) with a light irradiance of approximately 33.6 mW/cm^2 and artificial solar light, using a solar box equipped with two 15 W tubular fluorescent lamps (maximum emission at 540 nm) with a light irradiance of approximately 13.6 mW/cm^2. For both irradiation types, the distance of each sample from the lamps was set to 7 cm.

The photocatalytic activity of TC25, TC50, TC62.5 and TC75 catalysts, compared to TC0, was evaluated by studying the photodegradation of MO solution in the presence of the appropriate catalyst under UV-A light, simulated solar light, as well as sunlight in outdoor weather conditions. For this purpose, the solid photocatalyst (40 mg, 0.08% w/v) was dispersed to a 5 ppm MO aqueous solution (50 mL), and the mixture was allowed to stir in darkness for 30 min in order to reach adsorption–desorption equilibrium of dye on the catalyst surface. Thereafter, the mixture was irradiated with the appropriate light source for a period of 120 min, with a sample removal (4 mL) every 15 min for reaction monitoring. Analytical determination of the reduction in MO concentration during the photocatalytic degradation process was carried out spectrophotometrically at 465 nm. A commercially available catalyst Au (1%)/TiO$_2$ (TAu) was also used in the MO degradation under sunlight conditions for comparison. Under the same conditions, MO degradation was tested in the absence of a catalyst, where no decomposition was observed.

Furthermore, the reusability of TC50 was tested under simulated sunlight irradiation. In this case, TC50 (0.16 g) was added into a 5 ppm aqueous MO solution (80 mL) and irradiated with visible light for 180 min. After each cycle, the solution was decanted, and the catalyst was washed with water and ethanol (30 mL each), dried at 60 °C overnight, and then reused in the next cycle. This process was repeated 9 times, while at the end of the 10th cycle, the catalyst was recycled (0.08 g) and could be further used in the next experiment.

2.8. Self-Cleaning Performance of Treated Cementitious Mortars

In order to determine the self-cleaning and air depolluting capabilities of the protective films enhanced with TiO$_2$/C-dots, the protective coating was brushed onto the cement mortar substrates, and the photodegradation of a drop of MB under UV-A and visible light was determined using a portable CM-2600d spectrophotometer (Konica Minolta, Tokyo, Japan). Control experiments were performed by exposing the untreated cement mortar and the cementitious material with FX-C, covered with the same amount of MB. The total color difference (ΔE) can be calculated using the following equation [45]:

$$\Delta E = \sqrt{\Delta L^{*2} + \Delta a^{*2} + \Delta b^{*2}} \tag{2}$$

where the differences in color (Δ) with time, for the treated with a protective coating and spotted with MB cement mortars, were expressed as L^*, a^* and b^* values of the CIELab color space. The L^* values range from 0 for black to +100 for white, while the negative and positive a^* values represent green and red, respectively. As for the negative and positive b^* values, these represent blue and yellow, respectively.

3. Results and Discussion

3.1. Structure and Morphology of C-Dots and TC0-75 Photocatalysts

Figure 3 illustrates the XRD patterns of the pure synthesized TiO$_2$ (TC0) and TC25–75, evidencing the crystalline phases present in the samples.

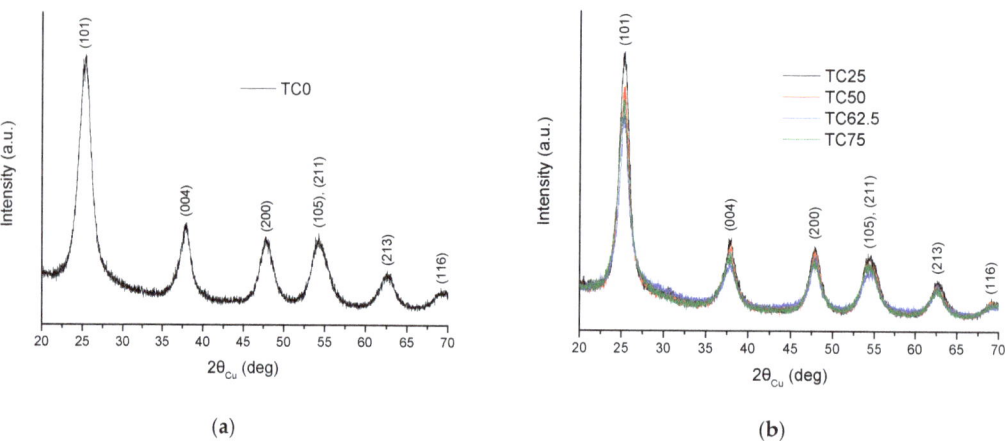

Figure 3. X-ray diffraction patterns of (**a**) TiO$_2$ (TC0) and (**b**) TiO$_2$ with different C-dot loading (TC25–75).

The diffractograms reveal the formation of anatase phase (JCPDS card no. 00-021-1272) exclusively in all the synthesized materials, excluding other possible phases of titania, such as rutile or brookite. The anatase crystalline phase is commonly accepted as a more efficient photocatalyst, compared to rutile and brookite, due to its high photocatalytic activity [46]. Crystallite average size of the TC0–75 nanoparticles was estimated using the following Scherrer equation [47]:

$$D = \frac{K\lambda}{\beta \cos\theta} \quad (3)$$

where D is the crystallite particle size (in nm), K is the Scherrer constant equal to ~0.9 in the case of spherical shapes, λ is the wavelength of the X-ray's radiation (0.15406 nm), β is the full width at half maximum (FWHM) height of the diffraction peaks and θ is the Bragg's diffraction angle. The diffraction peaks at 25.2°, 37.9°, 47.9° and 62.7° correspond to phases (101), (004), (200) and (213), respectively, while the peak appearing at 2θ 54.3° is attributed to both (105) and (211) phases. The crystallite size of the photocatalysts was calculated using the (101) peak and found to be 4.53, 6.06, 5,36, 6.35 and 5.07 nm for samples TC0, TC25, TC50, TC62.5 and TC75, respectively, showing only a slight increase in crystallite size with C-dot incorporation. The results are in good agreement with previous findings, showing the correlation between the low calcination temperature and the small crystal size of TiO_2 [48–50], as well as the stability of anatase when the particle size is below 14 nm [51].

TEM images of the as-synthesized C-dots (Figure S3) were similar to that observed in our previous work [26]. The average size of C-dots was 1.9 nm, ranging from 1.2 to 2.6 nm, and the shape appeared to be uniform and quasi-spherical (Figure S3c). TEM analysis of the TC50 and TC75 photocatalysts can be seen in Figure 4a,b. TC50 exhibited relatively uniform size nanoparticles, with an average size of 20–21 nm and the size distribution ranging approximately from 14 to 32 nm. In a previous work, a similar size distribution ranging from 10 to 40 nm for the TC25 photocatalyst was observed [26]. Agglomeration of crystallites resulting in nanocluster formation was observed in previous works for TiO_2 [52,53] and TiO_2 doped with C-dots [10] or with other species [54,55]. TEM images of TC75 nanoparticles in Figure 4c,d show a non-uniform but quasi-spherical shape in an average size of 64–68 nm and wide size distribution of 40 to 100 nm (Figure S4) due to the high C-dot loading in the TiO_2 surface. The XRD results and the crystal planes observed in the TEM images of our previous work for TC25 [26] were both attributed to the crystalline structure of the TiO_2/C-dots nanoparticles, further evidenced in the SEM images in Figure 4. Even though the synthesis of pure anatase requires calcinations at 500 °C within several hours [56], nevertheless, in our case, TEM images indicated that at a relatively low temperature, a C-dots-assisted crystalline TiO_2 could be formed [26].

Indeed, SEM was used to display the surface morphology of TC25 and TC62.5 nanocomposites (Figure 5). Both nanohybrids consist of nearly spherical in shape, rough and compact aggregates with crystal structures and various diameters. The spherical-like shape is typical for anatase crystallite structure [10].

The mineralogical and microstructural characterization of the studied photocatalysts revealed that anatase was formed in the presence of different C-dot loading upon a heating process up to 200 °C. By increasing the C-dot loading, as in the case of the TC75, the average size of the photocatalysts was also increased due to agglomeration phenomena.

The FTIR spectrum of C-dots (Figure 6) confirms the presence of polar groups derived from the incorporation of hydroxylamine into the carbon core. This observation is in accordance with the good dispersion properties that C-dots exhibit in water. More specifically, the broad peaks at 3450 and 3200 cm^{-1} correspond to the symmetric stretching vibration of O–H and N–H, respectively [23], while the peaks at 1605 and 1399 cm^{-1} can be attributed to the asymmetric and symmetric stretching vibration of COO^- [57]. The stretching vibration of C=O at 1697 cm^{-1} along with the bending vibration of N–H in amides at 1584 cm^{-1} [26,58] and the stretching vibration of C–O in carboxylates at 1183 cm^{-1} are also present in the C-dots spectrum. Other characteristic peaks of the main core in carbon dots can be considered: (a) the asymmetric and symmetric stretching vibration of C–H bonds with the broad peaks at 3041 and 2807 cm^{-1}, and (b) the stretching and bending vibrations of C=C at 1630 and 746 cm^{-1}. Finally, it is important to highlight that in the FTIR spectra of the TC25–75 photocatalysts (Figure 6), characteristic bands of C-dots at 1698, 1605 and 1183 cm^{-1} were shifted to higher and lower wavenumbers (e.g., 1630, 1552 and 1223 cm^{-1}, respectively), due to the C-dots' incorporation in the TiO$_2$ surface. In the sample TC75, the higher intensity of the C–H asymmetric and symmetric stretching vibrations at 2921 and 2852 cm^{-1} is directly correlated with the higher C-dots content.

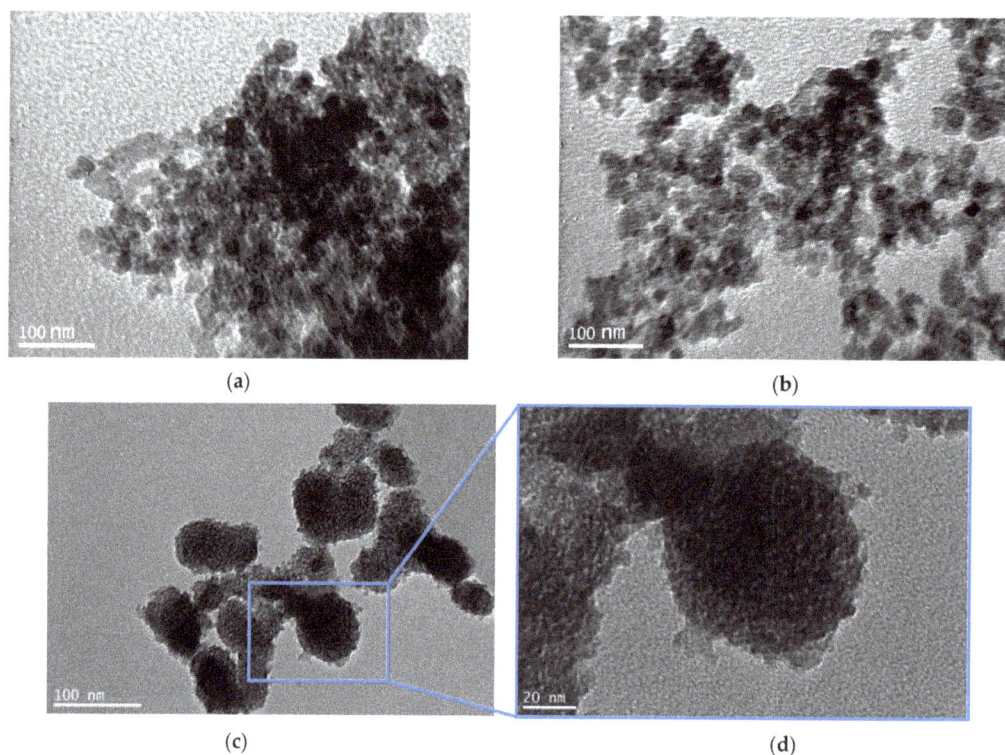

Figure 4. TEM images of (**a**,**b**) TC50 and (**c**,**d**) TC75 nanocomposites.

Figure 5. *Cont.*

Figure 5. SEM images of (**a**,**b**) TC25 and (**c**,**d**) TC62.5 nanocomposites.

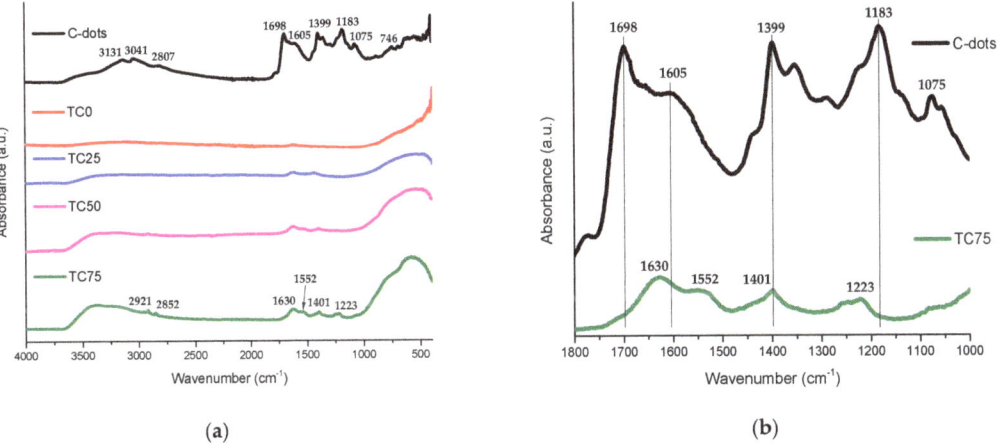

Figure 6. FTIR spectra of (**a**) C-dots, TC0, TC25, TC50 and TC75 photocatalysts and (**b**) the zoom area (1800–1000 cm^{-1}) of C-dots and TC75 photocatalyst.

The optical properties of the TC0–75 photocatalysts were studied by UV-Vis/NIR diffuse reflectance spectroscopic (DRS) measurement (Figure 7a). It is obvious that the absorption edge of TC25–75 was shifted towards the visible light region compared to TC0, suggesting significant absorption ability for solar light. The energy band gaps (Eg) of photocatalysts TC0–75 were determined from UV-Vis/NIR diffuse reflectance spectra (Figure 7b), using Tauc plot analysis for indirect allowed transition [59], due to the anatase crystallite form of TC structures. In this case, (f hv)$^{1/2}$ is expressed as a function of photon energy (hv), where f is the Kubelka–Munk function [60] of the measured reflectance (R), as shown in the following equation:

$$f(R) = \frac{(1-R)^2}{2R} \tag{4}$$

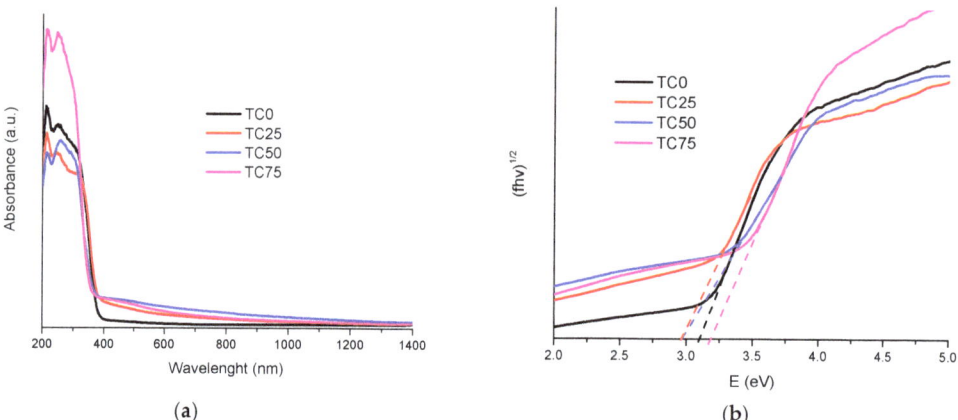

Figure 7. (**a**) UV-Vis/NIR diffuse reflectance spectra and (**b**) Tauc plots for TiO$_2$ (TC0), TC25, TC50 and TC75 photocatalysts.

This analysis led to Eg values of 3.10 eV for TC0, which is close to the bandgap of anatase. On the contrary, the bandgaps of TC25 and TC50 were measured at 2.97 eV each, while TC75 exhibited a bandgap of 3.18 eV. The findings are in correlation with the photocatalysis results, suggesting that the Eg of the nanocomposites appears to be an important parameter that accounts for the efficiency of the photocatalytic process. Thus, C-dots can reduce the energy bandgap of TiO$_2$ up to a point where C-dots dominate the structure of the photocatalyst, leading to an even higher Eg than TiO$_2$, as in the case of TC75 with higher C-dot content.

3.2. Photocatalytic Performance of TC25–75 Compared to TC0 and TAu

The results of the UV-A irradiation process were shown in Figure 8a, where TC25–62.5 composites provided a much higher photodegradation rate for MO compared to TiO$_2$ (TC0) and TC75, with both TC25 and TC50 photocatalysts showing the best performance. The light irradiance of the 36 W LED curing lamp used in our experiments, based on the sample distance, was approximately 33.6 mW/cm^2. The area of a 50 mL beaker that was used in the experiments can be defined as half the curved surface area of the cylinder, $\pi r h$, where r is the base radius, and h is the height of the cylinder. Therefore, the average continual radiation was ~1.42 W. After 60 min of UV-A light irradiation, the MO degradation rates in the presence of TC0, TC25, TC50, TC62.5 and TC75 were 20.8, 91.4, 92.0, 56.9 and 48.7%, respectively. C-dots had almost no photocatalytic ability for MO, as only a 4.6% of the dye was degraded after 60 min. The UV-Vis spectrum and macroscopic evaluation of MO degradation using TC50 under UV-A light are displayed in Figures S5 and S6. Similar results, in a slower degradation rate, were found during the simulated solar light irradiation, as shown in Figure 8b. The light irradiance of the two 15 W fluorescent lamps, based on the sample distance, was measured at 13.6 mW/cm^2; thus, the average continual radiation was ~0.58 W. It is evident that all of the TC25–75 composites exhibited higher photocatalytic activity for MO compared to TC0, whereas both TC25 and TC50 provided the highest degradation rate. More specifically, after 120 min of irradiation, the MO degradation rates in the presence of TC0, TC25, TC50, TC62 and TC75 were 45.4, 96.2, 90.9, 62.9 and 46.1%, respectively. The UV-Vis spectrum and macroscopic evaluation of MO degradation using TC50 under visible light are reported in Figures S7 and S8.

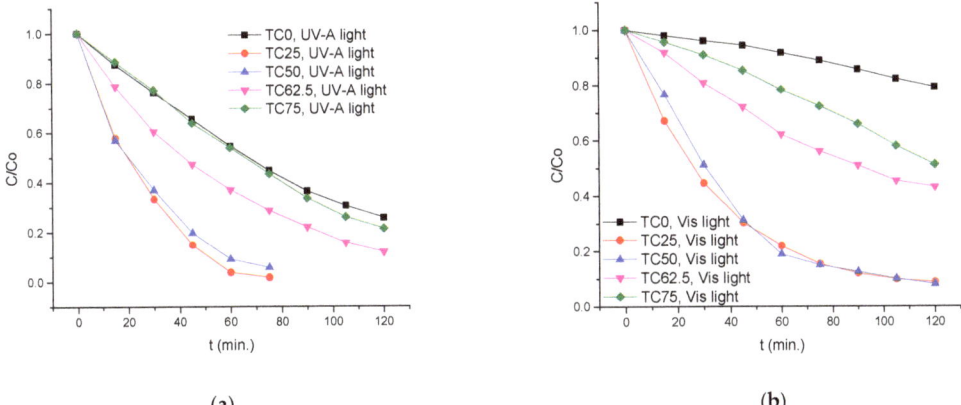

Figure 8. MO degradation rates in the presence of TC0–75 photocatalysts under (**a**) UV-A and (**b**) simulated solar irradiation.

In all photocatalytic experiments, the concentration of MO was fixed at 5 ppm. It is worth mentioning that for concentrations ranging from 5 to 25 ppm, the highest photocatalytic activity was found to be 20 ppm in an anatase titania suspension system, with 5 ppm being the least suitable MO concentration [61]. MO exhibits the lowest degradation rate compared to RhB and MB because an electron is more difficult to be transferred to TiO_2, as the Highest Occupied Molecular Orbital (HOMO) energy gap of MO is -5.624 eV, compared to -10.128 eV for RhB and -10.494 for MB [62]. For those reasons, the high photocatalytic activity of nanocomposites TC25 and TC50 in MO degradation can be described as a remarkable result.

Furthermore, the same photocatalytic group of TC nanocomposites was tested on the photodegradation of MO under solar light in outdoor weather conditions. According to accurate weather station data [63], during the days that the experiments took place in our department area, the daily all-sky surface shortwave downward irradiance was in the range of 2.46–7.49 $KWh/m^2/day$ (102.5–312.1 W/m^2). Thus, in our process, the average continual radiation was in the range of 0.43–1.32 W, and the degradation rates can be summarized in Figure 9. In this case, in the presence of the photocatalysts, TC25 and TC50 MO rapidly becomes degraded, as the rate increases to 96.0% and 90.2% for the first 30 min of irradiation (Figures S9 and S10). It must be pointed out that control experiments performed under the conditions of irradiation in the absence of the photocatalyst did not show any change in the azo-dye solution absorbance. It can also be noted that under similar reaction conditions, a lower degradation rate (similar to TC75) was obtained with a commercially available gold catalyst, supported by TiO_2 (TAu).

The results demonstrate that in all cases, the photocatalytic performance of the composite materials strongly depends on the C-dots percentage in TiO_2. A remarkable photocatalytic result in the case of low to moderate C-dot loading turns to reduced activity when the C-dot loading increases in the TiO_2 core. This could be attributed to the excess of C-dots on the surface of TiO_2 in TC75, which cause blockage of the pore channels, thus hindering light-harvesting and charge transfer [23,64].

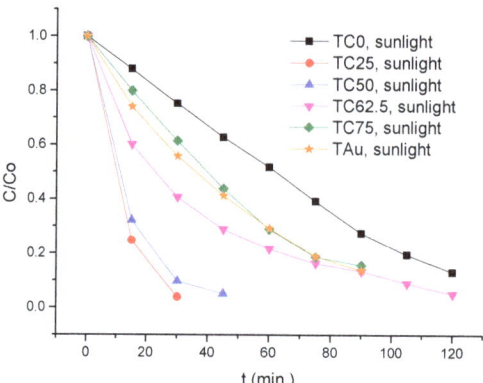

Figure 9. MO degradation rates in the presence of TAu and TC0–75 photocatalysts under sunlight irradiation.

Based on the heterogeneity of the catalytic system, the TC catalysts can be easily separated from the reaction mixture by careful decantation of the aqueous solution or, more accurately, via centrifugation, followed by washing with water and ethanol. Thus, the stability of the TC50 catalyst was examined by performing ten consecutive catalytic tests by irradiating 5 ppm of MO in the presence of 0.2% w/v photocatalyst for 180 min under simulated solar light. As it can be seen in Figure 10, even after 10 cycles, 89.3% of MO was degraded within 180 min, leading to complete degradation after 360 min in the 10th cycle. The UV-Vis spectra and macroscopic evaluation before and after ten consecutive cycles of MO degradation using TC50 under visible light are illustrated in Figures S11 and S12. To the best of our knowledge, this result is far superior compared to already published works on TiO_2/C-dots [23,25], showing a robust, almost eternal photocatalytic system for continuous pollution protection.

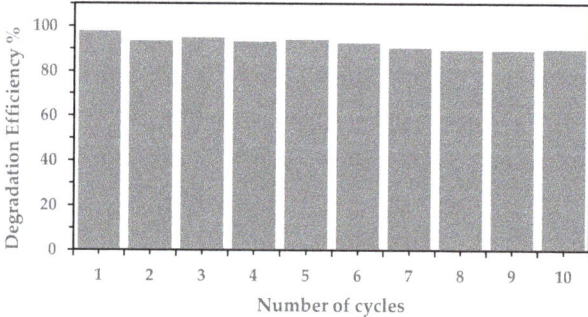

Figure 10. Reusability of TC50 photocatalyst in the MO degradation under simulated solar irradiation.

Since the photodegradation of MO in the presence of TiO_2/Cdots follows a pseudo-first-order kinetic model, photocatalytic efficiency of all TC25–75 nanocomposites, compared to TC0 and TAu, under different radiation sources can be summarized in Table 1. In all cases, the rate constants were obtained by plotting $\ln(C_0/C)$ to irradiation time. As it can be seen, the degradation rate of MO for TC25 and TC50 under visible light irradiation is 12.5 times higher than that of pure TiO_2 (TC0), while under UV-A and solar light irradiation, a 3.4–4.6 and 4.4–5.7 times higher rate was observed. Moreover, upon solar light irradiation, TC25 and TC50 provided a 3.3–4.3 higher degradation rate compared to commercially available Au/TiO_2 (TAu) catalyst.

Table 1. Comparison of photocatalytic efficiency between TC0–75 and TAu using the first order fitting kinetic data of MO degradation under different radiation sources.

Type of Irradiation	Photocatalyst	K_{app} (min^{-1})	R^2	Degradation Rate (%)/min
UV-A light	TC0	1.10×10^{-2}	0.997	45.4/60
	TC25	5.07×10^{-2}	0.982	96.2/60
	TC50	3.79×10^{-2}	0.997	90.9/60
	TC62.5	1.72×10^{-2}	0.999	62.9/60
	TC75	1.21×10^{-2}	0.991	46.1/60
Visible light	TC0	0.18×10^{-2}	0.984	20.8/120
	TC25	2.28×10^{-2}	0.991	91.4/120
	TC50	2.29×10^{-2}	0.991	92.0/120
	TC62.5	0.74×10^{-2}	0.998	56.9/120
	TC75	0.49×10^{-2}	0.980	48.7/120
Solar light	TC0	1.59×10^{-2}	0.971	48.1/60
	TC25	9.13×10^{-2}	0.983	96.0/30
	TC50	6.96×10^{-2}	0.992	90.2/30
	TC62.5	2.41×10^{-2}	0.994	78.2/60
	TC75	2.06×10^{-2}	0.993	71.0/60
	TAu	2.14×10^{-2}	0.998	70.6/60

3.3. Photocatalytic Activity of Multifunctional Protective Films

The photocatalytic ability of the catalysts was also tested after their incorporation into the silica-based consolidant FX-C and their application onto cementitious mortars. Mixtures containing 4% w/w of TC0, TC25 and TAu into the FX-C were prepared (Figure S1) and brushed until saturation onto the surface of the cementitious specimens, which were previously cleaned with isopropanol. TC25 was selected on the grounds of its bests photocatalytic activity both under solar and UV irradiation, as it was proved in the experiments with the MO degradation. For comparison, a part of the surface of the samples was left untreated, and another one was brushed with the consolidant without a catalyst. In order to explore the most favorable application method, another testing surface was created by first applying a layer of the FX-C until saturation and, afterward, a layer of the mixture of the consolidant with TC25.

In order to study the self-cleaning properties of the specimens with the different protective films, two drops (around 0.025 mL each) of methylene blue (MB) were released on each coating, and its degradation was evaluated by comparing the color difference of the stained surface with the color of the surface before the MB application (Figure S2), using a spectrophotometer. The goal is to observe a decline in this color difference (ΔE), reaching, in the best case, a value below 5. The macroscopic evaluation of MB degradation on treated and untreated cement mortars is reported in Figure S2. The specimens were tested under UV-A and visible light irradiation for 48 h, and the results do not significantly differ regarding the type of radiation used. As shown in Figure 11 and Table 2, the protective films containing TC0 and TAu were those with the best performance, reaching the desired value within 48 h.

In addition to the above-mentioned photocatalysts, the surface treated with a double layer consisting of FX-C and FX-C/TC25 also demonstrated considerable self-cleaning properties both under UV-A and visible light irradiation, reaching a value of 7.21 and 5.76, respectively. Moreover, in the case of the protective films containing TC25, the degradation of the pigment was also notable after 48 h of irradiation. The untreated surfaces and the surface treated with solely FX-C showed an insignificant decline in the color difference that can be attributed to the degradation of the dye because of the action of the irradiation. It should be mentioned that, despite the fact that the FX-C/TAu film demonstrated substantial photocatalytic properties, it cannot be proposed as a viable protective option for monumental surfaces, as it changes the color of the surface radically after its application (Figure S2).

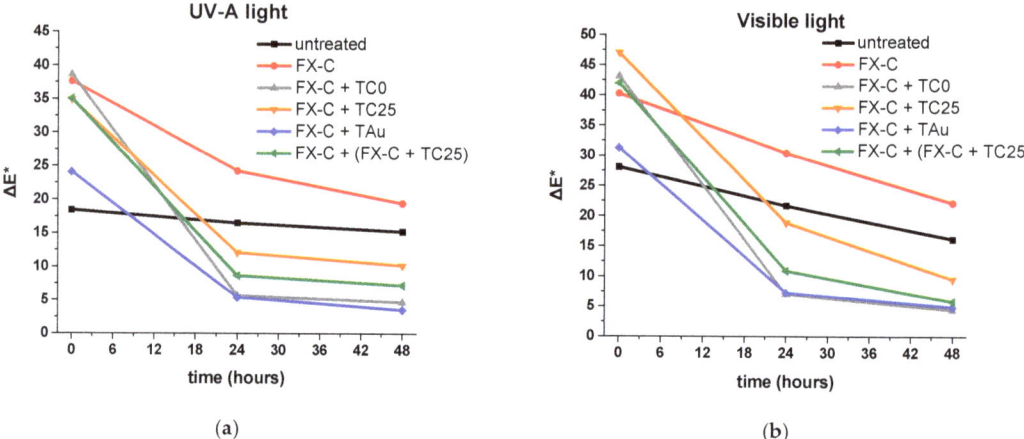

Figure 11. Variation graphs of the specimens' color difference (ΔE*) as a function of time for MB degradation under (**a**) UV-A and (**b**) simulated solar light irradiation.

Table 2. Comparison of the specimens' color difference (ΔE*) as a function of time for MB degradation under different radiation sources.

Light Irradiation	Time (h)	ΔE*					
		Untreated	FX-C	FX-C + TC0	FX-C + TC25	FX-C + TAu	FX-C + (FX-C + TC25)
UV-A light	0	18.41	37.60	38.62	34.88	24.08	35.02
	24	16.48	24.20	5.62	12.04	5.39	8.64
	48	15.21	19.43	4.67	10.16	3.54	7.21
Visible light	0	28.07	40.24	43.14	46.97	31.28	41.95
	24	21.67	30.39	7.00	18.85	7.24	10.88
	48	16.17	22.14	4.40	9.50	4.87	5.77

These in situ experiments clearly pointed out that significant differences arose in the efficiency of the pure photocatalysts and the subsequent incorporation within protective coatings [65,66]. Regardless of the best performed TC25 and TC50 photocatalysts compared to TC0 and TAu under UV-A and solar irradiation, however, the nanocomposites that faster degraded the MB resulted in being those without C-dot loading into TiO_2. Therefore, self-cleaning coatings applied to building materials require individual experiments, including a treatment process with both protective and self-cleaning agents. It was demonstrated that for cementitious surfaces, FX-C with pure TiO_2 and a double layer treatment of FX-C along with FX-C + TC25 could be considered promising solutions against pollutant decay. This study highlighted the need for in situ testing of the nanostructured protective coatings, in which the well-established photocatalysts were integrated for multifunctional building protection.

4. Conclusions

TiO_2/C-dots composites with different C-dot loading were synthesized using citric acid, hydroxylamine and titanium isopropoxide (TTIP) by a facile, green and large-scale hydrothermal strategy. The different mass ratios of C-dots solution to TTIP produced TiO_2

functionalized photocatalysts that were characterized by the anatase structure, formed at a low calcination temperature of 200 °C. The C-dot modified TiO_2 showed efficient photocatalytic MO and MB degradation activity under UV-A and, more specifically, under visible light and sunlight radiation. Under UV-A, visible and solar light irradiation, the degradation rate of MO for TC25 and TC50 is 3.4–12.5 times higher compared to pure TiO_2 (TC0), while solar light provides a 3.3–4.3 higher degradation rate of MO for TC25 and TC50 than commercially available Au/TiO_2 (TAu) catalyst. Additionally, the photocatalysts can be recycled at least 10 times without losing their activity, showing the potentiality of these nanocomposites for continuous pollution protection. Finally, it was demonstrated that the application of multifunctional coatings on the cementitious matrix played a significant role. In the in situ applications of the self-cleaning coatings onto cementitious materials, the protective film of the silica-based FX-C with the pure TiO_2 and the double layer treatment with FX-C and FX-C + TC25 resulted in being the best-performing coatings against pollutant degradation.

Supplementary Materials: The following supporting information can be downloaded at: https://www.mdpi.com/article/10.3390/coatings12050587/s1, Table S1: Mix design of cement mortars treated with the nanocomposites of FX-C with catalysts; Figure S1: Sols of FX-C and nanocomposites of FX-C with catalysts; Figure S2: Macroscopical evaluation of the gradual MB degradation on treated and untreated cement mortars, before and after MB application, under (a) UV-A and (b) visible irradiation; Figure S3: (a,b) TEM images and (c) diameter distribution histogram of C-dots; Figure S4: (a) TEM image and (b) diameter distribution histogram of TC75; Figure S5: UV-Vis spectra of gradual MO degradation using TC50 upon UV-A irradiation; Figure S6: Macroscopic image of gradual MO degradation using TC50 upon UV-A irradiation; Figure S7: UV-Vis spectra of gradual MO degradation using TC50 upon simulated solar light irradiation; Figure S8: Macroscopic image of gradual MO degradation using TC50 upon simulated solar light irradiation; Figure S9: UV-Vis spectra of gradual MO degradation using TC50 under sunlight irradiation; Figure S10: Macroscopic image of gradual MO degradation using TC50 under sunlight irradiation; Figure S11: UV-Vis spectra before (0 min) and after ten consecutive cycles (1–10) of MO degradation using TC50 upon simulated solar light irradiation; Figure S12: Macroscopic image before (0 min) and after ten consecutive cycles (1–10) of MO degradation using TC50 upon simulated solar light irradiation.

Author Contributions: Conceptualization, P.-N.M.; methodology, P.-N.M.; investigation, C.G. and T.K.; data curation, C.G., T.K. and P.-N.M.; writing—original draft preparation, C.G., T.K. and P.-N.M.; writing—review and editing, P.-N.M.; visualization, C.G., T.K. and P.-N.M.; supervision, P.-N.M.; project administration, P.-N.M. All authors have read and agreed to the published version of the manuscript.

Funding: This research received no external funding.

Institutional Review Board Statement: Not applicable.

Informed Consent Statement: Not applicable.

Data Availability Statement: Data are available in a publicly accessible repository.

Acknowledgments: The authors would like to thank the technicians of the "Vasilis Galanopoulos" electron microscopy laboratory (Department of Biology, University of Crete) for providing the SEM and TEM images and Evangelos K. Andreou (Department of Materials Science, University of Crete) for providing the UV-vis/near-IR diffuse reflectance spectra.

Conflicts of Interest: The authors declare no conflict of interest.

References

1. GBD 2016 Risk Factors Collaborators. Global, regional, and national comparative risk assessment of 84 behavioural, environmental and occupational, and metabolic risks or clusters of risks, 1990–2016: A systematic analysis for the Global Burden of Disease Study 2016. *Lancet* **2017**, *390*, 1345–1422. [CrossRef]
2. Masson-Delmotte, V.; Zhai, P.; Pirani, A.; Connors, S.L.; Péan, C.; Berger, S.; Caud, N.; Chen, Y.; Goldfarb, L.; Gomis, M.I.; et al. (Eds.) *IPCC, 2021: Climate Change 2021: The Physical Science Basis. Contribution of Working Group I to the Sixth Assessment Report of the Intergovernmental Panel on Climate Change*; Cambridge University Press: Cambridge, UK, 2021; Available online: https://www.ipcc.ch/report/ar6/wg1/#FullReport (accessed on 27 February 2022).
3. Vandyck, T.; Keramidas, K.; Kitous, A.; Spadaro, J.V.; Van Dingenen, R.; Holland, M.; Saveyn, B. Air quality co-benefits for human health and agriculture counterbalance costs to meet Paris Agreement pledges. *Nat. Commun.* **2018**, *9*, 4939. [CrossRef] [PubMed]
4. Kapridaki, C.; Pinho, L.; Mosquera, M.J.; Maravelaki-Kalaitzaki, P. Producing photoactive, transparent and hydrophobic SiO_2-crystalline TiO_2 nanocomposites at ambient conditions with application as self-cleaning coatings. *Appl. Catal. B Environ.* **2014**, *156–157*, 416–427. [CrossRef]
5. Ganesh, V.A.; Raut, H.K.; Nair, A.S.; Ramakrishna, S. A review on self-cleaning coatings. *J. Mater. Chem.* **2011**, *21*, 16304–16322. [CrossRef]
6. Sakar, M.; Prakash, R.M.; Do, T.-O. Insights into the TiO_2-Based Photocatalytic Systems and Their Mechanisms. *Catalysts* **2019**, *9*, 680. [CrossRef]
7. Fujishima, A.; Honda, K. Electrochemical photolysis of water at a semiconductor electrode. *Nature* **1972**, *238*, 37–38. [CrossRef] [PubMed]
8. Schneider, J.; Matsuoka, M.; Takeuchi, M.; Zhang, J.; Horiuchi, Y.; Anpo, M.; Bahnemann, D.W. Understanding TiO_2 Photocatalysis: Mechanisms and Materials. *Chem. Rev.* **2014**, *114*, 9919–9986. [CrossRef]
9. Linsebigler, A.L.; Lu, G.; Yates, J.T. Photocatalysis on TiO_2 Surfaces: Principles, Mechanisms, and Selected Results. *Chem. Rev.* **1995**, *95*, 735–758. [CrossRef]
10. Oseghe, E.O.; Msagati, T.A.M.; Mamba, B.B.; Ofomaja, A.E. An efficient and stable narrow bandgap carbon dot-brookite composite over other CD-TiO_2 polymorphs in rhodamine B degradation under LED light. *Ceram. Int.* **2019**, *45*, 14173–14181. [CrossRef]
11. Zhang, J.; Zhou, P.; Liu, J.; Yu, J. New understanding of the difference of photocatalytic activity among anatase, rutile and brookite TiO_2. *Phys. Chem. Chem. Phys.* **2014**, *16*, 20382–20386. [CrossRef] [PubMed]
12. Xu, X.; Ray, R.; Gu, Y.; Ploehn, H.J.; Gearheart, L.; Raker, K.; Scrivens, W.A. Electrophoretic Analysis and Purification of Fluorescent Single-Walled Carbon Nanotube Fragments. *J. Am. Chem. Soc.* **2004**, *126*, 12736–12737. [CrossRef] [PubMed]
13. Liu, J.; Li, R.; Yang, B. Carbon Dots: A New Type of Carbon-Based Nanomaterial with Wide Applications. *ACS Cent. Sci.* **2020**, *6*, 2179–2195. [CrossRef]
14. Lim, S.Y.; Shen, W.; Gao, Z. Carbon quantum dots and their applications. *Chem. Soc. Rev.* **2015**, *44*, 362–381. [CrossRef]
15. Li, H.; He, X.; Kang, Z.; Huang, H.; Liu, Y.; Liu, J.; Lian, S.; Tsang, C.H.A.; Yang, X.; Lee, S.-T. Water-Soluble Fluorescent Carbon Quantum Dots and Photocatalyst Design. *Angew. Chem. Int. Ed.* **2010**, *49*, 4430–4434. [CrossRef]
16. Hazarika, D.; Karak, N. Photocatalytic degradation of organic contaminants under solar light using carbon dot/titanium dioxide nanohybrid, obtained through a facile approach. *Appl. Surf. Sci.* **2016**, *376*, 276–285. [CrossRef]
17. Chen, J.; Shu, J.; Anqi, Z.; Juyuan, H.; Yan, Z.; Chen, J. Synthesis of carbon quantum dots/TiO_2 nanocomposite for photo-degradation of Rhodamine B and cefradine. *Diam. Relat. Mater.* **2016**, *70*, 137–144. [CrossRef]
18. Miao, R.; Luo, Z.; Zhong, W.; Chen, S.-Y.; Jiang, T.; Dutta, B.; Nasr, Y.; Zhang, Y.; Suib, S.L. Mesoporous TiO_2 modified with carbon quantum dots as a high-performance visible light photocatalyst. *Appl. Catal. B Environ.* **2016**, *189*, 26–38. [CrossRef]
19. Syafei, D.; Sugiarti, S.; Darmawan, N.; Khotib, M. Synthesis of TiO_2/Carbon Nanoparticle (C-dot) Composites as Active Catalysts for Photodegradation of Persistent Organic Pollutant. *Indones. J. Chem.* **2017**, *17*, 37–42. [CrossRef]
20. Wongso, V.; Chung, H.K.; Sambudi, N.S.; Sufian, S.; Abdullah, B.; Wirzal, M.D.H.; Ang, W.L. Silica–carbon quantum dots decorated titanium dioxide as sunlight-driven photocatalyst to diminish acetaminophen from aquatic environment. *J. Photochem. Photobiol. A Chem.* **2020**, *394*, 112436. [CrossRef]
21. Li, F.; Tian, F.; Liu, C.; Wang, Z.; Du, Z.; Li, R.; Zhang, L. One-step synthesis of nanohybrid carbon dots and TiO_2 composites with enhanced ultraviolet light active photocatalysis. *RSC Adv.* **2015**, *5*, 8389–8396. [CrossRef]
22. Kumar, M.S.; Yasoda, K.H.; Kumaresan, D.; Kothurkar, N.K.; Batabyal, S.K. TiO_2-carbon quantum dots (CQD) nanohybrid: Enhanced photocatalytic activity. *Mater. Res. Express* **2018**, *5*, 075502. [CrossRef]
23. Deng, Y.; Chen, M.; Chen, G.; Zou, W.; Zhao, Y.; Zhang, H.; Zhao, Q. Visible–Ultraviolet Upconversion Carbon Quantum Dots for Enhancement of the Photocatalytic Activity of Titanium Dioxide. *ACS Omega* **2021**, *6*, 4247–4254. [CrossRef] [PubMed]
24. Luo, H.; Dimitrov, S.; Daboczi, M.; Kim, J.-S.; Guo, Q.; Fang, Y.; Stoeckel, M.-A.; Samorì, P.; Fenwick, O.; Sobrido, A.B.J.; et al. Nitrogen-Doped Carbon Dots/TiO_2 Nanoparticle Composites for Photoelectrochemical Water Oxidation. *ACS Appl. Nano Mater.* **2020**, *3*, 3371–3381. [CrossRef]
25. Wang, A.; Xiao, X.; Zhou, C.; Lyu, F.; Fu, L.; Wang, C.; Ruan, S. Large-scale synthesis of carbon dots/TiO_2 nanocomposites for the photocatalytic color switching system. *Nanoscale Adv.* **2019**, *1*, 1819–1825. [CrossRef]
26. Stefanakis, D.; Krasoudaki, T.; Kaditis, A.-I.; Bakolas, A.; Maravelaki, P.-N. Design of Novel Photocatalytic Films for the Protection of Architectural Surfaces via the Incorporation of Green Photocatalysts. *Coatings* **2021**, *11*, 934. [CrossRef]

27. Barberena-Fernández, A.M.; Carmona-Quiroga, P.M.; Blanco-Varela, M.T. Interaction of TEOS with cementitious materials: Chemical and physical effects. *Cem. Concr. Compos.* **2015**, *55*, 145–152. [CrossRef]
28. Al-Qaradawi, S.; Salman, S.R. Photocatalytic degradation of methyl orange as a model compound. *J. Photochem. Photobiol. A Chem.* **2002**, *148*, 161–168. [CrossRef]
29. Kapetanaki, K.; Vazgiouraki, E.; Stefanakis, D.; Fotiou, A.; Anyfantis, G.C.; García-Lodeiro, I.; Blanco-Varela, M.T.; Arabatzis, I.; Maravelaki, P.N. TEOS Modified With Nano-Calcium Oxalate and PDMS to Protect Concrete Based Cultural Heritage Buildings. *Front. Mater.* **2020**, *7*, 16. [CrossRef]
30. Maravelaki, P.N.; Kapetanaki, K.; Stefanakis, D. TEOS-PDMS-Calcium Oxalate Hydrophobic Nanocomposite for Protection and Stone Consolidation. *Heritage* **2021**, *4*, 4068–4075. [CrossRef]
31. David, M.E.; Ion, R.-M.; Grigorescu, R.M.; Iancu, L.; Andrei, E.R. Nanomaterials used in conservation and restoration of cultural heritage: An up-to-date overview. *Materials* **2020**, *13*, 2064. [CrossRef] [PubMed]
32. Enesca, A.; Andronic, L.; Duta, A. The influence of surfactants on the crystalline structure, electrical and photocatalytic properties of hybrid multi-structured (SnO_2, TiO_2 and WO_3) thin films. *Appl. Surf. Sci.* **2012**, *258*, 4339–4346. [CrossRef]
33. Xu, H.; Hao, Z.; Feng, W.; Wang, T.; Fu, X. The floating photocatalytic spheres loaded with weak light-driven TiO_2-based catalysts for photodegrading tetracycline in seawater. *Mater. Sci. Semicond. Process* **2022**, *144*, 106610. [CrossRef]
34. Mugundan, S.; Praveen, P.; Sridhar, S.; Prabu, S.; Mary, K.L.; Ubaidullah, M.; Shaikh, S.F.; Kanagesan, S. Sol-gel synthesized barium doped TiO_2 nanoparticles for solar photocatalytic application. *Inorg. Chem. Commun.* **2022**, *139*, 109340. [CrossRef]
35. Rosales, A.; Esquivel, K. SiO_2@TiO_2 Composite Synthesis and Its Hydrophobic Applications: A Review. *Catalysts* **2020**, *10*, 171. [CrossRef]
36. Rosales, A.; Maury-Ramírez, A.; Mejía-De Gutiérrez, R.; Guzmán, C.; Esquivel, K. SiO_2@TiO_2 Coating: Synthesis, Physical Characterization and Photocatalytic Evaluation. *Coatings* **2018**, *8*, 120. [CrossRef]
37. Franzoni, E.; Fregni, A.; Gabrielli, R.; Graziani, G.; Sassoni, E. Compatibility of photocatalytic TiO_2-based finishing for renders in architectural restoration: A preliminary study. *Build. Environ.* **2014**, *80*, 125–135. [CrossRef]
38. Rosales, A.; Ortiz-Frade, L.; Medina-Ramirez, I.E.; Godínez, L.A.; Esquivel, K. Self-cleaning of SiO_2-TiO_2 coating: Effect of sonochemical synthetic parameters on the morphological, mechanical, and photocatalytic properties of the films. *Ultrason. Sonochem.* **2021**, *73*, 105483. [CrossRef] [PubMed]
39. Maravelaki-Kalaitzaki, P.; Agioutantis, Z.; Lionakis, E.; Stavroulaki, M.; Perdikatsis, V. Physico-chemical and mechanical characterization of hydraulic mortars containing nano-titania for restoration applications. *Cem. Concr. Compos.* **2013**, *36*, 33–41. [CrossRef]
40. Balliana, E.; Ricci, G.; Pesce, C.; Zendri, E. Assessing the value of green conservation for cultural heritage: Positive and critical aspects of already available methodologies. *Int. J. Conserv. Sci.* **2016**, *7*, 185–202. Available online: https://ijcs.ro/public/IJCS-16-SI01_Balliana.pdf (accessed on 14 October 2021).
41. Petronella, F.; Pagliarulo, A.; Truppi, A.; Lettieri, M.; Masieri, M.; Calia, A.; Curri, M.L.; Comparelli, R. TiO_2 Nanocrystal Based Coatings for the Protection of Architectural Stone: The Effect of Solvents in the Spray-Coating Application for a Self-Cleaning Surfaces. *Coatings* **2018**, *8*, 356. [CrossRef]
42. Kapridaki, C.; Maravelaki-Kalaitzaki, P. TiO_2–SiO_2–PDMS nano-composite hydrophobic coating with self-cleaning properties for marble protection. *Prog. Org. Coat.* **2013**, *76*, 400–410. [CrossRef]
43. Kapridaki, C.; Verganelaki, A.; Dimitriadou, P.; Maravelaki-Kalaitzaki, P. Conservation of Monuments by a Three-Layered Compatible Treatment of TEOS-Nano-Calcium Oxalate Consolidant and TEOS-PDMS-TiO_2 Hydrophobic/Photoactive Hybrid Nanomaterials. *Materials* **2018**, *11*, 684. [CrossRef]
44. Stefanakis, D.; Philippidis, A.; Sygellou, L.; Filippidis, G.; Ghanotakis, D.; Anglos, D. Synthesis of fluorescent carbon dots by a microwave heating process: Structural characterization and cell imaging applications. *J. Nanopart. Res.* **2014**, *16*, 12646. [CrossRef]
45. De Muynck, W.; Ramirez, A.M.; De Belie, N.; Verstraete, W. Evaluation of strategies to prevent algal fouling on white architectural and cellular concrete. *Int. Biodeterior. Biodegrad.* **2009**, *63*, 679–689. [CrossRef]
46. Allena, N.S.; Mahdjoubd, N.; Vishnyakovb, V.; Kellyb, P.J.; Kriekc, R.J. The effect of crystalline phase (anatase, brookite and rutile) and size on the photocatalytic activity of calcined polymorphic titanium dioxide (TiO_2). *Polym. Degrad. Stab.* **2018**, *150*, 31–36. [CrossRef]
47. Patterson, A.L. The Scherrer Formula for X-ray Particle Size Determination. *Phys. Rev.* **1939**, *56*, 978. [CrossRef]
48. Shah, A.H.; Rather, M.A. Effect of calcination temperature on the crystallite size, particle size and zeta potential of TiO_2 nanoparticles synthesized via polyol-mediated method. *Mater. Today Proc.* **2021**, *44*, 482–488. [CrossRef]
49. Esquivel, K.; Nava, R.; Zamudio-Méndez, A.; González, M.V.; Jaime-Acuña, O.E.; Escobar-Alarcón, L.; Peralta-Hernández, J.M.; Pawelec, B.; Fierro, J.L.G. Microwave-assisted synthesis of (S)Fe/TiO_2 systems: Effects of synthesis conditions and dopant concentration on photoactivity. *Appl. Catal. B Environ.* **2013**, *140–141*, 213–224. [CrossRef]
50. Górska, P.; Zaleska, A.; Kowalska, E.; Klimczuk, T.; Sobczak, J.W.; Skwarek, E.; Janusz, W.; Hupka, J. TiO_2 photoactivity in vis and UV light: The influence of calcination temperature and surface properties. *Appl. Catal. B Environ.* **2008**, *84*, 440–447. [CrossRef]
51. Zhang, H.; Banfield, J.F. Thermodynamic analysis of phase stability of nanocrystalline titania. *J. Mater. Chem.* **1998**, *8*, 2073–2076. [CrossRef]

52. Kibasomba, P.M.; Dhlamini, S.; Maaza, M.; Liu, C.-P.; Rashad, M.M.; Rayan, D.A.; Mwakikunga, B.W. Strain and grain size of TiO$_2$ nanoparticles from TEM, Raman spectroscopy and XRD: The revisiting of the Williamson-Hall plot method. *Results Phys.* **2018**, *9*, 628–635. [CrossRef]
53. Munafò, P.; Quagliarini, E.; Goffredo, G.B.; Bondioli, F.; Licciulli, A. Durability of nano-engineered TiO$_2$ self-cleaning treatments on limestone. *Constr. Build. Mater.* **2014**, *65*, 218–231. [CrossRef]
54. Zamudio-Méndez, A.; Sánchez-Palma, E.; Navarro-López, R.; Pérez-Lara, M.A.; Rivera Muñoz, E.M.; Godínez, L.A.; Velázquez-Castillo, R.; Nava, R.; Esquivel, K. Self-Cleaning Activity on Concrete Surfaces Coated with Fe- and S-Doped TiO$_2$ Synthesized by Sol–Gel Microwave Method. *J. Nanosci. Nanotechnol.* **2017**, *17*, 5637–5645. [CrossRef]
55. Shetty, R.; Chavan, V.B.; Kulkarni, P.S.; Kulkarni, B.D.; Kamble, S.P. Photocatalytic Degradation of Pharmaceuticals Pollutants Using N-Doped TiO$_2$ Photocatalyst: Identification of CFX Degradation Intermediates. *Indian Chem. Eng.* **2017**, *59*, 177–199. [CrossRef]
56. Wu, W.-Y.; Chang, Y.-M.; Ting, J.-M. Room-temperature synthesis of single-crystalline anatase TiO$_2$ nanowires. *Cryst. Growth Des.* **2010**, *10*, 1646–1651. [CrossRef]
57. Lu, W.; Qin, X.; Liu, S.; Chang, G.; Zhang, Y.; Luo, Y.; Asiri, A.M.; Al-Youbi, A.O.; Sun, X. Economical, green synthesis of fluorescent carbon nanoparticles and their use as probes for sensitive and selective detection of mercury(II) ions. *Anal. Chem.* **2012**, *84*, 5351–5357. [CrossRef] [PubMed]
58. Nie, X.; Jiang, C.; Wu, S.; Chen, W.; Lv, P.; Wang, Q.; Liu, J.; Narh, C.; Cao, X.; Ghiladi, R.A.; et al. Carbon quantum dots: A bright future as photosensitizers for in vitro antibacterial photodynamic inactivation. *J. Photochem. Photobiol. B Biol.* **2020**, *206*, 111864. [CrossRef]
59. Tauc, J. Optical properties of amorphous semiconductors. In *Amorphous and Liquid Semiconductors*; Tauc, J., Ed.; Springer: Boston, MA, USA, 1974; pp. 159–220. [CrossRef]
60. Kubelka, P. New contributions to the optics of intensely light-scattering materials. Part I. *J. Opt. Soc. Am.* **1948**, *38*, 448–457. [CrossRef]
61. Wang, J.; Guo, B.; Zhang, X.; Zhang, Z.; Han, J.; Wu, J. Sonocatalytic degradation of methyl orange in the presence of TiO$_2$ catalysts and catalytic activity comparison of rutile and anatase. *Ultrason. Sonochem.* **2005**, *12*, 331–337. [CrossRef] [PubMed]
62. Priyanshu, V.; Sujoy, K.S. Degradation kinetics of pollutants present in a simulated wastewater matrix using UV/TiO$_2$ photocatalysis and its microbiological toxicity assessment. *Res. Chem. Intermed.* **2017**, *43*, 6317–6341. [CrossRef]
63. NASA's Power Data Access Viewer. Available online: https://power.larc.nasa.gov/data-access-viewer/ (accessed on 18 February 2022).
64. Ke, J.; Li, X.; Zhao, Q.; Liu, B.; Liu, S.; Wang, S. Upconversion carbon quantum dots as visible light responsive component for efficient enhancement of photocatalytic performance. *J. Colloid Interface Sci.* **2017**, *496*, 425–433. [CrossRef] [PubMed]
65. Gherardi, F. Current and future trends in protective treatments for stone heritage. In *Conserving Stone Heritage*; Gherardi, F., Maravelaki, P.N., Eds.; Springer: Cham., Switzerland, 2022; pp. 137–176. [CrossRef]
66. Falchi, L.; Zendri, E.; Müller, U.; Fontana, P. The influence of water-repellent admixtures on the behaviour and the effectiveness of Portland limestone cement mortars. *Cem. Concr. Compos.* **2015**, *59*, 107–118. [CrossRef]

Article

Effects of Hydrolysis Parameters on AlN Content in Aluminum Dross and Multivariate Nonlinear Regression Analysis

Shuaishuai Lv [1], Hongjun Ni [1,2,*], Xingxing Wang [1], Wei Ni [1] and Weiyang Wu [1]

[1] School of Mechanical Engineering, Nantong University, Nantong 226019, China; lvshuaishuai@ntu.edu.cn (S.L.); wangxx@ntu.edu.cn (X.W.); niweiqa@163.com (W.N.); wuweiyang.woy@gmail.com (W.W.)
[2] Jiangsu Province Engineering Research Center of Aluminum Dross Solid Waste Harmless Treatment and Resource Utilization, Nantong 226019, China
* Correspondence: ni.hj@ntu.edu.cn

Citation: Lv, S.; Ni, H.; Wang, X.; Ni, W.; Wu, W. Effects of Hydrolysis Parameters on AlN Content in Aluminum Dross and Multivariate Nonlinear Regression Analysis. *Coatings* **2022**, *12*, 552. https://doi.org/10.3390/coatings12050552

Academic Editor: Giorgos Skordaris

Received: 3 March 2022
Accepted: 17 April 2022
Published: 19 April 2022

Publisher's Note: MDPI stays neutral with regard to jurisdictional claims in published maps and institutional affiliations.

Copyright: © 2022 by the authors. Licensee MDPI, Basel, Switzerland. This article is an open access article distributed under the terms and conditions of the Creative Commons Attribution (CC BY) license (https://creativecommons.org/licenses/by/4.0/).

Abstract: Aluminum dross, as a hazardous waste product, causes harm to the environment and humans, since the AlN it contains chemically reacts with water to produce ammonia. In the present study, a formula for modifying the AlN content in aluminum dross is proposed for the first time, by investigating the components of aluminum dross and changes in their respective contents during the hydrolysis process. Meanwhile, the effects of such hydrolysis parameters as time, temperature, and rotational speed on the hydrolysis rate of aluminum dross are explored. Furthermore, regression analysis is performed on the hydrolysis parameters and objective functions. The results show that as the reaction time increases, the variation in AlN content in aluminum dross decelerates gradually after modification. The hydrolysis rate is the fastest in the initial 4 h, which essentially stagnates after 20 h. The rise in temperature can significantly accelerate the AlN hydrolysis in aluminum dross, while the rotational speed has a non-obvious effect on the hydrolysis rate of AlN in aluminum dross. Regression analysis and secondary simplification are performed on the hydrolysis parameters and the modified AlN content, revealing that the relative error between the theoretical and experimental values is $\leq \pm 9.34\%$. The findings of this study have certain guiding significance for predicting and controlling modified AlN content in aluminum dross during hydrolysis.

Keywords: high-nitrogen-content aluminum dross; AlN; hydrolysis parameters; content prediction; modified formula

1. Introduction

Aluminum dross is a waste ash generated during aluminum production, which comes from infusible inclusions [1], oxides, and various additives that float on the aluminum melt surface during smelting [2]. As a typical solid waste product from the aluminum processing industry, it has been included in China's "*National Hazardous Waste List* (2021 edition)" due to its reactivity and leaching toxicity [3].

As of today, the only mature technique for utilizing aluminum dross is "ash frying", where a small amount of elemental aluminum is extracted from aluminum dross to obtain certain economic benefits [4]. However, the extensive remaining ash still cannot be treated effectively, and is often accumulated and buried as solid waste [5], polluting and occupying substantial land resources while seriously affecting the ecological environment and human health [6–8]. Some of the heavy metal elements contained in aluminum dross—such as Hg, Cd, and Pb—can cause severe pollution once they enter the soil or groundwater systems, tremendously impacting the ecosystem.

Additionally, all kinds of aluminum dross contain a certain amount of AlN. To purify the molten aluminum in the smelting process, nitrogen is often filled into the furnace to accelerate the purification and refinement [9]. During the evaporation process, part of the nitrogen is absorbed by the aluminum dross floating on the molten aluminum surface to form AlN [10]. AlN is a kind of functional ceramic belonging to diamond nitride, which has excellent thermal, mechanical, and electrical properties, including high thermal conductivity (theoretical value 320 W/(m·K)), linear expansivity (4.8×10^{-6}/K) comparable to silicon (Si), and outstanding electrical insulation (volume resistivity > 1012 Ω·m) [11]. The AlN in aluminum dross tends to be unstable overall due to its formation conditions, structural composition, and refinement difficulty [12]. Under wet or water-exposure conditions, it releases pungent ammonia gas, causing consequences such as imbalance in material performance and changes in solution pH during the utilization process [13]. Additionally, ammonia is a colorless gas with a strong, irritating odor, and can easily cause air pollution due to its strong diffusivity and harmfulness. The ammonia produced by aluminum dross hydrolysis is highly toxic to the human body [14]. Chronic intoxication can cause such respiratory diseases as bronchitis and emphysema, while acute intoxication can cause severe consequences such as persistent coughing, suffocation, and even death [15].

Clearly, the landfill of aluminum dross and related pollution have become hot issues in today's social development, and go against the current sustainability concept of zero emissions and pollution. The requirements of reducing, preventing, treating, and upcycling waste are imminent. There are various experimental methods for AlN, such as cathodoluminescence measurements (CL), X-ray absorption near-edge spectroscopy (XANES), and Fourier-transform infrared spectroscopy (FTIR) [16–18]. This study systematically analyzes the AlN (in aluminum dross) hydrolysis parameters such as temperature, time, and rotational speed, and performs their selection and optimization. Meanwhile, on the basis of extensive experimental data, each hydrolysis parameter and objective function (AlN content) is regressively analyzed, in order to achieve prediction and control of AlN in aluminum dross.

2. Materials and Methods

2.1. Raw Material and Pretreatment

The ultimate aluminum dross used in this experimentation came from a secondary aluminum company in Jiangsu, China, and was analyzed and tested after pretreatment. The pretreatment process was as follows: The aluminum dross was preliminarily screened via a standard large-mesh sieve to remove the aluminum metal blocks. Then, the remainder was ground, and screened using a standard small-mesh sieve to yield the experimental raw material [19]. Table 1 details the XRD results of the aluminum dross.

Table 1. XRD analysis results of aluminum dross.

Component	Al	Al_2O_3	AlN	AlO(OH)	SiO_2
Content (%)	5 ± 2	20 ± 2	13 ± 2	3 ± 1	1

2.2. Reagents and Instruments

The reagents used in the experimentation were NaOH (analytically pure, Xilong Scientific Co., Ltd., Shenzhen, China), methyl red (indicator, Yuanye Biotechnology), methylene blue (indicator, COOLABER SCIENCE & TECHNOLOGY), HCl (analytically pure, Lingfeng Chemical Reagent, Shanghai, China), and boric acid (chemically pure, Meryer Chemical Technology, Shanghai, China). Experimental instruments were as follows: ultra water purifier (EPED-10TH, EPED Technology, Nanjing, China), universal electric furnace (DK-98 type, Chushui Electrothermal Appliance Co., Ltd., Chushui, China), digital stirrer (JB60-SH, Lichen Instrument, Shaoxing, China), water bath (WB-4, Zhengrong Experimental Instrument), and electro-thermostatic blast oven (101A-2, Hengchang Instrument, Haimen, Nantong, China).

2.3. Analytical Test Methods

The aluminum dross was hydrolyzed, filtered, and dried under different hydrolysis parameters, and a distillation separation unit was built by utilizing the aforementioned instruments. Figure 1 illustrates the schematic of the unit.

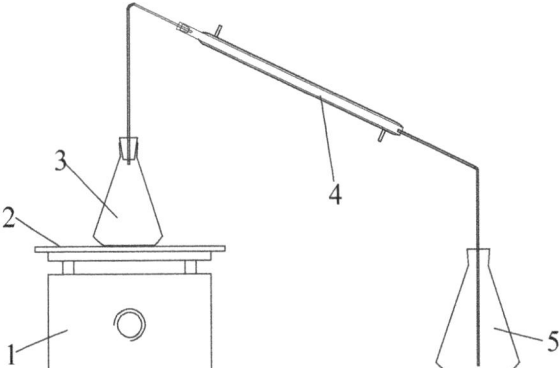

Figure 1. Schematic diagram of distillation separation plant (1—electric furnace; 2—asbestos net; 3,5—cone bottle; 4—condenser tube).

After weighing 2 g of aluminum dross, it was added to an Erlenmeyer flask containing 150 mL of 20%wt NaOH solution. The flask was quickly stoppered, and then heated and kept boiling for approximately 2 h to allow for distillation. Meanwhile, the distilled ammonia gas was absorbed with 200 mL of 40 g/L boric acid solution. Following distillation, titration was performed with 0.05 mol/L dilute HCl solution, where the standard methyl red–methylene blue served as an indicator. The endpoint was marked by a blue-to-purple change in solution color [20,21].

A blank experiment was conducted according to the aforementioned procedure, and the relative volume of HCl consumed in the experiment was determined, based on which the AlN content was calculated as follows:

$$Z = \frac{0.041C(V_2 - V_1)}{M} \times 100\% \quad (1)$$

where C denotes the concentration of dilute HCl (mol/L), V_2 and V_1 represent the volumes of diluted HCl consumed in the titration process and the blank experiment, respectively (mL), M denotes the sample mass (g), and Z denotes the AlN content in aluminum dross directly measured after hydrolysis (%).

3. Effects of Hydrolysis Parameters on the Nitrogen Removal Efficiency from Aluminum Dross

3.1. Prediction of AlN Content and Its Modification Formula

The AlN contents in different aluminum dross varied distinctly. To validate the accuracy of subsequent tests, predicting the AlN content in hydrolyzed products was necessary. Figure 2 depicts the XRD spectra of aluminum dross. Clearly, the hydrolyzed products contained Al, Al_2O_3, AlN, and other crystalline substances (amorphous components were excluded from the test results). The semi-quantitative XRD results of aluminum dross are detailed in Table 2. Since the aluminum dross used herein was already processed by ash frying, the AlN content was low, while the contents of Al_2O_3 and AlN were about 20% and 10%, respectively. Figure 3 displays the local EDS spectra at two material sites. Since almost all nitrogen contained in the aluminum dross was AlN, after conversion, the AlN contents in Figure 3a,b were estimated to be 12.97% and 16.37%, respectively.

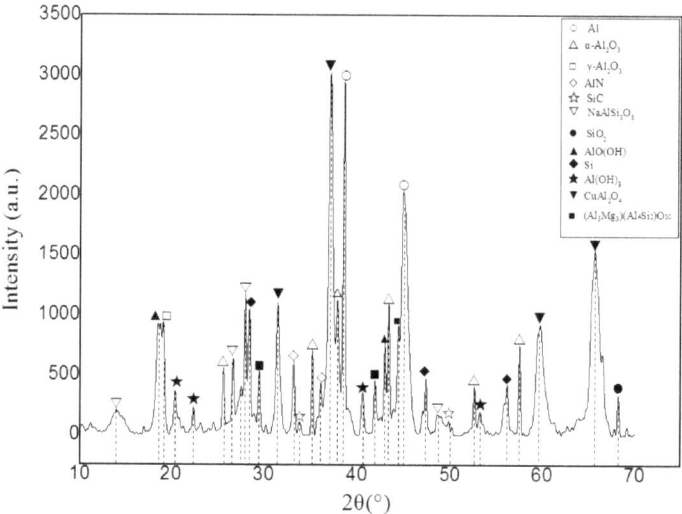

Figure 2. XRD patterns of aluminum dross.

Table 2. Semi-quantitative XRD results.

No.	Phase Name	Molecular Formula	Semi-Quantitative (%)
1	Aluminum nitride	AlN	9 ± 2
2	Alumina	Al_2O_3	19 ± 2
3	Elemental aluminum	Al	3 ± 2
4	Diaspore	AlO(OH)	1
5	Some low-content minerals		8 ± 2

The above analysis revealed that the main components of hydrolyzed products and their respective contents conformed to the definition of "secondary aluminum dross" in the previous paper, and that the hydrolyzed products were the dross residues from ash frying, showing good agreement with the actual situation. The AlN contents in the hydrolyzed products were predicted to range between 10 and 17%.

The hydrolytic reaction of AlN in aluminum dross is described in Equation (2). During this process, the N in AlN was separated from the aluminum dross in the form of NH_3, and its elemental position was occupied by OH^- ions to form $Al(OH)_3$ with a molecular mass of 78, which was considerably greater than the original AlN's 41. Additionally, apart from insoluble substances such as Al and Al_2O_3, the aluminum dross also contained small amounts of soluble salts, such as NaCl and KCl [22,23].

$$AlN + 3H_2O = Al(OH)_3 + NH_3$$
$$41 : 54 = 78 : 17 \quad (2)$$

If the above determinations were used directly as the AlN contents in aluminum dross, not only would the changes in other components and contents be ignored, but the AlN content standard also could not be unified. In particular, when measuring AlN-rich aluminum dross, the error would be larger. To address this problem, the compositional and content variations during the hydrolysis process in Equation (2) were investigated, thereby deriving the following formula:

$$Z = \frac{KM - m}{M - Y + \frac{78}{41}m} \quad (3)$$

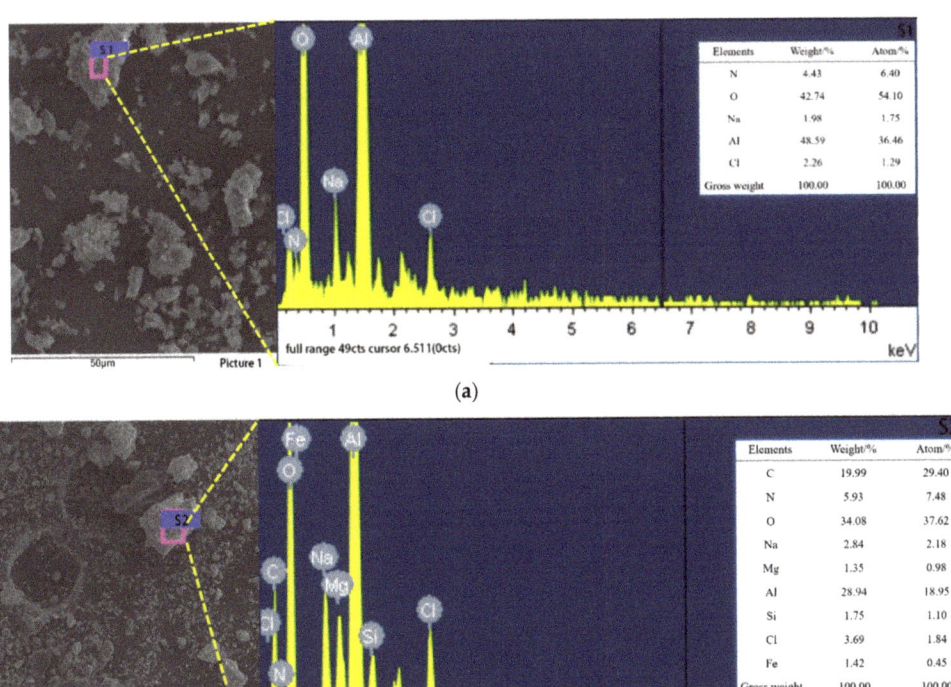

Figure 3. EDS results: (**a**) Spectrogram 1; (**b**) Spectrogram 2.

Transforming Equation (3) yields the *m* value, which was substituted into the formula below to modify the directly measured AlN content *Z*:

$$Z' = \frac{KM-m}{M} = K - \frac{m}{M} = \frac{78KZ + 41Z - \frac{41ZY}{M}}{78Z + 41} = \frac{78K + 41 - \frac{41Y}{M}}{78 + \frac{41}{Z}} \quad (4)$$

In Equation (3), *K* represents the original AlN content (%), *M* denotes the mass of the original sample (g), *m* denotes the mass of AlN reacted during hydrolysis (g), and *Y* is the mass of the salt in the hydrolyzed filtrate (g). Meanwhile, *Z* in Equation (4) denotes the AlN content in the aluminum dross after modification (%).

3.2. Effects of Time and Rotational Speed on the AlN Content

Under different rotational speeds, the aluminum dross was hydrolyzed at 60 °C. Figure 4 depicts the AlN variation trends in the aluminum dross after modification. Clearly, under mechanical stirring at both low speeds (0, 100, and 300 r/min) and high speeds (600 and 1000 r/min), the variation trends of modified AlN contents remained generally consistent. Moreover, at various time points, the modified AlN contents in aluminum dross differed little, suggesting an insignificant effect of rotational speed on the aluminum dross hydrolysis rate.

Meanwhile, the modified AlN content in aluminum dross decreased gradually with the increasing hydrolysis time, decreasing by about half (from 12.68% to 6.50%) under this hydrolysis condition 24 h after reaction. More obviously, at the 0–4 h stage, the modified

AlN content decreased drastically from the initial 12.68% to 9.00–10.40%, showing considerably sharper reduction than that at other reaction time periods. With the prolongation of hydrolysis time, the aluminum dross hydrolysis rate was gradually lowered, and the variation trends of modified AlN contents in aluminum dross were generally consistent after 20 h.

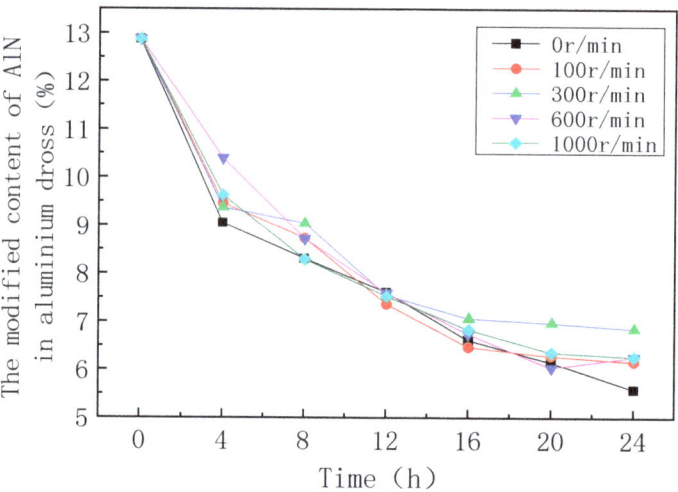

Figure 4. The modified contents of AlN in aluminum dross under different rotational speeds (60 °C).

The causes of the above phenomenon were as follows: On the one hand, the particle size and surface area of aluminum dross were large at the initial reaction stage, and the water contact area was also correspondingly large. Since more AlN participated in the reaction under identical hydrolysis conditions, the hydrolytic process was faster. On the other hand, given the basic hydrolysis properties of AlN in aluminum dross, the reaction product $Al(OH)_3$ was a white insoluble gelatinous precipitate. With the prolongation of hydrolysis time, this product and its attachments gradually blocked the pores of aluminum dross particles, which hindered the contact channel between AlN and water, ultimately affecting the hydrolysis rate of aluminum dross. Meanwhile, stirring facilitated the suspension of aluminum dross particles in the hydrolysis system, prevented the accumulation of dross at the system's bottom, and promoted the diffusion of generated NH_3 outside the hydrolysis system. However, at identical times and temperatures, the rotational speed generally did not affect the AlN–water contact area in the aluminum dross, which also had only a small effect on the hydrolysis process.

3.3. Effects of Temperature on the AlN Content

Under different temperature conditions, the aluminum dross hydrolysis experiments were carried out at 600 r min^{-1}. Figure 5 illustrates the relevant variation trends. As is clear, at identical hydrolysis times, the modified AlN content in aluminum dross was markedly lower in the hydrolysis system at 100 °C than that at 60 °C and 80 °C. After 24 h of hydrolysis, the modified AlN content at a 100 °C hydrolysis temperature decreased from the original 12.88% to 2.25%, and there was no obvious ammonia smell upon further continuation of the hydrolysis. Meanwhile, after 24 h of hydrolysis at 80 °C, the modified AlN content dropped to 4.23%, and ammonia was slightly smelt upon further hydrolysis. In contrast, at 60 °C, the AlN content only dropped to 6.28%, and apparent bubbles and ammonia smell were produced upon further hydrolysis, indicating that the hydrolysis process had not ended at 60 °C. Clearly, the increase in temperature could remarkably promote the hydrolysis of aluminum dross. It could be speculated that with the prolongation of hydrolysis time, the variation of modified AlN content in aluminum dross gradually slowed down, so it inevitably took a long time to achieve complete hydrolysis.

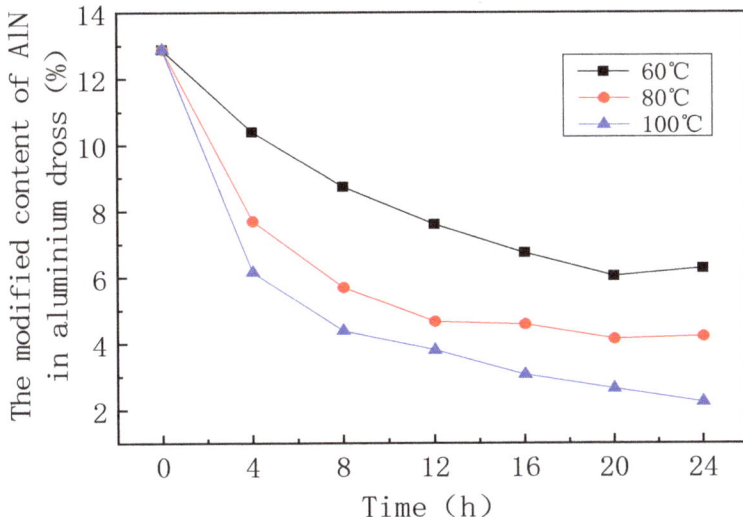

Figure 5. The modified contents of AlN in aluminum dross under different temperatures and speeds (600 r·min^{-1}).

The rate of a chemical reaction varies by temperature. The effect of temperature on the reaction rate is manifested primarily in the rate coefficient of reaction. Normally, the higher the temperature, the greater the reaction rate coefficient, which is directly manifested by the acceleration of the reaction rate [24]. At the end of the 19th century, the Swedish chemist Arrhenius proposed the famous Arrhenius equation after investigating the relationship between the hydrolysis rate of sucrose and the temperature:

$$\ln(k) = \ln(k_0) - \frac{Ea}{RT} \tag{5}$$

where k represents the reaction rate coefficient (dimensionless), k_0 represents the pre-exponential parameter (dimensionless), Ea denotes the reaction activation energy (kJ·mol^{-1}), T denotes the temperature (K), and R is the molar gas constant (J (mol·K)$^{-1}$).

Both k_0 and Ea are empirical parameters, which can be regarded as temperature-independent constants if the temperature fluctuates within a small range. At this time, the Arrhenius equation has excellent applicability. In different-temperature reaction systems, the correlation between the rate coefficients at two varying temperatures could be derived by Equation (5):

$$\ln(k_1) = \ln(k_0) - \frac{E_a}{RT_1} \tag{6}$$

$$\ln(k_2) = \ln(k_0) - \frac{E_a}{RT_2} \tag{7}$$

Supposing that $T_1 < T_2$ within an interval $T_1 - T_2$, and that k_0 and Ea could be deemed as constants, Equation (8) could be derived based on Equations (6) and (7). Since $Ea > 0$ and $T_1 < T_2$, the right side of Equation (8) was greater than 0, so $k_2 \cdot k_1 - 1 > 1$. In other words, the reaction rate coefficient at high temperatures was generally greater than that at low temperatures. Hence, the rise in temperature could increase the hydrolysis rate of AlN in the aluminum dross.

$$\ln \frac{k_2}{k_1} = \frac{Ea}{R} \left(\frac{T_2 - T_1}{T_1 T_2} \right) \tag{8}$$

Additionally, the microscopic essence of reaction lay in the breakage of old chemical bonds and the formation of new chemical bonds caused by effective particle collision. The rise in temperature increased the percentage and motion speed of activated molecules, and the increase in the number of activated molecules further promoted the effective intermolecular collision. Accordingly, the numbers of broken old bonds and formed new bonds per unit of time also increased.

To sum up, the hydrolysis rate of AlN in aluminum dross could be effectively accelerated by increasing the hydrolysis temperature and prolonging the hydrolysis time. As the hydrolysis time increased, the AlN hydrolysis rate gradually decreased, and after 20 h, the hydrolytic process essentially stopped. Meanwhile, the rotational speed produced a non-obvious effect on the hydrolysis rate of AlN in aluminum dross.

3.4. Effects of Hydrolysis Conditions on the pH

3.4.1. N Migration Model of Aluminum Dross during Hydrolysis

The N-containing compound in the aluminum dross is mostly AlN, and all of the N during the hydrolysis process is generally believed to come from this phase. After in-depth research and analysis, we proposed the N migration model of aluminum dross during its hydrolysis, as shown in Figure 6.

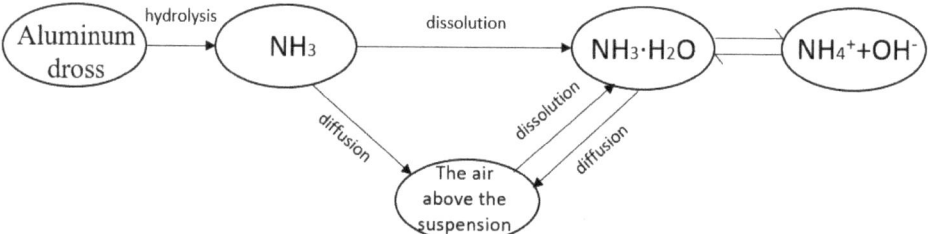

Figure 6. N migration model of aluminum dross during hydrolysis.

On the one hand, the NH_3 produced during the hydrolysis of AlN in aluminum dross was dissolved in suspension as a soluble gas, and then hydrated to form $NH_3 \cdot H_2O$. On the other hand, part of the undissolved NH_3 diffused above the suspension. Moreover, like other gas–liquid two-phase processes, the gas and liquid phases were dynamically interacting rather than mutually isolated. In macroscopic terms, the $NH_3 \cdot H_2O$ in suspension was decomposed by heat, so that the ammonia gas was diffused above the suspension. Meanwhile, the ammonia gas in the air above the suspension was also continuously dissolved into the suspension. In the meantime, the formed $NH_3 \cdot H_2O$ ionized into NH_4+ and OH^-, which also continued to reversely combine to form $NH_3 \cdot H_2O$ [25,26]. The above reactions interacted to establish a stable dynamic equilibrium system. By measuring the pH of the suspension in the equilibrium state, and referring to the "direct characterization" procedure for measuring the AlN content in aluminum dross, we assessed the feasibility of the "indirect characterization" procedure for determining the hydrolysis rate of aluminum dross.

3.4.2. Effects of Time and Rotational Speed on the Suspension pH

Figure 7 depicts the variation trends of the suspension pH under different rotational speeds at 60 °C. At the 0–2 h reaction stage, the suspension pH values under various rotational speeds increased sharply. With the prolongation of reaction time, the suspension pH values at low rotational speeds (0 and 100 r min^{-1}) were significantly greater than those at high rotational speeds (300 and 600 r min^{-1}). The primary cause of this phenomenon was that under identical conditions, a low rotational speed was conducive to the NH_3 retention in the suspension, which led to a higher concentration of OH^- ions ionized from the formed $NH_3 \cdot H_2O$, so that the pH value of the suspension was higher. Additionally, neither low nor high rotational speeds produced particularly obvious effects on the suspension pH. Overall, the rotational speed and the suspension pH appeared to be uncorrelated.

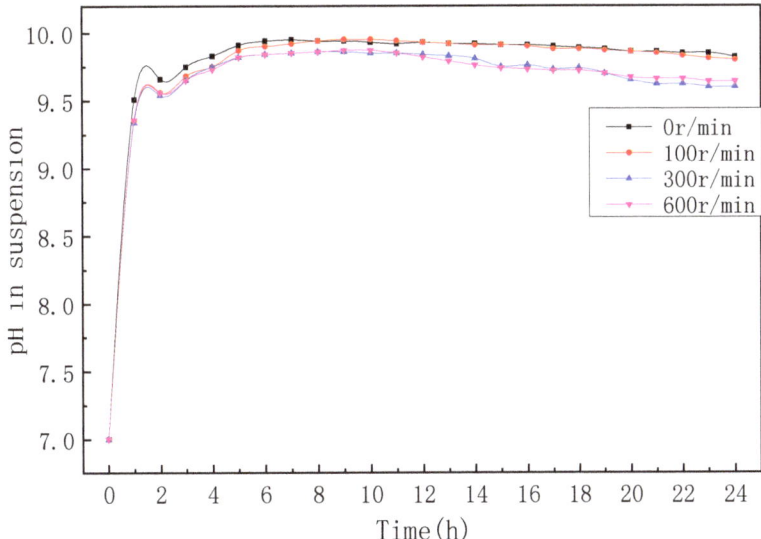

Figure 7. Suspension pH values under different rotational speeds (60 °C).

3.4.3. Effects of Temperature on the Suspension pH

Figure 8 depicts the variation trends of the suspension pH under different temperatures at 300 r min^{-1}. Clearly, at the 0–2 h reaction stage, the suspension pH values at various temperatures increased sharply. With the prolongation of reaction time, the maximal pH values could be maintained at different temperatures for certain periods of time, since the reaction rate was sufficient to keep the system saturated. Eventually, however, the pH values dropped slowly due to the decrease in the reaction rate and the slow release of ammonia. The suspension pH reading was the highest (approximately 9.8) at a temperature of 60 °C, whereas it was the lowest (only about 9.0) at 100 °C.

With the "direct characterization" method for AlN content in aluminum dross, the higher the temperature, the faster the hydrolysis. In contrast, with the present "indirect characterization" procedure, the higher the temperature, the lower the pH of suspension, which dropped extremely slowly per unit of time. This stood in contrast to the temporal variation result of AlN content in aluminum dross by the direct characterization method. According to the solution ionization equation, the following formulae could be derived:

$$K_1 = \frac{C(NH_4^+) \cdot C(OH^-)}{C(NH_3 \cdot H_2O)} = \frac{C^2(OH^-)}{C(NH_3 \cdot H_2O)} \quad (9)$$

$$K_2 = \frac{C(H^+) \cdot C(OH^-)}{C(H_2O)} \quad (10)$$

where K_1 represents the ionization constant of ammonia water (dimensionless), K_2 represents the ionization constant of water (dimensionless), $C(NH_4^+)$ denotes the concentration of NH_4^+ ionized from ammonia water (mol L^{-1}), $C(OH^-)$ denotes the concentration of OH^- ionized from ammonia water (mol L^{-1}), $C(H^+)$ denotes the concentration of H^+ ionized from water (mol L^{-1}), $C(NH_3 \cdot H_2O)$ is the concentration of unionized $NH_3 \cdot H_2O$ in the solution (mol L^{-1}), and $C(H_2O)$ is the concentration of unionized H_2O in the solution (mol L^{-1}).

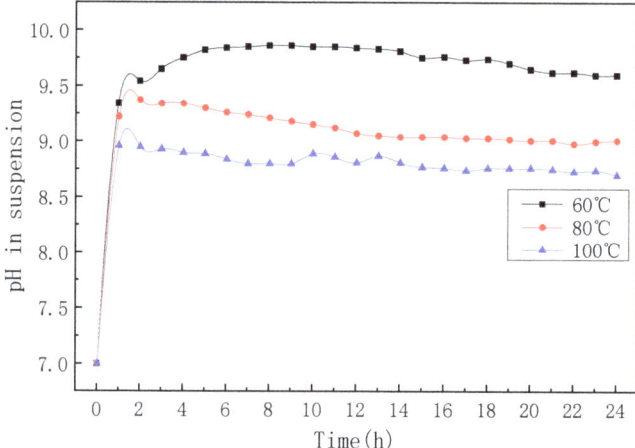

Figure 8. Suspension pH values at different temperatures (300 r min^{-1}).

Combining Equations (9) and (10) gives:

$$\text{pH} = -\log(\text{CH}^+) = \log \frac{\sqrt{K_1 \cdot C(\text{NH}_3 \cdot \text{H}_2\text{O})}}{K_2 \cdot C(\text{H}_2\text{O})} \quad (11)$$

As is clear from Equation (11), assuming that the reaction system was at the saturation stage of ammonia, the $C(\text{NH}_3 \cdot \text{H}_2\text{O})$ and $C(\text{H}_2\text{O})$ values in the system were constant, and the pH value of the suspension was related mainly to the ammonia ionization constant K_1 and the water ionization constant K_2. Since these two constants were directly correlated with the temperature T, the resultant values were essentially nonlinear, so the suspension pH could not accurately characterize the hydrolysis rate of aluminum dross under different temperature conditions. Meanwhile, according to Figure 8, when "indirect measures" such as suspension pH were used to assess the hydrolysis rate of aluminum dross, it was impossible to intuitively comprehend the effects of various parameters on the dross hydrolysis rate.

4. Regression Analysis of Hydrolysis Parameters

Based on the above experiments, during the aluminum dross hydrolysis, temperature, rotational speed, and time were independent variables that affected the AlN content (dependent variable). Since linear regression is a type of nonlinear regression, and there were multiple variables in the present experimentation, the "Analysis" module in Origin was utilized to perform relevant nonlinear regression analysis based on ternary quadratic polynomials. The preliminary regression model was formulated as follows:

$$y = ax_1^2 + bx_2^2 + cx_3^2 + dx_1x_2x_3 + ex_1x_2 + fx_1x_3 + gx_2x_3 + hx_1 + ix_2 + jx_3 + k \quad (12)$$

where x_1 denotes the hydrolysis temperature of aluminum dross (°C), x_2 denotes the hydrolysis speed of aluminum dross (r min^{-1}), x_3 denotes the hydrolysis time of aluminum dross (h), y represents the modified content of AlN (%), and others were undetermined constants (dimensionless).

As suggested by the above analysis, the independent variable "rotational speed" had an insignificant effect on the variation of AlN content in aluminum dross. To simplify the model, the rotational speed x_2 was "zeroed". The simplified regression model is as follows:

$$y = ax_1^2 + cx_3^2 + +ex_1x_2 + fx_1x_3 + hx_1 + jx_3 + k \quad (13)$$

Table 3 lists the values of various regression coefficients in Equations (12) and (13). The coefficient of multiple determination R_2, F-statistic, and p-value were 0.93035, 508.32976, and 0 in the preliminary model, respectively, while they were 0.93616, 1016.49681, and 0 in the simplified model, respectively. This indicates that the correlations between dependent and independent variables were higher with the simplified model, which yielded more significant results and achieved a better fitting effect [27,28].

Table 3. Numerical statistics of regression model.

Regression Coefficients	The Values of the Regression Coefficients		Standard Deviation of Regression Coefficient	
	Preliminary Model	Simplified Model	Preliminary Model	Simplified Model
a	0.00122	9.57857×10^{-4}	8.38698×10^{-4}	7.73811×10^{-4}
b	-1.36763×10^{-6}	—	1.24974×10^{-6}	—
c	0.01812	0.01812	0.00193	0.00185
d	0.27207	—	—	—
e	0.2718	—	—	—
f	−163.24244	−0.00267	—	8.8037×10^{-4}
g	−16.32398	—	—	—
h	−163.32872	−0.20235	—	0.12209
i	−16.30633	—	—	—
j	9793.85899	−0.52673	—	0.07604
k	9807.86654	21.45126	—	4.61293

Preliminary model: $R_2 = 0.93035$, $F = 508.32976$, $p = 0$; preliminary model: $R_2 = 0.93616$, $F = 1016.49681$, $p = 0$

The preliminary model is specifically expressed as follows:

$$yy = 0.00122x_1^2 - 1.36763 \times 10^{-6}x_2^2 + 0.01812x_3^2 + 0.27207x_1x_2x_3 \\ + 0.2718x_1x_2 - 163.24244x_1x_3 \\ - 16.32398x_2x_3 - 163.32872x_1 - 16.30633x_2 \\ + 9793.85899x_3 + 9807.86654 \quad (14)$$

The simplified model is specifically expressed as follows:

$$y = 9.57857 \times 10^{-4}x_1^2 + 0.01812x_3^2 - 0.00267x_1x_3 - 0.20235x_1 - 0.52673x_3 + 21.45126 \quad (15)$$

Table 4 compares the experimental and theoretical values between the preliminary and simplified models. Clearly, the simplified treatment could maintain good accuracy, with relative errors controlled generally within ±9.34%, which could somewhat guide the prediction and control of the AlN content in aluminum dross during hydrolysis.

Table 4. Comparison between experimental modified values and theoretical modified values of the AlN hydrolysis process in aluminum dross (part).

Number	Experimental Value	Theoretical Value		Relative Error (%)	
		Preliminary Model	Simplified Model	Preliminary Model	Simplified Model
1	7.07	6.53	6.41	−7.64	−9.34
2	6.86	6.83	6.71	−0.44	−2.19
3	12.88	12.97	12.76	0.70	−0.93
4	4.59	4.16	4.19	−9.37	−8.71
5	4.15	3.83	3.83	−7.71	−7.71
6	4.23	4.06	4.06	−4.02	−4.02
7	7.53	7.07	7.12	−6.11	−5.44
8	3.82	3.88	3.88	1.57	1.57
9	2.25	2.18	2.18	−3.11	−3.11

5. Conclusions

(1) By investigating the components of aluminum dross and changes in their respective contents during the hydrolysis process, a formula for modifying the AlN content in aluminum dross is proposed for the first time. Capable of reflecting the absolute content of AlN in aluminum dross, the formula provides a new way to scientifically characterize the hydrolysis rate of aluminum dross.

(2) The experimental study of aluminum dross hydrolysis under different times, temperatures, and rotational speeds found that after 24 h of hydrolysis, the modified AlN content decreased from the original 12.88% to 2.25% at a 100 °C hydrolysis temperature, dropped to 4.23% at 80 °C, and only dropped to 6.28% at 60 °C. The hydrolytic process essentially stopped 20 h later. The aluminum dross hydrolysis can be effectively promoted by the prolongation of time and the increase in temperature. Meanwhile, rotational speed produces an insignificant effect on the dross hydrolysis process.

(3) Through the hydrolysis experiments under different conditions, we found that the suspension pH increased sharply during the 0–2 h period. With the prolongation of reaction time, the suspension pH decreased slowly, exhibiting greater values at low rotational speeds (0 and 100 r min^{-1}) than at high rotational speeds (300 and 600 r min^{-1}). Meanwhile, the rise in temperature can significantly lower the maximum pH of the suspension in the system.

(4) Comparison of two methods for characterizing aluminum dross hydrolysis rate—namely, the AlN content detection, and the suspension pH measurement—reveals that the suspension pH is greatly affected by temperature and NH_3 release. Moreover, the results obtained at identical temperatures are obviously inconsistent with the direct characterization results. Overall, the direct detection of AlN content in aluminum dross is versatile and highly accurate.

(5) Multivariate nonlinear regression was performed between the modified AlN content and hydrolysis parameters such as time, temperature, and rotational speed. After removing the outliers in the model, the hydrolysis parameters and AlN content were subjected to multivariate nonlinear regression and quadratic simplification to derive a simplified model. With the simplified model, the relative errors between the theoretical and experimental values of modified AlN content were within ±9.34%, which can somewhat guide the prediction and control of AlN content in aluminum dross during hydrolysis.

Author Contributions: Conceptualization, H.N.; methodology, W.W.; validation, W.W.; formal analysis, S.L. and H.N.; investigation, X.W. and S.L.; resources, H.N. and S.L.; writing—original draft, H.N.; writing—review and editing, S.L. and W.W.; visualization, W.N. and W.W.; supervision, S.L. All authors have read and agreed to the published version of the manuscript.

Funding: This research was supported by the Priority Academic Program Development of Jiangsu Higher Education Institutions (PAPD), the Jiangsu Province Policy Guidance Program (International Science and Technology Cooperation) Project (BZ2021045), and Key R&D Projects of Jiangsu Province (BE2019060).

Institutional Review Board Statement: Not applicable.

Informed Consent Statement: Not applicable.

Data Availability Statement: Not applicable.

Conflicts of Interest: The authors declare no conflict of interest.

References

1. Foo, C.T.; Salleh, M.A.M.; Ying, K.K.; Matori, K.A. Mineralogy and thermal expansion study of mullite-based ceramics synthesized from coal fly ash and aluminum dross industrial wastes. *Ceram. Int.* **2019**, *45*, 7488–7494. [CrossRef]
2. Choo, T.F.; Mohd Salleh, M.A.; Kok, K.Y.; Matori, K.A.; Abdul Rashid, S. Characterization of high-temperature hierarchical Porous Mullite washcoat synthesized using Aluminum Dross and Coal fly ash. *Crystals* **2020**, *10*, 178. [CrossRef]

3. Elseknidy, M.H.; Salmiaton, A.; Nor Shafizah, I.; Saad, A.H. A Study on Mechanical Properties of Concrete Incorporating Aluminum Dross, Fly Ash, and Quarry Dust. *Sustainability* **2020**, *12*, 9230. [CrossRef]
4. Kikuchi, R. Recycling of municipal solid waste for cement production: Pilot-scale test for transforming incineration ash of solid waste into cement clinker. *Resour. Conserv. Recycl.* **2001**, *31*, 137–147. [CrossRef]
5. Mohammadzadeh, K.; Mahinroosta, M.; Allahverdi, A.; Dong, P.; Bassim, N. Non-supercritical drying synthesis and characterization of monolithic alumina aerogel from secondary aluminum dross. *Ceram. Int.* **2022**, *48*, 13154–13162. [CrossRef]
6. Ni, H.; Zhang, J.; Lv, S.; Gu, T.; Wang, X. Performance Optimization of Original Aluminum Ash Coating. *Coatings* **2020**, *10*, 831. [CrossRef]
7. Mailar, G.; Sreedhara, B.M.; Manu, D.S.; Hiremath, P.; Jayakesh, K. Investigation of concrete produced using recycled aluminium dross for hot weather concreting conditions. *Resour.-Effic. Technol.* **2016**, *2*, 68–80. [CrossRef]
8. Lee, J.R.; Lee, I.; Shin, H.Y.; Ahn, J.G.; Kim, D.J.; Chung, H.S. Nitride-related compounds preparation from waste aluminum dross by self-propagating high-temperature process. In *Materials Science Forum*; Trans Tech Publications Ltd.: Schwyz, Switzerland, 2005; Volume 486, pp. 297–300. [CrossRef]
9. Lukita, M.; Abidin, Z.; Riani, E.; Ismail, A. Utilization of hazardous waste of black dross aluminum: Processing and application-a review. *J. Degrad. Min. Lands Manag.* **2022**, *9*, 3265–3271. [CrossRef]
10. Zhang, F.; Zhang, J.; Zhu, Y.; Wang, X.; Jin, Y. Microstructure and Properties of Polytetrafluoroethylene Composites Modified by Carbon Materials and Aramid Fibers. *Coatings* **2020**, *10*, 1103. [CrossRef]
11. Sooksaen, P.; Puathawee, P. Properties of Unglazed Ceramics Containing Aluminum Dross as a Major Component. In *Solid State Phenomena*; Trans Tech Publications Ltd.: Schwyz, Switzerland, 2017; Volume 266, pp. 182–186. [CrossRef]
12. Benkhelif, A.; Kolli, M. Synthesis of Pure Magnesium Aluminate Spinel ($MgAl_2O_4$) from Waste Aluminum Dross. *Waste Biomass Valorization* **2022**, *13*, 2637–2649. [CrossRef]
13. Kale, M.; Yılmaz, I.H.; Kaya, A.; Çetin, A.E.; Söylemez, M.S. Pilot-scale hydrogen generation from the hydrolysis of black aluminum dross without any catalyst. *J. Energy Inst.* **2022**, *100*, 99–108. [CrossRef]
14. Tang, J.; Liu, G.; Qi, T.; Zhou, Q.; Peng, Z.; Li, X.; Yan, H.; Hao, H. Two-stage process for the safe utilization of secondary aluminum dross in combination with the Bayer process. *Hydrometallurgy* **2022**, *209*, 105836. [CrossRef]
15. Guo, J.; Zhou, Z.; Ming, Q.; Huang, Z.; Zhu, J.; Zhang, S.; Xu, J.; Xi, J.; Zhao, Q.; Zhao, X. Recovering precipitates from dechlorination process of saline wastewater as poly aluminum chloride. *Chem. Eng. J.* **2022**, *427*, 131612. [CrossRef]
16. Bellucci, S.; Popov, A.I.; Balasubramanian, C.; Cinque, G.; Marcelli, A.; Karbovnyk, I.; Savchyn, V.; Krutyak, N. Luminescence, vibrational and XANES studies of AlN nanomaterials. *Radiat. Meas.* **2007**, *42*, 708–711. [CrossRef]
17. Balasubramanian, C.; Bellucci, S.; Castrucci, P.; De Crescenzi, M.; Bhoraskar, S. Scanning tunneling microscopy observation of coiled aluminum nitride nanotubes. *Chem. Phys. Lett.* **2004**, *383*, 188–191. [CrossRef]
18. Zhukovskii, Y.F.; Pugno, N.; Popov, A.I.; Balasubramanian, C.; Bellucci, S. Influence of F centres on structural and electronic properties of AlN single-walled nanotubes. *J. Phys. Condens. Matter* **2007**, *19*, 395021. [CrossRef]
19. Shuaishuai, L.; Jiaqiao, Z.; Hongjun, N.; Xingxing, W.; Yu, Z.; Tao, G. Study on Preparation of Aluminum Ash Coating Based on Plasma Spray. *Appl. Sci.* **2019**, *9*, 4980.
20. Ni, H.; Zhang, J.; Lv, S.; Wang, X.; Zhu, Y.; Gu, T. Preparation and Performance Optimization of Original Aluminum Ash Coating Based on Plasma Spraying. *Coatings* **2019**, *9*, 770. [CrossRef]
21. Zhang, F.; Zhang, J.; Ni, H.; Zhu, Y.; Wang, X.; Wan, X.; Chen, K. Optimization of $AlSi_{10}MgMn$ Alloy Heat Treatment Process Based on Orthogonal Test and Grey Relational Analysis. *Crystals* **2021**, *11*, 385. [CrossRef]
22. Zhang, Y.; Ni, H.; Lv, S.; Wang, X.; Li, S.; Zhang, J. Preparation of Sintered Brick with Aluminum Dross and Optimization of Process Parameters. *Coatings* **2021**, *11*, 1039. [CrossRef]
23. Calignano, F.; Manfredi, D.; Ambrosio, E.P.; Iuliano, L.; Fino, P. Influence of process parameters on surface roughness of aluminum parts produced by DMLS. *Int. J. Adv. Manuf. Technol.* **2013**, *67*, 2743–2751. [CrossRef]
24. Felix, G.S.; Sellin, N.; Marangoni, C. Reduction of dross in galvanized sheets through automatic control of snout positioning in continuous operation. *Int. J. Adv. Manuf. Technol.* **2017**, *89*, 2345–2353. [CrossRef]
25. Zhou, C.; Wang, Q.; Zhang, W.; Zhao, W. Recovery of ammonia nitrogen from aluminum slag ash. *Conserv. Util. Miner. Resour.* **2012**, *3*, 38–41. [CrossRef]
26. Yoldi, M.; Fuentes-Ordoñez, E.G.; Korili, S.A.; Gil, A. Efficient recovery of aluminum from saline slag wastes. *Miner. Eng.* **2019**, *140*, 105884. [CrossRef]
27. Yang, Q.; Li, Q.; Zhang, G.; Shi, Q.; Feng, H. Investigation of leaching kinetics of aluminum extraction from secondary aluminum dross with use of hydrochloric acid. *Hydrometallurgy* **2019**, *187*, 158–167. [CrossRef]
28. Shabashov, V.A.; Kozlov, K.A.; Lyashkov, K.A.; Litvinov, A.V.; Dorofeev, G.A.; Titova, S.G.; Fedorenko, V.V. Effect of aluminum on mechanical solid-state alloying of iron with nitrogen in ball mill. *Phys. Met. Metallogr.* **2012**, *113*, 992–1000. [CrossRef]

Article

A Novel SERS Substrate Based on Discarded Oyster Shells for Rapid Detection of Organophosphorus Pesticide

Chi-Yu Chu [1], Pei-Ying Lin [2], Jun-Sian Li [2], Rajendranath Kirankumar [2], Chen-Yu Tsai [2], Nan-Fu Chen [3,4], Zhi-Hong Wen [5] and Shuchen Hsieh [2,6,7,*]

[1] Department of Pathology, Kaohsiung Armed Forces General Hospital, Kaohsiung 80284, Taiwan; cych11869@gmail.com
[2] Department of Chemistry, National Sun Yat-Sen University, Kaohsiung 80424, Taiwan; phoebe00315@yahoo.com.tw (P.-Y.L.); wwe874067@gmail.com (J.-S.L.); r7.kirankumar@gmail.com (R.K.); m870309@gmail.com (C.-Y.T.)
[3] Department of Surgery, Division of Neurosurgery, Kaohsiung Armed Forces General Hospital, Kaohsiung 80284, Taiwan; chen06688@gmail.com
[4] Institute of Medical Science and Technology, National Sun Yat-Sen University, Kaohsiung 80424, Taiwan
[5] Department of Marine Biotechnology and Resources, National Sun Yat-Sen University, Kaohsiung 80424, Taiwan; wzh@mail.nsysu.edu.tw
[6] Center for Stem Cell Research, Kaohsiung Medical University, Kaohsiung 80708, Taiwan
[7] School of Pharmacy, College of Pharmacy, Kaohsiung Medical University, Kaohsiung 80708, Taiwan
* Correspondence: shsieh@faculty.nsysu.edu.tw; Tel.: +886-7-525-2000 (ext. 3931)

Abstract: Over the past few years, the concern for green chemistry and sustainable development has risen dramatically. Researchers make an effort to find solutions to difficult challenges using green chemical processes. In this study, we use oyster shells as a green chemical source to prepare calcium oxide nanoparticles (CaO-NPs). Transmission electron microscopy (TEM) results showed the CaO-NPs morphology, which was spherical in shape, 40 ± 5 nm in diameter, with uniform dispersion. We further prepared silver/polydopamine/calcium-oxide (Ag/PDA/CaO) nanocomposites as the surface-enhanced Raman scattering (SERS) substrates and evaluated their enhancement effect using the methyl parathion pesticide. The effective SERS detection limit of this method is 0.9 nM methyl parathion, which is much lower than the safety limits set by the Collaborative International Pesticides Analytical Council for insecticide in fruits. This novel green material is an excellent SERS substrate for future applications and meets the goal of green chemistry and sustainable development.

Keywords: calcium oxide; nanoparticles; surface-enhanced Raman scattering; oyster shells; biocompatible

Citation: Chu, C.-Y.; Lin, P.-Y.; Li, J.-S.; Kirankumar, R.; Tsai, C.-Y.; Chen, N.-F.; Wen, Z.-H.; Hsieh, S. A Novel SERS Substrate Based on Discarded Oyster Shells for Rapid Detection of Organophosphorus Pesticide. *Coatings* 2022, 12, 506. https://doi.org/10.3390/coatings12040506

Academic Editor: Giorgos Skordaris

Received: 14 March 2022
Accepted: 6 April 2022
Published: 8 April 2022

Publisher's Note: MDPI stays neutral with regard to jurisdictional claims in published maps and institutional affiliations.

Copyright: © 2022 by the authors. Licensee MDPI, Basel, Switzerland. This article is an open access article distributed under the terms and conditions of the Creative Commons Attribution (CC BY) license (https://creativecommons.org/licenses/by/4.0/).

1. Introduction

The great advantages that surface-enhanced Raman scattering (SERS) spectroscopy offer in detection applications include fingerprint recognition of chemical molecules, non-destructive testing, high sensitivity, rapid detection, quantification, etc. [1]. Therefore, SERS can be applied to a wide range of applications, such as the detection of antibiotic and pesticide residues in agricultural products [2], pathogenic bacteria testing in aquatic environments [3], and identification of plastic particles released from consumer products [4], environmental exhaust fumes, etc. [5]. The significant impact of SERS techniques on molecular detection and plasmon catalysis are quite promising and due to the fact that these materials are of significant interest for a wide range of applications [6]. Since SERS-based techniques rely on the well-organized nanostructures, capable of generating plasmonic events, thus, SERS substrates based on the newly designed and more controlled nanostructures are highly demanded for further advancements. For the efficacy and to obtain the enhanced signals, simultaneous occurrence of a long-range electromagnetic effect (EM) and a short-range chemical effect, i.e., charge transfer (CT), are well needed during SERS measurements [7]. In this context, simultaneous CT can be seen by the use of semiconductor materials, which predominantly include metal oxides [8]. Therefore, the incorporation of semiconductor materials into SERS substrates would be able to generate a CT mechanism as a dominant contribution to the surface-enhanced Raman signal. To this end, efforts have been made with semiconductor oxide materials, such as ZnO, TiO_2 and Cu_2O, which are reported to be capable of generating weak and/or strong SERS signals via chemical enhancement [4,9,10]. The versatile application of metal oxides in various areas are well known. As a major drawback, after use, these materials are released to soil and surface water that passes over lakes, rivers, and seas. Nanomaterials, due to their shape and size, are prominent to bind to a wide variety of biomolecules, including proteins, drugs and nucleotides. Therefore, humans become exposed to nanoparticles (NPs) in medical and industrial settings quite commonly. The existence of metal nanoparticles (MNPs) in the ecological system lead to bioaccumulations, causing potential effects on human health and living organisms. Especially the induced toxicity of CuO, ZnO, TiO_2, and other metal oxides, which are well known and their adverse effects have been well studied in the past few decades [11–15]. Therefore, it is urgently required to incorporate some non-toxic, biocompatible and ecological friendly metal oxides for the construction of SERS substrates for societal welfare and care.

In our previous work, we reported a simultaneous occurrence of CT and EM mechanisms into a Ag/CuO nanocomposite system at the metal–semiconductor interface [16]. However, realizing the scenario of toxicity, we further present the utilization of non-toxic CaO MNPs into SERS substrates; i.e., a metal-molecular-semiconductor composite system consisting of Ag NPs, PDA and CaO NPs to construct a silver/polydopamine/calcium-oxide (Ag/PDA/CaO) system. For the generation of the SERS substrate, CaO NPs were mainly obtained from a green resource (oyster shells) via a simple chemical process. The strong adhesion of the PDA layer provided a uniform and well-packed composite system [17]. Simultaneously, in-situ preparation of Ag NPs and, thus, effective charge redistributions among Ag NPs and CaO, via a conducting PDA layer to induce a chemical effect during the SERS measurement, was obtained [18]. In our previous work, the Ag/CuO nanocomposite demonstrated an approximately 10^5 times SERS signal enhancement, mainly through the combination of CT and EM mechanisms [19]. In the present study, such an enhancement factor (4.7×10^5) in 10 h with LOD in a nano-regime for 4-aminothiophenol (4-ATP) was obtained for the Ag/CaO nanocomposite and, thus, the incorporation of CaO can be validated. To date, there are no reports demonstrating the incorporation of CaO into a SERS substrate to obtain enhanced SERS signals. Therefore, we believe our study would be able to introduce a newer and non-toxic metal oxide into the process of advancement for newer ecologically friendly SERS substrates and nanomaterials.

2. Materials and Methods

Oyster shells, having a high $CaCO_3$ content, were used as the green source of CaO in our study. After calcination, the CaO content was highest followed by CO_2; the percentages of Na_2O, MgO, SO_3, SiO_2, SrO, and H_2O were very low [20,21]. The shells were scrubbed to remove sand and salt from the surface and were then placed in an oven to remove moisture and thus to reduce impurities. The agents used in this experiment include dopamine hydrochloride (Sigma-Aldrich, Saint Louis, MO, USA), tris buffer (Sigma-Aldrich, Saint Louis, MO, USA), silver nitrate (Sigma-Aldrich, Saint Louis, MO, USA), sodium hydroxide (Sigma-Aldrich, Saint Louis, MO, USA), ethanol (ECHO Chemical CO., LTD, Kaohsiung, Taiwan), hydrochloric acid (J.T. Baker, Radnor, PA, USA), sodium carbonate (Sigma-Aldrich, Saint Louis, MO, USA), 4-aminothiophenol and methyl parathion (Sigma-Aldrich, Saint Louis, MO, USA). All agents were used directly in the experiment without further purification. The pure water required for samples synthesized and dispensed was obtained from pure water (18.2 MΩ-cm at 25 °C) Milli-Q reagent-grade (type I).

Scheme 1 shows the Ag/PDA/CaO NPs preparation process. After the oyster shells were ground at room temperature, 1 M NaOH solution was added and stirred for 2 h. It was then centrifuged and washed with deionized water until neutral. Hydrochloric acid was added, and the mixture was stirred for 6 h. The filtrate was filtered to obtain a CaCl$_2$ solution. CaCl$_2$ and sodium carbonate were mixed into the water-heating device to heat to 100 °C. After filtering, the powder was placed into a high-temperature furnace heated to 900 °C for 2 h, and spherical calcium oxide nanoparticles (CaO NPs) were synthesized. The reaction mechanism is given by Equation (1):

$$CaCO_{3(s)} + 2HCl_{(aq)} \rightarrow CaCl_{2(aq)} + CO_{2(g)} + H_2O_{(l)} \quad (1)$$

$$CaCl_{2(aq)} + NaCO_{3(s)} \rightarrow CaCO_{3(s)} + 2NaCl_{(aq)} \quad (2)$$

$$CaCO_{3(s)} \xrightarrow{\Delta} CaO_{(s)} + CO_{2(g)} \quad (3)$$

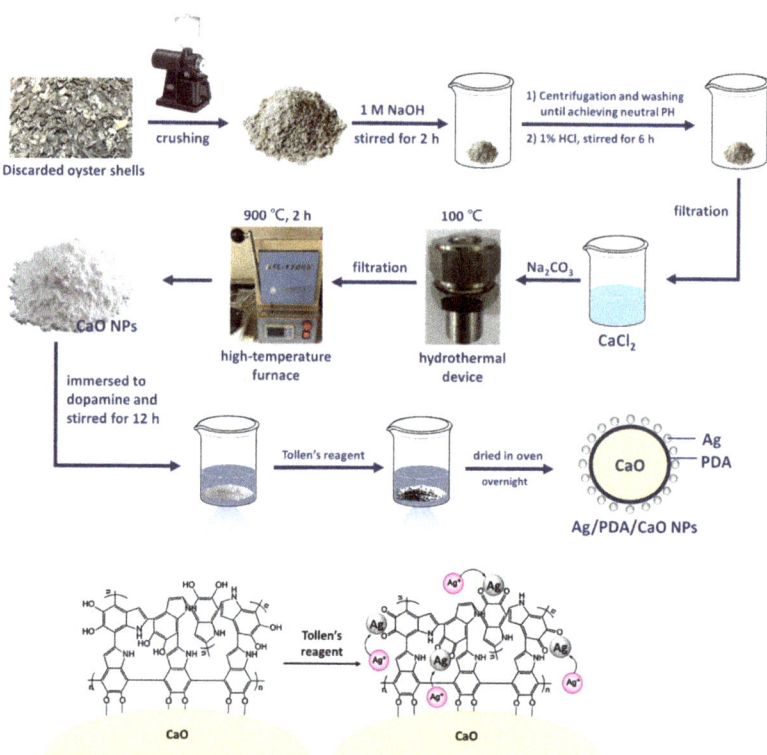

Scheme 1. Preparation process of Ag/PDA/CaO nanoparticles.

For the PDA and Ag NPs surface modification method of Ag/PDA/CaO NPs, refer to Lin et al. [5]. Utilizing the catechol-amine chemistry of dopamine (DA) as an adhesive layer between CaO NPs and Ag NPs, PDA coating layers were formed on the surface of CaO NPs by oxidative polymerization in an alkaline environment [22]. Surface-exposed PDA functional groups, including catechol, amine, and imine [23] can be used as reducing agents for Ag NPs. We prepared a solution of synthetic PDA, the tris buffer (C$_8$H$_{11}$NO$_2$•HCl) adjusted to pH 8.5 as a solvent and configured the 10 mM dopamine solution. We soaked the synthesized CaO NPs in the dopamine solution and stirred for 12 h. Then, we soaked the PDA-coated CaO NPs in a silver amine solution (Tollen's reagent), dried it in the oven overnight, and obtained the Ag/PDA/CaO NPs.

2.1. Transmission Electron Microscopy (TEM)

After the nanoparticle powder was dissolved in alcohol, 30 µL of the sample was dripped onto a carbonized copper mesh sheet and dried to complete the test substrate, shooting the image with a transmission electron microscope (JEOL, JEM-2100, Tokyo, Japan). Magnification: 2000–1,500,000×; Resolution: Point image: 0.27 nm and Lattice image: 0.14 nm; Dual tilt bases: X-axis ± 60°; Electron gun: field emission acceleration voltage of 200 kv.

2.2. X-ray Diffraction (XRD)

The samples were prepared by drying and grinding calcium oxide nanoparticles into a uniform powder. The powder was flattened on a test substrate for X-ray diffraction. The diffraction pattern of the calcium oxide nanoparticles was used to identify the crystal grid parameters. We measured calcium oxide and calcium carbonate powder samples at room temperature using an X-ray powder diffractometer (Bruker D8 Advance, Karlsruhe, Germany). The Cu Kα1 with a wavelength of 1.5418 Å and energy of 8.05 KeV was used as the X-ray source and the X-ray tube was a Cu target. The operating voltage was 40 kV and the operating current was 40 mA. The detector is a high-speed, high-sensitivity PSD detector with the best resolution of ~0.03° (FWHM). The 2θ angle was scanned at a scanning speed of 0.5 s at each pitch from 20 °C to 80 °C, and the spacing was 0.1°.

2.3. X-ray Photoelectron Spectroscopy (XPS)

Samples were prepared for analysis using X-ray photoelectron spectroscopy (XPS) by depositing an aliquot of the sample solution onto clean silicon wafer at room temperature and drying in air. XPS spectra were acquired using a JAMP-9500F (JEOL, Tokyo, Japan) equipped with a monochromatic Mg Kα X-ray (1253.6 eV) radiation source. The spectra were calibrated using the C 1s peak at 285 eV from adventitious surface carbon as an internal reference. The Cu and Ag peak curve fitting was performed with a Gaussian/Lorentzian ratio of 0.7 using peak fitting software by JEOL SpecSurf XPS (version 1.7.3.9), after a Shirley-type background subtraction.

2.4. Raman Spectroscopy

The modified metal/polymer/CaO NPs were deposited on a silicon substrate (1 cm × 1 cm) as a SERS substrate. Next, 5 µL 4-ATP and methyl parathion at different concentrations were dropped on the surface of SERS substrate, and its SERS enhancement effect was detected by Raman spectroscopy (WITec alpha300R, WITec, Ulm, Germany) at a laser wavelength of 532 nm and an intensity of 30.5 µW, and further detected the representative pesticide in water.

3. Results and Discussion

The surface morphologies of the composites were observed using transmission electron microscopy (TEM), as shown in Figure 1a,d, showing that CaO NPs, 40 ± 5 nm in diameter, were synthesized successfully with a spherical shape. Figure 1b and e demonstrate that after synthesizing the PDA onto CaO NPs, its surface became rough, and a coating was observed. The inset in Figure 1d and e shows high-resolution TEM (HRTEM) images of CaO NPs and PDA/CaO nanocomposites. The lattice spacings were 0.24 and 0.27 nm through calculation, corresponding to the crystal plane of (200) and (111), which proves a calcium oxide composition [24]. Figure 1c,f show that Ag NPs were successfully deposited on each calcium oxide nanoparticle. The inset in Figure 1f shows the HRTEM result. The spacing of Ag (111) was 0.237 nm [25], which confirms that Ag NPs were reduced on the surface of the CaO NPs by about 12 ± 5 nm.

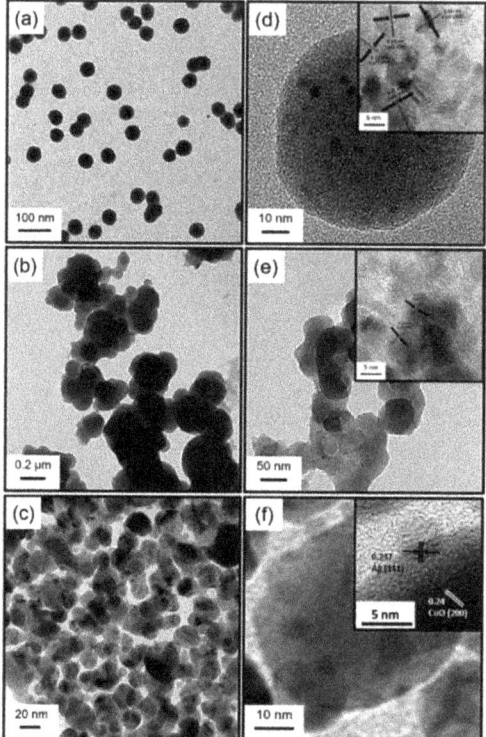

Figure 1. TEM images of various nanoparticles: (**a,d**) CaO NPs at different magnifications; (**b,e**) PDA/CaO nanocomposites at different magnifications; the insets in (**d**) and (**e**) are HRTEM images of random CaO nanoparticles and PDA/CaO nanocomposites, showing lattice spacing between CaO (111) and (200) planes. (**c,f**) Ag/PDA/CaO nanocomposites at different magnifications; the inset in (**f**) is an HRTEM image of a random Ag/PDA/CaO nanocomposite showing lattice spacing between Ag (111) and CaO (200) planes as well.

The X-ray diffraction (XRD) results of the synthesized composites are shown in Figure 2. In addition to the CaO signal, the Ag/PDA/CaO (blue line) also had Ag (111) and (200) signals, in accordance with the standard diffraction pattern of Ag NPs (JCPDS: 04-0783) [26]. These results prove that we successfully synthesized Ag NPs on CaO NPs.

We analyzed each step of the Ag/PDA/CaO SERS substrate formation by using high-resolution X-ray photo-electron spectroscopy (XPS). The Ag 3D XPS spectra in Figure 3a for Ag NPs exhibited peaks at 368.1 eV and 371.2 eV, which are assigned to Ag $3d_{5/2}$ and Ag $3d_{3/2}$, respectively. As shown in Figure 3b, the high-resolution XPS of the Ca 2p revealed a signal of calcium oxide at 350.8 eV and 354.4 eV. Moreover, dopamine molecules were polymerized on the CaO NP surfaces, and the Ca 2p signal shifted 0.4 eV from 350.8 eV to the lower binding energy of 350.4 eV. These results indicate that the push electron effects of PDA resulted in a shift in Ca 2p signal to a lower binding energy because the chemical environment around Ca changed. After the PDA reduced the Ag NPs on the calcium oxide, the signal of Ca 2p shifted to a lower binding energy, from 350.4 eV to 350.1 eV. According to the literature, when semiconductors combine with metals, electrons flow from higher to lower Fermi levels, allowing both Fermi levels to reach a new equilibrium [27–29]. Because the work function is 1.69 eV of calcium oxide and 4.1 eV of silver [30], the electrons transferred from calcium oxide to silver when they were combined, as shown in Figure 3c.

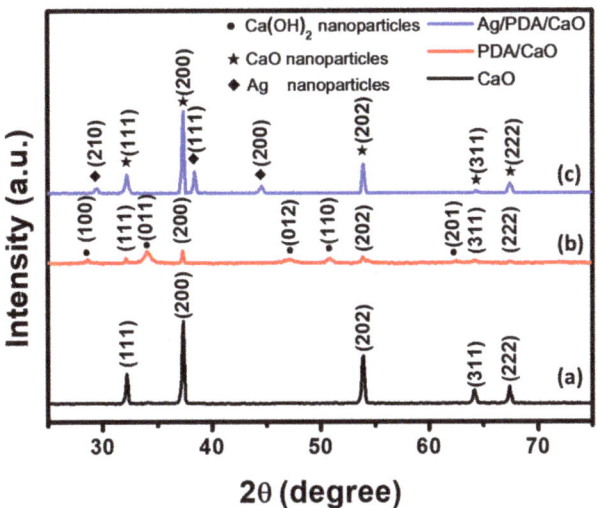

Figure 2. XRD patterns of synthesized (**a**) CaO NPs, (**b**) PDA/CaO and (**c**) Ag/PDA/CaO nanocomposites.

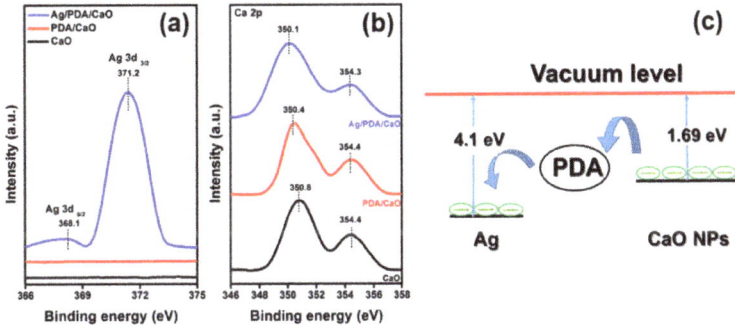

Figure 3. XPS spectra of (**a**) Ag 3d and (**b**) Ca 2p peak regions of the CaO NPs, PDA/CaO and Ag/PDA/CaO nanocomposites. (**c**) Scheme of charge transfer from CaO NPs to Ag.

To evaluate the SERS activity of the Ag/PDA/CaO substrate, we selected 4-ATP as model probe molecules and configured 4-ATP using ethanol to 10^{-3} M. We examined pure 4-ATP 10^{-2} M on a silicon substrate as normal Raman spectra (Figure 4a). The labeled peaks in the normal Raman spectra correspond well to those reported for pure 4-ATP in earlier reports. The peaks at 1154, 1200, and 1448 cm^{-1} are assigned to the δ(C−H) stretching vibration. The peaks at 1088, 1154, 1400, and 1588 cm^{-1} are assigned to the ν(C−S), δ(C−H), [ν(C−C) + δ(C−H)], and ν(C−C) stretching vibration, respectively. We used different Ag NP deposition times of 2, 4, 6, 8, 10, and 24 h to prepare the SERS substrates and analyzed 4-ATP, in order to find the most suitable preparation condition. Figure A1 in the Appendix A shows significant enhancement in intensity, beginning 2 h after deposition. We calculated the effective enhancement factors (EFs) by comparing the SERS peak intensity at 1448 cm^{-1} of 4-ATP on Ag/PDA/CaO nanocomposites to that of 4-ATP on CaO NPs, according to the formula $I_{SERS}/I_{substrate}$ × (concentration factor) [31,32] I_{SERS} is the Raman intensity of the probe molecule on the Ag/PDA/CaO substrate, $I_{substrate}$ is the Raman intensity of the probe molecule on the silicon substrate. The concentration factor is the corresponding concentration of the different concentrations of the probe molecule in ethanol. The calculated EFs were 3.8×10^5, 5.1×10^5, 5.2×10^5, 3.2×10^5, 4.7×10^5, and 4.6×10^5 at 2, 4, 6, 8, 10, and 24 h respectively. The results indicate that EF reached

5.1×10^5 after 4 h of deposition. Therefore, we found that 4 h of deposition time for Ag NPs is the most appropriate for preparing a Ag/PDA/CaO nanocomposite as the SERS substrate for detecting 4-ATP. Under these conditions, less deposition time was needed for the best EF result.

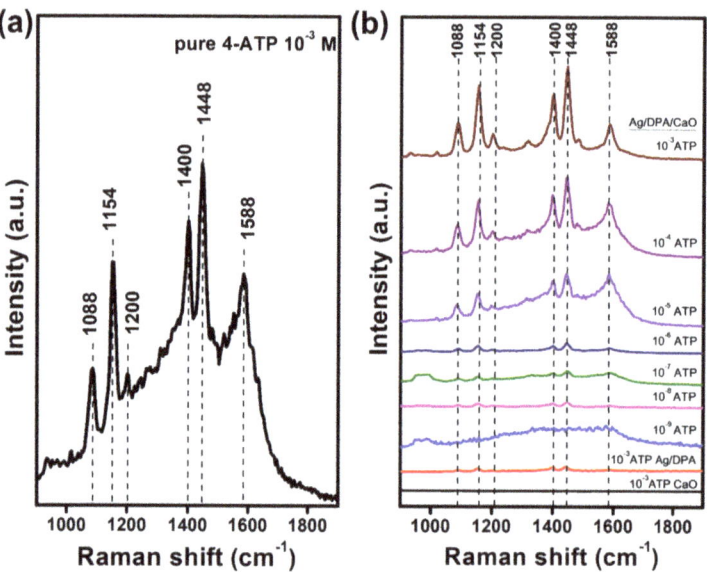

Figure 4. (**a**) Raman spectra of pure 4-ATP 10^{-2} M. (**b**) SERS spectra as a function of 4-ATP concentration while being adsorbed on Ag/PDA/CaO-4h nanocomposites.

We further investigated the detection limit of 4-ATP with the Ag/PDA/CaO-4h substrate. We formulated 4-ATP to 10^{-3}, 10^{-4}, 10^{-5}, 10^{-6}, 10^{-7} and 10^{-8} M in ethanol. As shown in Figure 4, CaO and Ag/PDA materials showed no SERS effects; however, after detection of the Ag/PDA/CaO-4 h substrates, an obvious SERS effect was observed. The calculated EFs were 1.1×10^6, 7.8×10^6, 3.6×10^7, 1.0×10^8, 9.6×10^8, and 6.5×10^9, for 10^{-3}, 10^{-4}, 10^{-5}, 10^{-6}, 10^{-7} and 10^{-8} M, respectively, and the detection limit reached $\sim 10^{-8}$ M.

We compared these results with those reported in other studies. Sun et al. [33] used a silver nanoparticle colloidal solution to enhance the 4-ATP molecular signal, and the detection limit was $\sim 10^{-7}$ M. Recently, scientists have used different substrates to reduce the silver nanoparticles to further enhance the 4-ATP signal, with a detection limit of $\sim 10^{-7}$ M for a single molecule [34–36]. In our study, Ag/PDA/CaO nanocomposites synthesized through simple steps successfully enhanced the SERS effect and improved the detection limit to $\sim 10^{-8}$ M for 4-ATP single-molecule detection. The SERS intensity is determined by several factors, such as particle size and shape and the nature of the aggregate [37,38]. In our results, the self-assembled Ag/PDA/CaO nanocomposites had uniform particle sizes and shapes of Ag NPs. This characteristic, resulting in the formation of numerous "hot spots", offers great signal improvement in SERS through strong electromagnetic enhancement.

The spectra of methyl parathion of differing concentrations (5 µL of 10 mM, 1 mM, 0.1 mM, 0.01 mM, 1 µM, 0.01 µM, 1 nM, and 0.9 nM) on the Ag/PDA/CaO nanocomposites are shown in Figure 5. The major characteristic bands in the SERS spectra of the methyl parathion could be identified at 1353 and 1596 cm^{-1}, which correspond with the variation in the C–H bending and phenyl stretching, respectively [39]. The calibration curve of 1353 cm^{-1} peak intensity and the corresponding concentration displayed a good linear relationship, ranging from 0.01 M to 0.9 nM, showing a linear correlation coefficient R^2 of 0.96 for methyl parathion (Figure 5c). The detection limit was 0.9 nM, which is much lower than the safety limits set by the Collaborative International Pesticides Analytical Council (CIPAC) for insecticide in fruits.

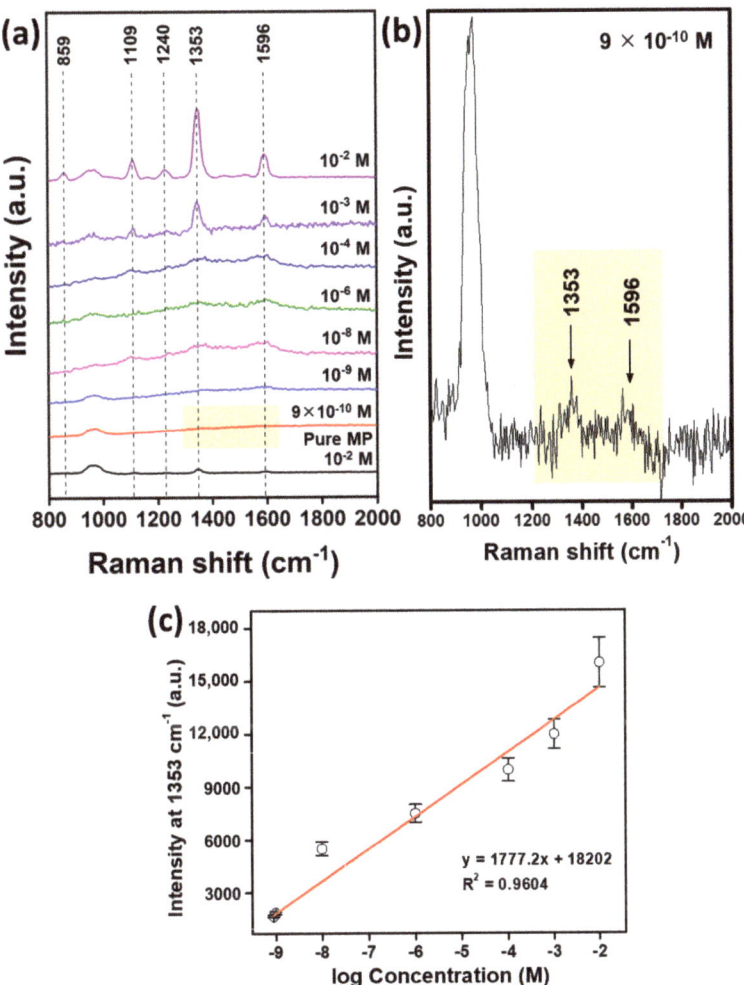

Figure 5. (a) SERS spectra of methyl parathion at different concentrations. (b) A zoomed-in view of the rectangle in (a), showing the intensity signal of 9×10^{-10} M methyl parathion. (c) Linear calibration plot between the SERS intensity (peak at 1353 cm^{-1}) and methyl parathion concentration.

In addition, Table 1 compares the analytical performances of various SERS nanocomposites for the detection of methyl parathion, which is comparable or even superior to the other SERS nanocomposites reported in the literature. The above results demonstrate that Ag/PDA/CaO nanocomposites have the potential for the highly sensitive detection and quantitative analysis of methyl parathion residue.

Table 1. Comparison of various SERS substrates in methyl parathion detection.

Materials	Method	Chemicals	Detection Limit (nM)	References
AgNPs-Ge/Si	Raman	Methyl Parathion	100	[39]
Au-Ag alloy	Raman	Methyl Parathion	5	[40]
Au/Ag-nanosphere/Al$_2$O$_3$-layer/Ag-nanoparticles	Raman	Methyl Parathion	1	[41]
Ag NPs on 3D PDMS nanotentacle array	Raman	Methyl Parathion	78	[42]
Silver nanoparticles/graphene	Raman	Methyl Parathion	600	[43]
Ag/AuNWs/PDMS	Raman	Methyl Parathion	3.8	[44]
polyaniline nanoparticles	Raman	Methyl Parathion	11.3	[45]
triangular gold nanoparticles	Raman	Methyl Parathion	500	[46]
Ag/PDA/CaO	Raman	4-ATP	10	This work
		Methyl Parathion	0.9	This work

4. Conclusions

A SERS substrate was successfully designed, constructed, and established by utilizing non-toxic CaO as a semiconductor material for the first time, over other toxic, traditionally used metal oxides (CuO, Cu$_2$O, ZnO, Fe$_2$O$_3$, etc.). The construction process involved the construction of CaO NPs from green resource oyster shells, followed by PDA coating on CaO NPs and further in-situ construction of Ag NPs onto a PDA/CaO support. The obtained LOD in the nanomolar regime and significant enhancement factor in SERS signals for sensing methyl parathion with the constructed substrate demonstrate the effectiveness of the SERS-constructed substrate by using CaO.

Author Contributions: C.-Y.C., P.-Y.L. and J.-S.L. contributed equally to this work. C.-Y.C.: funding acquisition, project administration, and supervision. P.-Y.L.: methodology, data curation, formal analysis, original draft writing. J.-S.L.: methodology, formal analysis, and original draft writing. R.K.: original draft writing. C.-Y.T.: methodology and data curation. N.-F.C. and Z.-H.W.: resources. S.H.: conceptualization, funding acquisition, project administration and supervision. All authors have read and agreed to the published version of the manuscript.

Funding: This work was financially supported by grants from Kaohsiung Armed Forces General Hospital (KAFGH-A-111007), the Ministry of Science and Technology of Taiwan (MOST 109-2113-M-110-001 and 110-2113-M-110-018) and the National Sun Yat-Sen University Center for Nanoscience and Nanotechnology.

Institutional Review Board Statement: Not applicable.

Informed Consent Statement: Not applicable.

Data Availability Statement: No data availability.

Acknowledgments: We would like to thank Hsien-Tsan Lin of the Regional Instruments Center at National Sun Yat-Sen University for his help in TEM experiments.

Conflicts of Interest: The authors declare no conflict of interest.

Appendix A

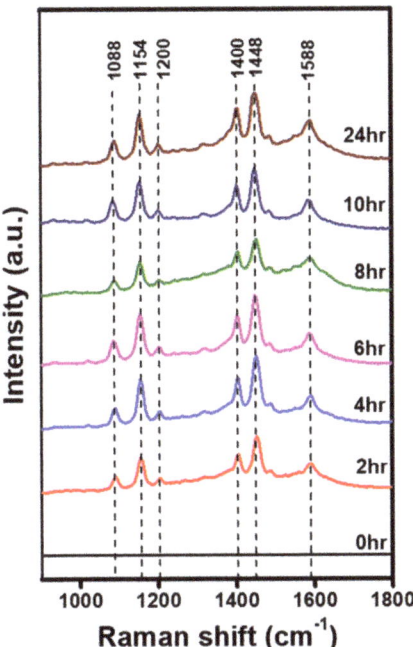

Figure A1. SERS spectra of 4-ATP acquired on Ag/PDA/CaO substrates prepared under different Ag-NPs deposition times (0, 2, 4, 6, 8, 10 and 24 h).

References

1. Han, X.X.; Rodriguez, R.S.; Haynes, C.L.; Ozaki, Y.; Zhao, B. Surface-enhanced Raman spectroscopy. *Nat. Rev. Methods Primers* **2022**, *1*, 87. [CrossRef]
2. Mikac, L.; Kovacevic, E.; Ukic, S.; Raic, M.; Jurkin, T.; Maric, I.; Gotic, M.; Ivanda, M. Detection of multi-class pesticide residues with surface-enhanced Raman spectroscopy. *Spectrochim. Acta A Mol. Biomol. Spectrosc.* **2021**, *252*, 119478. [CrossRef] [PubMed]
3. Ho, C.S.; Jean, N.; Hogan, C.A.; Blackmon, L.; Jeffrey, S.S.; Holodniy, M.; Banaei, N.; Saleh, A.A.E.; Ermon, S.; Dionne, J. Rapid identification of pathogenic bacteria using Raman spectroscopy and deep learning. *Nat. Commun.* **2019**, *10*, 4927. [CrossRef] [PubMed]
4. Lin, P.-Y.; Wu, I.H.; Tsai, C.-Y.; Kirankumar, R.; Hsieh, S. Detecting the release of plastic particles in packaged drinking water under simulated light irradiation using surface-enhanced Raman spectroscopy. *Anal. Chim. Acta* **2022**, *1198*, 339516. [CrossRef]
5. Lin, P.-Y.; He, G.; Chen, J.; Dwivedi, A.K.; Hsieh, S. Monitoring the photoinduced surface catalytic coupling reaction and environmental exhaust fumes with an Ag/PDA/CuO modified 3D glass microfiber platform. *J. Ind. Eng. Chem.* **2020**, *82*, 424–432. [CrossRef]
6. Koo, K.M.; Wang, J.; Richards, R.S.; Farrell, A.; Yaxley, J.W.; Samaratunga, H.; Teloken, P.E.; Roberts, M.J.; Coughlin, G.D.; Lavin, M.F.; et al. Design and Clinical Verification of Surface-Enhanced Raman Spectroscopy Diagnostic Technology for Individual Cancer Risk Prediction. *ACS Nano* **2018**, *12*, 8362–8371. [CrossRef] [PubMed]
7. Rajput, V.; Gupta, R.K.; Prakash, J. Engineering metal oxide semiconductor nanostructures for enhanced charge transfer: Fundamentals and emerging SERS applications. *J. Mater. Chem. C* **2022**, *10*, 73–95. [CrossRef]
8. Cui, Y.; Nilius, N.; Freund, H.-J.; Prada, S.; Giordano, L.; Pacchioni, G. Controlling the charge state of single Mo dopants in a CaO film. *Phys. Rev. B* **2013**, *88*, 205421. [CrossRef]
9. Li, L.; Hutter, T.; Finnemore, A.S.; Huang, F.M.; Baumberg, J.J.; Elliott, S.R.; Steiner, U.; Mahajan, S. Metal Oxide Nanoparticle Mediated Enhanced Raman Scattering and Its Use in Direct Monitoring of Interfacial Chemical Reactions. *Nano Lett.* **2012**, *12*, 4242–4246. [CrossRef]
10. Rhodes, C.; Franzen, S.; Maria, J.-P.; Losego, M.; Leonard, D.N.; Laughlin, B.; Duscher, G.; Weibel, S. Surface plasmon resonance in conducting metal oxides. *J. Appl. Phys.* **2006**, *100*, 054905. [CrossRef]
11. Ates, M.; Arslan, Z.; Demir, V.; Daniels, J.; Farah, I.O. Accumulation and toxicity of CuO and ZnO nanoparticles through waterborne and dietary exposure of goldfish (*Carassius auratus*). *Environ. Toxicol.* **2015**, *30*, 119–128. [CrossRef] [PubMed]

12. Hua, J.; Vijver, M.G.; Ahmad, F.; Richardson, M.K.; Peijnenburg, W.J.G.M. Toxicity of different-sized copper nano- and submicron particles and their shed copper ions to zebrafish embryos. *Environ. Toxicol. Chem.* **2014**, *33*, 1774–1782. [CrossRef] [PubMed]
13. Jia, X.; Wang, S.; Zhou, L.; Sun, L. The Potential Liver, Brain, and Embryo Toxicity of Titanium Dioxide Nanoparticles on Mice. *Nanoscale Res. Lett.* **2017**, *12*, 478. [CrossRef] [PubMed]
14. Rizk, M.Z.; Ali, S.A.; Hamed, M.A.; El-Rigal, N.S.; Aly, H.F.; Salah, H.H. Toxicity of titanium dioxide nanoparticles: Effect of dose and time on biochemical disturbance, oxidative stress and genotoxicity in mice. *Biomed. Pharmacother.* **2017**, *90*, 466–472. [CrossRef] [PubMed]
15. Das, S.; Thiagarajan, V.; Chandrasekaran, N.; Ravindran, B.; Mukherjee, A. Nanoplastics enhance the toxic effects of titanium dioxide nanoparticle in freshwater algae Scenedesmus obliquus. *Comp. Biochem. Physiol. Part C Toxicol. Pharmacol.* **2022**, *256*, 109305. [CrossRef]
16. Hsieh, S.; Lin, P.-Y.; Chu, L.-Y. Improved Performance of Solution-Phase Surface-Enhanced Raman Scattering at Ag/CuO Nanocomposite Surfaces. *J. Phys. Chem. C* **2014**, *118*, 12500–12505. [CrossRef]
17. Huang, Z.-H.; Peng, S.-W.; Hsieh, S.-L.; Kirankumar, R.; Huang, P.-F.; Chang, T.-M.; Dwivedi, A.K.; Chen, N.-F.; Wu, H.-M.; Hsieh, S. Polydopamine Ultrathin Film Growth on Mica via In-Situ Polymerization of Dopamine with Applications for Silver-Based Antimicrobial Coatings. *Materials* **2021**, *14*, 671. [CrossRef]
18. Cheng, M.; Li, C.; Li, W.; Liu, Y. Trace Cd(2+) Ions Detection on the Flower-Like Ag@CuO Substrate. *Nanomaterials* **2020**, *10*, 1664. [CrossRef]
19. Ye, F.; Ju, S.; Liu, Y.; Jiang, Y.; Chen, H.; Ge, L.; Yan, C.; Yuan, A. Ag-CuO Nanocomposites: Surface-Enhanced Raman Scattering Substrate and Photocatalytic Performance. *Cryst. Res. Technol.* **2019**, *54*, 1800257. [CrossRef]
20. Alvarenga, R.A.F.d.; Galindro, B.M.; Helpa, C.d.F.; Soares, S.R. The recycling of oyster shells: An environmental analysis using Life Cycle Assessment. *J. Environ. Manag.* **2012**, *106*, 102–109. [CrossRef]
21. Smith, R.A.; Wright, E.R. Elemental composition of oyster shell. *Tex. J. Sci.* **1962**, *14*, 222–224.
22. Ryu, J.H.; Messersmith, P.B.; Lee, H. Polydopamine Surface Chemistry: A Decade of Discovery. *ACS Appl. Mater. Interfaces* **2018**, *10*, 7523–7540. [CrossRef] [PubMed]
23. Lee, H.; Dellatore, S.M.; Miller, W.M.; Messersmith, P.B. Mussel-Inspired Surface Chemistry for Multifunctional Coatings. *Science* **2007**, *318*, 426–430. [CrossRef] [PubMed]
24. Prathap, A.; Shaijumon, M.M.; Sureshan, K.M. CaO nanocrystals grown over SiO_2 microtubes for efficient CO_2 capture: Organogel sets the platform. *Chem. Commun.* **2016**, *52*, 1342–1345. [CrossRef] [PubMed]
25. Gusev, A.; Sadovnikov, S. Acanthite–argentite transformation in nanocrystalline silver sulfide and the Ag_2S/Ag nanoheterostructure. *Semiconductors* **2016**, *50*, 682–687. [CrossRef]
26. Park, H.-H.; Zhang, X.; Choi, Y.-J.; Park, H.-H.; Hill, R.H. Synthesis of Ag Nanostructures by Photochemical Reduction Using Citrate-Capped Pt Seeds. *J. Nanomater.* **2011**, *2011*, 265287. [CrossRef]
27. Taherpour, A.A.; Rizehbandi, M.; Jahanian, F.; Naghibi, E.; Mahdizadeh, N.-A. Theoretical study of electron transfer process between fullerenes and neurotransmitters; acetylcholine, dopamine, serotonin and epinephrine in nanostructures [neurotransmitters].C n complexes. *J. Chem. Biol.* **2016**, *9*, 19–29. [CrossRef]
28. Feng, J.-J.; Zhang, P.-P.; Wang, A.-J.; Liao, Q.-C.; Xi, J.-L.; Chen, J.-R. One-step synthesis of monodisperse polydopamine-coated silver core-shell nanostructures for enhanced photocatalysis. *New J. Chem.* **2012**, *36*, 148–154. [CrossRef]
29. Akin, M.S.; Yilmaz, M.; Babur, E.; Ozdemir, B.; Erdogan, H.; Tamer, U.; Demirel, G. Large area uniform deposition of silver nanoparticles through bio-inspired polydopamine coating on silicon nanowire arrays for practical SERS applications. *J. Mater. Chem. B* **2014**, *2*, 4894–4900. [CrossRef]
30. Hopkins, B.J.; Vick, F.A. Thermionic and Related Properties of Calcium Oxide. *Br. J. Appl. Phys.* **1958**, *9*, 257–264. [CrossRef]
31. Hsieh, C.-W.; Lin, P.-Y.; Hsieh, S. Improved surface-enhanced Raman scattering of insulin fibril templated colloidal gold nanoparticles on silicon. *J. NanoPhotonics* **2012**, *6*, 063501. [CrossRef]
32. Le Ru, E.C.; Blackie, E.; Meyer, M.; Etchegoin, P.G. Surface Enhanced Raman Scattering Enhancement Factors: A Comprehensive Study. *J. Phys. Chem. C* **2007**, *111*, 13794–13803. [CrossRef]
33. Sun, L.; Song, Y.; Wang, L.; Guo, C.; Sun, Y.; Liu, Z.; Li, Z. Ethanol-Induced Formation of Silver Nanoparticle Aggregates for Highly Active SERS Substrates and Application in DNA Detection. *J. Phys. Chem. C* **2008**, *112*, 1415–1422. [CrossRef]
34. Cha, M.G.; Kim, H.M.; Kang, Y.L.; Lee, M.; Kang, H.; Kim, J.; Pham, X.H.; Kim, T.H.; Hahm, E.; Lee, Y.S.; et al. Thin silica shell coated Ag assembled nanostructures for expanding generality of SERS analytes. *PLoS ONE* **2017**, *12*, e0178651. [CrossRef]
35. Huang, J.; Ma, D.; Chen, F.; Chen, D.; Bai, M.; Xu, K.; Zhao, Y. Green in Situ Synthesis of Clean 3D Chestnutlike Ag/WO_{3-x} Nanostructures for Highly Efficient, Recyclable and Sensitive SERS Sensing. *ACS Appl. Mater. Interfaces* **2017**, *9*, 7436–7446. [CrossRef]
36. Yang, L.; Bao, Z.; Wu, Y.; Liu, J. Clean and reproducible SERS substrates for high sensitive detection by solid phase synthesis and fabrication of Ag-coated Fe_3O_4 microspheres. *J. Raman Spectrosc.* **2012**, *43*, 848–856. [CrossRef]
37. Pal, J.; Ganguly, M.; Dutta, S.; Mondal, C.; Negishi, Y.; Pal, T. Hierarchical Au-CuO nanocomposite from redox transformation reaction for surface enhanced Raman scattering and clock reaction. *CrystEngComm* **2014**, *16*, 883–893. [CrossRef]
38. Hu, J.-W.; Li, J.-F.; Ren, B.; Wu, D.-Y.; Sun, S.-G.; Tian, Z.-Q. Palladium-Coated Gold Nanoparticles with a Controlled Shell Thickness Used as Surface-Enhanced Raman Scattering Substrate. *J. Phys. Chem. C* **2007**, *111*, 1105–1112. [CrossRef]

39. Liu, J.; Meng, G.; Li, Z.; Huang, Z.; Li, X. Ag-NP@Ge-nanotaper/Si-micropillar ordered arrays as ultrasensitive and uniform surface enhanced Raman scattering substrates. *Nanoscale* **2015**, *7*, 18218–18224. [CrossRef]
40. Zhu, C.; Meng, G.; Zheng, P.; Huang, Q.; Li, Z.; Hu, X.; Wang, X.; Huang, Z.; Li, F.; Wu, N. A Hierarchically Ordered Array of Silver-Nanorod Bundles for Surface-Enhanced Raman Scattering Detection of Phenolic Pollutants. *Adv. Mater.* **2016**, *28*, 4871–4876. [CrossRef]
41. Hu, X.; Zheng, P.; Meng, G.; Huang, Q.; Zhu, C.; Han, F.; Huang, Z.; Li, Z.; Wang, Z.; Wu, N. An ordered array of hierarchical spheres for surface-enhanced Raman scattering detection of traces of pesticide. *Nanotechnology* **2016**, *27*, 384001. [CrossRef] [PubMed]
42. Wang, P.; Wu, L.; Lu, Z.; Li, Q.; Yin, W.; Ding, F.; Han, H. Gecko-Inspired Nanotentacle Surface-Enhanced Raman Spectroscopy Substrate for Sampling and Reliable Detection of Pesticide Residues in Fruits and Vegetables. *Anal. Chem.* **2017**, *89*, 2424–2431. [CrossRef] [PubMed]
43. Wang, X.; Zhu, C.; Hu, X.; Xu, Q.; Zhao, H.; Meng, G.; Lei, Y. Highly sensitive surface-enhanced Raman scattering detection of organic pesticides based on Ag-nanoplate decorated graphene-sheets. *Appl. Surf. Sci.* **2019**, *486*, 405–410. [CrossRef]
44. Ma, Y.; Du, Y.; Chen, Y.; Gu, C.; Jiang, T.; Wei, G.; Zhou, J. Intrinsic Raman signal of polymer matrix induced quantitative multiphase SERS analysis based on stretched PDMS film with anchored Ag nanoparticles/Au nanowires. *Chem. Eng. J.* **2020**, *381*, 122710. [CrossRef]
45. Liang, Y.; Wang, H.; Xu, Y.; Pan, H.; Guo, K.; Zhang, Y.; Chen, Y.; Liu, D.; Zhang, Y.; Yao, C.; et al. A novel molecularly imprinted polymer composite based on polyaniline nanoparticles as sensitive sensors for parathion detection in the field. *Food Control* **2022**, *133*, 108638. [CrossRef]
46. Wu, H.; Chen, S.; Wang, Y.; Tan, J.; Feng, Y.; Hou, L.; Wang, H. Spiny gold nanoparticles colloids as substrate for sensing of methyl parathion based on surfaced-enhanced Raman scattering. *Mater. Lett.* **2022**, *313*, 131687. [CrossRef]

Article

Polytetrafluoroethylene Modified Nafion Membranes by Magnetron Sputtering for Vanadium Redox Flow Batteries

Jun Su [†], Jiaye Ye [†], Zhenyu Qin and Lidong Sun *

State Key Laboratory of Mechanical Transmission, School of Materials Science and Engineering, Chongqing University, Chongqing 400044, China; sujun@cqu.edu.cn (J.S.); jiayesir@163.com (J.Y.); qzy014@gmail.com (Z.Q.)
* Correspondence: lidong.sun@cqu.edu.cn
† These authors contributed equally to this work.

Abstract: Commercial Nafion membranes have been widely used for vanadium redox flow batteries (VRFB) but with relatively low ion selectivity. A chemical method is commonly employed to modify the organic membranes, whereas physical approaches are rarely reported in view of less compatibility with the organic species. In this study, an ultrathin polytetrafluoroethylene (PTFE) film of less than 30 nm is deposited onto the Nafion substrates by radio frequency magnetron sputtering to form PTFE@Nafion composite membranes. The PTFE layer of hydrophobic and inert feature enhances the dimensional stability and the ion selectivity of the Nafion membranes. The VRFB single cell with an optimized composite membrane exhibits a better self-discharge property than that of the Nafion 212 (i.e., 201.2 vs. 18.6 h), due to a higher ion selectivity (i.e., 21.191×10^4 vs. 11.054×10^4 S min cm^{-3}). The composite membranes also show better discharge capacity retention than the Nafion 212 over the entire 100 cycles. The results indicate that the magnetron sputtering is an alternative and feasible route to tailor the organic membranes via surface modification and functionalization.

Keywords: ion exchange membrane; PTFE; Nafion; magnetron sputtering; vanadium redox flow battery

Citation: Su, J.; Ye, J.; Qin, Z.; Sun, L. Polytetrafluoroethylene Modified Nafion Membranes by Magnetron Sputtering for Vanadium Redox Flow Batteries. *Coatings* **2022**, *12*, 378. https://doi.org/10.3390/coatings12030378

Academic Editor: Torsten Brezesinski

Received: 8 February 2022
Accepted: 9 March 2022
Published: 14 March 2022

Publisher's Note: MDPI stays neutral with regard to jurisdictional claims in published maps and institutional affiliations.

Copyright: © 2022 by the authors. Licensee MDPI, Basel, Switzerland. This article is an open access article distributed under the terms and conditions of the Creative Commons Attribution (CC BY) license (https://creativecommons.org/licenses/by/4.0/).

1. Introduction

The utilization of renewable energies (e.g., wind, solar, and tidal energy) has been the global focus towards sustainable development [1–5]. The intermittent features of renewable energy make it difficult to achieve a reliable energy storage. Vanadium redox flow battery (VRFB) is regarded as one of the most promising systems for large-scale energy storage, considering its long lifetime, fast response, high reliability, low cost, and decoupling of power and capacity [6–8]. An ion exchange membrane is a core component in a flow battery to separate the catholyte and anolyte and to conduct protons. The commercial Nafion membranes have been widely employed due to their good conductivity and high chemical and mechanical stability [9–11]. However, Nafion membranes generally exhibit large permeability for vanadium ions and thus hinder VRFB development and deployment [12–15].

To reduce ion permeability and enhance ion selectivity, chemical strategies have been the prevailing approach to modify and improve the Nafion-based membranes, such as sol–gel modification, interfacial polymerization, oxidation polymerization, surfactant treatment, solution casting, and electrodeposition. Sol–gel modification is an effective way to incorporate different additives into the membranes. Xi et al. [16] employed SiO$_2$ nanoparticles to modify Nafion membranes by sol–gel reaction with tetraethyl-orthosilicate. The pores of Nafion membrane were filled with the SiO$_2$ nanoparticles, thus resulting in reduced permeability of vanadium ions. Polymerization is another useful method to introduce organic species. Luo and co-workers [17] synthesized a cationic charged layer on the surface of the Nafion membrane by interfacial polymerization, and Dai et al. [18,19]

designed a composite membrane with zwitterionic groups by grafting the sulfobetaine methacrylate onto Nafion membrane via polymerization. The charge repulsion effect and covalent bonds between the additives and matrix suppressed the vanadium ion penetration and enhanced ion selectivity. Solution casting is also feasible to modify the membranes with either organic or inorganic additives. Ye and co-workers [20] used the lignin to blend with Nafion by solution casting, and also [21] incorporated TiO_2 nanotubes into the Nafion membranes to achieve high-performance VRFBs. In addition, electrodeposition is also an alternative to tailor the membranes. Zeng et al. [22] modified the Nafion membranes with pyrrole by electrodeposition and revealed an improved property of membrane resistance and vanadium permeability. It is noteworthy that the chemical methods usually require harsh environments and multiple reactions with delicate conditions. To this end, a physical deposition approach is an alternative which is, however, rarely reported due to less compatibility with the organic species.

A physical deposition is a facile and cost-effective way for large-scale surface modification with high-quality thin films, which are generally used for metallic and inorganic species. In this study, a composite membrane of PTFE@Nafion is prepared by radio frequency magnetron sputtering. An ultrathin PTFE layer is deposited onto the Nafion surface and performs as a blocking layer to reduce the permeation of vanadium ions and enhance the dimensional stability of the Nafion membranes for VRFB application.

2. Experimental

The Nafion 212 membranes (DuPont, Wilmington, NC, USA) and PTFE targets (Teflon, purity 99.99%, Shanghai, China) were commercially obtained. The PTFE@Nafion composite membranes were prepared by radio frequency (RF) magnetron sputtering (JPGF-480, Beijing, China) under an argon atmosphere (purity 99.999%). The Nafion membranes were immediately put into the chamber after removing the surface protective layer. The chamber was then pumped down to a vacuum level of less than 3×10^{-3} Pa. The distance between the membrane and the target was set as 10 cm. The following major parameters were optimized to adjust the film thickness and surface roughness, i.e., the sputtering power (90, 100, 110, 120, and 130 W), the chamber pressure (0.6, 0.8, 1.0, 1.2, and 1.4 Pa), the substrate temperature (80, 90, 100, 110, and 120 °C), and the deposition time (30, 60, 90, and 120 s). The PTFE films were also sputtered onto the glass substrates (CITOGAS, REF10127105P-G) to determine the film thickness and roughness (R_a) by surface profiler (Dektak 150, Veeco Instruments, Tucson, AZ, USA). Although the film roughness on glasses might exhibit some deviations from that on Nafion membranes, it could be used to optimize the sputtering conditions.

The morphologies of composite membranes were examined by field emission scanning electron microscopy (FESEM, FEI Nova 400, Hillsboro, OR, USA). The contact angle measurement was performed with an optical tensiometer (Attension Theta, KSVMRI instruments Ltd., Helsinki, Finland) to evaluate the wetting property of the composite membranes. The static contact angle was measured with water droplet (about 5 μL) on the as-deposited composite membranes or the pristine Nafion after removing the protective layer on the surface. The water uptake (WU) was determined by the weight difference between the wet and dry membranes, as follows:

$$WU = \frac{W_{wet} - W_{dry}}{W_{dry}} \times 100\% \qquad (1)$$

where W_{wet} and W_{dry} are the weight of the wet and dry membranes, respectively. Before the water absorption test, the membranes were cut into a size of 2×2 cm^2 and dried for more than 24 h, and then the weight of dry membrane was measured. Thereafter, the membranes were soaked in deionized water for 12 h, taken out and quickly wiped with absorbent paper to remove the water on the surface, and then the weight of the wet membrane was measured. The swelling ratio (SR) was obtained by calculating the length difference between the wet and dry membranes. The swelling ratio was calculated by:

$$SR = \frac{L_{wet} - L_{dry}}{L_{dry}} \times 100\% \qquad (2)$$

where L_{wet} and L_{dry} are the length of the wet (soaked in deionized water for 12 h) and dry (the as-deposited composite membrane or the pristine Nafion membrane after removing the protective layer on surface) membranes, respectively. The permeability measurement was carried out using a home-made cell, as detailed in our previous report [23]. In brief, the two isolated compartments were filled with 70 mL of 1.5 M $VOSO_4$ in 3.0 M H_2SO_4 solution and 70 mL of 1.5 M $MgSO_4$ in 3.0 M H_2SO_4 solution, respectively, which were separated by a composite membrane. The concentration of VO^{2+} in the $MgSO_4$ compartment was examined with a UV-vis spectrometer (UV-2450PC, Shimadzu, Tokyo, Japan) every 12 h. The permeability (P) can be calculated by [12,17,18]:

$$P = \frac{VL}{A(C_0 - C_t)} \times \frac{dC_t}{dt} \qquad (3)$$

where V is the solution volume in the $MgSO_4$ compartment, L and A are the thickness and active area (2.01 cm^2) of the membrane, respectively, C_0 and C_t are the initial concentration of VO^{2+} (1.5 M) and the concentration of VO^{2+} in the $MgSO_4$ compartment as a function of time (t).

The conductivity of the membrane was examined with an electrochemical workstation (CHI660E, Chenhua, Shanghai, China) in a frequency range of 1 Hz–100 KHz. The membrane conductivity (σ) can be calculated by [24–26]:

$$\sigma = \frac{L}{A(R_1 - R_2)} \qquad (4)$$

where R_1 and R_2 are the cell resistance with and without membrane, respectively, which were determined by the intercept at the real axis in a Nyquist plot. The active area was 13.5 cm^2 for the conductivity measurement.

A VRFB single cell was assembled as per our previous report [20]. The 15 mL of 1.5 M VO^{2+} in 3 M H_2SO_4 and 15 mL of 1.5 M V^{3+} in 3 M H_2SO_4 were employed as the catholyte and anolyte, respectively. The PTFE side of a composite membrane was faced to the catholyte to confront the highly corrosive environment. The cell tests were conducted with a battery testing system (CT2001A, LANHE, Wuhan, China). A voltage window of 0.7–1.75 V was used for the cyclic test. An open circuit voltage (OCV) decay of single cell at 75% state of charge (SOC) was recorded beyond 0.85 V.

3. Results and Discussion

3.1. Characterization of Composite Membranes

In general, the polytetrafluoroethylene bears a hydrophobic and inert nature. Once formed on the Nafion surface, it behaves as a blocking layer to expel the water molecules and retard the ion permeation. However, the PTFE is also an insulator which, on the other hand, enlarges the membrane's resistance. As such, the film thickness should be optimized to reach a compromise between the permeability and conductivity. Four important parameters for the RF magnetron sputtering are considered herein, i.e., the sputtering power, chamber pressure, substrate temperature and deposition time. Figure 1 shows that the film thickness generally exhibits a linear relationship with the key parameters. The PTFE thickness increases with the power, as shown in Figure 1a). This is attributed to an enhanced energy of Ar^+ ions that generate more PTFE particles with a high deposition rate. Meanwhile, the sputtered particles also bear high energy and tend to form islands on the membrane surface, thus increasing the roughness. An abrupt enhancement arises beyond 110 W for the surface roughness. Accordingly, a sputtering power of 110 W is used for an optimal film deposition. The PTFE thickness displays a similar varying trend with the pressure, as shown in Figure 1b. The quantity of Ar$^+$ ions is promoted under high pressure, resulting in a high deposition rate. However, the average energy of each Ar$^+$ ion is reduced under this scenario because of shortened mean free path for collision. The size of PTFE particles bombarded by Ar$^+$ ions is decreased accordingly, which in turn decreases the film roughness. An optimal pressure of 1.2 Pa is hence selected based on the roughness step presented in Figure 1b. By contrast, the PTFE thickness decreases with the elevated temperature in light of an enhanced evaporation rate of the resulting films, as displayed in Figure 1c. Once reaching the membrane surface, the sputtered PTFE particles can obtain more energy at a high substrate temperature. This facilitates the translational movement of the particles, thereby giving rise to a reduced surface roughness over 110 °C. Consequently, an optimal substrate temperature of 110 °C is chosen herein. With these optimized parameters, the PTFE thickness is tailored to be about 5, 10, 20, and 30 nm by controlling the deposition time of 30, 60, 90, and 120 s, respectively, as shown in Figure 1d. Subsequently, the composite film is referred to as 30PTFE@Nafion, 60PTFE@Nafion, 90PTFE@Nafion, and 120PTFE@Nafion based on the deposition time. The surface roughness remains almost unchanged under the optimal conditions.

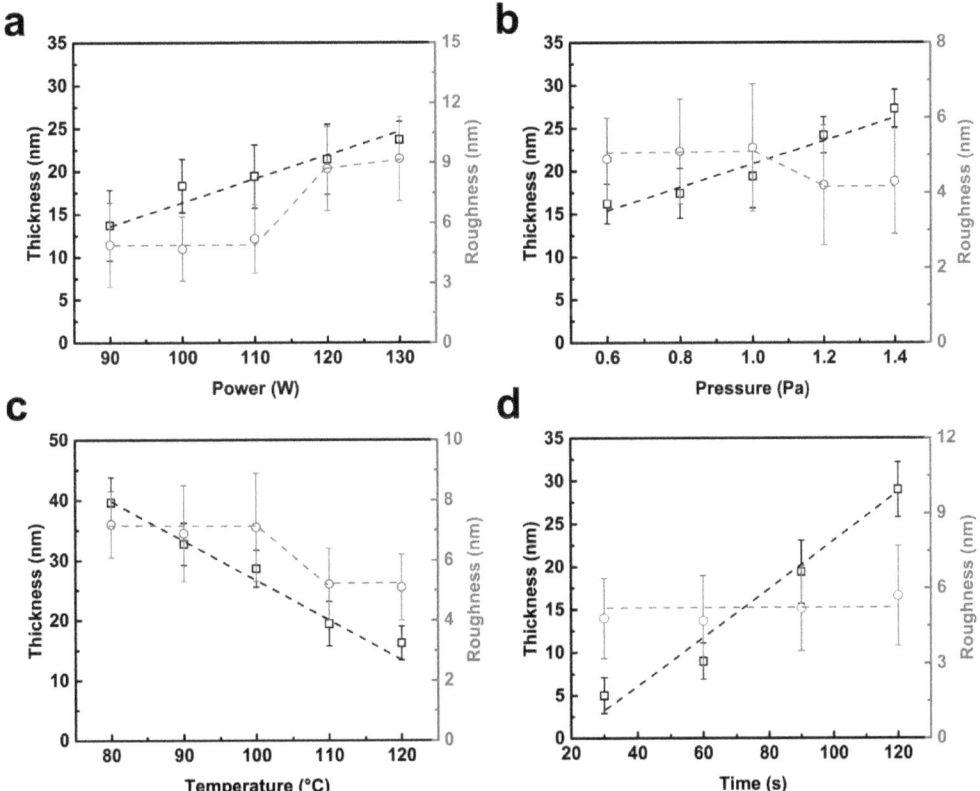

Figure 1. Optimization of sputtered PTFE films. Effect of sputtering power (**a**), chamber pressure (**b**), substrate temperature (**c**), and deposition time (**d**) on film thickness and surface roughness. Each of the data points is based on at least 5 measurements (see Figures S1–S4 in the Supplemental Information for typical surface profiles). The broken lines are used to assist view only.

Figure 2a illustrates the microstructure of a PTFE@Nafion composite membrane, where an ultrathin layer of PTFE film is deposited onto the Nafion surface. The PTFE film is expected to suppress the permeation of different vanadium ions and to maintain the transport of protons. Figure 2b displays the photographs of Nafion 212 and PTFE@Nafion membranes. All of the membranes are transparent and show less differences because of the ultrathin films. The corresponding FESEM images of the original and composite membranes reveal that a uniform and porous, thin film is developed on the Nafion substrate, as compared in Figure 2c,d. The porous feature provides the diffusion pathway for both ion permeation and water migration (proton transport) in the operation of VRFB cells, which can be tailored by varying the film thickness. Some small clusters are observed on the composite membrane surface, which stem from the aggregation of PTFE particles. Further examination discloses that the morphologies are almost the same for 60PTFE@Nafion, 90PTFE@Nafion, and 120PTFE@Nafion membranes, while that for 30PTFE@Nafion is not well developed because of the small thickness, as shown in Figure S5.

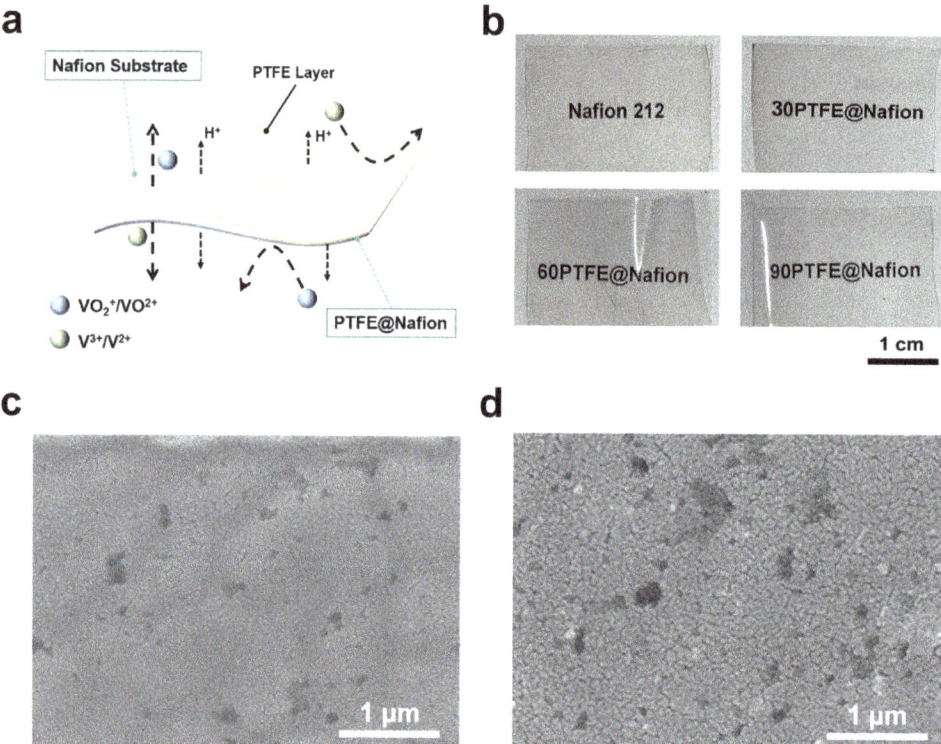

Figure 2. Morphologies of sputtered PTFE films. (**a**) Schematic illustration of a composite membrane and its influence on ion permeation, (**b**) photographs of different membranes, and top-view FESEM images of Nafion 212 (**c**) and 60PTFE@Nafion composite membrane (**d**). The dark contaminants in (**c**) and (**d**) are induced by gold sputtering to facilitate FESEM characterization.

3.2. Property of Composite Membranes

Water contact angle is used to evaluate the surface wettability of the membranes. Figure 3a shows that the contact angle increases with the PTFE thickness due to the hydrophobic nature of polytetrafluoroethylene. Such a hydrophobicity is beneficial to the water repulsion and thus enhances the dimensional stability of membranes. Figure 3b exhibits that the swelling ratio (SR) and water uptake (WU) decline gradually with the increased PTFE thickness, i.e., from 8.4% to 6.2% for SR and from 14.2% to 12.6% for WU. This indicates that the PTFE films can stabilize the Nafion membrane in view of the hydrophobicity. A low swelling ratio also facilitates the PTFE@Nafion membranes to reduce the permeation of vanadium ions and enhance the cyclic stability of flow cells. Nonetheless, the insulating character of PTFE films promotes the membrane resistance, as displayed in Figure 3c. The area resistance increases with the PTFE thickness and is determined to be 0.432, 0.837, 1.026, and 2.012 Ω cm^2 with an active area of 13.5 cm^2 for Nafion 212, 30PTFE@Nafion, 60PTFE@Nafion, and 90PTFE@Nafion membranes, respectively. (see Figure S6 for corresponding Nyquist plots and detailed interpretation).

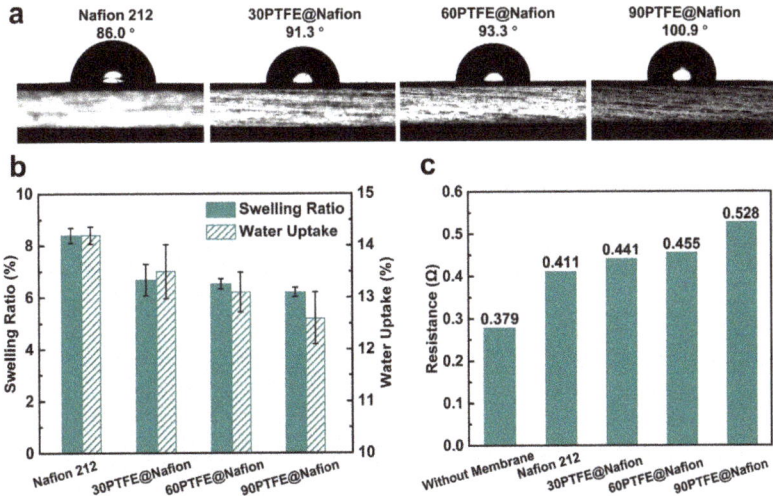

Figure 3. Surface wettability and dimensional stability of composite membranes. (**a**) Water contact angle, (**b**) swelling ratio and water uptake, and (**c**) membrane resistance.

Figure 4a displays the diffusion rate of VO^{2+} ions for different membranes. Upon coating with the PTFE films, the ion permeation is considerably suppressed with the composite membranes, and the permeation rate decreases with the increased PTFE thickness. This is due to the blocking effect of PTFE layer. The permeability thus declines accordingly, as shown in Figure 4b. On the other hand, the conductivity is also reduced with the increasing film thickness. This suggests that the proton transport is hindered with the PTFE films. Both can be ascribed to the increased diffusion pathway for ion permeation and water migration (proton transport) as the thickness of the porous film increases. In this regard, the ion selectivity (σ/P) is widely used to reach a compromise between the ion permeability and proton conductivity. A high ion selectivity is beneficial to the cell performance. The ion selectivity is hence determined to be 11.054 × 10^4, 21.109 × 10^4, 21.191 × 10^4, and 15.635 × 10^4 S min cm^{-3} for Nafion 212, 30PTFE@Nafion, 60PTFE@Nafion, and 90PTFE@Nafion, respectively. Considering the dimensional stability and ion selectivity of the composite membranes, the 60PTFE@Nafion is employed for vanadium redox flow battery.

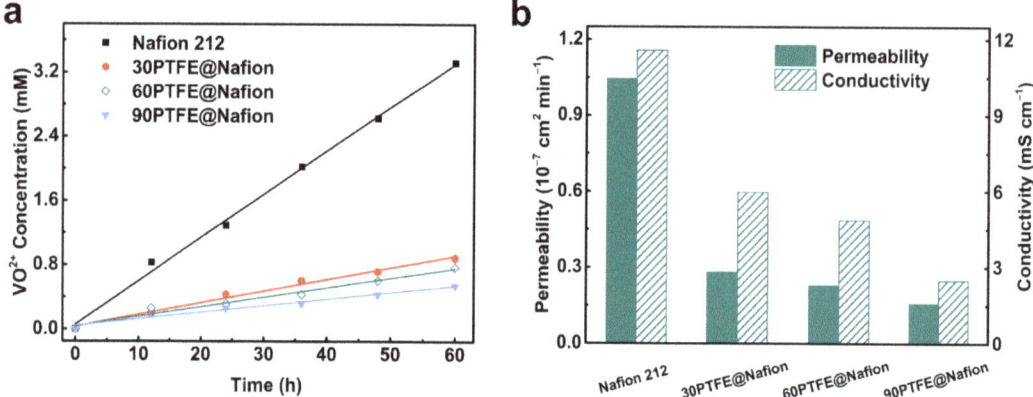

Figure 4. Ion permeation properties of composite membranes. (**a**) Time-dependent concentration of VO^{2+} ions, (**b**) permeability and conductivity.

3.3. Cell Performances

Figure 5a presents the charge–discharge curves of VRFB cells with the Nafion 212 and 60PTFE@Nafion membranes at 80 mA cm^{-2}. Although the charge capacity being a little lower, the single cell with 60PTFE@Nafion membrane exhibits comparable discharge capacity with the Nafion 212. This is attributed to the suppressed permeation of vanadium ions in the presence of PTFE films. The average charge and discharge voltage of 60PTFE@Nafion is not as good as that of the Nafion 212, in light of the enlarged resistance with PTFE layer (see Figure 3c). The OCV decay is used to further investigate the self-discharge property induced by the permeation of vanadium ions. Figure 5b discloses that the OCV decay maintains 201.2 h for the 60PTFE@Nafion membrane, being more than ten-times that for the Nafion 212 counterpart (i.e., 18.6 h). This is in good agreement with the high ion selectivity (i.e., 21.191 × 10^4 vs. 11.054 × 10^4 S min cm^{-3}). It demonstrates that the sputtered PTFE films are effective in retarding the crossover of vanadium ions.

Figure 5c shows a comparison of coulombic, voltage and energy efficiencies between the single cells with 60PTFE@Nafion and Nafion 212 membranes. The coulombic efficiency keeps about 98% for 60PTFE@Nafion membrane and is higher than of the Nafion 212 (i.e., ~94%). This is ascribed to the good ion selectivity with the sputtered PTFE films. The coulombic efficiency even performs better than most of the Nafion-based bulk membranes synthesized by chemical methods, as shown in Figure 6. However, the voltage efficiency of the cell with 60PTFE@Nafion membrane is relatively lower than the Nafion 212 (i.e., 82% vs. 87%), as a result of increased area resistance. Eventually, a comparable energy efficiency is achieved for both membranes. Nevertheless, the discharge capacity retention of the cell with 60PTFE@Nafion membrane performs better than the Nafion 212 over the entire 100 cycles, as displayed in Figure 5d. This originates from the hydrophobic and inert nature of PTFE films that give rise to high dimensional stability and ion selectivity. The major findings herein suggest that the magnetron sputtering approach provides an alternative to modify the organic membranes for VRFB application.

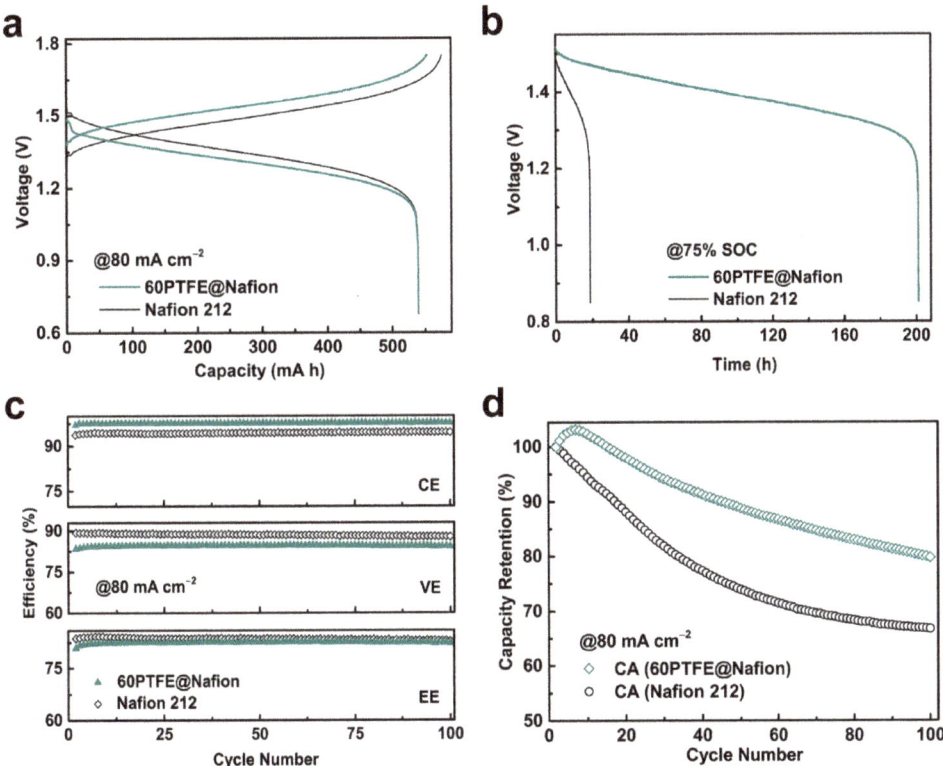

Figure 5. Cell performances employing Nafion and 60PTFE@Nafion membranes at 80 mA cm^{-2}. (**a**) Charge–discharge curves, (**b**) self-discharge curves under 75% SOC, (**c**) coulombic, voltage, and energy efficiencies, and (**d**) discharge capacity retention.

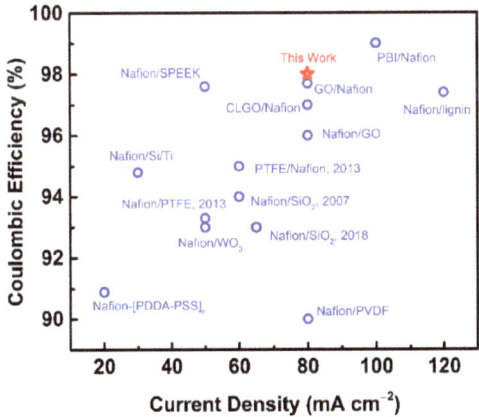

Figure 6. Comparison of coulombic efficiency for different Nafion-based membranes in VRFB at different current densities [16,20,27–38].

4. Conclusions

A PTFE@Nafion composite membrane is prepared by depositing an ultrathin PTFE film (less than 30 nm) on a Nafion substrate with radio frequency magnetron sputtering.

The composite membrane exhibits good dimensional stability and high ion selectivity, due to the hydrophobic and inert nature of the PTFE layer that suppresses the vanadium ion permeation. The 60PTFE@Nafion membrane renders a better self-discharge property than the Nafion 212 counterpart (i.e., 201.2 vs. 18.6 h) in light of higher ion selectivity (i.e., 21.191×10^4 vs. 11.054×10^4 S min cm^{-3}). The VRFB single cell with 60PTFE@Nafion membrane shows better discharge capacity retention than the Nafion 212 over the entire 100 cycles and maintains about 80% at 100 cycles. The major findings indicate that it is feasible and promising to use physical approaches to modify the organic membranes for VRFB application, as compared to the conventional chemical methods.

In addition, there are also some limitations for the method reported herein. (1) The property of the composite membranes is highly dependent on the quality of the sputtering targets. (2) The diversity of organic films is limited by the available sputtering targets. In this regard, more efforts are required to develop sputtering targets of high quality and diversity.

Supplementary Materials: The following supporting information can be downloaded at: https://www.mdpi.com/article/10.3390/coatings12030378/s1, Figure S1: Typical surface profiles of PTFE films prepared at different sputtering powers; Figure S2: Typical surface profiles of PTFE films prepared at different chamber pressures; Figure S3: Typical surface profiles of PTFE films prepared at different substrate temperatures; Figure S4: Typical surface profiles of PTFE films prepared at different deposition times; Figure S5: Top-view FESEM images of different composite membranes; Figure S6: The Nyquist plots for the cells with different membranes.

Author Contributions: Conceptualization, J.S., J.Y. and L.S.; methodology, J.S., J.Y. and Z.Q.; software, J.S.; validation, J.S., J.Y. and L.S.; formal analysis, J.S., J.Y. and L.S.; investigation, J.S. and J.Y.; resources, L.S.; data curation, J.S., J.Y. and Z.Q.; writing—original draft preparation, J.S. and L.S.; writing—review and editing, L.S.; visualization, J.S. and L.S.; supervision, L.S.; project administration, L.S.; funding acquisition, L.S. All authors have read and agreed to the published version of the manuscript.

Funding: The authors are grateful for the financial support from the National Natural Science Foundation of China (51871037), the Natural Science Foundation of Chongqing, China (cstc2021jcyj-jqX0020), the Chongqing Talents: Exceptional Young Talents Project (cstc2021ycjh-bgzxm0063, CQYC201905023), and the National Key Research and Development Program of China (2020YFF0421893).

Institutional Review Board Statement: Not applicable.

Informed Consent Statement: Not applicable.

Data Availability Statement: Data sharing is not applicable to this article.

Conflicts of Interest: The authors declare no conflict of interest.

References

1. Xiang, C.; Zhao, X.; Tan, L.; Ye, J.; Wu, S.; Zhang, S.; Sun, L. A solar tube: Efficiently converting sunlight into electricity and heat. *Nano Energy* **2019**, *55*, 269–276. [CrossRef]
2. Tan, R.; Wang, A.; Malpass-Evans, R.; Williams, R.; Zhao, E.W.; Liu, T.; Ye, C.; Zhou, X.; Darwich, B.P.; Fan, Z.; et al. Hydrophilic microporous membranes for selective ion separation and flow-battery energy storage. *Nat. Mater.* **2020**, *19*, 195–202. [CrossRef] [PubMed]
3. Chen, I.S.; Luo, T.; Moon, G.H.; Ogieglo, W.; Kang, Y.S.; Wessling, M. Untra-high proton/vanadium selectivity for hydrophobic polymer membranes with intrinsic nanopores for redox flow battery. *Adv. Energy Mater.* **2016**, *6*, 1600517.
4. Yuan, Z.; Duan, Y.; Zhang, H.; Li, X.; Zhang, H.; Vankelecom, I. Advanced porous membranes with ultra-high selectivity and stability for vanadium flow batteries. *Energy Environ. Sci.* **2016**, *9*, 441–447. [CrossRef]
5. Kim, J.Q.; So, S.; Kim, H.-T.; Choi, S.Q. Highly ordered ultrathin perfluorinated sulfonic acid Ionomer membranes for vanadium redox flow battery. *ACS Energy Lett.* **2021**, *6*, 184–192. [CrossRef]
6. Wu, C.; Lu, S.; Wang, H.; Xu, X.; Peng, S.; Tan, Q.; Xiang, Y. A novel polysulfone–polyvinylpyrrolidone membrane with superior proton-to-vanadium ion selectivity for vanadium redox flow batteries. *J. Mater. Chem. A* **2016**, *4*, 1174–1179. [CrossRef]
7. Wu, C.; Bai, H.; Lv, Y.; Lv, Z.; Xiang, Y.; Lu, S. Enhanced membrane ion selectivity by incorporating graphene oxide nanosheet for vanadium redox flow battery application. *Electrochim. Acta* **2017**, *248*, 454–461. [CrossRef]
8. Xia, L.; Long, T.; Li, W.; Zhong, F.; Ding, M.; Long, Y.; Xu, Z.; Lei, Y.; Guan, Y.; Yuan, D.; et al. Highly Stable Vanadium Redox-Flow Battery Assisted by Redox-Mediated Catalysis. *Small* **2020**, *16*, 2003321. [CrossRef] [PubMed]

9. Li, Y.; Lin, X.; Wu, L.; Jiang, C.; Hossain, M.; Xu, T. Quaternized membranes bearing zwitterionic groups for vanadium redox flow battery through a green route. *J. Membr. Sci.* **2015**, *483*, 60–69. [CrossRef]
10. Winardi, S.; Raghu, S.C.; Oo, M.O.; Yan, Q.; Wai, N.; Lim, T.M.; Skyllas-Kazacos, M. Sulfonated poly (ether ether ketone)-based proton exchange membranes for vanadium redox battery applications. *J. Membr. Sci.* **2014**, *450*, 313–322. [CrossRef]
11. Wang, N.; Yu, J.; Zhou, Z.; Fang, D.; Liu, S.; Liu, Y.-N. SPPEK/TPA composite membrane as a separator of vanadium redox flow battery. *J. Membr. Sci.* **2013**, *437*, 114–121. [CrossRef]
12. Semiz, L.; Sankir, N.D.; Sankir, M. Influence of the basic membrane properties of the disulfonated poly(arylene ether sulfone) copolymer membranes on the vanadium redox flow battery performance. *J. Membr. Sci.* **2014**, *468*, 209–215. [CrossRef]
13. Gindt, B.P.; Tang, Z.; Watkins, D.L.; Abebe, D.G.; Seo, S.; Tuli, S.; Ghassemi, H.; Zawodzinski, T.A.; Fujiwara, T. Effects of sulfonated side chains used in polysulfone based PEMs for VRFB separator. *J. Membr. Sci.* **2017**, *532*, 58–67. [CrossRef]
14. Ren, J.; Dong, Y.; Dai, J.; Hu, H.; Zhu, Y.; Teng, X. A novel chloromethylated/quaternized poly(sulfone)/poly(vinylidene fluoride) anion exchange membrane with ultra-low vanadium permeability for all vanadium redox flow battery. *J. Membr. Sci.* **2017**, *544*, 186–194. [CrossRef]
15. Pu, Y.; Huang, X.; Yang, P.; Zhou, Y.; Xuan, S.; Zhang, Y. Effect of non-sulfonated diamine monomer on branched sulfonated polyimide membrane for vanadium redox flow battery application. *Electrochim. Acta* **2017**, *241*, 50–62. [CrossRef]
16. Xi, J.; Wu, W.Z.; Qin, X.; Chen, L. Nafion/SiO$_2$ hybrid membrane for vanadium redox flow battery. *J. Power Sources* **2007**, *166*, 531–536. [CrossRef]
17. Luo, Q.; Zhang, H.; Chen, J.; Qian, P.; Zhai, Y. Modification of Nafion membrane using interfacial polymerization for vanadium redox flow battery applications. *J. Membr. Sci.* **2008**, *311*, 98–103. [CrossRef]
18. Dai, J.; Dong, Y.; Yu, C.; Liu, Y.; Teng, X. A novel Nafion-g-PSBMA membrane prepared by grafting zwitterionic SBMA onto Nafion via SI-ATRP for vanadium redox flow battery application. *J. Membr. Sci.* **2018**, *554*, 324–330. [CrossRef]
19. Dai, J.; Zhang, H.; Sui, Z.; Hu, H.; Gao, P.; Zhu, Y.; Dong, Y.; Teng, X. Study on Nafion/Nafion-g-poly (sulfobetaine methacrylate)-blended amphoteric membranes for vanadium redox flow battery. *Ionics* **2020**, *26*, 801–811. [CrossRef]
20. Ye, J.; Yuan, D.; Ding, M.; Long, Y.; Long, T.; Sun, L.; Jia, C. A cost-effective nafion/lignin composite membrane with low vanadium ion permeation for high performance vanadium redox flow battery. *J. Power Sources* **2021**, *482*, 229023. [CrossRef]
21. Ye, J.; Zhao, X.; Ma, Y.; Su, J.; Xiang, C.; Zhao, K.; Ding, M.; Jia, C.; Sun, L. Hybrid Membranes Dispersed with Superhydrophilic TiO$_2$ Nanotubes Toward Ultra-Stable and High-Performance Vanadium Redox Flow Batteries. *Adv. Energy Mater.* **2020**, *10*, 1904041. [CrossRef]
22. Zeng, J.; Jiang, C.; Wang, Y.; Chen, J.; Zhu, S.; Zhao, B.; Wang, R. Studies on polypyrrole modified nafion membrane for vanadium redox flow battery. *Electrochem. Commun.* **2008**, *10*, 372–375. [CrossRef]
23. Ye, J.; Lou, X.; Wu, C.; Wu, S.; Ding, M.; Sun, L.; Jia, C. Ion Selectivity and Stability Enhancement of SPEEK/Lignin Membrane for Vanadium Redox Flow Battery: The Degree of Sulfonation Effect. *Front. Chem.* **2018**, *6*, 549. [CrossRef] [PubMed]
24. Yang, R.; Cao, Z.; Yang, S.; Michos, I.; Xu, Z.; Dong, J. Colloidal silicalite-nafion composite ion exchange membrane for vanadium redox-flow battery. *J. Membr. Sci.* **2015**, *484*, 1–9. [CrossRef]
25. Niu, R.; Kong, L.; Zheng, L.; Wang, H.; Shi, H. Novel graphitic carbon nitride nanosheets/sulfonated poly(ether ether ketone) acid-base hybrid membrane for vanadium redox flow battery. *J. Membr. Sci.* **2017**, *525*, 220–228. [CrossRef]
26. Zheng, L.; Wang, H.; Niu, R.; Zhang, Y.; Shi, H. Sulfonated poly (ether ether ketone)/sulfonated graphene oxide hybrid membrane for vanadium redox flow battery. *Electrochim. Acta* **2018**, *282*, 437–447. [CrossRef]
27. Xi, J.; Wu, Z.; Teng, X.; Zhao, Y.; Chen, L.; Qiu, X. Self-assembled polyelectrolyte multilayer modified Nafion membrane with suppressed vanadium ion crossover for vanadium redox flow batteries. *J. Mater. Chem.* **2008**, *18*, 1232–1238. [CrossRef]
28. Teng, X.; Zhao, Y.; Xi, J.; Wu, Z.; Qiu, X.; Chen, L. Nafion/organic silica modified TiO$_2$ composite membrane for vanadium redox flow battery via in situ sol–gel reactions. *J. Membr. Sci.* **2009**, *341*, 149–154. [CrossRef]
29. Mai, Z.; Zhang, H.; Li, X.; Xiao, S.; Zhang, H. Nafion/polyvinylidene fluoride blend membranes with improved ion selectivity for vanadium redox flow battery application. *J. Power Sources* **2011**, *196*, 5737–5741. [CrossRef]
30. Teng, X.; Dai, J.; Su, J.; Zhu, Y.; Liu, H.; Song, Z. A high performance polytetrafluoroethene/Nafion composite membrane for vanadium redox flow battery application. *J. Power Sources* **2013**, *240*, 131–139. [CrossRef]
31. Teng, X.; Sun, C.; Dai, J.; Liu, H.; Su, J.; Li, F. Solution casting Nafion/polytetrafluoroethylene membrane for vanadium redox flow battery application. *Electrochim. Acta* **2013**, *88*, 725–734. [CrossRef]
32. Yu, L.; Lin, F.; Xu, L.; Xi, J. A recast Nafion/graphene oxide composite membrane for advanced vanadium redox flow batteries. *RSC Adv.* **2016**, *6*, 3756–3763. [CrossRef]
33. Ahn, S.M.; Jeong, H.Y.; Jang, J.-K.; Lee, J.Y.; So, S.; Kim, Y.J.; Hong, Y.T.; Kim, T.-H. Polybenzimidazole/Nafion hybrid membrane with improved chemical stability for vanadium redox flow battery application. *RSC Adv.* **2018**, *8*, 25304–25312. [CrossRef]
34. Su, L.; Zhang, D.; Peng, S.; Wu, X.; Luo, Y.; He, G. Orientated graphene oxide/Nafion ultra-thin layer coated composite membranes for vanadium redox flow battery. *Int. J. Hydrogen Energy* **2017**, *42*, 21806–21816. [CrossRef]
35. Luo, Q.; Zhang, H.; Chen, J.; You, D.; Sun, C.; Zhang, Y. Preparation and characterization of Nafion/SPEEK layered composite membrane and its application in vanadium redox flow battery. *J. Membr. Sci.* **2008**, *325*, 553–558. [CrossRef]
36. Zhang, D.; Wang, Q.; Peng, S.; Yan, X.; Wu, X.; He, G. An interface-strengthened cross-linked graphene oxide/Nafion212 composite membrane for vanadium flow batteries. *J. Membr. Sci.* **2019**, *587*, 117189. [CrossRef]

37. Wang, N.; Peng, S.; Lu, D.; Liu, S.; Liu, Y.-N.; Huang, K. Nafion/TiO2 hybrid membrane fabricated via hydrothermal method for vanadium redox battery. *J. Solid State Electrochem.* **2012**, *16*, 1577–1584. [CrossRef]
38. Lee, K.J.; Chu, Y.H. Preparation of the graphene oxide (GO)/Nafion composite membrane for the vanadium redox flow battery (VRB) system. *Vacuum* **2014**, *107*, 269–276. [CrossRef]

Article

A Novel Method for Calcium Carbonate Deposition in Wood That Increases Carbon Dioxide Concentration and Fire Resistance

Vicente Hernandez [1,2,*], Romina Romero [3], Sebastián Arias [2] and David Contreras [2,4,5]

1 Facultad de Ciencias Forestales, Universidad de Concepción, Victoria 631, Concepción 40730386, Chile
2 Centro de Biotecnología, Universidad de Concepción, Barrio Universitario S/N, Concepción 40730386, Chile; searias@udec.cl (S.A.); dcontrer@udec.cl (D.C.)
3 Laboratorio de Investigaciones Medioambientales de Zonas Áridas (LIMZA), Depto. Ingeniería Mecánica, Facultad de Ingeniería, Universidad de Tarapacá, Arica 100001, Chile; rominaromero@udec.cl
4 Facultad de Ciencias Químicas, Universidad de Concepción, Edmundo Larenas 129, Concepción 4070371, Chile
5 Millennium Nuclei on Catalytic Processes towards Sustainable Chemistry, Santiago 7820436, Chile
* Correspondence: vhernandezc@udec.cl; Tel.: +56-41-2661449

Abstract: In this study, a novel method for calcium carbonate deposition in wood that increases carbon dioxide concentration and fire resistance is proposed. The method promoted the mineralization of radiata pine wood microstructure with calcium carbonate by using a process consisting in the vacuum impregnation of wood with a calcium chloride aqueous solution and the subsequent sequential diffusion of gaseous ammonium and carbon dioxide. In the most favorable conditions, the method yielded a weight gain of about 20 wt.% due to mineralization, which implied the accumulation of 0.467 mmol·g^{-1} of carbon dioxide in the microstructure of wood. In addition, a weight gain of about 8% was sufficient to provide fire resistance to a level similar to that achieved by a commercially available fire-retardant treatment. The feasibility of retaining carbon dioxide directly inside the wood microstructure can be advantageous for developing wood products with enhanced environmental characteristics. This method can be a potential alternative for users seeking materials that could be effective at supporting a full sustainable development.

Keywords: carbon dioxide accumulation; wood fire-protection; calcium carbonate; in situ mineralization; wood protection

Citation: Hernandez, V.; Romero, R.; Arias, S.; Contreras, D. A Novel Method for Calcium Carbonate Deposition in Wood That Increases Carbon Dioxide Concentration and Fire Resistance. *Coatings* 2022, 12, 72. https://doi.org/10.3390/coatings12010072

Academic Editor: Giorgos Skordaris

Received: 10 December 2021
Accepted: 4 January 2022
Published: 7 January 2022

Publisher's Note: MDPI stays neutral with regard to jurisdictional claims in published maps and institutional affiliations.

Copyright: © 2022 by the authors. Licensee MDPI, Basel, Switzerland. This article is an open access article distributed under the terms and conditions of the Creative Commons Attribution (CC BY) license (https://creativecommons.org/licenses/by/4.0/).

1. Introduction

The increasing interest in wood as a building material is driven by its outstanding physical and mechanical behavior and versatility, but also by its biodegradable nature and ability to act as a carbon reservoir [1]. For these characteristics, many final users, seeking materials supporting sustainable development, have turned their attention to wood and wood products [2]. The use of wood in construction ensures that an important portion of the carbon incorporated by trees during their growth will remain sequestered in wood fibers for a long time [3]. Unfortunately, wood's ability to act as a carbon reservoir ends when it is degraded, and carbon is released to the atmosphere [4]. Among the agents of deterioration, fire stands out because of its devastating effects, which can almost wholly degrade wood cells in a very short time. Thus, fire is a significant risk in any construction containing wooden elements.

Fire retardancy in wood products is regularly achieved by superficial or integral impregnation with chemical formulations. In this sense, a wide range of very effective formulations, mostly based on the action of phosphorus, nitrogen, and boron compounds, have been developed [5]. However, nowadays many of these formulations have been questioned because of their possible toxic effects on living organisms [6–8]. For instance, the release of chlorinated and brominated dioxins and dibenzofurans during accidental fires or waste incineration is a recurrent concern [8]. Consequently, the development of new and innovative technologies for wood's fire protection, capable of meeting the current environmental demands, is an evident necessity.

Mineralization treatments can certainly be accounted as an option to protect wood against fire while reducing the risk of releasing toxic compounds during its thermal degradation. Examples of innovations in such area can be found in the development of treatments promoting an accelerated petrification of wood [9,10] and others based on the formation of a wide range of minerals with inherent fire resistance properties inside the wood cells [11,12]. Several authors have proposed calcium carbonate ($CaCO_3$) as a fire protection agent for wood products [13–16]. During the endothermic decomposition of $CaCO_3$, the release of carbon dioxide (CO_2) apparently dilutes and cools combustion gases, reducing the effectiveness of the combustion [14]. The thermal degradation of $CaCO_3$ does not generate toxic compounds and therefore is considered an environmentally friendly treatment [14,17–19]. Cone calorimetry tests on wood treated with $CaCO_3$ have shown up to a 65% decrease in heat release capacity, demonstrating the great potential of the treatment. In addition, the intrinsic properties of wood are not significantly affected, expanding its reliability in construction uses [14,17].

In this work, we synthesized $CaCO_3$ inside the microstructure of radiata pine samples using a novel process that consisted in vacuum impregnation of an aqueous solution of calcium chloride ($CaCl_2$) and subsequent sequential diffusion of gaseous ammonium (NH_3) and CO_2. The main objective of the study was to verify the feasibility of an innovative treatment to accumulate CO_2 inside wood that simultaneously could improve the response of wood against the action of fire, as the CO_2 involved in the reaction becomes part of the salt crystals of $CaCO_3$ accumulated in the wood microstructure. Specifically, we planned to verify the level of fire protection offered by the best conditions of the treatment, if any, and the amount of CO_2 accumulated inside the wood in such conditions.

2. Materials and Methods

2.1. Wood Samples

Radiata pine (*Pinus radiata* D. Don) wood pieces, acquired at a local lumber store, were used to prepare two sets of samples. The first set consisted of cubic sapwood pieces with perfectly tangential, radial, and transverse faces (20 mm × 20 mm × 20 mm). The second set consisted of 6 mm × 150 mm × 300 mm wood pieces with the tangential section on the main face and the end grain oriented along the longest axis. These samples were not screened for the presence of sapwood or heartwood. All samples were dried to anhydrous conditions in a convection oven at 50 °C, and their anhydrous mass was recorded. After that, the samples were conditioned at 65% relative humidity (R.H.) and 21 °C for two weeks, prior to the impregnation treatment.

2.2. Mineralization Treatment

$CaCO_3$ accumulation inside the wood microstructure was conducted on the 20 × 20 × 20 mm (n = 100) conditioned samples. The samples were vacuum-impregnated (−1 atm, 120 min) with $CaCl_2$ (7.5%, 15%, 30%, and 60% w/v) in a stainless-steel vessel (30 L, approx.). After that, the samples were removed and then sequentially exposed to gaseous ammonia (40 °C for 60 min, atmospheric pressure) inside a vaporing chamber and to CO_2 (99%, 60 min, 4 bars, room temperature) inside the stainless-steel vessel. The aim of using gaseous ammonia was to promote a rapid change of pH throughout the wood section, creating an alkaline environment, in which the presence of CO_2 could initiate the reaction of mineralization of $CaCl_2$ into $CaCO_3$. Similarly, the use of gaseous CO_2 was expected to improve the penetration of the mineralization treatment. After the treatment, the samples were washed overnight in distilled water at neutral pH to remove residual salts and byproducts. The whole process was consecutively repeated, producing cycles of treatment from 1 to 5. The treated samples were oven-dried at 50 °C until constant weight

(±0.002 g), and the anhydrous mass was recorded and afterwards used to calculate the weight gain of the treatment according to Equation (1):

$$\text{Weight gain (\%)} = [(m_2 - m_1)/m_1] \times 100 \qquad (1)$$

where m_1 and m_2 are the anhydrous mass of the samples before and after the mineralization treatment, respectively.

2.3. Scanning Electron Microscopy and Energy-Dispersive X-ray Spectroscopy

Scanning electron microscopy (SEM) and energy-dispersive X-ray spectroscopy (EDS) analyses were used to observe the accumulation of salt crystals inside the wood microstructure and assess their chemical composition. SEM and EDS evaluations were performed on samples that achieved the higher weight gains. Primarily, observations were made at the surface and then at 5 and 10 mm depth, using a JEOL-JSM 6380 (Tokyo, Japan) scanning electron microscope and an acceleration voltage of 20 kV.

2.4. Carbon Dioxide Accumulation Due to the Mineralization Treatment

The concentration of CO_2 accumulated inside the wood microstructure due to the mineralization treatment was indirectly determined by assessing the gain of calcium in the samples. According to the chemical reaction associated with the mineralization treatment, each molecule of calcium acquired by the wood due to the accumulation of $CaCO_3$ reacted with one molecule of CO_2 (Equation (2)). Thus, for each mol of calcium accumulated in the wood, at least one mol of CO_2 was captured due to the treatment.

$$CaCl_{2(s)} + 2\,NH_{3(g)} + H_2O_{(l)} + CO_{2(g)} \rightarrow CaCO_{3(s)} + 2NH_4Cl_{(s)} \qquad (2)$$

Calcium concentration in the samples was measured by Atomic Absorption Spectroscopy (AAS) by using 20 mm × 20 mm × 40 mm wood samples ($n = 5$). These samples were sawn in half to obtain 10 paired samples of 20 mm × 20 mm × 20 mm. Five of the paired samples remained as controls without treatment, and the other five were mineralized until reaching a 20% weight gain. After treatment, the samples were grounded in a small grinder to 200 mesh, and then the ash content was determined according to ASTM 1755-01(2020). The grounded samples were digested using a mixture of 18 mL HNO_3 and 7 mL HCl (final volume 25 mL). The samples were boiled for 2 h in covered beakers on a hot plate. Then, the digested samples were transferred and diluted to 100 mL with distilled water. The resultant products were measured and analyzed in an iCE 3000 Series Atomic Absorption Spectrometer (Thermo Fisher Scientific, Waltham, MA, USA). The initial content of calcium in the samples was subtracted from the content of calcium measured after the mineralization treatment. This information was used to calculate the mol of calcium per gram of anhydrous wood and hence the content of CO_2.

2.5. Infrared and Thermogravimetric Analyses

The Fourier transform infrared spectroscopy (FTIR) spectra of the untreated and mineralized samples, as well as of the pure $CaCO_3$ standard, were measured by direct transmittance using the KBr pellet technique. The spectra were recorded in the range of 4000–400 cm^{-1} using a Thermo-Nicolet Nexus 670 FTIR (Thermo Fischer Scientific, Waltham, MA, USA). Pretreatment was carried out by grinding the samples in a mill to 200 mesh and, subsequently, by compressing the mixture of each sample and KBr (where KBr has a ratio of 0.5 wt.%–1 wt.%).

In addition, 2.5 mg of each grinded sample was used to conduct a thermogravimetric analysis (TGA) in a TG 209F3 Tarsus from NETZSCH Instruments (Selb, Germany) thermogravimetric analyzer. These tests were conducted in an ultra-high purity nitrogen atmosphere with a heating rate of 20 $°C·min^{-1}$, from 50 to 600 °C. The weight loss relative to the temperature values was used to perform the derivative thermogravimetric analysis (DTG). All TGA tests were performed in triplicate.

2.6. Fire Resistance Test

The fire resistance of the wood samples was tested according to Chilean Standard NCh1974 (Figure 1). The samples were ignited in controlled conditions inside a 400 mm × 700 mm × 810 mm cabinet, under a constant air flux (0.2 m·s^{-1}). In the test, the samples are individually placed in the middle of the cabinet laying at 45°. Then, in a stainless-steel container, a known volume of absolute ethanol is burned directly under the samples. The test lasts until the total combustion of the fuel is achieved and the flames in the sample are completely off.

Figure 1. Fire resistance test according to the Chilean Standard NCh1974. The samples are ignited inside a 400 mm × 700 mm × 810 mm cabinet under a constant air flux (0.2 m·s^{-1}). Blurriness of the image due to the template glass cover on the front of the cabinet.

Wood samples, 6 mm × 150 mm × 300 mm (n = 15), were mineralized using the conditions that resulted in the highest accumulation of calcium carbonate, treated with the commercial fire-retardant agent AF7000 (AF7000 Fire Protection, Industrial y Comercial Ciprés Ltda, Santiago, Chile), or left without treatment to act as control samples. The fire-retardant agent was selected due to its commercial availability and chemical composition based on boric and phosphoric acid and ammonium sulfate [20]. The fire-retardant agent was applied by brushing, according to the manufacturer's instructions, on samples previously conditioned at 65% R.H. and 21 °C for 2 weeks.

Before the combustion test, all samples were oven-dried at 50 °C until constant weight (±0.002 g), and their mass recorded. After the combustion, the samples were allowed to cool down, and their mass was recorded again and used to calculate the weight loss due to combustion (Equation (3)). In addition, the maximum extension of the carbonization in each axis of the samples was recorded in mm and used to calculate the carbonization index according to Equation (4).

$$\text{Weight loss} = [(m_1 - m_2)/m_1] \times 100 \qquad (3)$$

where m_1 and m_2 are the mass of the samples before and after the combustion, respectively.

$$\text{Carbonization index} = L_1 \times L_2 \times L_3 \qquad (4)$$

where L_1, L_2, and L_3 are the maximum extension of the carbonization at the length, width, and depth of the sample, respectively.

2.7. Data Analysis

Analysis of variance (ANOVA) was used to evaluate the mineralization and fire resistance of the wood samples. The mineralization experiment was conducted according to a factorial design that tested the concentration of $CaCl_2$ and the number of cycles of treatment as factors, while the weight gain of the samples after treatment was considered as the main response. Similarly, ANOVA was used to test the effect of the fire-retardant treatment on the wood samples. Such analysis considered as responses the weight loss due to combustion and the carbonization index. In all cases, after ANOVA ($p < 0.05$), significant differences were estimated using Fisher's least significant test (l.s.d.) and expressed as error bars in the different result charts.

Statistical analyses were performed using the Software Design-Expert 11 (Stat-ease, Inc., Minneapolis, MN, USA, version 11).

3. Results

3.1. Mineralization Treatment

The results of the mineralization experiments are depicted in Figure 2. The ANOVA results, available as Supplementary Data (Figure S1), indicated a significant effect (p-value < 0.05) of $CaCl_2$ concentration, the number of treatment cycles, and the interaction of these two factors. Figure 2 shows that the increment of the treatment cycles had a significant effect only until the fourth cycle. Similarly, 30% $CaCl_2$ concentration yielded significantly higher weight gains values compared to other concentrations. Nevertheless, relevant interactions detected among factors showed that the highest weight gain, 22% in samples treated with five cycles using $CaCl_2$ at a concentration of 30%, was not significantly different than those obtained in samples treated with five cycles; 15% $CaCl_2$, and 4 cycles; 15% and 30% $CaCl_2$. With the goal to establish adequate conditions for the treatment of larger samples, four cycles and 30% $CaCl_2$ were chosen as treatment conditions to mineralize the fire-resistant test samples.

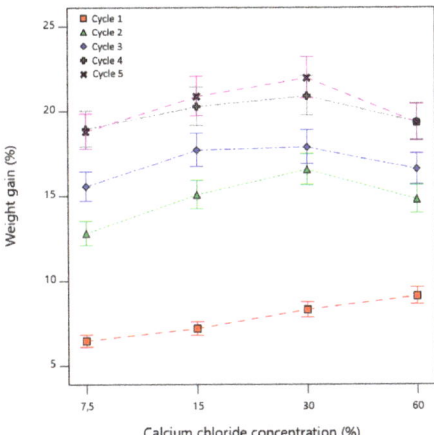

Figure 2. Weight gain of radiata pine samples mineralized for calcium carbonate accumulation. Error bars depict the least significant difference (l.s.d.) after analysis of variance ($p < 0.05$).

SEM images of samples treated with 30% $CaCl_2$, five cycles, show that the treatment induced an extended mineralization of the external area of the samples (Figure 3). Mineralization was less prevalent at 5 mm depth, forming semispherical particles, and apparently absent at 10 mm depth. EDS analysis of the samples showed that the amount of calcium present at the surface of the observed samples was 35.6% (Figure 3a). This amount decreased to 3.6% at 5 mm depth and to 1.08% at 10 mm depth (Figure 3b,c), which was almost equal to that measured in an untreated sample. The EDS analysis of the semispherical particles found in the interior and at the surface of the samples (Figure 3d) and at 5 and 10 mm depth in the treated samples was stoichiometrically sufficient to determine $CaCO_3$ presence. This clearly contrasts with the results obtained for the untreated surface, where the content of C, O, and Ca were 52.85%, 46.39%, and 0.79%, respectively.

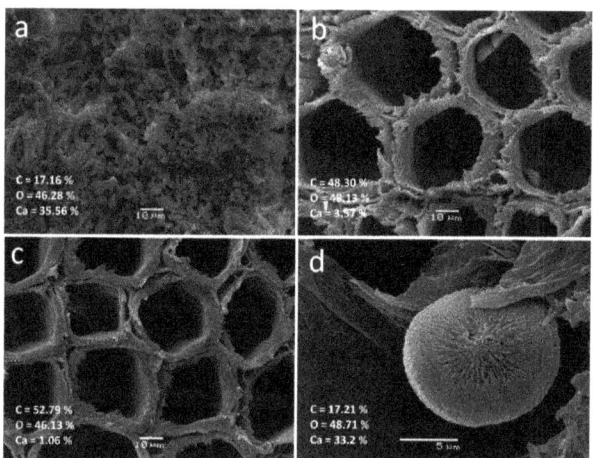

Figure 3. SEM images of a sample treated with 30% calcium chloride, five cycles. (**a**) Surface of the sample; EDS chemical determination: C = 17.16%, O = 46.28%, and Ca = 36.56%, bar size 10 µm; (**b**) 5 mm depth from the surface of the sample; EDS chemical determination: C = 48.30%, O = 48.13%, and Ca = 3.57%, bar size 10 µm; (**c**) 10 mm depth from the surface of the sample; EDS chemical determination: C = 52.79%, O = 46.13%, and Ca = 1.06%, bar size 10 µm; (**d**) semispherical particle formed due to the mineralization process; EDS chemical determination: C = 17.21%, O = 48.71%, and Ca = 33.2%, bar size 5 µm.

3.2. CO_2 Concentration

The gains of weight, calcium, and CO_2, of selected wood samples are shown in Table 1. The average carbon dioxide gain was 0.467 (mmol of $CO_2 \cdot g^{-1}$ of anhydrous wood) for samples that achieved about 20% weight gain (wt.%) due to the mineralization treatment. These samples were treated with an initial $CaCl_2$ concentration of 30% in four consecutive cycles.

Table 1. Gains of weight, calcium, and carbon dioxide in mineralized samples.

Sample	Weight Gain (%)	Calcium Gain (%)	CO_2 Gain (m·mol·g^{-1} Anhydrous Wood)
1	20.40	1.55	0.490
2	20.50	1.96	0.602
3	19.01	1.19	0.381
4	20.20	1.56	0.489
5	20.10	1.15	0.374

3.3. Infrared Analysis

The FTIR spectrum of pure $CaCO_3$ showed the presence of a strong band centered around 1453 cm^{-1}, characteristic of the C–O stretching mode of carbonate, together with a narrow band around 873 cm^{-1} characteristic of the bending mode (Figure 4B). These bands were also observed in the $CaCO_3$-treated wood sample, showing an increased absorbance in both wavenumbers (Figure 4A). Additionally, when both spectra were compared, it was observed that the treated wood sample showed differences at 1050 cm^{-1} that corresponded to C=C stretching vibrations in the aromatic structure. Interestingly, a variation was seen at 1735 cm^{-1} associated with the stretching vibrations of unconjugated C=O and related to carbonyl groups in the hemicellulose structure [21,22].

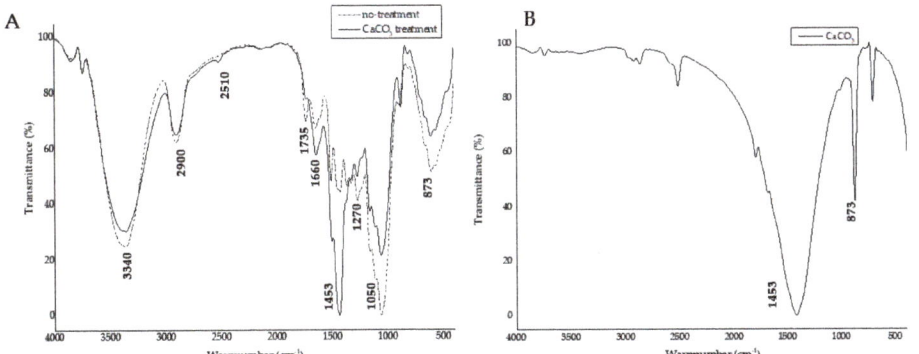

Figure 4. FTIR analysis, (**A**) untreated and treated wood samples and (**B**) CaCO$_3$ standard.

3.4. Thermogravimetric Analysis

TGA of the untreated wood samples showed a large signal at 367 °C, which implied a mass loss of almost 80% (Figure 5a). In contrast, the samples previously treated with CaCO$_3$ showed a different thermal degradation profile. It is possible to observe in Figure 5b the important increase of the signal (shoulder) near 300 °C that appeared in the untreated samples. This signal in CaCO$_3$-treated samples appears at lower temperatures (283 °C) with a DTG of −3.87%/min. The DTG curve of CaCO$_3$-treated samples presents two important decomposition stages: the first one appears as a signal in the form of a slight curve after 200 °C, and the second stage presents the highest peak of decomposition near 350 °C. In all replicates, the mineralized samples showed a maximum decomposition peak before that of untreated wood. Therefore, these TGA results suggest that, at the beginning, CaCO$_3$-treated wood lost weight faster than untreated wood; however, after 350 °C, the opposite behavior was observed, and the untreated wood lost weight faster (Figure 5a), reaching a loss of about 92% compared with a weight loss of only 64% of the CaCO$_3$-treated samples.

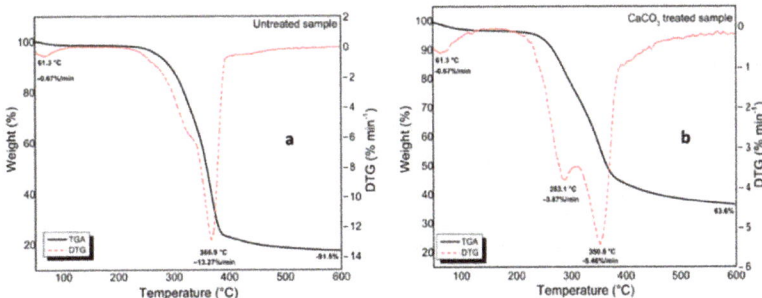

Figure 5. TGA and DTG analysis of (**a**) an untreated wood sample and (**b**) a sample treated with CaCO$_3$.

3.5. Fire Resistance of the Wood Samples

The weight gain of the mineralized samples in the fire-resistance test is shown in Table 2. The results of the fire resistance tests are summarized in Figure 6. The ANOVA results, available as Supplementary Data (Figures S2 and S3) show differences between the treatments (p-values < 0.05). Thus, a significant improvement in fire resistance was observed for the mineralized samples treated with four cycles of $CaCl_2$ 30% compared to the control samples left without treatment. However, differences between the mineralized samples and those treated with the commercial fire-retardant agent AF7000 were not significant. In terms of weight losses, the mineralized samples reached an average value that was about 20 percentage points lower than that of the samples treated with the fire-retardant agent and more than 40 percentage points lower than that of control samples. A similar trend was observed for the carbonization index, which was almost 30 percentage points lower for the mineralized samples compared to the samples treated with the retardant agent and approximately 60 percentage points lower compared to the control samples.

Table 2. Weight gain of wood samples mineralized for the fire-resistance tests.

Sample	Weight Gain (%)
1	7.86
2	9.07
3	10.47
4	7.75
5	8.67

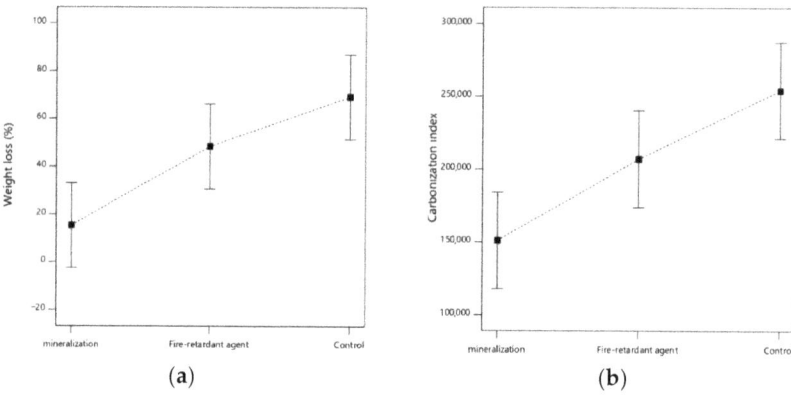

Figure 6. (a) Average weight loss and (b) carbonization index of mineralized wood samples, samples treated with the fire-retardant agent AF7000, and controls without treatment. Error bars depict the least significant difference (l.s.d.) after analysis of variance ($p < 0.05$).

The appearance of the wood samples after the fire-resistance test is shown in Figure 7. It can be observed that the mineralized samples maintained their structural integrity, which was not the case for some of the samples treated with the fire-resistant-agent and for the control samples. The integrity of the mineralized samples supports their reliability if used in a wooden structure subjected to fire, pointing out the great potential of the mineralization treatment for wood fire protection.

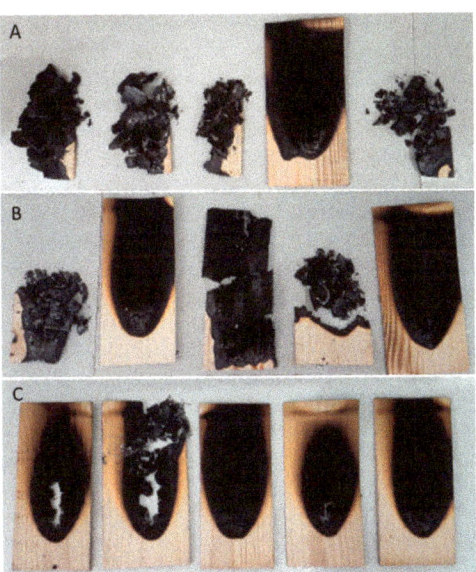

Figure 7. Wood samples after the fire-resistance test. (**A**) control samples; (**B**) samples treated with the fire-retardant agent AF7000; (**C**) $CaCO_3$ mineralized samples.

4. Discussion

The results obtained in this work support the hypothesis that a mineralization treatment based on wood impregnation with an aqueous solutions of $CaCl_2$, followed by sequential diffusion of ammonia and carbon dioxide, promotes $CaCO_3$ deposition into the wood microstructure, resulting in improved fire resistance and in the accumulation of CO_2. In the process of mineralization, the main product synthesized on the surface and inside the wood was $CaCO_3$. In the process, CO_2 reacted in presence of $CaCl_2$ in an alkaline environment (NH_3) to form $CaCO_3$. The best conditions of mineralization reached for radiata pine sapwood led to a weight gain of 20 wt.%, with an average of 0.467 (mmol·g^{-1} wood) of CO_2 captured into the wood.

A significant increment of weight due to the mineralization after cycle 4 was apparently made difficult by the saturation of the wood surfaces and penetration pathways with $CaCO_3$. These pathways may include pits, rays, and cell lumens in a transverse section. In support of this, SEM images and EDS analysis showed a compact layer of minerals on the wood surface and a scarce presence of $CaCO_3$ at 10 mm depth. The ideal process of wood mineralization requires that the reaction occurs from the center of the sample towards the surface. In most mineralization processes, this is difficult to achieve, as the precursors of the chemical reaction are all first available at the surface, and consequently, the synthesis of minerals and byproducts will inevitably saturate the surface after several cycles of replications, if not during the first. In this case, diffusion of gaseous CO_2 apparently occurs in parallel with the formation of $CaCO_3$. In such conditions, it is likely that gaseous CO_2 would follow only free passages in the microstructure of the wood, avoiding the already coated surfaces and clogged lumens. Further testing, increasing CO_2 pressure or reaction times and using methods to promote the liberation of pathways before CO_2 injection into the system (such as sonication), may be valid strategies to improve the efficiency of the treatment.

Changes detected by FTIR, specifically, the decrease in absorbance of the band 1735 cm^{-1}, suggest a chemical interaction between $CaCO_3$ impregnated in the wood and the hemicellulose matrix. Evidence of this has been attributed to the absorption of calcium cations into pectin and lignin surfaces, which are rich in metal-complexing groups and are negatively charged [14]. The TGA results showed that the mineralized samples degraded faster at the beginning of the test, but after 350 °C, this behavior reversed, and finally the untreated samples were more extensively degraded. These results are in agreement with those reported previously in the literature for different $CaCO_3$-treated samples [15,23]. It has been proposed that a fast initial degradation of $CaCO_3$-treated wood samples was due to the early degradation of $CaCO_3$, H_2O, and CO_2 [24] before reaching 100 °C (Figure 5). However, at higher temperatures (>350 °C), $CaCO_3$-treated samples become more stable to combustion due to aragonite transforming into calcite at 387 °C [25], which afterwards is decomposed into CaO and CO_2 at temperatures above 850 °C [26].

In terms of fire resistance, it is relevant to remark that the conditions of mineralization—$CaCl_2$ 30% and 4 cycles—only produced a weight gain of 8 wt.% for the fire-resistant test samples. This may be due to the presence of heartwood within large samples, which is more difficult to impregnate than sapwood. Despite this, the results showed that the mineralized samples performed to a similar level as the samples treated with a commercially available fire-retardant agent, while maintaining their structural conformation. This suggest the great potential of the treatment. The experimental results indicated that the main portion of the synthesized $CaCO_3$ was located at the surface of the samples, which is also the main area coming in contact with fire. Thus, it appears that, in terms of fire resistance, the specific concentration of $CaCO_3$ at the wood surface is more relevant than the total amount of $CaCO_3$ accumulated within the sample.

The mineralization of wood with $CaCO_3$ has been proposed by several authors to increase wood's fire resistance, but to our knowledge, a method comprising sequential gas diffusion has not been reported. Most of the reported treatments use liquid diffusion of $CaCl_2$ in combination with a number of agents, such as sodium hydroxide and supercritical carbon dioxide [23], aqueous sodium carbonate with dodecanoic acid [11], sodium bicarbonate [14,15,27], ammonium carbonate [28], sodium carbonate, alkaline hydrolysis of dimethyl carbonate [17], and calcium dimethylcarbonate in methanol [13]. In addition, a novel method was recently proposed consisting in impregnating wood with calcium acetoacetate [16]. The mineralization treatment used in this work yielded weight gains that were comparable to those obtained by using sodium bicarbonate and calcium acetoacetate as precursors [14,16], generating weight gains close to 35% for spruce samples and of about 28% and 8.3% for beech samples.

Common drawbacks of $CaCO_3$ mineralization treatments include the impossibility to achieve a deep deposition of minerals, the toxicity of certain precursors, and the complexity of the treatments. Although this was partially confirmed for the treatment proposed in this study, as certainly gaseous ammonia can be categorized as toxic, the diffusion of ammonia, in practical terms, can be achieved in a closed system in which the residual gases can be recovered and recycled. In addition, it is expected that the increment of reaction time with NH_3 and CO_2 may also increase the weight gain and the depth of mineralization.

The development of new and innovative technologies for fire protection capable of meeting the current environmental demands is a necessity. The method presented in this paper indicates that it is possible to promote the deposition of $CaCO_3$ inside the wood microstructure using a hybrid process that includes the use of liquid and gaseous reagents, resulting in the retention of CO_2 and improving wood's fire resistance. This opens the way for the development of wood products with enhanced performance and environmental characteristics that can be very attractive for many final users.

5. Conclusions

A mineralization treatment consisting in the impregnation of wood with $CaCl_2$ and subsequent sequential gaseous diffusion of NH_3 and CO_2 resulted in the accumulation of CO_2 in the form of $CaCO_3$ and in the enhancement of the fire-resistance properties of the treated wood samples. In the best conditions, the treatment yielded weight gains of about 20 wt.%, corresponding to an accumulation of 0.467 mmol·g^{-1} (anhydrous wood) of CO_2 in the wood structure. In less favorable treatment conditions, the same methodology yielded a weight gain of about 8 wt.%, which was sufficient to provide fire resistance similar to that of a commercially available fire-retardant agent. Further research, increasing CO_2 pressure and reaction times and using physical and mechanical methods to promote the liberation of circulation pathways inside wood before CO_2 injection, is advisable in order to improve the efficiency of the treatment. The application of in situ $CaCO_3$ deposition in wood can result in wood products with enhanced performance and environmental characteristics, supporting their sustainable development for wood construction and manufacture.

Supplementary Materials: The following supporting information can be downloaded at: https://www.mdpi.com/article/10.3390/coatings12010072/s1, Figure S1: Analysis of variance of $CaCO_3$ mineralized samples; Figure S2: Analysis of variance of weight losses after the fire-resistance test; Figure S3: Analysis of variance of the carbonization index after the fire-resistance test.

Author Contributions: Conceptualization, V.H., S.A., R.R. and D.C.; methodology, S.A., V.H. and R.R.; formal analysis, V.H., D.C. and R.R.; writing—original draft preparation, V.H. and R.R.; writing—review and editing, V.H., R.R. and D.C.; supervision, V.H. All authors have read and agreed to the published version of the manuscript.

Funding: This research was funded by ANID, PAI Convocatoria Nacional Subvención a Instalación en la Academia 2018, 77180054.

Institutional Review Board Statement: Not applicable.

Informed Consent Statement: Not applicable.

Data Availability Statement: Data sharing is not applicable to this article.

Acknowledgments: V.H. acknowledges the support from FONDECYT 11180030. D.C. and R.R. thank the support from FONDECYT 1201895, FONDAP Solar Energy Research Center (SERC) ANID/FONDAP/15110019 and Millennium Science Initiative, Grant NCN2021_090, Nuclei on Catalytic Processes towards Sustainable Chemistry (CSC).

Conflicts of Interest: The authors declare no conflict of interest.

References

1. Brischke, C. Timber. In *Long-Term Performance and Durability of Masonry Structures: Degradation Mechanisms, Health Monitoring and Service Life Design*; Woodhead Publishing: Sawston, UK, 2018.
2. Zubizarreta, M.; Cuadrado, J.; Orbe, A.; García, H. Modeling the environmental sustainability of timber structures: A case study. *Environ. Impact Assess. Rev.* **2019**, *78*, 106286. [CrossRef]
3. Hill, C. *Wood Modification: Chemical, Thermal and other Processes*; John Wiley & Sons: Hoboken, NJ, USA, 2007; ISBN 978-0-470-02173-6.
4. Dietenberger, M.; Hasburgh, L. Wood Products: Thermal Degradation and Fire. In *Reference Module in Materials Science and Materials Engineering*; Elsevier: Amsterdam, The Netherlands, 2016.
5. Wang, F.; Wang, Q.; Wang, X. Progress in research on fire retardant–treated wood and wood-based composites: A Chinese perspective. *For. Prod. J.* **2010**, *60*, 668–678. [CrossRef]
6. van der Veen, I.; de Boer, J. Phosphorus flame retardants: Properties, production, environmental occurrence, toxicity and analysis. *Chemosphere* **2012**, *88*, 1119–1153. [CrossRef] [PubMed]
7. Yu, G.; Bu, Q.; Cao, Z.; Du, X.; Xia, J.; Wu, M.; Huang, J. Brominated flame retardants (BFRs): A review on environmental contamination in China. *Chemosphere* **2016**, *150*, 479–490. [CrossRef] [PubMed]
8. Purser, D. Toxicity of fire retardants in relation to life safety and environmental hazards. In *Fire Retardant Materials*; Horrocks, A.R., Price, D., Eds.; Woodhead Publishing: Cambridge, UK, 2001; pp. 69–127.
9. Stein, J. Accelerated Petrifaction Process for Lignocellulose Materials. U.S. Patent 20040105938, 29 November 2002.
10. Hellberg, M.; Öhrn, A. Environmentally Friendly Wood Treatment Process. WO Patent 2012072592, 29 November 2010.
11. Wang, C.; Liu, C.; Li, J. Preparation of hydrophobic $CaCO_3$-wood composite in situ. *Adv. Mater. Res.* **2010**, *113–116*, 1712–1715.
12. Guo, H.; Luković, M.; Mendoza, M.; Schlepütz, C.M.; Griffa, M.; Xu, B.; Gaan, S.; Herrmann, H.; Burgert, I. Bioinspired Struvite Mineralization for Fire-Resistant Wood. *ACS Appl. Mater. Interfaces* **2019**, *11*, 5427–5434. [CrossRef] [PubMed]
13. Klaithong, S.; Van Opdenbosch, D.; Zollfrank, C.; Plank, J. Preparation of $CaCO_3$ and CaO replicas retaining the hierarchical structure of sprucewood. *Z. Für Nat. B* **2013**, *68*, 533–538. [CrossRef]
14. Merk, V.; Chanana, M.; Gaan, S.; Burgert, I. Mineralization of wood by calcium carbonate insertion for improved flame retardancy. *Holzforschung* **2016**, *70*, 867–876. [CrossRef]

15. Moya, R.; Gaitán-Alvarez, J.; Berrocal, A.; Araya, F. Effect of $CaCO_3$ on the wood properties of tropical hardwood species from fast-growth plantation in costa rica. *BioResources* **2020**, *15*, 4802–4822. [CrossRef]
16. Pondelak, A.; Škapin, A.S.; Knez, N.; Knez, F.; Pazlar, T. Improving the flame retardancy of wood using an eco-friendly mineralisation process. *Green Chem.* **2021**, *23*, 1130–1135. [CrossRef]
17. Merk, V.; Chanana, M.; Keplinger, T.; Gaan, S.; Burgert, I. Hybrid wood materials with improved fire retardance by bio-inspired mineralisation on the nano- and submicron level. *Green Chem.* **2015**, *17*, 1423–1428. [CrossRef]
18. Kong, H.-S.; Kim, B.-J.; Kang, K.-S. Synthesis of $CaCO_3$–SiO_2 composite using CO_2 for fire retardant. *Mater. Lett.* **2019**, *238*, 278–280. [CrossRef]
19. Tao, Y.; Li, P.; Cai, L.; Shi, S.Q. Flammability and mechanical properties of composites fabricated with $CaCO_3$-filled pine flakes and Phenol Formaldehyde resin. *Compos. Part B Eng.* **2019**, *167*, 1–6. [CrossRef]
20. Miranda, R. Composición Química Ignífuga Con Propiedades Biocidas Contra Termitas, Roedores y Hongos Que Comprende Una Solución de Sulfato de Amonio, Ácido Bórico, Una Pelicula Acuosa Formadora de Espuma y Ácido Fosfórico: Procedimiento de Obtención y Procedimiento de Aplicación de Dicha Composición. CL Patent 49107, 8 June 2004.
21. Bodirlau, R.; Teaca, C. Fourier transform infrared spectroscopy and thermal analysis of lignocellulose fillers treated with organic anhydrides. *Rom. J. Phys.* **2009**, *54*, 93–104.
22. Pucetaite, M. Archaeological Wood from the Swedish Warship Vasa Studied by Infrared Microscopy. Master's Thesis, Lund University, Lund, Sweden, 2012.
23. Tsioptsias, C.; Panayiotou, C. Thermal stability and hydrophobicity enhancement of wood through impregnation with aqueous solutions and supercritical carbon dioxide. *J. Mater. Sci.* **2011**, *46*, 5406–5411. [CrossRef]
24. Dash, S.; Kamruddin, M.; Ajikumar, P.; Tyagi, A.; Raj, B. Nanocrystalline and metastable phase formation in vacuum thermal decomposition of calcium carbonate. *Thermochim. Acta* **2000**, *363*, 129–135. [CrossRef]
25. Brown, M.; Gallagher, P. (Eds.) Handbook of Thermal Analysis and Calorimetry. In *Handbook of Thermal Analysis and Calorimetry*; Elsevier Science B.V.: Amsterdam, The Netherlands, 2003; Volume 2.
26. Singh, N.; Singh, N. Formation of CaO from thermal decomposition of calcium carbonate in the presence of carboxylic acids. *J. Therm. Anal. Calorim.* **2007**, *89*, 159–162. [CrossRef]
27. Yang, T.; Yuan, G.; Xia, M.; Mu, M.; Chen, S. Kinetic analysis of the pyrolysis of Wood/Inorganic composites under non-isothermal conditions. *Eur. J. Wood Prod.* **2021**, *79*, 273–284. [CrossRef]
28. Jia, Z.; Li, G.; Li, J.; Wang, C.; Zhang, Z. Preparation of nano-$CaCO_3$/fast-growing poplar composite by the diffusion method. *Appl. Mech. Mater.* **2012**, *184–185*, 1129–1133. [CrossRef]

Article

Formulation of Non-Fired Bricks Made from Secondary Aluminum Ash

Hongjun Ni [1,2], Weiyang Wu [1], Shuaishuai Lv [1,*], Xingxing Wang [1] and Weijia Tang [1]

1. School of Mechanical Engineering, Nantong University, Nantong 226019, China; ni.hj@ntu.edu.cn (H.N.); wuweiyang.woy@gmail.com (W.W.); 1910310003@stmail.ntu.edu.cn (X.W.); twj101680631@outlook.com (W.T.)
2. Jiangsu Province Engineering Research Center of Aluminum Dross Solid Waste Harmless Treatment and Resource Utilization, Nantong 226019, China
* Correspondence: lvshuaishuai@ntu.edu.cn

Abstract: The secondary aluminum ash is the black slag left after the primary aluminum ash is extracted from the metal aluminum. To address the environmental pollution and resource waste caused by the accumulation and landfill of aluminum ash, this study fabricated non-fired bricks by using secondary aluminum ash as the principal raw material, which was supplemented by cement, slaked lime, gypsum and engineering sand. The effects of mix proportions of various admixtures on the mechanical properties of non-fired bricks were investigated, and on this basis, the hydration mechanism was analyzed. The results showed that the mix proportions were 68.3% aluminum ash, 11.4% cement, 6.4% slaked lime, 4.2% gypsum and 9.7% engineering sand. The compressive strength of the fabricated bricks reached 22.19 MPa, and their quality indicators were in line with the MU20 requirements for *Non-fired Rubbish Gangue Bricks*. Evident hydration reaction occurred inside the non-fired bricks, with main products being calcium silicate hydrate (CSH), calcium aluminate hydrate (CAH) and ettringite (AFt). Besides, a dense structure was formed, which enhanced the brick strength.

Keywords: high nitrogen content aluminum ash; non-fired brick; mix proportion; mechanical property

1. Introduction

Aluminum ash is a kind of solid waste produced during industrial aluminum production. According to statistics, approximately 30–250 kg of aluminum ash is produced per ton of aluminum during electrolysis, melting and casting, as well as scrap recycling processes [1]. Depending on source, aluminum ash can be classified into the primary and secondary types. Primary aluminum ash refers to the waste slag produced during the electrolysis of primary aluminum or casting of aluminum, which has high elemental aluminum content and can be used as the raw material for ash frying [2,3]. Meanwhile, secondary aluminum ash refers to the hardly-disposable waste slag left after the ash frying with primary aluminum ash. Although the secondary aluminum ash contains less aluminum than the primary type, the total mass of aluminum-containing phase still exceeds 50%, which thus has great utilization value [4]. Currently, secondary aluminum ash is disposed in China mainly by accumulation and landfill, which causes environmental pollution and resource waste [5].

To date, research on the comprehensive utilization of aluminum ash is concentrated in the areas of building, refractory and other inorganic materials. Xu et al. [6] fabricated composite cement by utilizing waste aluminum ash from the production of aluminum sulfate water purifier, where the aluminum ash amounted to 8%–10%. Li et al. [7] utilized aluminum ash to fabricate a high-alumina refractory with $MgAl_2O_4$ as the main phase and $CaAl_2O_4$ as the secondary phase. Adeosun et al. [8] prepared insulating refractory bricks by incorporating aluminum ash into plastic clay and kaolin, which could withstand up to 1200 °C. Murayama et al. [9] prepared Mg-Zn, Ca-Al and Zn-Al hydrotalcites by dissolving aluminum ash (raw material) in HCl and NaOH solvents, which could serve as steelmaking deoxidizers. Zhang et al. [10] prepared sintered bricks with aluminum ash as raw materials, and obtained the best formula as follows: aluminum ash 50%, engineering soil 37.50%, coal gangue 12.50%, and the compressive strength of the sintered brick can reach 16.21 MPa. Ni and Zhang [11,12] prepared aluminum ash coating by plasma spraying. The aluminum ash coating prepared by plasma spraying has excellent properties and the strength of the coating can reach 606.54 HV. However, the existing techniques for secondary aluminum ash treatment consume only small amounts aluminum ash, which cannot thoroughly manage the long-accumulated aluminum ash or the newly produced aluminum ash every year. Hence, finding a comprehensive utilization method that enables substantial consumption of secondary aluminum ash is of profound significance to environmental and resource protection.

Since 2003, the production and use of solid clay bricks have been banned in China. Non-fired brick is an efficient substitute for solid clay brick, which contributes to protecting the environment and saving land resources. In this work, non-fired bricks were fabricated by using secondary aluminum ash from an aluminum company in Jiangsu as the principal raw material, which was supplemented by cement, slaked lime, gypsum and engineering sand. The effects of the four admixtures on the mechanical properties of the non-fired bricks were investigated, and fairly optimal mix proportions for making non-fired bricks from secondary aluminum ash were obtained. Further, the relevant hydration mechanism was analyzed. The findings of this study provide a theoretical basis for the application of aluminum ash for making non-fired bricks.

2. Materials and Methods

2.1. Raw Materials

Secondary aluminum ash was collected from Haiguang Metal Co., Ltd. in Suqian, China, which was produced after frying treatment of primary aluminum ash. Slaked lime ($Ca(OH)_2$) and gypsum ($CaSO_4 \cdot 2H_2O$) (analytical grade) were both from Zhiyuan Chemical Reagent Co., Ltd. in Tianjin, China. Engineering sand (fineness modulus: 2.6) was produced by Jiuqi Building Materials Co., Ltd. in Weifang, China and Portland cement PO 42.5 was produced by Yangchun Cement Co., Ltd. in Weifang, China.

2.2. Experimental Procedure

As a first step, the secondary aluminum ash was homogenized and sieved with a 50-mesh screen, and the aluminum ash that does not pass through the screen is regarded as impurity removal. Then, the primarily sieved aluminum ash was treated for 24 h with ultrapure water to remove nitrogen and salt, followed by thorough drying. Next, the dried aluminum ash and the admixtures were mixed uniformly according to proportions in the table, and then poured into an independently-prepared mold with a molding pressure of 15 MPa to form 40 mm × 40 mm × 160 mm specimens. Finally, the specimens were steam-cured at 80 °C for 18 h, and then sprinkle cured naturally for 3 d to obtain the experimental samples. Five samples were made in each group, and after the maximum and minimum values were removed, the average values of the remaining three data were taken as the experimental data. Figure 1 depicts the sample preparation flow.

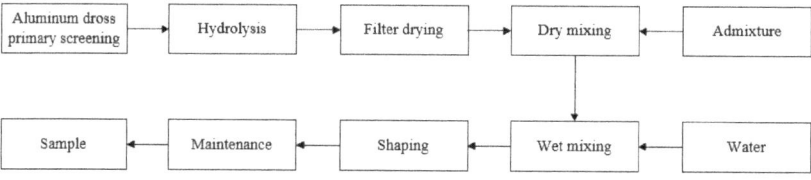

Figure 1. Sample preparation flow chart.

2.3. Characterization Method

In accordance with the Chinese standard GB/T 17671-1999 "Test Method for Cement Mortar Strength" (ISO Method) [13], the compressive strength and flexural strength of the non-fired bricks was examined with a microcomputer-controlled compressive and flexural tester. The frost resistance and water absorption index were determined according to the JC/T 422-2007 standard "Non-fired Rubbish Gangue Bricks" [14]. The leaching toxicity test was carried out according to the Chinese standard GB 5085.3-2007 "Hazardous waste Identification Standard leaching toxicity Identification" [15]. Scanning electron microscope (SEM, HITACHI S-3400N, Hitachi Corporation, Tokyo, Japan) was employed to observe the microstructures of secondary aluminum ash and non-fired brick samples, while X-ray diffractometer (XRD, Rigaku D/max2550V, Rigaku Corporation, Tokyo, Japan) was utilized to analyze their phase compositions.

3. Results and Discussion

3.1. Aluminum Ash and Its Characteristics

Table 1 details the chemical composition of aluminum ash, whereas Figures 2 and 3 display its XRD pattern and particle morphology.

Table 1. Major components of aluminum ash (wt.%).

Al_2O_3	AlN	Al	SiO_2	NaCl
44 ± 2	13 ± 2	5 ± 2	4 ± 2	2 ± 1

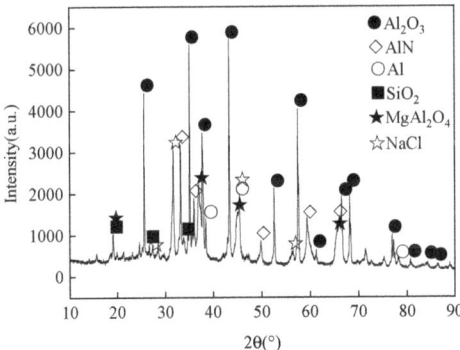

Figure 2. XRD pattern of aluminum ash.

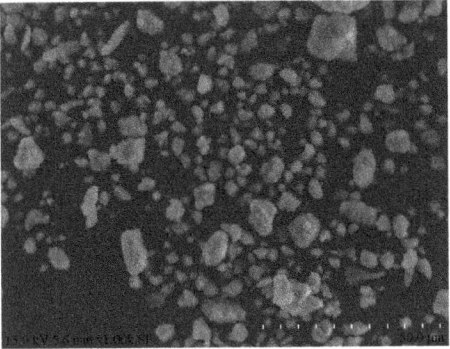

Figure 3. Morphology of aluminum ash.

As is clear from Table 1 and Figure 2, the major chemical components of secondary aluminum ash were alumina (Al_2O_3), aluminum nitride (AlN) and Aluminum (Al), among which the content of AlN was very high. This was attributed to the chemical reaction between molten aluminum and nitrogen at high temperatures during the aluminum casting, which led to substantial generation of AlN. Figure 3 displays the morphology of aluminum ash, which was mostly scattered and loose, approximately spherical, and the particles were relatively independent. Such particle morphology could facilitate the subsequent hydration reaction.

3.2. Effects of Mix Proportions on Non-Fired Bricks

3.2.1. Effect of Cement–Aluminum Ash Mix Proportion

Using aluminum ash as the adjusting component, the cement was incorporated at different ratios for experimentation. Table 2 details the specific mix proportions, and Figure 4 illustrates the experimental results.

Table 2. Cement and aluminum ash mix design. (wt.%).

No.	Cement	Slaked Lime	Gypsum	Engineering Sand	Aluminum Ash
A1	0	10	5	15	70
A2	5	10	5	15	65
A3	10	10	5	15	60
A4	15	10	5	15	55
A5	20	10	5	15	50

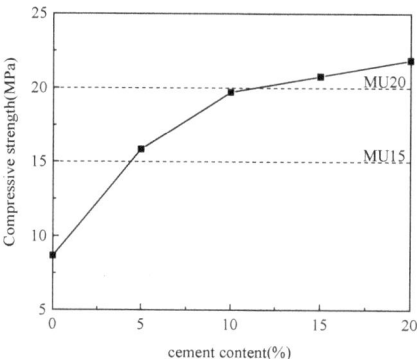

Figure 4. Effect of cement content on compressive strength.

As is clear from Figure 4, with the increase in cement content and the decrease in aluminum ash content, the compressive strength of non-fired bricks increased gradually, which peaked (21.86 MPa) at a cement content of 20%. The cement itself underwent hydration reaction to produce calcium silicate hydrate (CSH) and other products. Meanwhile, $Ca(OH)_2$ was produced during cement hydration, which further deepened the alkalinity to generate calcium aluminate hydrate (CAH), thereby enhancing the brick strength. The chemical reactions involved are formulated in (1)–(4) [16–18]. However, the amount of cement directly affected the production cost of non-fired bricks while reducing the utilization rate of aluminum ash. Accordingly, considering the effect of cement content on the compressive strength, the appropriate mix proportion between aluminum ash and cement was determined to be 60:10.

$$3CaO \cdot SiO_2 + 2H_2O \rightarrow 2CaO \cdot SiO_2 \cdot H_2O + Ca(OH)_2 \tag{1}$$

$$2CaO \cdot SiO_2 + H_2O \rightarrow 2CaO \cdot SiO_2 \cdot H_2O \tag{2}$$

$$CaO + H_2O \rightarrow Ca(OH)_2 \quad (3)$$

$$nCa(OH)_2 + Al_2O_3 \rightarrow nCaO \cdot Al_2O_3 \cdot nH_2O \quad (4)$$

3.2.2. Effect of Slaked Lime–Aluminum Ash Mix Proportion

Using aluminum ash as the adjusting component, the slaked lime was incorporated at different ratios for experimentation. Table 3 details the specific mix proportions, and Figure 5 illustrates the experimental results.

Table 3. Slaked lime and aluminum ash mix design. (wt.%).

No.	Cement	Slaked Lime	Gypsum	Engineering Sand	Aluminum Ash
B1	10	0	5	15	70
B2	10	2	5	15	68
B3	10	4	5	15	66
B4	10	6	5	15	64
B5	10	8	5	15	62
B6	10	10	5	15	60

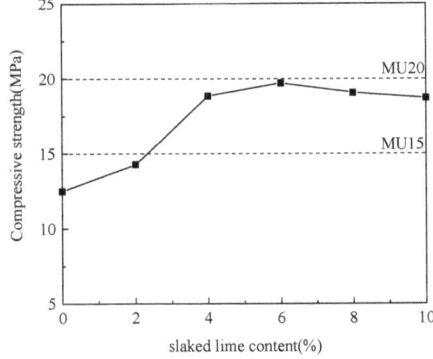

Figure 5. Effect of slaked lime content on compressive strength.

According to Figure 5, with the increase in slaked lime content and the decrease in aluminum ash content, the compressive strength of non-fired bricks increased initially and then decreased, which peaked (19.71 MPa) at a slaked lime content of 6%. The use of slaked lime as an activator enhanced the brick strength, and the reaction occurred was formulated in (4) [19]. However, given the insufficient activity of slaked lime, its further increase could not participate timely in the reaction, which was dispersed in the system and reacted with CO_2 in the air to form expansive $CaCO_3$, resulting in strength decrease. Thus, the mix proportion between aluminum ash and slaked lime was determined to be 64:6.

3.2.3. Effect of Gypsum–Aluminum Ash Mix Proportion

Using aluminum ash as the adjusting component, the gypsum was incorporated at different ratios for experimentation. Table 4 details the specific mix proportions, and Figure 6 illustrates the experimental results.

As is clear from Figure 6, with the increase in gypsum content and the decrease in aluminum ash content, the compressive strength of non-fired bricks increased initially and then decreased, which peaked (21.31 MPa) at a gypsum content of 4%. Gypsum, also as an activator, worked jointly with slaked lime to stimulate the active Al_2O_3 in the aluminum ash, thereby forming ettringite (AFt). The relevant reactions were formulated in (4) and (5) [19,20]. In the case of excessively large amount of gypsum, the calcium sulfate residue after the reaction was disorderly distributed in the system, reducing the strength of the non-fired bricks [21]. Thus, the mix proportion between aluminum ash and gypsum was determined to be 65:4.

$$3CaO \cdot Al_2O_3 \cdot H_2O + 3CaSO_4 \cdot 2H_2O + nH_2O \rightarrow 3CaO \cdot Al_2O_3 \cdot 3CaSO_4 \cdot (n+3)H_2O \quad (5)$$

Table 4. Gypsum and aluminum ash mix design. (wt.%).

No.	Cement	Slaked Lime	Gypsum	Engineering Sand	Aluminum Ash
C1	10	6	0	15	69
C2	10	6	2	15	67
C3	10	6	4	15	65
C4	10	6	6	15	63
C5	10	6	8	15	61
C6	10	6	10	15	59

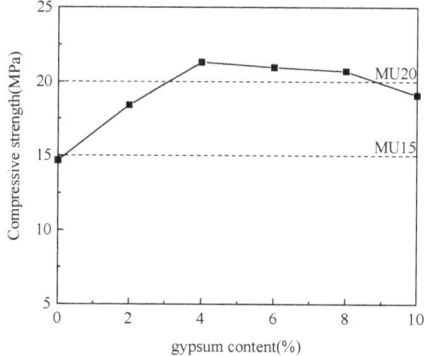

Figure 6. Effect of gypsum content on compressive strength.

3.2.4. Effect of Engineering Sand–Aluminum Ash Mix Proportion

Using aluminum ash as the adjusting component, the engineering sand was incorporated at different ratios for experimentation. Table 5 details the specific mix proportions, and Figure 7 illustrates the experimental results.

Table 5. Engineering sand and aluminum ash mix design. (wt.%).

No.	Cement	Slaked Lime	Gypsum	Engineering Sand	Aluminum Ash
D1	10	6	4	0	80
D2	10	6	4	5	75
D3	10	6	4	10	70
D4	10	6	4	15	65
D5	10	6	4	20	60

According to Figure 7, with the increase in engineering sand content and the decrease in aluminum ash content, the compressive strength of non-fired bricks increased initially and then decreased, which peaked (21.97 MPa) at an engineering sand content of 10%. With further increase in engineering sand content, the brick compressive strength tended to decline. Engineering sand did not participate in the hydration reaction, which acted as aggregates to support the framework. Non-fired bricks are essentially a kind of concrete product. Adequate addition of aggregates could reduce the brick shrinkage and effectively prevent brick cracking. Meanwhile, the hydration products were attached to aggregates with larger particle sizes, which facilitated the enhancement of brick strength [22]. Thus, the mix proportion between aluminum ash and engineering sand was determined to be 70:10.

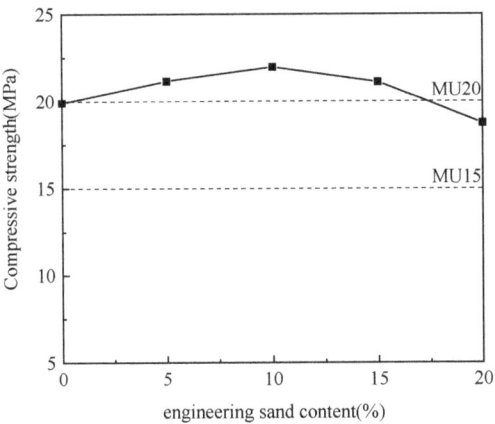

Figure 7. Effect of engineering sand content on compressive strength.

3.2.5. Validation of Optimal Mix Proportions

According to the aforedetermined mix proportions of aluminum ash and other components, the fairly optimal ratios of aluminum ash, cement, slaked lime, gypsum and engineering sand were calculated to be 68.3%, 11.4%, 6.4%, 4.2% and 9.7% (as percentages of total mass). Table 6 details the performance test results of the non-fired bricks (E1) made by the optimal mix proportions. Clearly, the compressive strength was 22.19 MPa, which was superior to the best experimental results of A5, B4, C3 and D3. The compressive strength, flexural strength and frost resistance of E1 samples conformed to the MU20 grade requirements, and the water absorption and leaching toxicity met the corresponding indicator.

3.3. Sample Texture and Structure

XRD analysis was performed on the e1 samples, which had the highest compressive strength, and the results are displayed in Figure 8. Meanwhile, SEM analysis was carried out on the E1 samples, as well as the A1 samples, which had the lowest compressive strength, and the results are depicted in Figures 9 and 10. According to Figure 8, substantial hydration products were generated in the system following reaction, mainly AFt, CSH and CAH. Besides, some products were present as Al_2O_3 in the non-fired bricks, which were further hydrated during later curing. There was also some amount of $CaCO_3$ in the samples, which was probably generated by the reaction of residual $Ca(OH)_2$ with CO_2 in the air. According to Figure 9, in the absence of cement and other cementitious materials, obvious voids or cracks were present on the sample surfaces, and the inter-particle binding was not tight. As is clear from Figure 10, the brick structure was denser, without obvious voids, where a hydraulically hardened slurry with a dense spatial structure was formed, thereby enhancing the sample strength [23].

Table 6. Performance test results of E1 samples.

No.	Test Item	Standard	Measured Value	Result
1	Compressive strength MU20/MPa	Average compressive strength ≥ 20.0	22.19	Qualified
2	Frost resistance MU20/MPa	Average compressive strength after frost ≥ 16	18.95	Qualified
3	Water absorption rate/%	≤18	7.76	Qualified
4	Flexural strength/MPa	GB/T 17671-1999 Test Method for Cement Mortar Strength (ISO Method)	4.23	Qualified
5	Leaching toxicity/mgL^{-1}	GB 5085.3-2007 Inorganic fluoride ≤ 100	11	Qualified

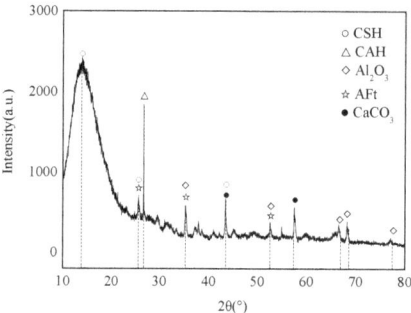

Figure 8. XRD pattern of E1 samples.

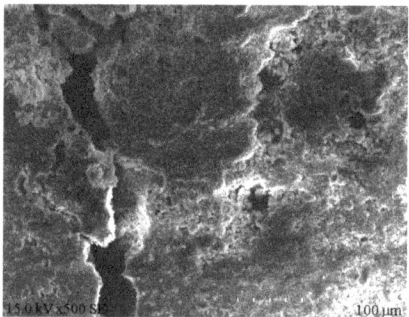

Figure 9. Microstructure of A1 samples.

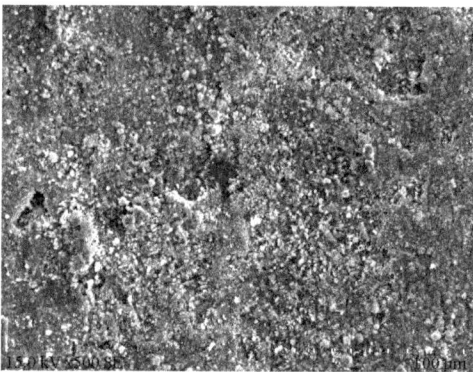

Figure 10. Microstructure of E1 samples.

4. Conclusions

(1) Preferred mix ratios for making non-fired bricks from secondary aluminum ash were 68.3% aluminum ash, 11.4% cement, 6.4% slaked lime, 4.2% gypsum and 9.7% engineering sand. The compressive strength, frost resistance and water absorption index of the resulting non-fired bricks all conform to the MU20 grade requirements in JC/T 422-2007 "Non-fired Rubbish Gangue Bricks". The findings of this study provide a theoretical basis for the application of aluminum ash for making non-fired bricks, which can effectively reduce the inventory of aluminum ash and reduce environmental pollution.

(2) Major hydration products of non-fired aluminum ash bricks are AFt, CSH and CAH, which are cemented together to form a dense interwoven and interfilled structure, thereby enhancing the brick strength.

Author Contributions: Conceptualization, H.N. and W.T.; methodology, W.W.; validation, W.W.; formal analysis, S.L. and H.N.; investigation, X.W. and S.L.; resources, H.N. and S.L.; writing—original draft, H.N.; writing—review and editing, S.L. and W.W.; visualization, H.N. and W.W.; super-vision, S.L. All authors have read and agreed to the published version of the manuscript. **Funding:** This research was Supported by the Priority Academic Program Development of Jiangsu Higher Education Institutions (PAPD), Jiangsu Province Policy Guidance Program (International Science and Technology Cooperation) Project (BZ2021045), Key R&D Projects of Jiangsu Province (BE2019060).

Institutional Review Board Statement: Not applicable.

Informed Consent Statement: Not applicable.

Data Availability Statement: Not applicable.

Conflicts of Interest: The authors declare no conflict of interest.

References

1. Zhang, N.Y.; Ning, P.; Xie, T.J.; Duan, G. research progress on recycling and comprehensive utilization of aluminum dross. *Bull. Chin. Ceram. Soc.* **2017**, *6*, 23.
2. Yang, Q.; Li, Q.; Zhang, G.F.; Shi, Q. Research and prospect of comprehensive utilization of aluminum dross. *Light Met.* **2019**, *6*, 1–5.
3. Guo, R.; Liu, X.Z.; Li, Q.D.; Yi, X.M. Present situation of high value recycling technology of aluminum ash. *Inorg. Chem. Ind.* **2017**, *49*, 12–15, 25.
4. Yang, H.; Shen, S.F.; Liu, H.Y.; Zheng, X.J.; Li, W.G.; Zhao, Q.C.; Wang, J.L.; Luo, Y.F. Study on mineralogical characteristics of secondary aluminum ash. *Nonferrous Met. Eng.* **2019**, *9*, 117–124.
5. Hong, J.P.; Wang, J.; Chen, H.Y.; Sun, B.D.; Li, J.J.; Chen, C. Process of aluminum dross recycling and life cycle assessment for Al-Si alloys and brown fused alumina. *Trans. Nonferrous Met. Soc. China* **2010**, *20*, 2155–2161. [CrossRef]
6. Xu, Z.F.; Song, W.G.; Xu, G.C. Successful practice of producing composite cement with aluminum slag. *China Cem.* **2004**, *12*, 76–77.
7. Li, A.; Zhang, H.; Yang, H. Evaluation of aluminum dross as raw material for high-alumina refractory. *Ceram. Int.* **2014**, *40*, 12585–12590. [CrossRef]
8. Adeosun, S.O.; Akpan, E.I.; Dada, M.O. Refractory characteristics of aluminum dross-kaolin composite. *JOM J. Miner. Met. Mater. Soc.* **2014**, *66*, 2253–2261. [CrossRef]
9. Murayama, N.; Maekawa, I.; Ushiro, H.; Miyoshi, T.; Shibata, J.; Valix, M. Synthesis of various layered double hydroxides using aluminum dross generated in aluminum recycling process. *Int. J. Mineral. Process.* **2012**, *110*, 46–52. [CrossRef]
10. Zhang, Y.; Ni, H.J.; Lv, S.S.; Wang, X.X.; Li, S.Y.; Zhang, J.Q. Preparation of sintered brick with aluminum dross and optimization of process parameters. *Coatings* **2021**, *11*, 1039. [CrossRef]
11. Ni, H.J.; Zhang, J.Q.; Lv, S.S.; Gu, T.; Wang, X.X. Performance optimization of original aluminum ash coating. *Coatings* **2020**, *10*, 831. [CrossRef]
12. Ni, H.J.; Zhang, J.Q.; Lv, S.S.; Wang, X.X.; Zhu, Y.; Gu, T. Preparation and performance optimization of original aluminum ash coating based on plasma spraying. *Coatings* **2019**, *9*, 770. [CrossRef]
13. GB/T 17671-1999 *Test Method for Cement Mortar Strength*, General Administration of Quality Supervision, Inspection and Quarantine of the People's Republic of China: Beijing, China, 1999.
14. JC/T 422-2007 *Non-Fired Rubbish Gangue Bricks*, General Administration of Quality Supervision, Inspection and Quarantine of the People's Republic of China: Beijing, China, 2007.
15. GB 5085.3-2007 *Hazardous Waste Identification Standard Leaching Toxicity Identification*, General Administration of Quality Supervision, Inspection and Quarantine of the People's Republic of China: Beijing, China, 2007.
16. Wang, Y.; Ni, W.; Zhang, S.Q.; Wang, Y.J.; Li, J. Research status of hydration mechanism of alumino-based cementitious system. *Metal. Mines* **2019**, *4*, 194–198.
17. Kashef-Haghighi, S.; Shao, Y.X.; Ghoshal, S. Mathematical modeling of CO_2 uptake by concrete during accelerated carbonation curing. *Cem. Concr. Res.* **2015**, *67*, 1–10. [CrossRef]
18. Tang, M.Y.; Wang, Y.C.; Wang, Z.S. Production of building materials bricks with barium containing waste residue. *Inorg. Chem. Ind.* **1994**, *2*, 34–38.
19. Hu, P.H.; Zhang, G.Z. Research on lightweight energy-saving unburned brick with large mixed fly ash. *Bull. Chin. Ceram. Soc.* **2012**, *31*, 984–987.

20. Xing, J.; Hu, J.W.; Li, C.; Qiu, J.P.; Sun, X.G. The effect of gypsum on the cementitious performance of blast furnace slag stimulated by calcium oxide. *China Min. Mag.* **2019**, *28*, 166–171.
21. Wang, D.Z.; He, Y.X. Study on effect of gypsum whisker on properties and structure of cement composite material. *Inorg. Chem. Ind.* **2015**, *47*, 65–68, 72.
22. Feng, N.Q. *New Practical Collection of Cement Concrete*; Science Press: Beijing, China, 2005; pp. 78–125.
23. Wu, H.; Lu, X.Y.; Luo, Z.J.; Yang, Y. Preparation and mechanism analysis of activated coal gangue unburned bricks. *Non-Met. Mines* **2018**, *41*, 30–33.

Article

Optical Characterization of H-Free *a*-Si Layers Grown by rf-Magnetron Sputtering by Inverse Synthesis Using Matlab: Tauc–Lorentz–Urbach Parameterization

Emilio Márquez [1,*], Juan J. Ruíz-Pérez [2], Manuel Ballester [3], Almudena P. Márquez [4], Eduardo Blanco [1], Dorian Minkov [5], Susana M. Fernández Ruano [6] and Elias Saugar [6]

1. Department of Condensed-Matter Physics, Faculty of Science, University of Cadiz, 11510 Puerto Real, Spain; eduardo.blanco@uca.es
2. Royal Institute and Observatory of the Navy, E-11100 San Fernando, Spain; jjruizperez@gmail.com
3. Department of Computer Sciences, Northwestern University, 633 Clark St, Evanston, IL 60208, USA; manuel.ballester@fau.de
4. Department of Mathematics, Faculty of Science, University of Cadiz, 11510 Puerto Real, Spain; almudena.marquez@uca.es
5. College of Energy and Electronics, Technical University of Sofia, 2140 Botevgrad, Bulgaria; d.minkov@tu-sofia.bg
6. Photovoltaic Solar Energy Unit, Energy Department, CIEMAT, Avenida Complutense 40, 28040 Madrid, Spain; susanamaria.fernandez@ciemat.es (S.M.F.R.); elias.saugar@csic.es (E.S.)
* Correspondence: emilio.marquez@uca.es

Abstract: Several, nearly-1-µm-thick, pure, unhydrogenated amorphous-silicon (*a*-Si) thin layers were grown at high rates by non-equilibrium rf-magnetron Ar-plasma sputtering (RFMS) onto room-temperature low-cost glass substrates. A new approach is employed for the optical characterization of the thin-layer samples, which is based on some new formulae for the normal-incidence transmission of such a samples and on the adoption of the inverse-synthesis method, by using a devised Matlab GUI environment. The so-far existing limiting value of the thickness-non-uniformity parameter, Δd, when optically characterizing wedge-shaped layers, has been suppressed with the introduction of the appropriate corrections in the expression of transmittance. The optical responses of the H-free RFMS-*a*-Si thin films investigated, were successfully parameterized using a single, Kramers–Krönig (KK)-consistent, Tauc–Lorentz oscillator model, with the inclusion in the model of the Urbach tail (TLUC), in the present case of non-hydrogenated *a*-Si films. We have also employed the Wemple–DiDomenico (WDD) single-oscillator model to calculate the two WDD dispersion parameters, dispersion energy, E_d, and oscillator energy, E_{so}. The amorphous-to-crystalline mass-density ratio in the expression for E_d suggested by Wemple and DiDomenico is the key factor in understanding the refractive index behavior of the *a*-Si layers under study. The value of the porosity for the specific rf-magnetron sputtering deposition conditions employed in this work, with an Ar-pressure of ~4.4 Pa, is found to be approximately 21%. Additionally, it must be concluded that the adopted TLUC parameterization is highly accurate for the evaluation of the UV/visible/NIR transmittance measurements, on the H-free *a*-Si investigated. Finally, the performed experiments are needed to have more confidence of quick and accurate optical-characterizations techniques, in order to find new applications of *a*-Si layers in optics and optoelectronics.

Keywords: amorphous semiconductors; optical properties; dielectric function; thin-film characterization; Tauc–Lorentz model; Tauc–Lorentz–Urbach model

1. Introduction

Amorphous silicon (*a*-Si) still remains at the center of attention of the modern amorphous/glassy solid-state physics community for two main reasons. First, this disordered material, in its hydrogenated and doped forms, is technologically very important. Thin-film transistors made from *a*-Si:H are commonly found in today's electronic devices, as for instance, liquid crystal at panel displays. Other relevant applications of *a*-Si include light sensors, microchips, and solar cells [1–7]. The second reason, which is of great interest to condensed matter physics theory and computer simulation scientists, is that *a*-Si is one of the simplest and archetypal systems readily available, in order to be tested new theoretical and simulation techniques, specifically developed for non-crystalline materials in general. Regarding the various non-equilibrium methods for metastable *a*-Si fabrication, it has to be mentioned, among others, ion implantation, ion-beam sputtering, dc and rf magnetron sputtering, and pulsed-laser deposition. We have specifically chosen rf magnetron sputtering in order to prepare the *a*-Si layers, because is a low-cost technique which allows us to reasonably control, up to a point, the deposition parameters.

Multi-layered, hydrogen-less *a*-Si thin films, on the other hand, has nowadays become one of the most prominent and promising candidates for Li-ion battery anodes, since it was found to be a reversible host material for Li intercalation, and, very importantly, it possesses an extremely high theoretical specific capacity of nearly 4200 mAh/g, which is certainly the highest known value among all materials in nature, so far [8]. In addition, it was recently reported by Karabacak and Demirkan [9] a simple, low-cost, and scalable technique of producing non-hydrogenated *a*-Si-thin-layer anodes, with high specific-capacity values and with high numbers of charging/discharging cycles. H-free *a*-Si thin films were deposited for such a particular purpose, with a mass-density-modulated multi-layer structure, and these *a*-Si thin layers were grown by a high/low working-gas-pressure sputtering deposition technique.

Normal-incidence optical transmittance spectrophotometry is, indeed, an physically appealing tool for accurately determining the optical properties of thin semiconducting films upon thick non-absorbing substrates, because it is relatively simple, non-destructive, and noninvasive [10]. The optical constants, n and k, are obviously relevant physical quantities, as they ultimately control the optical behavior of the thin layer [11,12]. Even though the room-temperature measurement of the normal-incidence transmission spectrum by a commercially available spectrophotometer is an easy task, highly accurate calculation of the optical and geometrical parameters n, k, and d for a thin layer, from its experimentally-measured transmission spectrum, turns out to be a very challenging problem. It has to be pointed out that there is a very abundant literature describing methods for the determination of such an optical constants of both uniform- and non-uniform-thickness thin layers. Therefore, there are various transmittance formulae being proposed, corresponding to diverse approaches to this really complex optical problem [13,14].

In this paper, a method is employed for the optical characterization of thin *a*-Si semiconductor layers onto thick transparent substrates, which is based upon newly derived formulae for the normal-incidence transmission of these kind of samples. These novel equations are not limited to the usual cases, where the real refractive index of the thin layer, n, and that of the non-absorbing substrate, s, must necessarily obey the two conditions: $n^2 \gg k^2$ and $s^2 \gg k^2$, where k is the extinction coefficient of the semiconductor film. Importantly, a non-limited value of the homogeneity parameter, Δd, will be additionally computed, as the optical characterization of *real*, amorphous semiconductor thin films is carried out. It has to be emphasized at this point that the main newness of the present approach is based upon the combination of three contributions: (i) a new formulae for the normal-incidence transmission, taking into account the existing lack of thickness uniformity; (ii) the use of the reverse or inverse synthesis approach for the calculation of the optical properties of the *a*-Si thin films; and finally (iii) the adoption of the Tauc–Lorentz-Urbach-Continuous (TLUC) optical-dispersion model [15].

We will investigate in-depth in this work, from the optical standpoint, several hydrogen-less *a*-Si thin layers, grown at high deposition rates (around 72 nm/min) and with different degrees of thickness non-uniformity, onto room-temperature inexpensive transparent glass substrates, by employing the non-equilibrium rf-magnetron Ar-plasma sputtering (RFMS) deposition technique. Besides, it has to be mentioned that a created Matlab GUI, 'Adjust-

TransIS', has allowed us to accurately compute the optical constants, n and k; the average layer thickness, $\bar{d}$; and the homogeneity parameter, Δd, of films even thicker than up to approximately 4 µm, a value of the average thickness larger than the recommendable maximum average-film-thickness limit of the alternative, and, undoubtedly, more complex optical characterization technique of variable-angle spectroscopic ellipsometry (VASE).

2. Experimental Procedure and Preliminary Structural Analysis

2.1. Preparation of H-Free a-Si Thin-Film Samples by RFMS

A set of several, nearly 1 µm thick, pure, non-hydrogenated and fully amorphous Si thin layers were deposited onto intentionally unheated low-cost 1 mm thick Corning Glass Eagle XG substrates, having a reported value of the refractive index of 1.5078 at $\lambda = 643.8$ nm. To that end, a commercially available MVSystem RFMS deposition system was employed. In this particular deposition system there is only one cathode vertically adjustable, operated by rf power of frequency 13.56 MHz. The Si (p-type) sputtering target, i.e., the source, from Kurt J. Lesker Company, has a size of 3.00-in. diameter × 0.250-in. thickness, a purity of 99.999%, a bulk electrical resistivity of 0.005–0.020 Ω cm, and a theoretical mass density for this Si material of 2.32 g/cm^3. A type-K reference thermocouple was used to carry out the substrate temperature measurement during the whole RFMS deposition process.

Before the thin-layer deposition at around room temperature was performed, the glass substrates were ultrasonically cleaned. The target-to-substrate distance was appropriately set to 6.1 cm, in order to be able to grow closely-to-uniform a-Si thin films. The working gas used in the present sputtering process was argon; its purity was higher than 99.9999%, and its flow was regulated by a MKS mass-flow controller. This Ar-gas flow in the present case was about 70 sccm, which gave place to an Ar-pressure of approximately 4.4 Pa. All the RFMS depositions were carried out with a specific rf power of 525 W, and a power density applied to the Si sputtering target of about 3.0 W/cm^2. The resulting large average growth rate was ~72 nm/min (the maximum deposition time was nearly 15 min), and such an average growth rate was calculated ex situ from the *a posteriori* optically determined values of the average a-Si layer thickness and the corresponding deposition time.

2.2. Surface, Structural and Optical Characterizations of the a-Si Films

The surface morphology, i.e., the root mean square value of the surface roughness, R_q, of the a-Si films under study was examined by atomic force microscopy (AFM microscope, Bruker Multimode Nanoscope IIIA). The example AFM image shown in Figure 1a demonstrates that RFMS-a-Si samples have a reasonably smooth, shiny, and flat surface. AFM measurements indicate that the value of R_q is approximately 1.3 nm; on the other hand, the value of R_q for the glass substrate employed was found to be around 0.60 nm.

It has also to be emphasized the surface effect, commonly described as 'broccoli', 'orange peel', or 'dry mud' cracks, observed in the top-view SEM micrograph (SEM: FEI Nova NanoSEM 450), corresponding to the representative RFMS-a-Si sample, Si#1, shown in Figure 1b. There is to date no clear explanation whatsoever of the possible mechanisms giving rise to the formation of this particular surface effect. Nevertheless, it may be suggested that it is related to the natural tendency of shrinkage of a thin coating layer, in order to diminish its surface energy, and would be caused by the tensile internal stresses in the films generated all of the them by the present RFMS-coating process.

Moreover, the refractive index, n, and the extinction coefficient, k, were determined from measurements of the normal-incidence optical transmittance spectra, exclusively. These optical spectra were collected by employing a Perkin-Elmer Lambda 1050 UV/visible/NIR, double-beam, ratio-recording spectrophotometer. The illuminated sample area was 10 mm × 1 mm, and the spectral slit width used was 2 nm; the transmission data were measured with a data interval of 4 nm. All the transmission spectra were recorded over the wavelength range from 300 up to 2500 nm. Various different spots on each a-Si sample have been measured, and we have certainly found consistency between the results from the different spot-to-spot transmission measurements. Besides, the optically calculated film thickness was systematically cross-checked by cross-sectional SEM images, and additionally corroborated for some selected a-Si thin-film samples with a Veeco Dektak 150 mechanical surface profiler; it must be emphasized that the difference found with the average layer thickness obtained from the transmission spectrum only was, for all the cases studied, less than approximately 3.0%.

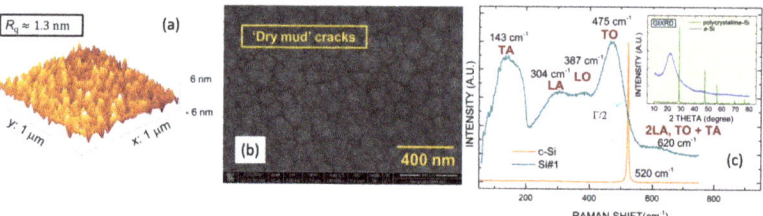

Figure 1. 3D AFM (**a**) and top-view SEM (**b**) images of a 4.4 Pa RFMS-a-Si sample (Si#1). (**c**) Room-temperature first-order Raman spectrum of the representative sample, Si#1, and also that for a poly-Si (crystallized by the use of CW-diode laser), respectively; in its inset, the GIXRD-patterns of the a-Si layer, together with that of the laser-crystallized Si.

In order to examine the atomic structure of the present sputtered samples, the first-order Raman spectra, recorded at room temperature, of the a-Si thin films investigated were excited with a 633 nm laser of 20 nW nominal power, by using a Horiba LabRAM HR Raman spectrometer. In Figure 1c, we display the Raman spectrum for the representative a-Si specimen, Si#1.

Let us start now with the interpretation of this measured Raman spectrum. Note that in a solid, a small part of an incoming photon energy can be used to excite a lattice vibration (phonon). The remaining energy escapes as a photon with slightly smaller energy compared to the incoming photon, and this energy shift is denoted as the Raman shift. In the case of a-Si solid, the momentum selection rule is relaxed, in comparison with the monocrystalline Si, and a variety of phonon modes and energies are permitted: A broad peak centered at around 480 cm^{-1} now almost dominates the whole Raman spectrum. It is observed that the Raman spectrum for the particular sample Si#1, shown in Figure 1c, exhibits a main peak at approximately 465 cm^{-1}, corresponding to the transverse-optic (TO) phonon band, and another main peak at about 152 cm^{-1}, belonging to the transverse-acoustic (TA) phonon band [16]. In addition to these two clearly dominant Raman peaks, two additional peaks associated with the longitudinal-optic (LO) phonon band, located at about 380 cm^{-1}, and with the longitudinal-acoustic (LA) phonon band, located at nearly 300 cm^{-1}, are also found. Moreover, note that the broad and small feature noticed at the high-energy side in the representative a-Si spectrum, illustrated in Figure 1c, at a value of the Raman shift of around 620 cm^{-1}, can be attributed to two plausible two-phonon excitations, such as TO + TA and 2LA, respectively [16].

The intensity of the TA-peak, I_{TA}, taken with respect to the TO-peak intensity, I_{TO}, is commonly interpreted as a degree of the extent of the intermediate range order within the atomic structure of the non-crystalline material. Thus, the larger the Raman–intensity ratio, $\gamma_{Ram}(= I_{TA}/I_{TO})$, the smaller the degree of intermediate-range order. The obtained value of the parameter γ_{Ram} for the specimen of a-Si, Si#1, is found to be 0.89; the reported value of the γ_{Ram} ratio for the sputtered (in this case, without magnetron enhancement) a-Si [17,18] is reported to be 0.80. Orapunt et al. [19], on the other hand, prepared a novel form of a-Si by ultra-high-vacuum evaporation of Si atoms onto a room-temperature quartz substrate, with 98% of the mass density of the c-Si, and with a smaller value of the γ_{Ram} ratio of 0.70. Therefore, it is consequently concluded that our RFMS-a-Si samples are the most structurally-disordered of the previously considered and differently prepared forms of a-Si solid.

We can now undertake an indirect quantification of the bond-angle dispersion or structural disorder of the material, $\Delta\theta$, using the measured Raman spectra. Experiments and theory [20] clearly show that the transverse-optical TO band broadens as $\Delta\theta$ (expressed in degrees) increases. The general relationship proposed by Beeman-Tsu [21] has the following form:

$$(\Gamma/2) = A_{\text{Raman}} + 3\Delta\theta, \tag{1}$$

with the parameter A_{Raman} verifying that $7.5\,\text{cm}^{-1} \lesssim A_{\text{Raman}} \lesssim 12\,\text{cm}^{-1}$, and where $(\Gamma/2)$ (in cm^{-1}) is defined as the half-width at the high-energy side of the TO band (see Figure 1c). Therefore, the bond angle dispersion obtained in this case is found to be 15.0 degrees $\lesssim \Delta\theta \lesssim 16.5$ degrees, which can be consider a range of large values of $\Delta\theta$, indicating again the highly-disordered atomic structure of the RFMS-a-Si material under study.

Finally, the grazing incidence X-ray diffraction (GIXRD) measurements made on our a-Si samples corroborate, beyond any doubt, that these sputtered samples are fully amorphous, and none of the characteristic sharp X-ray diffraction peaks of its c-Si counterpart being evident, at all (these two clearly distinct diffraction patterns are shown in the inset of Figure 1c). The GIXRD and EDAX measurements performed also indicate the complete absence of impurities in the present RFMS-a-Si films.

3. Thin-Film Optics Theory and Model Dielectric Function

3.1. Some Basic Theoretical Considerations

Figure 2 displays the bi-layered specimen geometry, being made up of a weakly-absorbing layer of optical constants, n and k, on top of a transparent substrate. Thin-layer a-Si materials can be grown onto a thick substrate by using several physical or chemical vapor deposition techniques [22]. The investigated a-Si film has ideally a thickness, d. The glass substrate has smooth surfaces, and it is thick enough so that all the possible Fabry–Perot (FP) interference effects related to the non-absorbing glass substrate, totally disappear. The substrate refractive index, s, is independently obtained from normal-incidence transmittance measurements on the transparent substrate alone. This bi-layered sample is usually immersed in air with index of refraction $n_{\text{air}} = 1$.

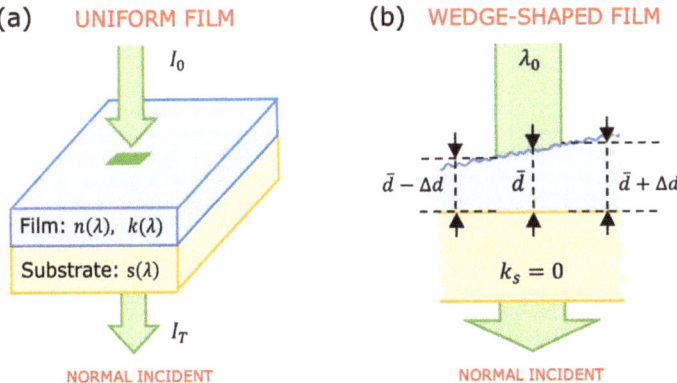

Figure 2. A schematic representation of the geometry of (**a**) a uniform, and (**b**) a nonuniform weakly-absorbing thin film, onto a thick non-absorbing glass substrate.

The complex refractive index, $n = n - ik$, of a thin layer is wavelength-dependent, or, in other words, optically dispersive. Its real part, $n(\lambda)$, is the refractive index, whereas its imaginary part, $k(\lambda)$, is the extinction coefficient, which accounts for the existing optical absorption of the semiconductor material. Furthermore, it is advisable to quantify such an absorption by the alternative absorption coefficient, $\alpha(\lambda)$, and also by the dimensionless absorbance, $x(\lambda)$, which are both interrelated to the also dimensionless extinction coefficient by the equations, $k(\lambda) = \alpha(\lambda)\lambda/(4\pi)$, and $x(\lambda) = \exp(-\alpha(\lambda)d)$, respectively.

The model dielectric function, as a function of the photon energy, for non-crystalline materials, used in this study, is based upon both the Tauc joint density of states [23], and the Lorentz electron oscillator model [12]. The optical quantity to be now considered is the complex dielectric function, $\epsilon = \epsilon_1 - i\epsilon_2$, and it should be mentioned that its dispersive real and imaginary parts are *not* independent, but, instead, they are correlated by the Kramers–Krönig (KK) equations [12]. For the present non-magnetic materials the interrelationships between the respective real and imaginary parts of this complex dielectric constant, ϵ, and the respective real and imaginary parts of its associated complex index of refraction, n, are given by

$$\epsilon_1 = n^2 - k^2,$$
$$\epsilon_2 = 2nk, \tag{2}$$

or equivalently expressed,

$$n = \sqrt{\frac{1}{2}\left(\sqrt{\epsilon_1^2 + \epsilon_2^2} + \epsilon_1\right)},$$
$$k = \sqrt{\frac{1}{2}\left(\sqrt{\epsilon_1^2 + \epsilon_2^2} - \epsilon_1\right)}. \tag{3}$$

3.2. Spectral Transmittance for a Nonuniform Semiconductor Film onto a Transparent Glass Substrate

Let us first consider the normal-incidence monochromatic light illumination on the ideal surface of a uniform semiconductor thin layer, grown onto a thick transparent glass substrate, as illustrated in Figure 2a. Taking into consideration the infinite reflections occurring at the three existing interfaces between the three constituent media of this particular optical system: air–layer, layer–substrate, and substrate–air, respectively, it is found that the spectral transmittance is given by the exact expression [14,24,25] (we present here again these already reported formulae for the sake of self-completeness of the present manuscript),

$$T_{\text{uniform}}(\lambda; n, k, s, d) = \frac{Ax}{B + x(C\sin\varphi - D\cos\varphi) + Ex^2}, \tag{4}$$

where

$$A = 16(n^2 + k^2)s,$$
$$B = \left((n+1)^2 + k^2\right)\left((n+1)(n+s^2) + k^2\right),$$
$$C = -2k\left(2(n^2 + k^2 - s^2) + (n^2 + k^2 - 1)(s^2 + 1)\right), \tag{5}$$
$$D = 2\left((n^2 + k^2 - 1)(n^2 + k^2 - s^2) - 2k^2(s^2 + 1)\right)\left((n-1)(n-s^2) + k^2\right),$$
$$E = (n-1)^2 + k^2,$$
$$\varphi = 4\pi nd/\lambda.$$

In addition, in the spectral range corresponding to the strong optical absorption within the thin film, where the FP interference pattern starts disappearing, the spectral transmittance formula basically depends only on the exponential term appearing in the numerator of such equation. Therefore, in this specific spectral range of high absorption, the previous exact formula of the transmittance is reasonably approximated as follows:

$$T_{\text{approx}}(\lambda; n, k, s, d) \approx \frac{Ax}{B} \approx \frac{16n^2 sx}{(n+1)^3(n+s^2)} \quad (6)$$

The thickness of an idealized perfect layer is certainly constant, but in a real specimen, this is rarely the situation, and, thus, significant thickness nonuniformity and/or surface roughness are often present in these real thin-film samples. The necessary simplifying assumption to be made in order to model the geometry of a real nonuniform thin layer will be to suppose such a layer, having a wedge-shaped profile as that depicted in Figure 2b. In order to be able to determine this film-thickness variation through the light spot, a homogeneity or wedging parameter, Δd, is introduced, and also its associated average film thickness, $\bar{d}$, in such a way that the actual layer thickness in the illuminated area of the sample increases linearly from a minimum value of $d = \bar{d} - \Delta d$, up to a maximum value of $d = \bar{d} + \Delta d$ (see Figure 2b).

Under this proposed working hypothesis of a linear dependence of the variable layer thickness through the spectrophotometer light spot, a much more realistic and accurate expression for the transmission, accounting for its varying thickness, will be derived by integrating upon the two optical variables which depend upon the variable thickness, that is, the optical phase, φ, and the absorbance, x, both previously introduced in Equation (5). Nevertheless, the influence of the variable thickness upon the absorbance is clearly negligible, as compared with the much stronger influence of the variable thickness upon the phase. Taking into account this suggested simplification, the integral for the transmittance will then be written as follows:

$$T_{\Delta d}^{\text{non-uniform}}(\lambda; n, k, s, \bar{d}, \Delta d) =$$

$$\int_{\varphi_1}^{\varphi_2} \frac{d\varphi}{\varphi_2 - \varphi_1} \int_{x_1}^{x_2} \frac{dx}{x_2 - x_1} \left[\frac{Ax}{B + (C \sin \varphi - D \cos \varphi)x + Ex^2} \right] \quad (7)$$

$$\approx \frac{1}{\varphi_2 - \varphi_1} \int_{\varphi_1}^{\varphi_2} \left[\frac{Ax}{B + (C \sin \varphi - D \cos \varphi)x + Ex^2} \right] d\varphi,$$

with

$$\begin{aligned} x_1 &= \exp(-\alpha(\bar{d} - \Delta d)), \\ x_2 &= \exp(-\alpha(\bar{d} + \Delta d)), \\ \varphi_1 &= 4\pi n(\bar{d} - \Delta d)/\lambda, \\ \varphi_2 &= 4\pi n(\bar{d} + \Delta d)/\lambda. \end{aligned} \quad (8)$$

Note that the *analytical* integration of Equation (7) gives rise to inverse hyperbolic functions; it is indeed a complex expression, derived in the present study by employing the useful tool of Mathematica software system (version 12.3), and it is now expressed after doing various algebraic manipulations [25],

$$T_{\Delta d}^{\text{hyper}}(\lambda; n, k, s, \bar{d}, \Delta d) = -\frac{2Ax}{(\varphi_2 - \varphi_1)\sqrt{-H}} \left[\tanh^{-1}\left(\frac{G}{\sqrt{-H}}\right) - \tanh^{-1}\left(\frac{F}{\sqrt{-H}}\right) \right], \quad (9)$$

where

$$\begin{aligned} F &= x(C + B(D + Ex)) \tan(\varphi_1/2), \\ G &= x(C + B(D + Ex)) \tan(\varphi_2/2), \\ H &= B^2 - x^2 \left(C^2 + D^2 - 2BE - E^2 x^2\right). \end{aligned} \quad (10)$$

It is convenient next to express the spectral transmittance for the layer-on-substrate specimen, using angle or circular functions, instead of the previously employed hyperbolic functions, for reasons that later will be absolutely evident. Again, after some additional algebraic manipulations, a more compacted expression for the transmittance is derived,

$$T_{\Delta d}^{uncorrected}(\lambda; n, k, s, \bar{d}, \Delta d) = \frac{2Ax}{(\varphi_2 - \varphi_1)\sqrt{H}} \left[\tan^{-1}\left(\frac{G}{\sqrt{H}}\right) - \tan^{-1}\left(\frac{F}{\sqrt{H}}\right) \right]. \quad (11)$$

It has to be stressed that Equation (11), unfortunately, cannot be yet used in the optical characterization of real semiconducting thin layers, by only using the transmission spectrum. The existence of a multi-valued inverse goniometric function in Equation (11) is the cause of the presence of discontinuities around the minima of the transmission spectrum if the necessary correcting angles, multiples of π radians, are not correctly introduced. The existence of such discontinuities in the transmittance expression makes Equation (11) certainly useless for our final purpose.

All of these invalidating effects corresponding to the expression of the transmission are clearly evidenced in Figure 3, where such a previous mathematical expression is graphed for a hypothesized hydrogenated a-Si thin film, whose homogeneity parameter Δd has the proposed exact value of $\Delta d = 30$ nm, with a typical average thickness, $\bar{d} = 1500$ nm, a plotted wavelength range $\lambda = 500$–900 nm, and, finally, with the optical constants given by the two following semiempirical dispersion relationships [24], with the wavelength in nm,

$$n_{a\text{-Si:H}}(\lambda) = 2.60 + \frac{3.0 \times 10^5}{\lambda^2},$$
$$k_{a\text{-Si:H}}(\lambda) = \frac{\lambda}{4\pi} 10^{(1.5 \times 10^6/\lambda^2) - 8}. \quad (12)$$

The refractive index of the non-absorbing glass substrate is assumed to be for this simulation $s = 1.51 = $ constant. It is further noticed in Figure 3 that the uncorrected transmittance curve of the gedanken film completely matches the corresponding numerical integration of Equation (11), except around the minima of the transmission curve, as was already referred to above.

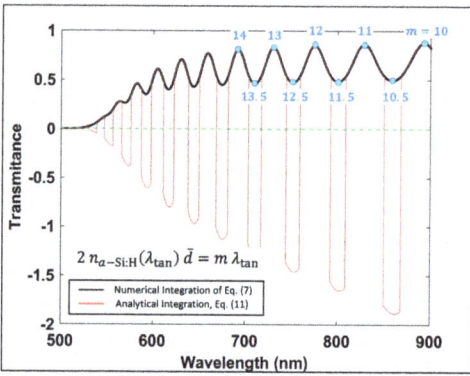

Figure 3. Numerical integration of Equation (7), for the spectral transmittance corresponding to a specific thickness variation of $\Delta d = 30$ nm (see the text for the rest of the details shown in the figure).

3.3. Optical Phase Change: Correcting Integer Numbers $N_{c,1}$ and $N_{c,2}$

The two parameters F and G in Equation (10) do contain circular functions, whose two respective arguments φ_1 and φ_2 are expressed by Equation (8). Depending on the values of n, k, $\bar{d}$, Δd, and λ, the corresponding argument of the angle functions could exceed the critical value of 2π radians. Taking into account the fact that those two parameters F and G, are within inverse goniometric functions, it is, thereby, necessary to properly evaluate the exact number of multiples of 2π radians summed up to the corresponding optical phase, φ, within the thin layer. This newly proposed key step will, indeed, suppress the above-mentioned completely invalidating discontinuities in the transmission spectrum.

In order to account for the strong influence of the layer thickness nonuniformity upon the spectral transmittance, we now suggest the introduction of two absolutely necessary correcting integer numbers, $N_{c,1}$ and $N_{c,2}$, respectively, into Equation (11), in order to be able to properly evaluate the optical-phase change taking place within the thin film. Therefore, the *corrected* equation of the spectral transmittance for a real, non-uniform layer is finally derived:

$$T^{corrected}_{\Delta d}(\lambda; n, k, s, \bar{d}, \Delta d) = \frac{2Ax\left[\left(\tan^{-1}\left(\frac{G}{\sqrt{H}}\right)+N_{c,2}\pi\right)-\left(\tan^{-1}\left(\frac{F}{\sqrt{H}}\right)+N_{c,1}\pi\right)\right]}{(\varphi_2-\varphi_1)\sqrt{H}}, \quad (13)$$

where the two new correcting integer numbers, $N_{c,1}$ and $N_{c,2}$, respectively, are now given by the two following expressions:

$$N_{c,1} = \text{round}(\varphi_1/2\pi),$$
$$N_{c,2} = \text{round}(\varphi_2/2\pi). \quad (14)$$

This defined function 'round' does round off the argument to its closest integer value, and it is totally equivalent to the function with the same name, implemented in the mathematical software system Matlab (R2021a), which will be largely used later. The novel corrected Equation (13) presented in this work, very importantly, is a continuous function that can be used in order to optically characterize an ample variety of thin-film amorphous semiconductors, with high accuracy. Moreover, the so-far existing limiting maximum value for the wedging parameter, $\Delta d(=\lambda/4n)$, when optically characterizing strongly wedge-shaped layers, has been fortunately successfully suppressed with the introduction of the appropriate correcting numbers, $N_{c,1}$ and $N_{c,2}$, in the expression of the corrected spectral transmittance, $T^{corrected}_{\Delta d}$.

It should be added that the physical importance of Equation (13) is based upon the fact that the whole normal-incidence transmittance spectra of wedge-shaped semiconducting thin films, can be evaluated by using the 'inverse-synthesis method' proposed by Dobrowolski et al. [26], rather than the dispersion model-free characterization method, based exclusively upon the use of the upper and lower envelopes of the spectral transmittance, making use of only a portion of the transmission spectrum. In addition, the obliged exclusion for optical characterization purposes of nonuniform layers having a value of $\Delta d > \lambda/4n$ [27], is now, with the presented novel formulae, fully eliminated.

3.4. Determination of the Optical Constants by Applying the TLUC Dispersion Model

The glass substrate onto which the *a*-Si thin layer was deposited was previously checked for transparency in the complete wavelength region analyzed and, therefore, completely characterized from the optical standpoint by only its real refractive index, s (knowing that $k_s \approx 0$). This optical parameter has been obtained from independent normal-incidence transmission measurements on the bare transparent glass substrate, and by making use of the well-known expression for the transmittance of a non-absorbing slab, T_s [24]:

$$T_s(\lambda) = \frac{2s}{s^2+1}, \quad (15)$$

So, by solving for s we get

$$s(\lambda) = \frac{1}{T_s} + \sqrt{\frac{1}{T_s^2} - 1}. \quad (16)$$

The KK-consistent optical dispersion model that is adopted in this study is the Tauc–Lorentz–Urbach-Continuous parameterization, as suggested by Foldyna et al. [28]. This TLUC dispersion model was found to be appropriate for amorphous semiconductors in general, and is a realistic generalization of the very popular Tauc–Lorentz (TL) parametrization proposed by Jellison and Modine [29,30], with the certainly existing exponential Urbach band tail being added to the TL model. Thus, the more complete TLUC model employs six free-adjusting parameters: A_{mp}, E_0, C_{br}, E_g, E_c, and $\epsilon_1(\infty)$; that is, interestingly just only one more parameter than in the case of the TL model, specifically that one associated to the Urbach absorption tail spectral region, E_c.

The imaginary part of the complex dielectric constant, $\epsilon_2(E)$, in this TLUC parameterization is now given as a piecewise function:

$$\epsilon_2(E) = \frac{1}{E}\frac{A_{mp}E_0 C_{br}(E-E_g)^2}{(E^2-E_0^2)^2+C_{br}^2 E^2}, \quad E \geq E_c,$$

$$\epsilon_2(E) = \frac{A_u}{E}\exp\left(\frac{E}{E_u}\right), \quad 0 \leq E < E_c, \tag{17}$$

where the first expression of $\epsilon_2(E)$, for values of $E \geq E_c$, is identical to that corresponding to the TL parameterization, and the second expression of $\epsilon_2(E)$, for $0 \leq E < E_c$, indicates the contribution of the exponential Urbach band tail. In particular, the four free-fitting parameters E_g, A_{mp}, E_0, and C_{br} stand for the optical band-gap energy, the Lorentz-oscillator amplitude, the Lorentz-oscillator peak energy, and the broadening parameter or single-oscillator width, respectively.

Furthermore, the two additional TLUC model parameters A_u and E_u, the so-called Urbach amplitude and energy, respectively, have been added in order to guarantee the continuity of the imaginary part of the complex model dielectric function, and also its first derivative, and for that particular reason the model has been labeled 'continuous'. These two Urbach tail parameters are expressed this way,

$$E_u = \frac{E_c - E_g}{2 - 2E_c(E_c - E_g)\frac{C_{br}^2 + 2(E_c^2 - E_0^2)}{C_{br}^2 E_c^2 + (E_c^2 - E_0^2)^2}},$$

$$A_u = \exp\left(-\frac{E_c}{E_u}\right)\frac{A_{mp}E_0 C_{br}(E_c - E_g)^2}{(E_c^2 - E_0^2)^2 + C_{br}^2 E_c^2}. \tag{18}$$

The real part of the complex dielectric function, $\epsilon_1(E)$, is now derived by employing the analytical integration associated with the fundamental KK relations between $\epsilon_1(E)$ and $\epsilon_2(E)$. Thus, we can write next this real part of the complex dielectric constant as follows:

$$\epsilon_1(E) = \epsilon_1(\infty) + \frac{2}{\pi}\mathcal{P}\int_0^\infty \frac{\xi\,\epsilon_2(E)}{\xi^2 - E^2}d\xi, \tag{19}$$

where $\mathcal{P}$ denotes the Cauchy Principal value of the integral. In this fashion, we find

$$\epsilon_1(E) = \epsilon_1(\infty) + \epsilon_{1,TL}(E) + \epsilon_{1,UT}(E), \tag{20}$$

where the contribution of the TL model to the TLUC model dielectric function, for $E_c \leq E < \infty$, is expressed by the following equation, given here again for the sake of self-completeness of the present paper:

$$\epsilon_{1,TL}(E) = -A_{mp}E_0 C_{br}\frac{E^2+E_g^2}{\pi\sigma_4 E}\ln\left(\frac{|E_c-E|}{E_c+E}\right) + \frac{2A_{mp}E_0 C_{br}E_g}{\pi\sigma_4}\ln\left(\frac{|E_c-E|(E_c+E)}{\sqrt{(E_0^2-E_c^2)^2+C_{br}^2 E_c^2}}\right) +$$

$$+\frac{A_{mp}C_{br}a_L}{2\pi\sigma_4\alpha E_0}\ln\left(\frac{E_0^2+E_c^2+\alpha E_c}{E_0^2+E_c^2-\alpha E_c}\right) - \frac{A_{mp}a_A}{\pi\sigma_4 E_0}\left[\pi - \tan^{-1}\left(\frac{2E_c+\alpha}{C_{br}}\right) - \tan^{-1}\left(\frac{2E_c-\alpha}{C_{br}}\right)\right] +$$

$$+4A_{mp}E_0 E_g\frac{E^2-\gamma^2}{\pi\sigma_4\alpha}\left[\frac{\pi}{2} - \tan^{-1}\left(\frac{2(E_c^2-\gamma^2)}{\alpha C_{br}}\right)\right], \tag{21}$$

and where the intermediate variables previously introduced in the above equations are also given by the next expressions:

$$a_L = (E_g^2 - E_0^2)E^2 + E_g^2 C_{br}^2 - E_0^2(E_0^2 + 3E_g^2),$$
$$a_A = (E^2 - E_0^2)(E_0^2 + E_g^2) + E_g^2 C_{br}^2,$$
$$\gamma = \sqrt{E_0^2 - C_{br}^2/2},$$
$$\alpha = \sqrt{4E_0^2 - C_{br}^2},$$
$$\sigma_4 = (E^2 - E_0^2)^2 + C_{br}^2 E^2. \qquad (22)$$

Additionally, the Urbach band tail part of the model dielectric function, for $0 < E < E_c$, is written in this way,

$$\epsilon_{1,UT}(E) = \tfrac{A_u}{E\pi}\{\exp\left(-\tfrac{E}{E_u}\right)[Ei(\tfrac{E}{E_u}) - Ei(\tfrac{E_c+E}{E_u})]$$
$$+ \exp\left(\tfrac{E}{E_u}\right)[Ei(\tfrac{E_c-E}{E_u}) - Ei(-\tfrac{E}{E_u})]\} \qquad (23)$$

where the introduced function '$Ei(x)$' is the so-called 'generalized exponential integral function', as also implemented as a Matlab function, with exactly the same name, in the Matlab software used in this work, in such a fashion that

$$Ei(x) = \mathcal{P}\int_{-\infty}^{x} \frac{\exp(t)}{t}dt. \qquad (24)$$

3.5. 'AdjustTransIS': Matlab Computational Program for Optical Characterization of Amorphous Semiconductor Films

The computational program named 'AdjustTransIS', created in order to perform the accurate optical characterization of non-uniform semiconductor thin layers, was coded in Matlab and falls into the category of reverse-engineering, or inverse-synthesis method. The simplified flowchart of its algorithm was already presented in [31], and it ought to be noted that is fully configurable by the use of MS-Excel files. The Matlab toolbox developed is able to quickly fit a model-generated transmittance spectrum to the experimentally measured transmittance spectrum of a wedge-shaped semiconductor layer, by fitting up to a maximum number of *nine* free-adjusting model parameters. Up to *seven* of them associated with the implemented dispersion relations, plus *two* nonuniform film geometrical parameters, namely, the average film thickness and, interestingly, the homogeneity parameter, $\bar{d}$ and Δd, respectively.

The very basic idea behind 'AdjustTransIS', in order to accurately compute the TLUC-model parameters, is to calculate the set of values minimizing the figure-of-merit (FOM) metrics that follows,

$$\text{FOM} = 100 \times \text{RMSE} = 100 \times \sqrt{\frac{\sum_{i=1}^{N}(T_{i,\text{meas}} - T_{i,\text{simu}})^2}{N}}, \qquad (25)$$

where N stands for the total number of experimentally measured data points, $T_{i,\text{meas}}$ denotes the as-measured value of transmission, and $T_{i,\text{simu}}$ the corresponding TLUC model-generated value of transmission, for those particular wavelengths for which the glass substrate employed is non-absorbing. The chosen FOM metrics in order to be minimized is 100 times the root mean square error (RMSE) associated with the differences between the measured and simulated transmission data, that is, the square root of the average of the squared transmission differences, or residues. In the Matlab-coded program 'AdjustTransIS', the minimization routine employed was the Nelder–Mead (downhill) simplex algorithm, incorporated into a specific Matlab function. This is a nonlinear, direct search method, implemented in Matlab by its corresponding 'fminsearch' function, and it was adopted in order to compute the minimum of an unconstrained multivariable function.

4. Results and Discussion

4.1. Computation of the Optical Properties by 'AdjustTransIS'

The as-measured transmittance spectra for the RFMS-*a*-Si samples studied were all analyzed by the Matlab-coded program 'AdjustTransIS', and in Table 1 all the best-fit parameters belonging to the TLUC model used, obtained from two representative spectra, are listed. Figure 4a,b displays the comparison between the simulated and experimental transmittance spectra for the corresponding two representative *a*-Si samples, Si#1 and Si#2. The difference between the simulated and measured data, $\Delta T = T_{simu} - T_{meas}$, in the particular case of sample Si#1, is also plotted in Figure 4a. It is seen the remarkable agreement between the two generate and experimental transmittance spectra, with a very low value of FOM of 0.421, in the specific case of Si#1; a value slightly larger of the FOM of 0.565 has been found for the sample Si#2. For the sake of clarity, the *x*-axis of all the graphs represents photon energy, instead of vacuum wavelength, as given initially by the spectrophotometer, when plotting the transmission curves, and we have also employed open circles instead of solid lines, in order to plot the transmission data for these two representative *a*-Si specimens, Si#1 and Si#2, respectively.

Table 1. Values of all the Tauc–Lorentz–Urbach-Continuous oscillator model parameters for representative RFMS-*a*-Si thin films, sputtered with high Ar gas pressure. Furthermore, the values of the employed merit function, FOM = 100×RMSE, for those *a*-Si samples selected. The Tauc–Lorentz model parameters for implanted *a*-Si (*i*-*a*-Si) and evaporated *a*-Si (*e*-*a*-Si), are also listed in the table for the sake of comparison. The values of the Urbach energy and amplitude, E_u and A_u, respectively, obtained from the previous TLUC-model parameters, are also indicated in the present table. Besides, the values of the geometrical parameters, $\bar{d}$ and Δd, respectively, are presented in this table. For a greater completeness of the information reported in this work, we have also added, in the columns headed 'Si#3' and 'Si#4', respectively, the descriptive results associated to the fits of two more RFMS-*a*-Si samples studied.

a-Si Material	RFMS-*a*-Si (Si#1)	RFMS-*a*-Si (Si#2)	RFMS-*a*-Si (Si#3)	RFMS-*a*-Si (Si#4)	*e*-*a*-Si Jellison-Modine	*i*-*a*-Si Adichi-Mori
Data reference	Present work	Present work	Present work	Present work	[29]	[32]
Wavelength/ Energy range (nm/eV)	400–2500	400–2500	400–2500	400–2500	0.8–5.9	1.5–5.2
Figure-of-merit	0.421	0.565	0.445	0.446	N/A	N/A
E_g(eV)	1.19	1.19	1.26	1.27	1.20	1.11
Offset, $\epsilon_1(\infty)$	1.00 (fixed)	1.00 (fixed)	1.00 (fixed)	1.00 (fixed)	1.15	0.17
A_{mp}(eV)	102	95	96	104	122	150
E_0(eV)	3.72	3.73	3.70	3.65	3.45	3.40
C_{br}(eV)	2.54	2.26	2.29	2.35	2.54	2.55
E_c(eV)	1.76	1.73	1.78	1.79	N/A	N/A
E_u(meV)	253	240	245	245	N/A	N/A
A_u(meV)	2.19	1.28	1.2	1.3	N/A	N/A
$\bar{d}$(nm)	770	1123	730	795	N/A	N/A
d_{SEM}(nm)	777	1112	735	789	N/A	N/A
Δd(nm)	25	0	12	13	N/A	N/A

Table 1. Cont.

a-Si Material	RFMS-a-Si (Si#1)	RFMS-a-Si (Si#2)	RFMS-a-Si (Si#3)	RFMS-a-Si (Si#4)	e-a-Si Jellison-Modine	i-a-Si Adichi-Mori
$(\Delta d/\bar{d}) \times 100 (\%)$	3.2	0	1.6	1.6	N/A	N/A
$E_{g,\text{Tauc}}$ (eV)	1.38	1.37	1.43	1.43	N/A	N/A
β_{Tauc} (cm$^{-1/2}$ eV$^{-1/2}$)	591	557	574	604	N/A	N/A
Dispersion model	TLUC	TLUC	TLUC	TLUC	TL	TL

N/A: Data not given in the paper referenced in the table.

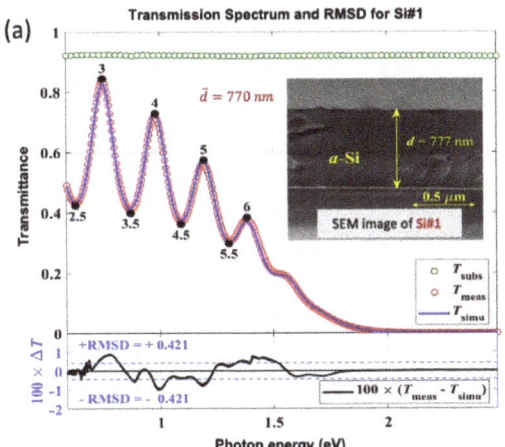

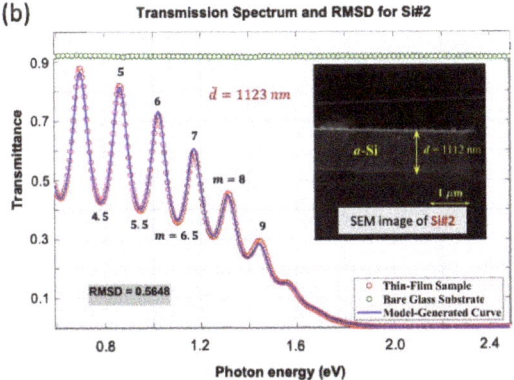

Figure 4. Experimental and best-fit transmittance spectra of the representative RFMS-a-Si thin-film samples, Si#1 (**a**) and Si#2 (**b**), respectively. Additionally, the difference between the TLUC-model-generated transmission spectrum and the as-measured transmission spectrum, ΔT, for both specimens. The excellent comparison between the simulated and experimental normal-incidence transmission spectra are clearly illustrated in this figure. The cross-sectional SEM micrographs for Si#1 and Si#2 are also displayed in this figure, as insets. The measured values of film thickness are marked in the insets. Moreover, the values of some FP-interference order numbers are indicated in the figure, together with the calculated values of $\bar{d}$. The very-low computed values of FOM are also inserted into the figure, for the two samples Si#1 and Si#2.

In addition, in order to clearly show the effectiveness and ease of use of the detailed-designed Matlab-based GUI (i.e., the graphical user interface or main window) of the toolbox 'AdjustTransIS', showing the representative values of all the free-adjusting and fixed parameters, associated with the simultaneous optical and geometrical characterizations, corresponding to the transmittance spectrum for the particular a-Si specimens Si#3 and Si#4, is displayed in the Appendix A. The full information regarding the free and fixed TLUC-model parameters was notified by convenient checkboxes. Furthermore, by using the appropriate radio buttons, we choose the TLUC parametrization selected in this work. The very low value of the FOM found for these other representative specimens analyzed, Si#3 and Si#4, of 0.445 and 0.446, respectively (see the Figure A1 in the Appendix A), have again unambiguously demonstrated the excellent fit between the experimental and simulated transmission data for the samples under investigation.

The computed complex refractive indices of the two RFMS-a-Si samples, Si#1 and Si#2, as a function of wavelength, are illustrated in Figure 5. The average thickness of the particular wedge-shaped specimen Si#1 was 770 nm, and its corresponding non-zero thickness variation was 25 nm. It should be stressed that this lack of film-thickness uniformity represents a percentage of 3.2 % of the calculated average layer thickness. Note that this particular result shows the great capability of the proposed optical characterization approach, highly sensible to the lack of parallelism between the faces of the thin-layer material, and based uniquely upon the shrunk transmittance spectrum of the non-homogeneous a-Si layer.

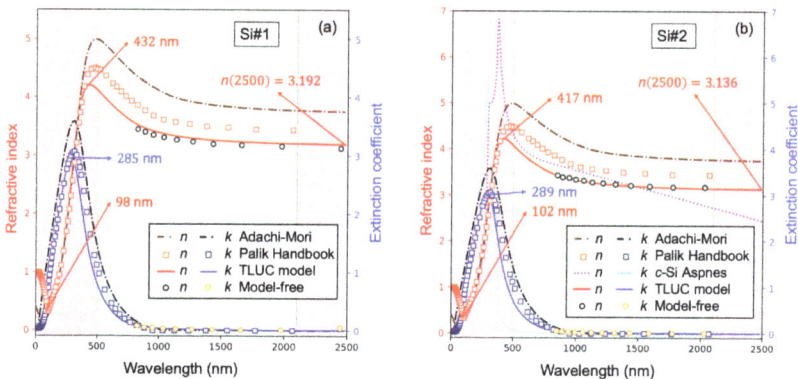

Figure 5. Complex refractive indices vs. vacuum wavelength, obtained from the two transmission spectra analyzed. Extrapolated refractive index, n, and extinction coefficient, k, of the two representative RFMS-a-Si specimens, (**a**) Si#1 and (**b**) Si#2, following the adopted TLUC-oscillator function (plotted for the spectral range 0.5–100 eV, or, equivalently, 12–2482 nm).

Continuing with the analysis of the optical properties of RFMS-a-Si, it is observed in Figure 5 that in the range of very low wavelengths, very close to zero, n is a decreasing function of the wavelength (i.e., $dn/d\lambda < 0$). The optical dispersion is then called normal dispersion. For values of the wavelength larger than approximately 100 nm, the refractive index starts increasing, instead, with increasing wavelength (i.e., $dn/d\lambda > 0$), and it is then called the regime of 'anomalous' dispersion. For wavelength values higher than 425 nm, covering now the whole visible spectral range, the regime of normal optical dispersion is again observed. From the KK relationships, the obtained wavelength dependence of the refractive index has to be necessarily correlated to the existing optical absorption of the a-Si layer, clearly shown by the behavior of its extinction coefficient, k (see Figure 5). There is, as expected, a noticeable narrow absorption band around the specific wavelength value of nearly 287 nm.

The best-fit single-TLUC oscillator parameters for the two representative layers, Si#1 and Si#2, respectively, indicated in Table 1, are next discussed. For these two specific cases, the amplitude of the TLUC oscillator A_{mp} is approximately 99 eV, and the optical band gap value E_g is nearly 1.19 eV. The values of the other three TLUC parameters E_0, C_{br}, and E_c, are around 3.73 eV, 2.40 eV, and 1.75 eV, respectively. Furthermore, following Ferlauto et al. [33], and also Foldyna et al. [28], the corresponding offset parameter, $\epsilon_1(\infty)$, has been always reasonable and successfully fixed to 1. Note at this point of the discussion that, in the present physical interpretation of the experimental results, the concept of energy band gap, E_g, is still maintained, even in the absence of crystallinity (there is no long-range order in the atomic structure) of the studied amorphous semiconductor material, through the existing influence of the preserved short-range ordering in the atomic structure of pure a-Si, upon its electronic density of states.

The values of the Urbach energy and amplitude parameters, E_u and A_u, respectively, determined from the previous TLUC parameters, are also listed in Table 1, for the two representative transmittance spectra analyzed. For the sake of comparison and greater completeness, we have added to Table 1 the values of the TLUC-parameters found for two more RFMS-a-Si samples, Si#3 and Si#4, respectively. The values of the so-called Tauc gap, $E_{g,Tauc}$, all also computed by applying the Matlab program 'AdjustTransIS' (it will be discussed in detail below), are also shown in Table 1. The reproducibility of the properties of RFMS a-Si films depends on the sputtering parameters having the required time stability, which is actually difficult to assure due to the complexity of the non-equilibrium deposition process. This could justify the relative small differences found, lower than about 4.0% (beyond the unavoidable statistical fluctuations), in the values of the TLUC parameters corresponding to the four metastable a-Si samples listed in Table 1. Moreover, the calculated value of the refractive index at the particular NIR wavelength of 2500 nm, $n(2500)$, approximately 3.164, and the two alternative iso-absorption gaps, E_{03}, and E_{04} (the values of energy corresponding to the values of α of 10^3 cm^{-1} and 10^4 cm^{-1}, respectively), are clearly marked in Figures 5 and 7, respectively.

Our computed TLUC parameters are consistent with those reported by some of the authors of this paper, in a previous work concerning with the study of a-Si semiconducting material, by the VASE technique [34]. Moreover, they are also comparable with the best-fit values of the TL-parameters, calculated from the evaporated amorphous silicon (e-a-Si) data compiled in the Palik's Handbook of Optical Properties [35], by Jellison and Modine [29]. Furthermore, note that Foldyna et al. [28] found a value of the band gap energy of 1.19 eV, as a result of applying its TLUC model to previous data belonging to e-a-Si, reported in the mentioned Palik's Handbook. This particular value of the optical gap of 1.19 eV, interestingly, is coincident with the value of E_g reported by Jellison and Modine (see Table 1), employing its limited TL parametrization, instead (it must be stressed that without the inclusion of the contribution of the Urbach absorption tail), to the same set of Palik's compiled optical data for e-a-Si. It has finally to be noted that the value of E_g calculated for the two representative samples Si#1 and Si#2, respectively, is, in both cases, exactly the value of 1.19 eV (see Table 1), in perfect agreement with the above-mentioned value of the optical gap reported in the literature.

4.2. Comparison with the Transmission Data Analysis by the Swanepoel Envelope Method

The complex refractive index, n, was also determined by the alternate transmission envelope method proposed by Swanepoel [24,27], by using our transmittance data from the weak-to-medium absorption region. The envelope method is relatively simple and, importantly, does not require assuming a dispersion model prior to the transmission data analysis, on the contrary to the inverse synthesis method. The basic principle of this method is to accurately draw the two top and bottom envelope curves, that are tangential interpolation curves between the interference maximum points, and between the interference minimum points, respectively. At those tangential points, the average film thickness, $\bar{d}$, the real refractive index, n, and the respective tangential wavelength, λ_{tan}, are related by the interference equation:

$$2n(\lambda_{\text{tan}})\bar{d} = m\lambda_{\text{tan}},\tag{26}$$

where m is the interference order number, which is an exact integer for top tangential points, and an exact half-integer for bottom tangential points (see Figure 3).

In this work, this envelope method is also used in order to calculate n and k, in the transparent-to-medium absorbing region, where the photon energy is correspondingly around or below the optical band-gap. From the transmittance spectra displayed in Figure 4, one can expect in the strong absorption region that the two upper and lower envelope curves merge into one single curve. Thus, it is not possible to find the optical constants, n and k, in this spectral range of high absorption, without assuming a particular dispersion model for the refractive index. Furthermore, for a sufficiently thin layer ($\bar{d} \lesssim 100$ nm) that has very few (or none of them) peaks and valleys, the interpolation is certainly inaccurate, but fortunately it is not at all the case in the present spectra. Instead, we have been able to make reliable interpolations between the tangential points because our a-Si layers are certainly thick enough. The two top and bottom envelopes have been smoothly constructed by the useful algorithm suggested by Minkov et al. [36,37].

After computer generating the two envelopes of the transmission spectrum, the refractive index, the extinction coefficient, the average layer thickness, and the homogeneity parameter have been calculated by using the dispersion model-free Swanepoel method, described in detail in [38–43]. Particularly, it must be emphasized that the differences between the values of $\bar{d}$ and Δd, determined by the envelope method and by inverse synthesis, respectively, are, in all the samples analyzed in this study, smaller than 1.0%; especially remarkable is the fact that the values of the thickness variation Δd for the tapered layer Si#1, calculated by the two different approaches, are coincident, within our range of accuracy. It is found that $\Delta d = 25$ nm with both methods, and as indicated before, it represents approximately the 3.2% of the value of $\bar{d}$. Besides, the values of $\bar{d}$ calculated by the TLUC parametrization and the Swanepoel method are, indeed, fairly close to the cross-checking film-thickness values directly measured by cross-sectional of SEM microscopy images, d_{SEM}'s, listed in Table 1. Superimposed as two insets to Figure 4a,b are displayed the respective two cross-sectional SEM images, belonging each of them to a representative a-Si sample, along with the comparison of the two generated and measured transmission spectra corresponding to each one of the specimens. The directly measured values of d_{SEM} are 777 nm and 1112 nm for the specimens, Si#1 and Si#2, respectively, which are remarkably near to their optically calculated values of $\bar{d}$. The values of the two optical constants n and k obtained by using the transmission envelope method in the allowed photon energy range of about 0.62–1.8 eV, and those from TLUC model fit, in the whole measured spectral range, are very close too, as shown in Figure 5.

4.3. Wemple–DiDomenico Framework: Insight into Material Porosity

The TLUC model yielded sub-gap n-values, and those other data for the refractive index dispersion independently calculated by the model-free envelope method, were systematically analyzed by using the Wemple–DiDomenico (WDD) single effective oscillator model [44,45]. These authors thoroughly investigated optical-dispersion data below the band gap for more than a hundred diverse materials, both covalent and ionic, and both crystalline and amorphous. They found that the whole set of dispersion data examined can be fitted, to an excellent approximation, by the following relationship:

$$\epsilon_1(\hbar\omega) = n^2(\hbar\omega) = 1 + \frac{E_{\text{so}}E_{\text{d}}}{E_{\text{so}}^2 - (\hbar\omega)^2},\tag{27}$$

where $\hbar\omega$ is the photon energy, E_{so} is the single-oscillator energy, and E_{d} is the single-oscillator strength or dispersion energy. Therefore, plotting $(n^2 - 1)^{-1}$ vs. $(\hbar\omega)^2$ lets us determine the two single-oscillator parameters, E_{so} and E_d, respectively, by fitting a linear function to the set of calculated values of the refractive index. The WDD plots for the two representative RFMS-a-Si samples Si#1 and Si#2, respectively, are shown in Figure 6. The obtained values of the WDD single-oscillator parameters E_{so} and E_d are presented in Table 2. Mention must be made that departures from the linear behavior in the WDD plots, at larger photon energies, very often reported in the literature [44,45], and being explained by the result of the proximity of the band edge (that is, the interband transitions), have not been observed in any of the present RFMS-a-Si specimens.

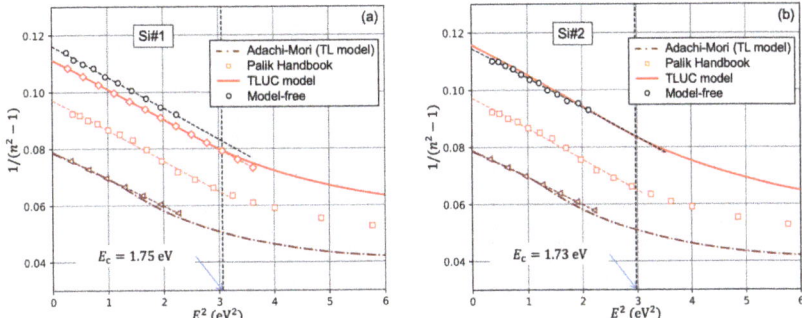

Figure 6. Plot of the refractive-index factor, $1/(n^2 - 1)$, versus the incident photon-energy squared (i.e., the so-called WDD plot), corresponding to the two example specimens of RFMS-a-Si under analysis; (**a**) Si#1 and (**b**) Si#2.

The extrapolated values of the static refractive index, $n(\hbar\omega = 0)$, for the representative a-Si samples are also listed in Table 2.

Table 2. Values of the height and position of the ϵ_2 peak, along with the refractive-index dispersion parameters revealed by the WDD single effective oscillator model, for various investigated a-Si materials. The different approaches used in order to determine the values of the presented optical parameters, are also listed in the table.

Sample ID	Approach	$\epsilon_{2,max}$	$E(\epsilon_{2,max})$ (eV)	E_{so} (eV)	E_d (eV)	$n(0)$	f_{void}^{WDD} (%)	f_{void}^{Bru} (%)
Si#1	Inverse Synthesis	18.6	3.71	3.24	29.2	3.158	20.6	20.8
Si#1	Envelope Method	N/A	N/A	3.26	28.0	3.099	21.2	21.4
Si#2	Inverse Synthesis	19.5	3.72	3.27	28.2	3.102	22.5	22.5
Si#2	Envelope Method	N/A	N/A	3.33	29.1	3.199	18.4	20.8
e-a-Si Palik [35]	Reflectance	20.8	3.50	3.01	30.9	3.359	15.1	13.2
i-a-Si Adachi-Mori [32]	VASE	26.6	3.45	2.87	36.4	3.697	0	0

N/A: Data not given in the paper referenced in the table.

Interestingly, a value of the TL model-yielded E_d parameter for dense and fully amorphous, self-implanted (i.e., non-hydrogenated) a-Si of 36.4 eV, reported by Chen and Shen [46], that was, in turn, calculated from the values previously reported by Adachi and Mori [32,47] of the complex dielectric constant, being this measured by the VASE technique, has been found in the literature. This result demonstrated that the popular literature data used a reference of 'dense' a-Si, measured by Bagley et al. [48,49], is actually influenced by the contribution of an existing component of microcrystalline Si, whose volume fraction was estimated to be, surprisingly, about 53%. It should be pointed out that the reference values of E_{so} and $n(0)$ reported by Chen and Shen [46] and, also, corroborated by us, for that Si-ion-implanted a-Si layer, are 2.87 eV and 3.697 eV, respectively.

The dispersion energy, E_d, measures the average strength of the interband electronic transitions, and obeys the widely-applicable empirical rule, expressed as follows [44,45]:

$$E_d = \beta N_c Z_a N_e (\rho_a/\rho_x), \quad (E_d \text{ in eV units}), \tag{28}$$

where $\beta_c = 0.37 \pm 0.04$ eV, in the case of 'covalent' materials, and $\beta_i = 0.26 \pm 0.03$ eV, in the case of the more 'ionic' materials. In Equation (28), N_c stands for the coordination number of the cation, that is, the nearest neighbor to the anion; Z_a denotes the formal chemical valency of the anion ($Z_a = 4$, for Si); N_e is the total number of valence electrons, where the cores are excluded, per anion ($N_e = 8$); and, last, ρ_a/ρ_x is the ratio of the mass density of the amorphous material to the mass density of its crystalline counterpart, with its 'compacted' and well-ordered atomic structure, instead.

This mass–density ratio, ρ_a/ρ_x, in Equation (28) is indeed the key factor in determining the refractive-index behavior of the studied RFMS-a-Si films. Therefore, we assume that E_d decreases linearly with the increase of the volume fraction of interconnected voids, i.e., the porosity, within the underdense a-Si layers; it must be also noted that E_d scales with the volume density of valence electrons in the material, involved in the transition at the single-oscillator energy of E_{so}. We have thereby used the already mentioned particular value of the dispersion energy, E_d, of 36.4 eV for the dense, or pore- or void-free, self-implanted a-Si material, obtained, in turn, from the data found by Adachi and Mori [32,47], as the correct reference value of E_d in order to be compared with. Therefore, we can express mathematically all those previously-proposed simplifying ideas about the calculation of the WDD void volume fraction, f_{void}^{WDD}, from the calculated values of E_d, by the simple approximate formula:

$$E_d^{underdense} = E_d^{dense}\left(1 - f_{void}^{WDD}\right), \tag{29}$$

and, by solving for f_{void}^{WDD}, we derive the following expression:

$$f_{void}^{WDD} = 1 - \frac{E_d^{underdense}}{E_d^{dense}}. \tag{30}$$

Finally, by using this approximate relation, the values of f_{void}^{WDD} for the representative underdense RFMS-a-Si samples were obtained from the calculated values of the WDD parameter, E_d, and are indicated in Table 2.

4.4. Bruggeman Effective Medium Approximation: Alternate Void-Volume-Fraction Determination

An additional cross-checking determination of the void volume fraction, f_{void}^{Bru}, for the RFMS-a-Si thin layers under investigation, by using the self-consistent Bruggeman effective medium approximation (B-EMA) [49,50], and the values of the complex dielectric function, ϵ, calculated by the extrapolation of the WDD dispersion formula to zero photon energy, $\epsilon(0) = 1 + E_d/E_{so}$, is next carried out.

According to this B-EMA approximation, a mixture of diverse materials can be considered as a homogeneous medium, possessing an effective complex dielectric constant, ϵ_{eff}, that can be obtained from the individual complex dielectric constants, ϵ_i's, and the respective volume fractions, f_i's, of the individual constituent components, whereas it is obeyed that $\sum_i f_i = 1$. Following the B-EMA approach, for a composite material of the various i-components, the following equation is verified:

$$\sum_i f_i \left(\frac{\epsilon_i - \epsilon_{eff}}{\epsilon_i + 2\epsilon_{eff}}\right) = 0 \tag{31}$$

By using the basic relation between the complex dielectric constant and the complex refractive index, $\epsilon = n^2$, and by also taking into account the two constituent components of our a-Si thin-film material, whose volume fractions are $1 - f_{void}^{Bru}$ (for the dense or pore- or void-free a-Si), and $f_{\text{void}}^{\text{Bru}}$ (for the air-filled pores or voids), respectively, along with setting, $n_{\text{eff}} \equiv n_{\text{underdense}}(0)$, we can express Equation (31) as

$$(1 - f_{\text{void}}^{\text{Bru}})\left(\frac{n_{\text{dense}}^2(0) - n_{\text{underdense}}^2(0)}{n_{\text{dense}}^2(0) + 2n_{\text{underdense}}^2(0)}\right) + f_{\text{void}}^{\text{Bru}}\left(\frac{1 - n_{\text{underdense}}^2(0)}{1 + 2n_{\text{underdense}}^2(0)}\right) = 0. \quad (32)$$

Hence, by solving for $f_{\text{void}}^{\text{Bru}}$, we have the final formula:

$$f_{\text{void}}^{\text{Bru}} = \frac{(1 + 2n_{\text{underdense}}^2(0))(n_{\text{dense}}^2(0) - n_{\text{underdense}}^2(0))}{3n_{\text{underdense}}^2(0)(n_{\text{dense}}^2(0) - 1)}. \quad (33)$$

The two necessary values of the refractive index, that is, the value of $n(0)$ for the dense, self-implanted a-Si, $n_{\text{dense}}(0) = 3.697$, and the value of the refractive index for air, $n_{\text{air}} = 1$, are now inserted in the previous Equation (33). The calculated values of B-EMA-based void volume fraction, $f_{\text{void}}^{\text{Bru}}$, are listed in Table 2. Note that the two respective values of the void volume fraction, $f_{\text{void}}^{\text{WDD}}$ and $f_{\text{void}}^{\text{Bru}}$, obtained by two different approaches, for our RFMS-a-Si specimens, are practically identical; the value of the porosity for the specific rf-magnetron sputtering deposition conditions employed in this work, with an Ar-pressure of about 4.4 Pa, is found to be approximately 21%. Karabacak and Demirkan [9] have reported values of the porosity for their sputter-deposited a-Si films produced at different Ar-gas pressures, as high as approximately 30%, corresponding to a mass density of 1.64 g/cm^3. Their reported void fractions as a function of working pressure are absolutely consistent with our obtained value of about 21%, which is surprisingly much higher than the reference value found for ion-implanted a-Si of around 3.0%.

4.5. Absorption Edge and Dielectric Function of RFMS-a-Si

In Figure 7, we plot the optical absorption coefficient, as a function of photon energy, of the two representative RFMS-a-Si samples, Si#1 and Si#2. From these optical absorption edges shown in Figure 7, the corresponding two iso-absorption gaps, E_{03} and E_{04}, respectively, for each representative sample, are now determined. The value of the gap parameter, E_{04}, is generally found to be around 0.15–0.30 eV larger than the value of the Tauc gap, $E_{g,\text{Tauc}}$, for Si-based non-crystalline materials. In the present a-Si films, the difference between the obtained value of the iso-absorption gap E_{04} and the fitted value of the optical gap, $E_{04} - E_g$, has been found to be approximately 0.28 eV, well within the previous gap-difference range. This gap difference, $E_{04} - E_g$, has been attributed to the width of the conduction-band tail, and, therefore, to the contribution associated with the electron states localized in the mobility pseudo-gap of the a-Si material. It is generally argued that the larger the structural randomness within the amorphous material, the wider its conduction band tail.

The exponential dependence of the absorption coefficient, α, of the form $\alpha(\hbar\omega) = \alpha_0 \exp(\hbar\omega/E_u)$, the well-known Urbach region equation, is clearly obeyed, as shown in Figure 7. The computed values of the Urbach energy parameter, E_u, listed in Table 1, were found to be about 250 meV, in all the samples under study. This very large value of the parameter E_u is indicative of the high degree of structural disorder of the present non-crystalline material, fully consistent with was above inferred from both X-ray diffraction and Raman spectroscopy measurements. The maximum value of α, α_{max}, found at the specific value of the photon energy of nearly 5.0 eV is 1.42×10^6 cm^{-1} (see Figure 8).

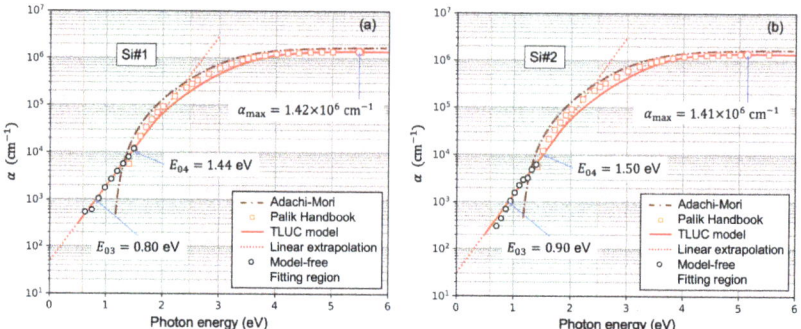

Figure 7. Spectral dependence of the logarithm of the absorption coefficient (that is, the optical-absorption edge), for the two RFMS-*a*-Si thin layers; (**a**) Si#1 and (**b**) Si#2. The values of the two iso-absorption gaps, E_{03} and E_{04}, are both conveniently marked in the corresponding absorption spectra.

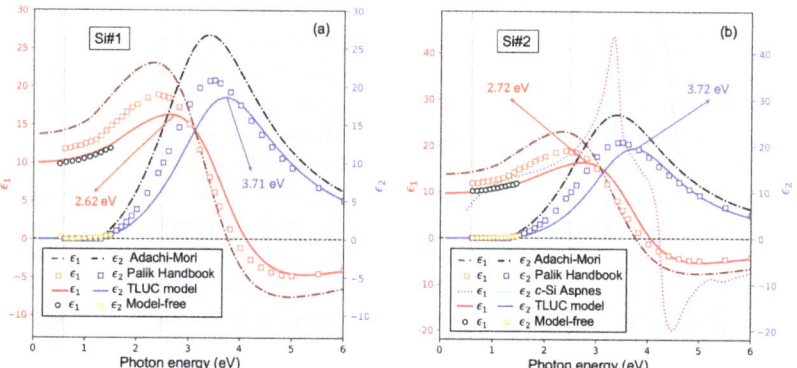

Figure 8. Real (ϵ_1) and imaginary (ϵ_2) parts of the complex dielectric functions versus photon energy, determined from the normal-incidence spectral transmittance, of the two representative RFMS-*a*-Si thin layers; (**a**) Si#1 and (**b**) Si#2.

Next, the real and imaginary parts, ϵ_1 and ϵ_2, respectively, of the complex dielectric constant, ϵ, as a function of the photon energy, for the two representative *a*-Si samples, Si#1 and Si#2, are shown in Figure 8. Note that the broad peak of the imaginary part of the dielectric function, ϵ_2, whose highest value, $\epsilon_{2,max}$, is associated with the splitting of bonding and antibonding electron states, and it is located at an energy position, $E(\epsilon_{2,max})$, of nearly 3.71 eV (see Table 2), almost coincident with the calculated value of the TLUC parameter E_0. This single smeared peak of ϵ_2 is usually shown by tetrahedrally bonded amorphous semiconductors [34], like the cases of elemental Si and Ge.

Generally speaking, the amorphous material exhibiting the largest value of $\epsilon_{2,max}$ (see Figure 8), is regarded as that belonging to 'intrinsic', dense, hydrogen-less, and fully amorphous silicon; i.e., the self-implanted *a*-Si, with the above-mentioned highest value of the WDD dispersion energy, E_d, of 36.4 eV. The largest value of the $\epsilon_{2,max}$ of 26.6 for this particular H-free *a*-Si material, measured by Adachi and Mori [32,47], and based upon VASE measurements, at a value of the photon energy, $E(\epsilon_{2,max})$, of 3.45 eV, must necessarily be the reference value in order to be compared with. Hence, the significantly lower value of the amplitude of the ϵ_2 peak for the case of our RFMS-*a*-Si samples can plausibly be explained by their proportionally lower mass density, attributed to their inherent micro-voided structure, under the present growth conditions.

The experimental $\epsilon_1(E)$ and $\epsilon_2(E)$ spectra of c-Si [49] are also displayed in Figure 8 by dashed lines. The prominent critical point (CP) features seen in these two $\epsilon_1(E)$ and $\epsilon_2(E)$ spectra of c-Si are the E_1 and E_2 structures at E approximately 3.4 and 4.3 eV, respectively. The disappearance of the CP features in the amorphous material is due to the breakdown of the crystal periodicity in the amorphous material. The comparative experimental $n(\lambda)$ and $k(\lambda)$ spectra of c-Si shown in Figure 5, on the other hand, have been directly obtained from those previous $\epsilon_1(E)$ and $\epsilon_2(E)$ spectra of c-Si reported by Aspnes [49].

4.6. Determination of the Band-Gap Energy by Tauc's Extrapolation Method

The absorption coefficient was also independently obtained directly from a portion of the transmission spectrum, by using *only* the spectral region of high absorption of such a spectrum, where the FP interference fringes start disappearing (see Figure 4 and the inset of Figure 9). Therefore, for those large values of α, α_{free}'s, where the absorbance obeys that $x \ll 1$, the interference effects can be totally ignored (interference-free transmission region), and the transmittance is then reasonably expressed, by taking into consideration Equation (5), this way,

$$T_{\text{meas}}(E) = \frac{16n^2 s \exp(-\alpha_{\text{free}} d)}{(n+1)^3(n+s^2)}. \tag{34}$$

Therefore, the absorption coefficient, α_{free}, is obtained by employing the following expression:

$$\alpha_{\text{free}}(E) = -\frac{1}{d} \ln\left(\frac{(n+1)^3(n+s^2)T_{\text{meas}}}{16n^2 s}\right), \tag{35}$$

where T_{meas} stands for the experimentally-measured transmittance, in the spectral interval of high absorption, where the FP fringes start disappearing.

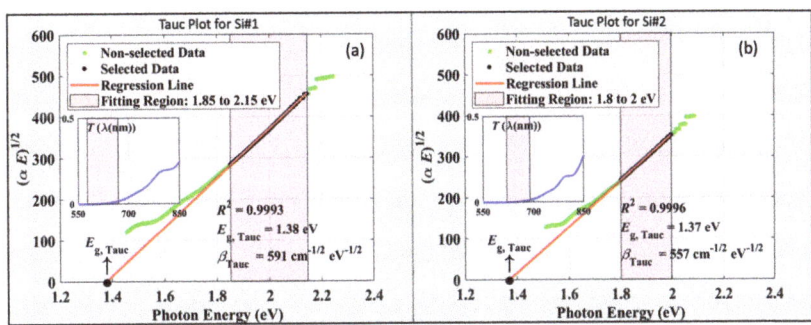

Figure 9. Tauc plots obtained from the room-temperature, normal-incidence transmission spectra belonging to the two representative RFMS-a-Si samples; (**a**) Si#1 and (**b**) Si#2.

We have just found an equation equivalent to Equation (6), derived by Swanepoel, in [24]. He also suggested the use of the two-term Cauchy empirical equation for the spectral dependence of the index of refraction in the interference-free region of high absorption. On the contrary, we have much more accurately employed the relationship for $n(E)$ derived from the adopted KK-consistent TLUC model.

Additionally, Tauc et al. [23] found that for values of $\alpha \gtrsim 10^4$ cm^{-1} (the Tauc's region), it is obeyed that

$$\sqrt{\alpha E} = \beta_{\text{Tauc}}(E - E_{g,\text{Tauc}}). \tag{36}$$

This is the well-known Tauc law, very often employed in order to determine the Tauc gap, $E_{g,\text{Tauc}}$, and the Tauc slope, β_{Tauc}, from the values of α_{free} calculated by Equation (35); in other words, the Tauc's extrapolation method. The so-called Tauc plots for the representative specimens Si#1 and Si#2 are shown in Figure 9; see in Table 1 the calculated values of both $E_{g,\text{Tauc}}$ and β_{Tauc}.

According to the ideas put forward by Chen and Shen [46], it has been finally drawn the conclusion that the three contributions present in the adopted TLUC parametrization, i.e., the Lorentz electron oscillator, the Tauc joint density of states, and the Urbach exponential tail, respectively, accurately give a complete account of both the below and above band gap absorption, in all the RFMS-a-Si samples analyzed. The 'trade-off' between these three contributions has caused the small decrease of the nominal (fitted) gap, E_g (TLUC gap), relative to the true Tauc (extrapolated) gap, $E_{g,\text{Tauc}}$, when the TLUC-oscillator model has been used in the fitting of the experimentally-measured transmission data for RFMS-a-Si; see again Table 1, where all the different values of all the introduced bandgaps, are listed. It can be speculated that part of the absorption of the material is, in a way, 'contained' within the additional photon energy range, from the true extrapolated gap, $E_{g,\text{Tauc}}$, down to the TLUC fitted gap, E_g. Thus, E_g could be reasonably considered, to some extent, a kind of 'mathematical gap', rather than an actually and purely 'physical gap' [46].

5. Summary and Concluding Remarks

The complex refractive index and dielectric constant of the RFMS-a-Si layers studied, grown onto glass substrates, were very accurately calculated as a function of the vacuum wavelength/photon energy, by using inverse synthesis in a Matlab GUI environment, which is based upon the use of the as-measured transmission spectrum. The wavelength region studied was from 400 to 2500 nm, and it has been clearly shown that the adopted TLUC parameterization is highly accurate for the evaluation of the UV/visible/NIR transmittance measurements, on the H-free a-Si investigated. The values of the seven best-fit TLUC parameters A_{mp}, E_0, C_{br}, E_g, E_c, A_u, and E_u, are computed, as well as the respective values of $\epsilon_{2,\text{max}}$ and $E(\epsilon_{2,\text{max}})$ associated to each peak of the ϵ_2 spectrum. It has been also demonstrated that the value of $\epsilon_{2,\text{max}}$ is related with the volume fraction of micro-voids, within the RFMS-a-Si layers. After ending the execution of 'AdjustTransIS', it produces several output files: (i) the Tauc and Cody [51] plots, where the Tauc and Cody, gaps and slopes, respectively, are determined, and (ii) the absorption spectrum, where the two iso-absorption gaps, E_{03} and E_{04}, are marked.

The remarkable agreement between the values of the refractive index and extinction coefficients, calculated by the model-free envelope method, and those determined by using inverse synthesis, and initially by the excellent comparison between the measured and generated transmittance curves, is a good base in order to fully validate our novel approach. It can also be asserted that our results very much encourage the use of the i-a-Si $\epsilon(E)$ data as an accurate reference of dense and fully a-Si, as claimed by Adachi and Mori [32,47].

Note that the Swanepoel envelope method is not practicable for layers thinner than around 100 nm; these specific layers are so thin that there are no interference fringes, or too few of them, in the spectra. However, this is not the case when using the inverse synthesis approach adopted in this investigation, which was also applicable to films thinner than approximately 100 nm, corresponding with the absence of fringes in the spectrum. We have also successfully run 'AdjustTransIS', in order to apply the Solomon dispersion model [52], aimed to the energy band structure calculation, and also the KK-consistent Cody–Lorentz–Urbach parameterization [33], that we recently implemented. The results obtained from these other fits are fully consistent with the results that were found by using the TLUC parameterization. We might point out a drawback or con when assessing the pros and cons of the proposed inverse synthesis method. This con would be the lack of determination of a potential refractive index profile, $n(z)$, and to overcome this con will be our next goal in the near future.

Last, but not least, we can end this work by emphasizing that the studied H-free, porous RFMS-a-Si, grown onto glass substrates, additionally shows a realistic potential for anode material in Li-ion batteries, appropriately increasing the Ar-sputtering pressure, in order to be able to control the mass density of the present RFMS-a-Si. The recently reported improvement of the cycling stability, by employing micro-voided a-Si electrodes (anodes) for Li-ion batteries [53], has been explained by the availability of pores in the a-Si material, necessary for the large volumetric expansion that occurs during the corresponding lithiation process, in the Si anode Li-ion batteries. The main perspective for the future to be highlighted is to study the influence of different deposition conditions upon the optical properties and porosity of the RFMS-*a*-Si, by using the devised inverse-synthesis method, based on the TLUC parametrization.

Author Contributions: Conceptualization, E.M. and J.J.R.-P.; methodology, E.M.; software, J.J.R.-P., A.P.M., and M.B.; validation, J.J.R.-P., M.B., and A.P.M.; formal analysis, E.M.; investigation, E.M. and J.J.R.-P.; resources, E.M.; data curation, S.M.F.R. and E.S.; writing—original draft preparation, E.M.; writing—review and editing, A.P.M. and M.B..; visualization, M.B., E.B., and J.J.R.-P.; supervision, E.M., D.M., and E.B.; project administration, E.M.; funding acquisition, E.M. All authors have read and agreed to the published version of the manuscript.

Funding: This research was funded by the SCALED project, grant number PID2019-109215RB-C42, provided by the Spanish Ministry of Science and Innovation.

Institutional Review Board Statement: Not applicable.

Informed Consent Statement: Not applicable.

Data Availability Statement: Data supporting the findings of this study are available from the corresponding author E.M. on request.

Acknowledgments: We would like to sincerely thank J. Gandía (CIEMAT, Madrid) and J. Cárabe (CIEMAT, Madrid), for their very insightful and constructive comments during the realization of the present investigation.

Conflicts of Interest: The authors declare no conflict of interest.

Appendix A

By presenting the designed Matlab GUI to the audience they come to a closer understanding of how the optical characterization of real non-crystalline semiconductor thin films is performed by using the created computational program, 'AdjustTransIS'.

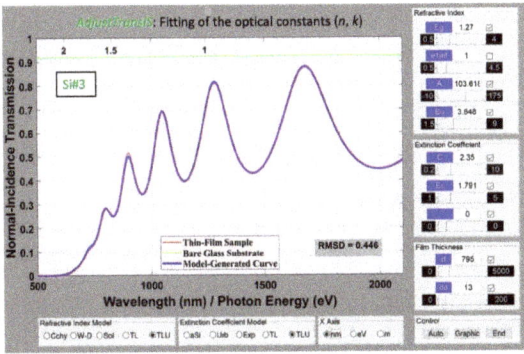

Figure A1. *Cont.*

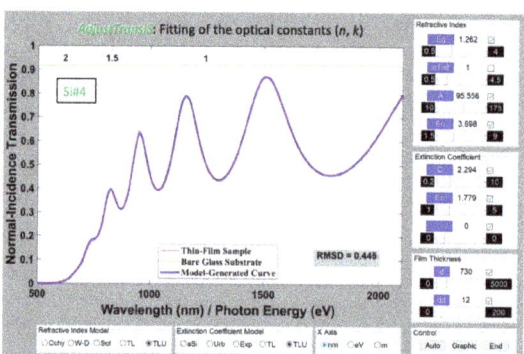

Figure A1. GUI screenshots for the determination of the optical constants of Si#3 and Si#4 thin-film samples, corresponding to the created Matlab-coded program, 'AdjustTransIS'. Appropriate checkboxes and radio buttons are employed for the quick and accurate transmission-data analysis.

References

1. Street, R.A. *Hydrogenated Amorphous Silicon*; Cambridge University Press: Cambridge, UK, 2005.
2. Street, R.A. *Technology and Applications of Amorphous Silicon*; Springer: Palo Alto, CA, USA, 1999.
3. Morigaki, K.; Kugler, S.; Shimakawa, K. *Amorphous Semiconductors: Structural, Optical, and Electronic Properties*; John Wiley & Sons: Hoboken, NJ, USA, 2017.
4. Powell, M.J. The physics of amorphous-silicon thin-film transistors. *IEEE Trans. Electron Devices* **1989**, *36*, 2753–2763. [CrossRef]
5. Bye, G.; Ceccaroli, B. Solar grade silicon: Technology status and industrial trends. *Sol. Energy Mater. Sol. Cells* **2014**, *130*, 634–646. [CrossRef]
6. Haschke, J.; Amkreutz, D.; Korte, L.; Ruske, F.; Rech, B. Towards wafer quality crystalline silicon thin-film solar cells on glass. *Sol. Energy Mater. Sol. Cells* **2014**, *128*, 190–197. [CrossRef]
7. Haschke, J.; Amkreutz, D.; Rech, B. Liquid phase crystallized silicon on glass: Technology, material quality and back contacted heterojunction solar cells. *Jpn. J. Appl. Phys.* **2016**, *55*, 04EA04. [CrossRef]
8. Boukamp, B.; Lesh, G.; Huggins, R. All-solid lithium electrodes with mixed-conductor matrix. *J. Electrochem. Soc.* **1981**, *128*, 725. [CrossRef]
9. Karabacak, T.; Demirkan, M.T. Density Modulated Thin Film Electrodes, Methods of Making Same, and Applications of Same. U.S. Patent 10,333,148, 25 June 2019.
10. Poelman, D.; Smet, P.F. Methods for the determination of the optical constants of thin films from single transmission measurements: A critical review. *J. Phys. D Appl. Phys.* **2003**, *36*, 1850. [CrossRef]
11. Capper, P.; Willoughby, A.; Kasap, S. *Optical Properties of Materials and Their Applications*; John Wiley & Sons: Hoboken, NJ, USA, 2020.
12. Fox, M. *Optical Properties of Solids*; Academic Press: Cambridge, MA, USA, 2010.
13. Stenzel, O. *The Physics of Thin Film Optical Spectra*; Springer: Berlin/Heidelberg, Germany, 2015.
14. Heavens, O.S. *Optical Properties of Thin Solid Films*; Dover books on physics; Courier Corporation: North Chelmsford, MA, USA, 1991.
15. Rodríguez-de Marcos, L.V.; Larruquert, J.I. Analytic optical-constant model derived from Tauc-Lorentz and Urbach tail. *Opt. Express* **2016**, *24*, 28561–28572. [CrossRef]
16. Smit, C.; Van Swaaij, R.; Donker, H.; Petit, A.; Kessels, W.; Van de Sanden, M. Determining the material structure of microcrystalline silicon from Raman spectra. *J. Appl. Phys.* **2003**, *94*, 3582–3588. [CrossRef]
17. O'Leary, S.; Fogal, B.; Lockwood, D.; Baribeau, J.M.; Noël, M.; Zwinkels, J. Optical dispersion relationships in amorphous silicon grown by molecular beam epitaxy. *J.-Non-Cryst. Solids* **2001**, *290*, 57–63. [CrossRef]
18. Fogal, B.; O'Leary, S.; Lockwood, D.; Baribeau, J.M.; Noël, M.; Zwinkels, J. Disorder and the optical properties of amorphous silicon grown by molecular beam epitaxy. *Solid State Commun.* **2001**, *120*, 429–434. [CrossRef]
19. Orapunt, F.; Tay, L.L.; Lockwood, D.J.; Baribeau, J.M.; Noël, M.; Zwinkels, J.C.; O'Leary, S.K. An amorphous-to-crystalline phase transition within thin silicon films grown by ultra-high-vacuum evaporation and its impact on the optical response. *J. Appl. Phys.* **2016**, *119*, 065702. [CrossRef]
20. Roura, P.; Farjas, J.; Roca i Cabarrocas, P. Quantification of the bond-angle dispersion by Raman spectroscopy and the strain energy of amorphous silicon. *J. Appl. Phys.* **2008**, *104*, 073521. [CrossRef]
21. Beeman, D.; Tsu, R.; Thorpe, M. Structural information from the Raman spectrum of amorphous silicon. *Phys. Rev. B* **1985**, *32*, 874. [CrossRef] [PubMed]

22. Smith, D. *Thin-Film Deposition: Principles and Practice*; McGraw-Hill Education: New York, NY, USA, 1995.
23. Tauc, J.; Grigorovici, R.; Vancu, A. Optical properties and electronic structure of amorphous germanium. *Phys. Status Solidi (b)* **1966**, *15*, 627–637. [CrossRef]
24. Swanepoel, R. Determination of the thickness and optical constants of amorphous silicon. *J. Phys. E Sci. Instruments* **1983**, *16*, 1214. [CrossRef]
25. Pérez, J.J.R. Nuevos metodos de caracterizacion optica de semiconductores basados en medidas espectroscopicas de reflexion. Ph.D. Thesis, University of Cadiz, Cadiz, Spain, 1997.
26. Dobrowolski, J.; Ho, F.; Waldorf, A. Determination of optical constants of thin film coating materials based on inverse synthesis. *Appl. Opt.* **1983**, *22*, 3191–3200. [CrossRef] [PubMed]
27. Swanepoel, R. Determination of surface roughness and optical constants of inhomogeneous amorphous silicon films. *J. Phys. E Sci. Instruments* **1984**, *17*, 896. [CrossRef]
28. Foldyna, M.; Postava, K.; Bouchala, J.; Pistora, J.; Yamaguchi, T. Model dielectric functional of amorphous materials including Urbach tail. In *Microwave and Optical Technology 2003*; International Society for Optics and Photonics: Bellingham, WA, USA, 2004; Volume 5445, pp. 301–305.
29. Jellison, G., Jr.; Modine, F. Parameterization of the optical functions of amorphous materials in the interband region. *Appl. Phys. Lett.* **1996**, *69*, 371–373; Erratum in **1996**, *69*, 2137–2137. [CrossRef]
30. Blanco, E.; Domínguez, M.; González-Leal, J.; Márquez, E.; Outón, J.; Ramírez-del Solar, M. Insights into the annealing process of sol-gel TiO2 films leading to anatase development: The interrelationship between microstructure and optical properties. *Appl. Surf. Sci.* **2018**, *439*, 736–748. [CrossRef]
31. Ruíz-Pérez, J.J.; Navarro, E.M. Optical Transmittance for Strongly-Wedge-Shaped Semiconductor Films: Appearance of Envelope-Crossover Points in Amorphous As-Based Chalcogenide Materials. *Coatings* **2020**, *10*, 1063. [CrossRef]
32. Adachi, S. *Optical Constants of Crystalline and Amorphous Semiconductors*; Springer: Berlin/Heidelberg, Germany, 1999.
33. Ferlauto, A.; Ferreira, G.; Pearce, J.M.; Wronski, C.; Collins, R.; Deng, X.; Ganguly, G. Analytical model for the optical functions of amorphous semiconductors from the near-infrared to ultraviolet: Applications in thin film photovoltaics. *J. Appl. Phys.* **2002**, *92*, 2424–2436. [CrossRef]
34. Márquez, E.; Blanco, E.; García-Vázquez, C.; Díaz, J.; Saugar, E. Spectroscopic ellipsometry study of non-hydrogenated fully amorphous silicon films deposited by room-temperature radio-frequency magnetron sputtering on glass: Influence of the argon pressure. *J.-Non-Cryst. Solids* **2020**, *547*, 120305. [CrossRef]
35. Palik, E.D. *Handbook of Optical Constants of Solids*; Academic Press: Cambridge, MA, USA, 1998; Volume 3.
36. Minkov, D.; Gavrilov, G.; Moreno, J.; Vázquez, C.; Márquez, E. Optimization of the graphical method of Swanepoel for characterization of thin film on substrate specimens from their transmittance spectrum. *Meas. Sci. Technol.* **2017**, *28*, 035202. [CrossRef]
37. Minkov, D.; Gavrilov, G.; Angelov, G.; Moreno, J.; Vázquez, C.; Ruano, S.; Márquez, E. Optimisation of the envelope method for characterisation of optical thin film on substrate specimens from their normal incidence transmittance spectrum. *Thin Solid Films* **2018**, *645*, 370–378. [CrossRef]
38. Márquez, .E.; Ramírez-Malo, J.; Villares, P.; Jiménez-Garay, R.; Ewen, P.; Owen, A. Calculation of the thickness and optical constants of amorphous arsenic sulphide films from their transmission spectra. *J. Phys. D Appl. Phys.* **1992**, *25*, 535. [CrossRef]
39. Ramírez-Malo, J.; Márquez, E.; Villares, P.; Jiménez-Garay, R. Determination of the Refractive Index and Optical Absorption Coefficient of Vapor-Deposited Amorphous As–S Films from Transmittance Measurements. *Phys. Status Solidi (a)* **1992**, *133*, 499–507. [CrossRef]
40. Márquez, E.; Ramírez-Malo, J.; Villares, P.; Jiménez-Garay, R.; Swanepoel, R. Optical characterization of wedge-shaped thin films of amorphous arsenic trisulphide based only on their shrunk transmission spectra. *Thin Solid Films* **1995**, *254*, 83–91. [CrossRef]
41. Márquez, E.; Díaz, J.; García-Vázquez, C.; Blanco, E.; Ruíz-Pérez, J.; Minkov, D.; Angelov, G.; Gavrilov, G. Optical characterization of amine-solution-processed amorphous AsS2 chalcogenide thin films by the use of transmission spectroscopy. *J. Alloy. Compd.* **2017**, *721*, 363–373. [CrossRef]
42. Márquez, E.; Saugar, E.; Díaz, J.; García-Vázquez, C.; Fernández-Ruano, S.; Blanco, E.; Ruíz-Pérez, J.; Minkov, D. The influence of Ar pressure on the structure and optical properties of non-hydrogenated a-Si thin films grown by rf magnetron sputtering onto room-temperature glass substrates. *J.-Non-Cryst. Solids* **2019**, *517*, 32–43. [CrossRef]
43. Ruíz-Pérez, J.; González-Leal, J.; Minkov, D.; Márquez, E. Method for determining the optical constants of thin dielectric films with variable thickness using only their shrunk reflection spectra. *J. Phys. D Appl. Phys.* **2001**, *34*, 2489. [CrossRef]
44. Wemple, S.; DiDomenico, M., Jr. Behavior of the electronic dielectric constant in covalent and ionic materials. *Phys. Rev. B* **1971**, *3*, 1338. [CrossRef]
45. Wemple, S. Refractive-index behavior of amorphous semiconductors and glasses. *Phys. Rev. B* **1973**, *7*, 3767. [CrossRef]
46. Chen, H.; Shen, W. Perspectives in the characteristics and applications of Tauc-Lorentz dielectric function model. *Eur. Phys. J. B-Condens. Matter Complex Syst.* **2005**, *43*, 503–507. [CrossRef]
47. Adachi, S.; Mori, H. Optical properties of fully amorphous silicon. *Phys. Rev. B* **2000**, *62*, 10158. [CrossRef]
48. Bagley, B.; Aspnes, D.; Celler, G.; Adams, A. Optical characterization of chemically vapor deposited and laser-annealed polysilicon. *MRS Online Proc. Libr. (OPL)* **1981**, *4*, 483. [CrossRef]
49. Aspnes, D.E. Optical properties of thin films. *Thin Solid Films* **1982**, *89*, 249–262. [CrossRef]

50. Bruggeman, V.D. Berechnung verschiedener physikalischer Konstanten von heterogenen Substanzen. I. Dielektrizitätskonstanten und Leitfähigkeiten der Mischkörper aus isotropen Substanzen. *Ann. Phys.* **1935**, *416*, 636–664. [CrossRef]
51. Cody, G. *Hydrogenated Amorphous Silicon*; Pankove, J.I., Ed.; Part B; Academic Press: New York, NY, USA, 1984.
52. Solomon, I. Band-structure determination by subgap spectroscopy in thin films of semiconductors. *Philos. Mag.* B **1997**, *76*, 273–280. [CrossRef]
53. Demirkan, M.; Trahey, L.; Karabacak, T. Low-density silicon thin films for lithium-ion battery anodes. *Thin Solid Films* **2016**, *600*, 126–130. [CrossRef]

Article

Deposition of Super-Hydrophobic Silver Film on Copper Substrate and Evaluation of Its Corrosion Properties

Fani Stergioudi [1], Aikaterini Baxevani [1], Azarias Mavropoulos [1] and Georgios Skordaris [2,*]

[1] ,Physical Metallurgy Laboratory, Mechanical Engineering Department, Aristotle University of Thessaloniki, 54124 Thessaloniki, Greece; fstergio@auth.gr (F.S.); ampaxeva@auth.gr (A.B.); azarias@auth.gr (A.M.)
[2] Laboratory for Machine Tools and Manufacturing Engineering, Mechanical Engineering Department, Aristotle University of Thessaloniki, 54124 Thessaloniki, Greece
* Correspondence: skordaris@auth.gr

Abstract: A simple and versatile chemical solution deposition process is reported to manipulate the wettability properties of copper sheets. The whole process has the advantage of being time-saving low cost and environment-friendly. An adherent silver coating was achieved under optimal conditions. Scanning electron microscopy and X-ray diffraction were used to examine the silver film structure. A confocal microscope was used to record the 3D topography and assess the film roughness of the surface. A dual morphology was revealed, consisting of broad regions with feather-like structured morphologies and some areas with spherical morphologies. Such silver-coated copper samples exhibited a sufficiently stable coating with superhydrophobicity, having a maximum water contact angle of 152°, along with an oleophilic nature. The corrosion behavior of the produced hydrophobic copper under optimal conditions was evaluated by means of potentiodynamic polarization and electrochemical impedance spectroscopy (EIS) using a 3.5% NaCl solution. The corrosion protection mechanism was elucidated by the proposed equivalent circuits, indicating that the superhydrophobic silver coating acted as an effective barrier, separating the Cu substrate from the corrosive solution. The superhydrophobic coating demonstrated enhanced anti-corrosion properties against NaCl aqueous solution in relation to the copper substrate as indicated from both EIS and potentiodynamic polarization experiments.

Keywords: superhydrophobic; deposition process; corrosion properties; copper; equivalent circuit modeling

Citation: Stergioudi, F.; Baxevani, A.; Mavropoulos, A.; Skordaris, G. Deposition of Super-Hydrophobic Silver Film on Copper Substrate and Evaluation of Its Corrosion Properties. Coatings 2021, 11, 1299. https://doi.org/10.3390/coatings11111299

Academic Editor: Heping Li

Received: 4 October 2021
Accepted: 25 October 2021
Published: 27 October 2021

Publisher's Note: MDPI stays neutral with regard to jurisdictional claims in published maps and institutional affiliations.

Copyright: © 2021 by the authors. Licensee MDPI, Basel, Switzerland. This article is an open access article distributed under the terms and conditions of the Creative Commons Attribution (CC BY) license (https://creativecommons.org/licenses/by/4.0/).

1. Introduction

Superhydrophobic surfaces have attracted a lot of attention recently. They have potential uses as corrosion protection systems, in oil–water separation processes, as self-cleaning, anti-icing surfaces and as modified surfaces in bio-applications [1–6]. Different technologies have been utilized to generate rough micro- and nanostructures on metallic surfaces so as to provide superhydrophobicity and simultaneously improve their corrosion resistance [7–9].

The corrosion protection of copper remains an active research area, while numerous researchers attest that the deposition of syperhydrophobic surfaces on copper substrates can act as an effective mean for corrosion protection [10–13]. Preparation of superhydrophobic copper surfaces has been reported in numerous papers by utilizing several techniques such as nanotextured surfaces by using laser beam machining [14] and nanocoatings by employing hydrothermal methods to fabricate CuO film on copper substrate [15]. Several chemical etching and hydrothermal methods, electrodeposition techniques and solution (wet) chemical reaction were also employed such as potentiostatic electrolysis in an ethanolic solution of tetradecanoic acid [16], oxidation in alkaline solutions [17], etching in ammonia solution combined with hydrothermal creation of copper oxide film [18], hydrothermal creation of superhydrophobic Cu_2S film [19], electrolysis in capsaicin ethanolic solution [20], electrodeposition in myristic acid solution [21], chemical etching in an ammonia solution and consequent calcination in air [22], fabrication of multiscale textured surfaces on copper via electrodeposition in $CuSO_4$ and H_2SO_4 [23], an electrochemical modification process in an ethanolic stearic acid solution [24] and electrodeposition to develop multifunctional zinc/polydopamine composite hydrophobic coatings [25].

Nevertheless, most of the production techniques reported in literature suffer certain drawbacks, such as utilization of expensive raw material, long-lasting production stages that involve complicated processes under tight operating conditions or utilization of specific equipment. Consequently, finding a technique to generate a superhydrophobic surface on a copper substrate with a facile and versatile process, which is simultaneously cost effective, requires minimal instrument needs and employs more environmentally friendly procedures, is critical.

An important issue is the mechanical stability and corrosion protection of the hydrophobic coatings. Corrosion protection depends on mechanical stability and is one of the biggest problems facing scientists dealing with hydrophobic surfaces. So far, no hydrophobic coating has been found that meets the requirements for mechanical stability, high corrosion resistance and at the same time is produced by methods that do not require time-consuming stages and specialized equipment. This combination is a challenge and this will be the main scope of this paper.

In this research work, immersion in chemical solutions was used to effectively produce superhydrophobic silver coatings on copper sheet substrates. The procedure is divided into three stages: the growth of a silver film to improve micro roughness, immobilization of the silver film and functionalization with a polydopamine film, and finally, modification using thiol groups. Polydopamine (PDA) coating is broadly used for functionalization of surfaces via wet solution methods since it provides the creation of a functional and homogeneous coating on a wide range of substrates in an one-step reaction [26,27]. An additional advantage is that the polydopamine coatings are environmentally friendly since they are produced by simple self-polymerization of dopamine. Static water contact angle measurements were used to describe the wetting characteristics of the generated hydrophobic silver-coated copper. Scanning electron microscopy (SEM) and X-ray diffraction (XRD) were used to examine the silver film structure (XRD). A confocal microscope was used to record the 3D topography and assess the film roughness of the surface. The corrosion behavior of the produced hydrophobic copper under optimum conditions was evaluated by means of potentiodynamic polarization and electrochemical Impedance Spectroscopy (EIS) using a 3.5% NaCl solution. The production method proposed has the advantage of a simple, low cost and environment-friendly process and aspires to facilitate mass and industrial production of superhydrophobic copper.

2. Materials and Methods
2.1. Synthesis of the Hydrophobic Coating

The synthesis of the hydrophobic silver film on copper substrate was achieved via a four-stage process that involves immersion in solutions with a view to silver coat the copper surface. The final stage involved the modification of the silver coating to enhance the hydrophobicity. The stages followed for the production of the hydrophobic silver coating on copper sheets are schematically shown in Figure 1.

The copper sheets (20 × 20 mm^2) used as a substrates were initially cleaned mechanically by grinding using SiC paper with grit size 1200 grit size. Thereafter, the copper substrates were cleaned chemically in an ultrasonic bath for 15 min using acetone for removing any grease residues. Finally, they were immersed in 0.1 M HCl aqueous solution for 5 min and rinsed by water and ethanol.

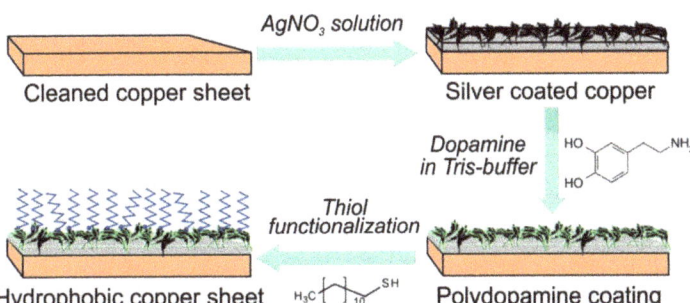

Figure 1. Schematic representation of the stages followed for the production of the hydrophobic silver coating on copper sheets.

The cleaned copper sheets were dipped in an aqueous solution of silver nitrate (AgNO$_3$) (4 mM) under different durations ranging from 5 to 50 min. The solution was continuously stirred (300 rpm) and the temperature was kept constant at 30 °C. The following step was the production of polydopamine coating which was achieved via the self-polymerization of dopamine as follows: 2 g/L dopamine (more specifically a dopamine derivative 3,4-dihydroxyphenylalanine (DOPA)), was introduced to a Tris-buffer solution at a temperature of 40 °C and constant stirring (300 rpm). The samples were then immersed for 20 min. For the preparation of the Tris-buffer solution 10 mM tris-hydroxymethyl-aminomethane (C$_4$H$_{11}$NO$_3$) was dissolved in distilled water acquiring a pH value of 10. The pH was adjusted from 10 to 7.5 with hydrochloric acid down to the desired value. The final step involved the reduction of surface energy which was accomplished by rinsing all specimens in an ethanolic solution of dodecanethiol (DDT) (0.1 M) for 1 h. All immersion stages for the coating process were carried out in a similar-sized beaker, allowing contact with air (oxygen) which is necessary for the self-polymerization of dopamine via oxidation.

2.2. Characterization of Modified Copper Surfaces

For smooth surfaces, static water contact angle (WCA) measurement can be used as an efficient tool to quantify the hydrophobicity. Therefore, static water contact angles were used to assess the wetting properties of the sample surfaces. Each measurement was taken using a 6 µL water droplet, to minimize the gravity impact and accurately determine the WCA.

Scanning electron microscopy (SEM, Bruker, Billerica, MA, USA) was used to evaluate the microstructure of the produced silver films. X-ray diffraction (XRD, 2-circle Rigaku Ultima+ X-ray diffractometer with radiation source of Cu (λ_{Ka} = 0.15406 nm), Rigaku Corporation, Tokyo, Japan) was performed to determine the crystal structure of the produced hydrophobic coatings. The 3D topography and measurement of the surface's film roughness were captured using a confocal microscope (nanofocus µsurf, of NANOFOCUS AG, Oberhausen, Germany).

An ultrasonic procedure was employed to examine the silver coating adhesion and stability. The silver-coated copper sheets were immersed in distilled water and subjected to ultrasound at 25 °C for 60 min. The static water contact angles were measured at regular intervals to evaluate any loss of hydrophobicity provoked by this ultrasound treatment.

2.3. Corrosion Behavior of Silver-Coated Copper Surfaces

Potentiodynamic polarization and Electrochemical Impedance Spectroscopy (EIS) tests were employed to investigate the corrosion properties of the silver-coated copper samples prepared under optimal conditions, thus exhibiting better hydrophobicity as indicated by the static contact angle measurements. As a reference material the corrosion properties of the copper substrate was used. All electrochemical experiments were performed in a three-electrode cell at room temperature (25 °C) using a 3.5% w/v NaCl aqueous solution. A platinum plate and a saturated calomel electrode (SCE) were used as counter and reference electrodes, respectively. The experimental installation of the electrochemical corrosion apparatus was in accordance with ASTM G69-20 [28] and ASTM G71-81 [29]. To ensure a steady state, the samples were immersed in the solution for at least 30 min at open circuit potential (OCP), before each polarization and EIS experiment. The polarization curves were obtained using a scan rate of 1 mV/s and a scan range from −600 to 800 mV (versus SCE) with respect to the OCP value. Corrosion density values were estimated and normalized to the surface area of each sample.

The EIS experiments were carried out at OCP (E_{OCP}) for a frequency range from 0.01 to 100 kHz using peak-to-peak voltage of 10 mV. A wide range of frequency was chosen so as to measure all possible corrosion and diffusion related phenomena taking place on the metal/solution interface. The EIS results were normalized to the surface area of each sample.

3. Results and Discussion

3.1. Influence of Various Parameters in the Synthesis Process of the Hydrophobic Coating

The process of producing the silver (Ag) coating is a crucial step that affects the hydrophobicity of the samples since different Ag film micro-morphologies are obtained. Therefore the deposition time in the $AgNO_3$ solution was investigated in this study. The role of the PDA coating is dual: to stabilize the uniform silver (Ag) coating and to serve as an anchor for the dodecanethiol ($CH_3(CH_2)_{11}SH$) hydrophobic molecule that forms a low-surface-energy self-assembled monolayer. More specifically, the thiol group adheres to the polydopamine coating, leaving the hydrophobic alkyl chain exposed to any solution. It is known that the process of dopamine-polymerization first involves oxidation of a catechol to a benzoquinone and that the formed PDA film comprises of free catechol groups available for further chemical surface modification [21,30,31].

The concentration of dopamine and the buffer were not varied. The "standard conditions", as referred in literature [26,27,32], of 2 g/L dopamine in a Tris buffer, were used for producing surface functionalization of the copper samples. Preliminary experiments using higher concentrations of dopamine were conducted. However, the appearance of white aggregates on the surface of the polydopamine coatings which were visual by bare eye indicated a non-uniform coating process. These aggregates have been noticed in other studies also and were attributed either to un-oxidised dopamine or the presence of oxidized polydopamine in the solution, before reaching the surface of the sample [26–30]. Therefore, the concentration limit was set to 2 g/L.

The pH of the Tris buffer solution was considered a parameter worthy to examine since it is proven that it strongly affects the kinetics of the deposition of dopamine [30,33]. Therefore, the hydrophobicity of the films produced as a function of pH value of Tris buffer solution at a constant dopamine concentration was investigated.

The simultaneous coating of silver and PDA in one beaker was also investigated in order to minimize the stages of the coating production. However, for ad-layer formations (such is our case) the conventional two-step silver and PDA coatings to obtain immobilization of silver coating showed the most promising results. In the case of simultaneous deposition of silver and PDA coatings using one beaker the produced hydrophobic film was unsuccessful, having a contact angle of 140°, possibly due to the disturbance of PDA polymerization and the appearance of aggregations on the surface of the polydopamine coating. Moreover, it is known that the O- and N-sites of PDA can reduce the Ag+ ions to Ag, promoting the formation of single Ag nanoparticles rather than the formation of leaf of dendritic Ag morphologies that are more useful in hydrophobic applications due to the micro-roughness increase that they offer [27].

3.2. Surface Morphology and Wettability of the Prepared Hydrophobic Coatings

Figure 2 depicts the WCA measurements for the produced silver-coated copper samples under different process conditions. Various hydrophobic (WCA > 90°) and superhydrophobic (WCA > 150°) copper surfaces were synthesized via immersion of the copper substrates in silver nitrate solution for durations of 5, 10, 15, 20, 30, 40 and 50 min. The pH effect of the Tris-buffer solution, used for the formation of the PDA coating, on the WCA is also illustrated in Figure 2. Three different pH values were examined. The initial contact angle received from the grinding process of the copper substrates was 128°. The direction of grinding did not influence the contact angle value for the copper samples.

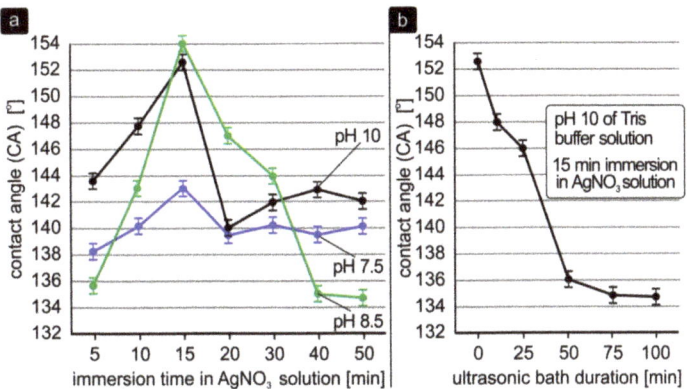

Figure 2. (a) Contact angle relative to the immersion time in silver nitrate solution and pH value of the Tris buffer solution, (b) contact angle after various durations in an ultrasonic bath for the optimum silver-coated superhydrophpobic copper sample.

All the samples exhibited superhydrophobic behaviour for pH 8.5 of the Tris-buffer solution and immersion duration between 10 to 20 min. The water drop after its deposition on the surface showed an inability to hold on to the surface and rolled over the surface immediately. These silver-coated copper samples presented a uniform coating, having a black color, which leads to a superhydrophobic surface angle of 180°. However, these coatings were unstable. After a short period of contact with the environment these surfaces lost the optimal property to repel the water drop, while the WCAs of the droplets stabilized at the values that are depicted in Figure 2a.

From Figure 2a it is clear that for all pH values of Tris buffer solution the WCA increases up to a maximum value, corresponding for all cases examined to 15 min immersion in silver nitrate solution. The slope of the curves is different indicating that different mechanisms and kinetics is probably taking place during PDA coating process at pH 7.5, 8.5 and 10 of the Tris-buffer solution. When the immersion time in silver nitrate solution is longer than 15 min the WCA decreases, irrespective of the pH of the Tris buffer solution until the value of 140–142° for pH 7.5 and 10 of the Tris buffer solution and at the lower value of 134° for pH 8.5 of Tris buffer solution.

Overall, it was deduced that silver-coated copper samples exhibit superhydrophobic behavior with a maximum value of WCA at 154° and 152.5° when produced under 15 min immersion in the silver nitrate solution and pH 8.5 and 10 of the Tris buffer solution, respectively. Because the hydrophobic coatings produced using pH 8.5 of the Tris buffer solution were unstable, it was considered that the optimum value of pH for the Tris buffer solution was pH = 10. Correspondingly, the optimal immersion time in AgNO$_3$ solution was 15 min.

The initial evaluation of the adhesion and stability of the silver coating was performed by immersing the silver-coated samples in an ultrasonic bath for various durations and subsequently measuring the WCA. Possible abruption of the silver film was also investigated by visual inspection and optical microscopy. The hydrophobicity loss, as indicated by the WCA reduction, for the optimal silver-coated copper samples (15 min silver deposition, pH 10 of the Tris buffer solution) is shown in Figure 1 Although the WCA decreased after

50 min in the ultrasonic bath, it stabilized at a constant value of about 135° suggesting no further hydrophobicity loss. It is noteworthy that throughout the experiment no coating abruption was observed even after 90 min in the ultrasonic bath, thus attesting that the stability and cohesion of the developed silver coatings on the copper substrate is sufficient.

The shape and morphology of the silver film obtained is considerably influenced by the immersion time in the nitrate silver solution. The silver salt is dissolved and the Ag cations are deposited via an electroless galvanic process (reaction (1)) to the copper substrate in order to produce a nano- or micro-roughened silver coating as follows [34]:

$$Cu\ (s) + 2AgNO_3\ (aq) \rightarrow Cu(NO_3)_2\ (aq) + 2Ag\ (s) \tag{1}$$

SEM images of the Ag coatings on copper substrate with different Ag deposition times are shown in Figure 3. The initial copper substrate used to develop the hydrophobic silver coatings is also presented, as a reference material.

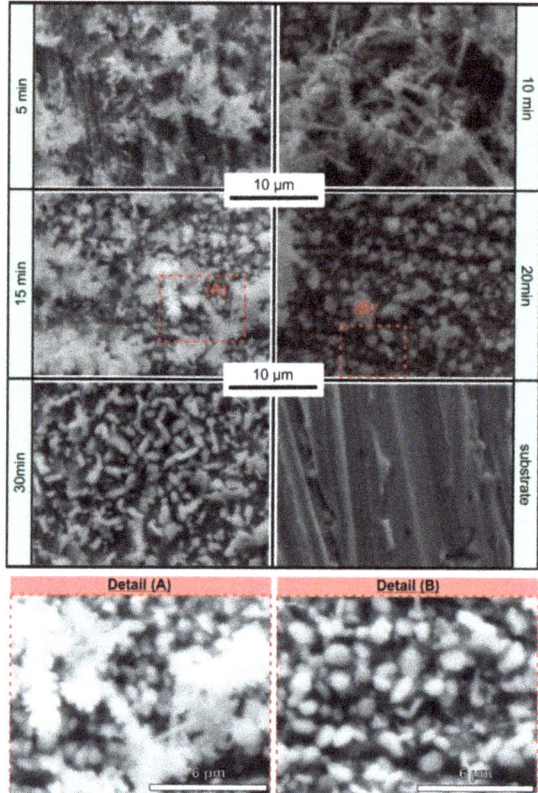

Figure 3. Scanning electron microscope (SEM) images of hydrophobic silver coatings produced using different immersion durations in silver nitrate solution and pH 10 for the Tris buffer solution.

It is known that the wettability properties of a solid surface depend on both the roughness and surface energy [35]. Roughness is a metric of surface topography which directly affects the contact angle values. Therefore, the parameters S_a, S_q, were calculated. S_a is the arithmetical mean height of the surface. S_q is the root mean square height of the surface. The 3D topography of the hydrophobic coatings is shown in Figure 4. The roughness increase was associated with the WCA growth following the same trend with respect to the immersion duration.

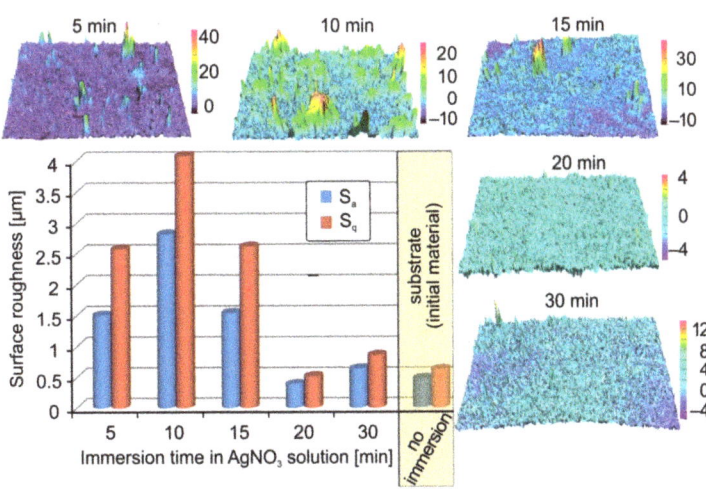

Figure 4. Surface roughness of silver-coated copper samples relative to the immersion duration in silver nitrate solution (using pH 10 for the Tris buffer solution) along with corresponding 3D imaging of surface topography.

When the immersion time in the silver nitrate solution was 5 min, the coating presented discontinuities. SEM images verified this observation since the copper substrate was visible (Figure 3). By increasing the immersion time to 10 min, the development of needle-like structures was observed, derived by self-organization of silver ions. However, the 3D topography of the surface revealed several clusters (of various sizes), indicating sufficient heterogeneity in the morphologically of the coating (Figure 4).

Further increase of the immersion time to 15 min, gave rise to new morphologies and different topography of the coating produced. More specifically, a collapse of the agglomerates was observed whilst the needle-like morphologies almost disappear. Instead, dendritic or feather-like morphologies (see detail A of Figure 3) were distinguished. A second morphology was also pronounced, characterized by spherical agglomerates of approximately 1 μm. The results from the 3D topography for the silver-coated copper surface that was immersed in the silver nitrate of solution for 15 min, indicated good homogeneity presenting only few large aggregates (Figure 4). When the immersion time increased to 20 min, the feather-like structures decomposed, whilst larger aggregates of Ag with polyhedral morphology, became visible. This morphology (polyhedral aggregates) was also visible for the samples that were immersed for 30 min in silver nitrate solution. As shown in Figure 4, for immersion duration higher than 20 min the silver coating was homogeneous and uniform. However, the roughness of this silver-coated copper surface decreased and resembled the roughness of the initial copper substrate which was attributed to the polyhedral and spherical structural morphologies of the coating. The hydrophobicity in this case diminished, with the WCA ranging from 140° to 143°.

The self-organization process for the Ag ions is well documented in literature [36–41]. Several mechanisms have been proposed while the reduction capacity of the substrate is stated as a key factor for the creation of dendrite-leaf morphologies [42]. In all cases the process begins with initial random reduction of Ag ions by the copper substrate and creation of nuclei that then grow to form primary nanoscale particles that further aggregate to form feather like structures or spherical morphologies. The immersion duration is also a significant factor that affects the 3D topography of the obtained silver coating (Figures 3 and 4) [36,40]. The SEM and micro-roughness results are consistent with the WCA measurements indicating that silver deposition time of 15 min is optimal to acquire a superhydrophobic silver film.

Figure 5a depicts a silver-coated copper sample produced under optimal conditions. The produced silver film exhibits good structural stability, superhydrophobic properties, (water contact angle of 152°) and displays an oleophilic nature.

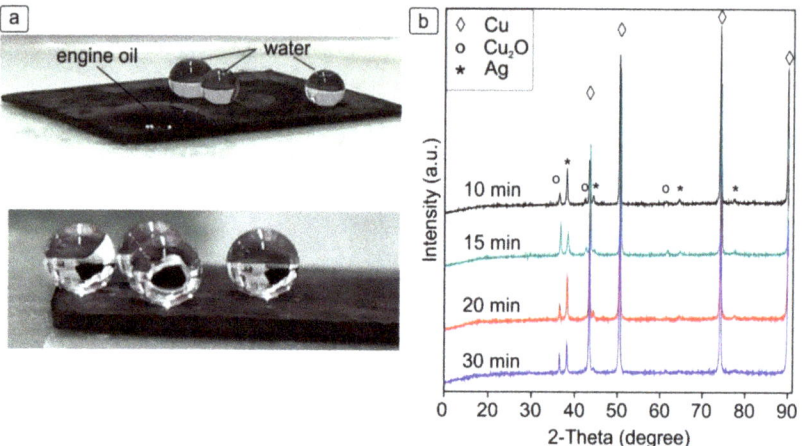

Figure 5. (**a**) Typical images of water droplets of about 6 µL on the surface of silver-silver-coated copper prepared under optimum conditions. (**b**) X-ray diffraction (XRD) results for hydrophobic silver-coated copper samples prepared under different immersion durations in silver nitrate solution and pH 10 for the Tris buffer solution.

XRD was used to examine the phases of the produced hydrophobic silver-coating under different immersion durations in silver nitrate solution and for pH 10 of the Tris-buffer solution. The XRD patterns revealed the presence of Ag for all immersion durations in AgNO$_3$ solution. Phases of Cu$_2$O oxide were also detected in all cases (Figure 5b), which probably can be ascribed to the decomposition of Cu(NO$_3$)$_2$ in copper oxide [34]. For all immersion durations the Ag diffraction peaks observed were the same, indexed as (111), (200), (220), and (311) planes of cubic Ag with a unit cell parameter equal to 4.0816 Å. This result is consistent with the XRD results of previous works [34,43,44]. However, it should be noted that for immersion times of 20 and 30 min the (220) and (311) diffraction peaks were less intense and almost disappear indicating different growth mechanism of the Ag film at those immersion durations.

3.3. Electrochemical Corrosion Results

3.3.1. Polarization Measurements

All corrosion experiments were performed at silver-coated copper samples prepared under optimal conditions, e.g., at pH 10 for the Tris buffer solution and immersion time of 15 min in silver nitrate solution.

The OCP plots for the silver-coated copper and the bare substrate are shown in Figure 6a. After approximately 10 min of exposure a stable state of the E_{OCP} is reached. The corrosion potential (E_{corr}), corrosion current (i_{corr}), Tafel slopes, polarization resistance R_p and corrosion rate are presented in Table 1 whist the polarization curves for the copper substrate and the silver-coated copper are shown in Figure 6b.

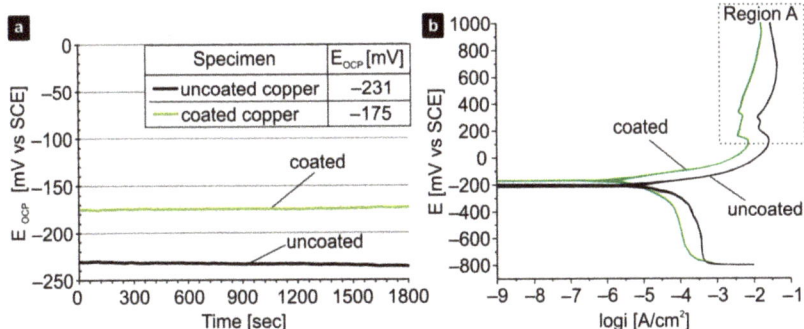

Figure 6. (a) Open circuit potential (OCP) and (b) polarization curves for silver-coated copper and copper substrate in 3.5 wt.% NaCl aqueous solution at a temperature of 25 °C.

Table 1. Polarization parameters.

Specimen	E_{corr} [mV]	i_{corr} [μA/cm^2]	βc [mV]	βa [mV]	R_p [kohm/cm^2]	Corrosion Rate μm/year
Copper substrate	−214	2	36.8	−53.7	0.82	24.06
Silver-coated copper	−170	0.5	38.3	−49.8	10.41	15.961

In terms of corrosion potential the silver-coated copper demonstrated a more electropositive (−174 mV vs. SCE) value than the uncoated copper (~−214 mV vs. SCE). Regarding the current density i_{corr}, the silver-coated copper (i_{corr} = 0.5 μA/cm^2) displayed almost an order of magnitude smaller value compared to that of the copper substrate (i_{corr} = 2 μA/cm^2). Similar values for the current density i_{corr} of superhydrophobic copper were found in literature [12,15,21]. The values of the polarization resistance are consistent with those of the current density showing increased resistance for the silver-coated copper.

Both the reduced value of corrosion current i_{corr} and larger polarization resistance R_p indicated superior corrosion properties of the silver-coated copper sample. The superhydrophobic silver film, which was developed on the copper substrate, sifted the anodic and cathodic polarization curve to lower current densities, indicating thus an increased corrosion resistance. Concerning the anodic polarization part of Figure 6b an active dissolution nature is observed in both samples. However, as the potential rises the anodic current stabilizes and its values is independent of the voltage, suggesting an attenuation of the dissolution process (region A in Figure 6b). The anodic current density, at this region of reduced dissolution rate, is 50 times higher for the uncoated copper than that of the silver-coated copper (about 0.50 A/cm^2 for the uncoated copper and 0.01 A/cm^2 for the silver-coated copper). For the uncoated copper the two current density peaks apparent at the less active dissolution are ascribed to formation of Cu (I) and Cu (II) corrosion products probably due to reaction with Cl$^-$ as reported in literature [45]. The same peaks are apparent for the silver-coated copper; however, they are not so intensive. This might be an indication that the hydrophobic coating retards the dissolution of copper through the interface of the silver surface and finally to NaCl solution.

The WCA of the silver-coated copper sample after the polarization tests amounted of 142°. Compared to the initial WCA value of 152° only a 7% decrease was observed, indicating that the hydrophobicity loss was negligible. This observation could verify the existence of a reasonably stabilized air layer within the roughness of the silver-coated samples that was maintained throughout the entire polarization test. The formation of a relatively stable air film is considered as a sufficient corrosion protection mean since it leads to a reduced contact surface of the specimen with the solution [11–13].

3.3.2. Electrochemical Impedance Spectroscopy (EIS)

Impedance plots (Nyquist and Bode) may be used to evaluate the overall corrosion properties of materials as well as to elucidate the acting corrosion mechanism. The development of localized corrosion can be recognized in the kHz regions of the Bode plot, the global corrosion rate is related to the Hz zone while any adsorption of corrosion products or diffusion of ions is related to the mHz region of the Bode plot. [46,47].

For silver-coated copper, the Nyquist plot (Figure 7a) reveals a semicircle that begins at high frequencies and terminates at low frequencies, with the maximum imaginary value of impedance lying at 0.158 Hz ($-33°$). The total impedance of silver-coated copper approaches a distinct DC limit (15.85 kohm cm^2) at low frequencies (0.01 Hz) while at frequencies exceeding 10^3 Hz reaches a value close to 80 ohm cm^2, according to the Bode plot (Figure 7b). The existence of a distinct DC limit may imply the absence of a diffusion process and pitting corrosion [47]. Two time constants for silver-coated copper are apparent at around 5.62 and 13.8 kHz, according to Bode plots. At phase angle–frequency curves, the existence of two or more time constants, together with a phase angle divergence from $-90°$, may suggest the development of inhomogeneous and irregularly dispersed reaction products on the sample surfaces [46,47].

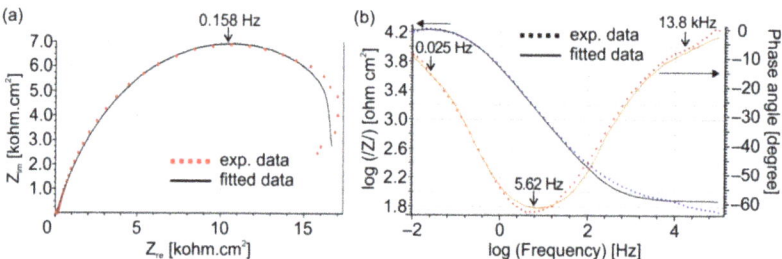

Figure 7. (a) Nyquist and (b) Bode plots of silver-coated copper in 3.5 wt.% NaCl aqueous solution.

For uncoated copper the Nyquist plots reveal a semicircle at high frequencies and a straight line at low frequencies (Figure 8a), which may be attributed to ion diffusion processes as well as any intermediate corrosion product absorbed on the specimen's surface [11,18,20]. High-frequency loops, in general, can be physically associated to the existence of corrosion products, whereas low-frequency loops can characterize the overall corrosion process [46,47].

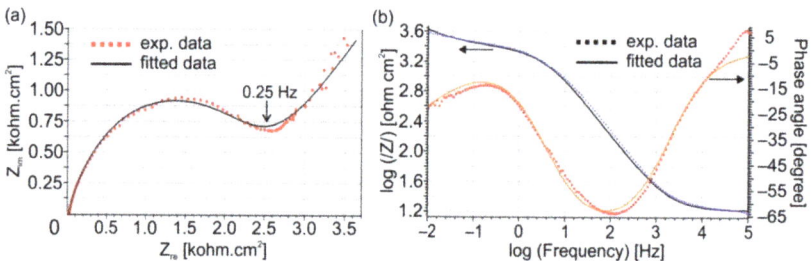

Figure 8. (a) Nyquist and (b) Bode plots of uncoated copper substrate in 3.5 wt.% aqueous solution NaCl.

At low frequencies (0.1 Hz), the slope of the impedance-frequency plot curve for uncoated copper (Figure 8b) remains stable, creating a continual rise in total impedance without establishing at an obvious DC limit. Furthermore, for frequencies higher than $10^{3.5}$ Hz, the impedance approaches zero value. The steady rise in impedance at low frequencies (below 0.1 Hz) confirms the presence of a diffusion mechanism that governs the electrochemical reactions as well as the initiation of pitting corrosion on the test piece surface [46].

The Nyquist plots reveal that the hydrophobic silver-coated copper has higher ohmic and capacitive resistance (real and imaginary Z, respectively) which is connected with a decrease in corrosion rates. This result is consistent with the ones obtained by polarization measurements. The total impedance values of silver-coated copper are larger than those of uncoated samples. Finally, the total impedance–frequency curve tendencies corroborate with the results of Nyquist plots.

3.3.3. Proposed Equivalent Electrical Circuit Models

The corrosion behavior of hydrophobic surfaces in various solutions remains a major research topic and is well-documented in literature [10–16,18–23]. However, the simulation of experimental EIS data using equivalent electrical circuits that have a coherent physical interpretation remains a controversial issue. Hereupon, diverse models have been proposed in the literature for identical hydrophobic materials under similar corrosive conditions [15,18,20–23]. In any event, the goal of these simulations is to connect various electrical circuit parts to electrochemical processes that take place at existing or generated interfaces (e.g., metal/'double' layer/solution), providing better knowledge of the corrosion mechanism.

Two equivalent electrical circuits models are proposed (see Figure 9a,b) based on the microstructures, XRD analysis (Figures 3 and 4b) and impedance curves (Figures 7 and 8) of the silver-coated and uncoated copper samples. The goal was to minimize the number of elements used and simultaneously provide an appropriate fitting of experimental and simulated results. Additionally, the proposed models also provide the most rational and consistent physical interpretation of the results obtained.

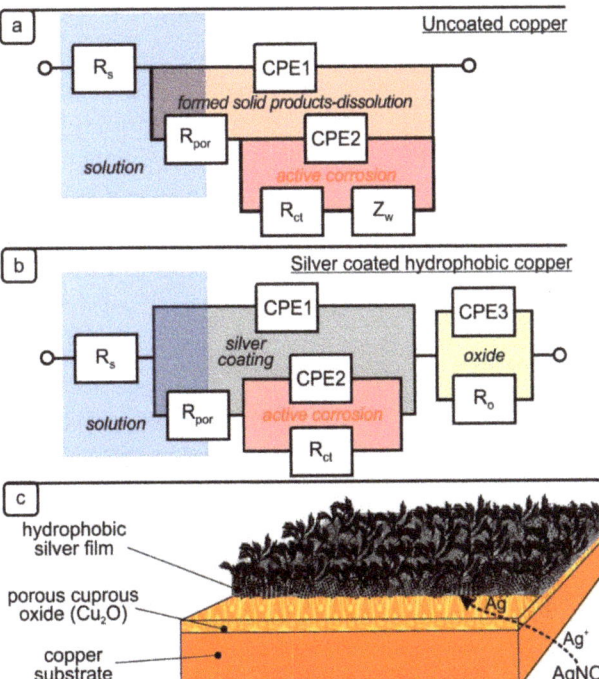

Figure 9. Proposed equivalent circuit representing the corrosion behavior of (**a**) uncoated copper and (**b**) silver-coated hydrophobic copper. (**c**) Schematic illustration of the silver coating structure used for representing the corrosion behavior of hydrophobic copper.

The proposed equivalent electrical circuit model for copper substrate (Figure 9a) comprises of: the electrolyte solution resistance R_s, a constant phase element CPE1 with an impedance Z_{CPE1} which represents the capacitance of the formed solid products on the surface of dense copper, the solution resistance R_{por} within a formed pore or pit on the solid surface related with solid product dissolution, a charged transfer resistance R_{ct} which characterizes the polarization of the copper due to the formation of the electric double layer on the interface between copper and the solution, a constant phase element CPE2 with impedance Z_{CPE2} which represents the capacitance contribution C_{dl} of the formed electric double layer on the interface between the copper substrate and the solution and a Warburg resistance Z_w related to the diffusion of the chloride ions and oxidants to the electric double layer on the interface between the copper and the solution.

Respectively, the proposed circuit for the silver-coated hydrophobic copper sample is that of the Figure 9b. It comprises of: the electrolyte solution resistance R_s, an constant phase element CPE1 with an impedance Z_{CPE1} which represents the capacitance C_C of the bulk silver coating (outside the pores), the pore resistance R_{por} represents the solution resistance within the formed pore and physically represents the ionic charge transfer through the aqueous or semi-aqueous pores and channels of the coating within the silver coating bulk, the C_C is the ability of the coating bulk to store charge, the R_{ct} charged transfer resistance which represents the polarization reaction between the copper–solution interface (inside the pore or channels that are formed) due to the electric double layer, an constant phase element CPE2 with an impedance Z_{CPE2} which represents the capacitance C_{dl} of the formed electric double layer between the copper–solution interface (inside the pore or channels that are formed). In terms of physical interpretation, this corresponds to current passing through the coating via ionic charge transfer, leading in either charge transfer across the interface (corrosion) or charge storage at the interface [21,22]. The R_o resistance characterizes the oxide layer of the silver-coated copper. The impedance Z_{CPE3} of the constant phase element CPE3 represents the capacitance C_O of the oxide layer which was detected through the XRD analysis.

Table 2 lists the parameters of the equivalent circuit elements of Figure 9. The fitting curves in Figures 7 and 8 are well adapted to the experimental ones, indicating that the proposed equivalent electric circuits are consistent with the experimental data (calculated standard error 6%).

Table 2. Best fitting parameters obtained using the proposed equivalent circuits, delineating the corrosion behavior of uncoated and silver silver-coated copper.

Material	R_s Ohm cm^2	R_{por} Ohm cm^2	CPE1 $F \times s^{\alpha-1}$	α_1	R_{ct} Ohm cm^2	CPE2 $F \times s^{\alpha-1}$	α_2	Z_w Ohm s$^{-0.5}$	R_o Ohm cm^2	CPE3 $F s^{\alpha-1}$	α_3
Copper substrate	6.57	721	67×10^{-6}	0.82	717	228×10^{-6}	0.99	148	-	-	-
Coated copper	22.31	5446	110×10^{-6}	0.89	2229	1.96×10^{-6}	0.53	-	18.76	155×10^{-6}	0.62

A relatively homogeneous silver film covering almost entirely the copper substrate is reported according to the equivalent circuit model for the hydrophobic silver-coated copper. The relatively small divergence of coefficient α_1 from 1 suggests that the silver coating is fairly homogeneous (Table 2). There are, however, locations where metastable micro-pits or pores emerge that may facilitate the creation of active corrosion zones. Nevertheless, there is no evidence of significant ion diffusion (lack of diffusion resistance), which is most likely due to only locally scaled deterioration of the silver-coated coating. This means that the degradation of the hydrophobic film has not evolved to an extent that would allow chorine ions to freely penetrate the silver coating and move towards the metal surface indicating that a fairly stable air film was formed [18,21]. Additionally, any electrochemical reactions of copper corrosion can be attributed only to an early stage of corrosion evolution and not to extended corrosion result.

These results are supported by the values shown of Table 2. More specifically, the value of Z_{CPE2} impedance is closely linked with active corrosion regions existing inside any created pore/pit. The decreased value for the silver-coated copper thus indicates enhanced corrosion resistance. The large divergence of α_2 from value 1 confirms the large instability of any formed corrosion products and the inhomogeneity of the electric double layer formed inside the pits.

For the silver-coated copper substrate the resistance R_{por}, associated with the solution properties, is one order of magnitude higher for the silver-coated copper indicating also that the interfacial coating properties in this case, play a dominant role in the evolution of corrosion. The higher values for R_{ct} for silver-coated copper indicate also superior resistance of the sample in general corrosion in relation to the copper substrate [11,18,20].

Finally, coefficient α_3 significantly deviates from 1 which indicates that the oxide layer is heterogeneous or has a porous nature (Figure 9c). The very low value of resistance Ro may indicate small thicknesses of the oxide thus allowing easy transfer of ions from the oxide layer through the copper substrate. Therefore it is suggested that the existence of the oxide layer does not contribute to any corrosion resistance.

4. Conclusions

Superhydrophobic silver coatings were successfully deposited on copper sheets substrates via immersion in chemical solutions. The surface morphologies were easily controlled by varying process conditions. The whole methodology is fast, simple and flexible. The influence of process conditions on wettability and the obtained silver film morphology and topography were examined and clarified. Under optimal conditions, a coherent silver coating with a dual morphology was produced, comprising of large regions with feather like structured and small regions with spherical morphologies. The produced silver-coated copper sheets were superhydrophobic and exhibited sufficient adhesion and stability. The superhydrophobicity of the silver coating derived from the synergistic effect of their surface microstructures and the modification with low energy groups of thiols. The superhydrophobic coating demonstrated enhanced anti-corrosion properties against NaCl aqueous solution in relation to the copper substrate as indicated from both EIS and potentiodynamic polarization experiments.

Author Contributions: Conceptualization, F.S. and G.S.; methodology, F.S., A.B. and A.M.; validation, F.S., A.B. and A.M.; formal analysis, F.S. and A.B.; investigation, A.B. and A.M.; resources, F.S.; data curation, F.S. and A.B.; writing—original draft preparation, F.S., A.B.; writing—review and editing, F.S. and G.S.; visualization, F.S. and A.B.; supervision, F.S.; project administration, F.S.; funding acquisition, F.S. All authors have read and agreed to the published version of the manuscript.

Funding: This research was funded by Research Committee, Aristotle University of Thessaloniki, grant number 89251.

Institutional Review Board Statement: Not applicable.

Informed Consent Statement: Not applicable.

Data Availability Statement: Data is contained within the article.

Acknowledgments: The authors would like to express its deepest gratitude to Georgios Vourlias and Eleni Pavlidou for technical support and assistance with the XRD and SEM studies.

Conflicts of Interest: The authors declare no conflict of interest.

References

1. Ellinas, K.; Tserepi, A.; Gogolides, E. Durable Superhydrophobic and superamphiphobic polymeric surfaces and their applications: A review. *Adv. Colloid Interface Sci.* **2017**, *250*, 132–157. [CrossRef]
2. Barati Darband, G.; Aliofkhazraei, M.; Khorsand, S.; Sokhanvar, S.; Kaboli, A. Science and engineering of superhydrophobic surfaces: Review of corrosion resistance, chemical and mechanical stability. *Arab. J. Chem.* **2020**, *13*, 1763–1802. [CrossRef]

3. Wang, S.; Liu, K.; Yao, X.; Jiang, L. Bioinspired surfaces with superwettability: New Insight on theory, design, and applications. *Chem. Rev.* **2015**, *115*, 8230–8293. [CrossRef] [PubMed]
4. Liu, K.; Jiang, L. Metallic Surfaces with Special Wettability. *Nanoscale* **2011**, *3*, 825–838. [CrossRef] [PubMed]
5. Vanithakumari, S.C.; Choubey, A.K.; Thinaharan, C.; Gupta, R.K.; George, R.P.; Kaul, R.; Bindra, K.S.; Philip, J. Laser patterned titanium surfaces with superior antibiofouling, superhydrophobicity, self-cleaning and durability: Role of line spacing. *Surf. Coat. Technol.* **2021**, *418*, 127257. [CrossRef]
6. Wu, S.-T.; Huang, C.-Y.; Weng, C.-C.; Chang, C.-C.; Li, B.-R.; Hsu, C.-S. Rapid prototyping of an open-surface microfluidic platform using wettability-patterned surfaces prepared by an atmospheric-pressure plasma jet. *ACS Omega* **2019**, *4*, 16292–16299. [CrossRef]
7. Bi, P.; Li, H.; Zhao, G.; Ran, M.; Cao, L.; Guo, H.; Xue, Y. Robust super-hydrophobic coating prepared by electrochemical surface engineering for corrosion protection. *Coatings* **2019**, *9*, 452. [CrossRef]
8. Dimitrakellis, P.; Gogolides, E. Hydrophobic and superhydrophobic surfaces fabricated using atmospheric pressure cold plasma technology: A review. *Adv. Colloid Interface Sci.* **2018**, *254*, 1–21. [CrossRef]
9. Hafiz, M.A.; Muhammad, A.Q.; Sullahuddin, M.; Ghulam, M. Techniques for the Fabrication of Super-Hydrophobic Surfaces and Their Heat Transfer Applications. In *Heat Transfer—Models, Methods and Applications*; Konstantin, V., Ed.; IntechOpen: London, UK, 2018. Available online: https://www.intechopen.com/chapters/58728 (accessed on 12 May 2021).
10. Fateh, A.; Aliofkhazraei, M.; Rezvanian, A.R. Review of corrosive environments for copper and its corrosion inhibitors. *Arab. J. Chem.* **2020**, *13*, 481–544. [CrossRef]
11. Liu, L.; Chen, R.; Liu, W.; Zhang, Y.; Shi, X.; Pan, Q. Fabrication of superhydrophobic copper sulfide film for corrosion protection of copper. *Surf. Coat. Technol.* **2015**, *272*, 221–228. [CrossRef]
12. Feng, L.; Zhao, L.; Qiang, X.; Liu, Y.; Sun, Z.; Wang, B. Fabrication of superhydrophobic copper surface with excellent corrosion resistance. *Appl. Phys. A Mater. Sci. Process.* **2015**, *119*, 75–83. [CrossRef]
13. Zhang, B.; Feng, H.; Lin, F.; Wang, Y.; Wang, L.; Dong, Y.; Li, W. Superhydrophobic surface fabricated on iron substrate by black chromium electrodeposition and its corrosion resistance property. *Appl. Surf. Sci.* **2016**, *378*, 388–396. [CrossRef]
14. Chun, D.M.; Ngo, C.V.; Lee, K.M. Fast Fabrication of superhydrophobic metallic surface using nanosecond laser texturing and low-temperature annealing. *CIRP Ann.-Manuf. Technol.* **2016**, *65*, 519–522. [CrossRef]
15. Fan, Y.; Chen, Z.; Liang, J.; Wang, Y.; Chen, H. Preparation of superhydrophobic films on copper substrate for corrosion protection. *Surf. Coat. Technol.* **2014**, *244*, 1–8. [CrossRef]
16. Wang, P.; Qiu, R.; Zhang, D.; Lin, Z.; Hou, B. Fabricated Super-Hydrophobic Film with Potentiostatic Electrolysis Method on Copper for Corrosion Protection. *Electrochim. Acta* **2010**, *56*, 517–522. [CrossRef]
17. Talesh Bahrami, H.R.; Ahmadi, B.; Saffari, H. Optimal Condition for Fabricating Superhydrophobic Copper Surfaces with Controlled Oxidation and Modification Processes. *Mater. Lett.* **2017**, *189*, 62–65. [CrossRef]
18. Wan, Y.; Chen, M.; Liu, W.; Shen, X.X.; Min, Y.; Xu, Q. The research on preparation of superhydrophobic surfaces of pure copper by hydrothermal method and its corrosion resistance. *Electrochim. Acta* **2018**, *270*, 310–318. [CrossRef]
19. Feng, L.; Wang, J.; Shi, X.; Chai, C. Superhydrophobic copper surface with mechanical, chemical, and UV durability along with corrosion resistance and self-cleaning effect. *Appl. Phys. A Mater. Sci. Process.* **2019**, *125*, 261. [CrossRef]
20. Wang, P.; Zhang, D.; Qiu, R.; Wan, Y.; Wu, J. Green approach to fabrication of a super-hydrophobic film on copper and the consequent corrosion resistance. *Corros. Sci.* **2014**, *80*, 366–373. [CrossRef]
21. Liu, Y.; Li, S.; Zhang, J.; Liu, J.; Han, Z.; Ren, L. Corrosion inhibition of biomimetic super-hydrophobic electrodeposition coatings on copper substrate. *Corros. Sci.* **2015**, *94*, 190–196. [CrossRef]
22. Liu, W.; Xu, Q.; Han, J.; Chen, X.; Min, Y. A novel combination approach for the preparation of superhydrophobic surface on copper and the consequent corrosion resistance. *Corros. Sci.* **2016**, *110*, 105–113. [CrossRef]
23. Mousavi, S.M.A.; Pitchumani, R. A Study of corrosion on electrodeposited superhydrophobic copper surfaces. *Corros. Sci.* **2021**, *186*, 109420. [CrossRef]
24. Huang, Y.; Sarkar, D.K.; Gallant, D.; Chen, X.G. Corrosion resistance properties of superhydrophobic copper surfaces fabricated by one-step electrochemical modification process. *Appl. Surf. Sci.* **2013**, *282*, 689–694. [CrossRef]
25. Wang, H.; Zhu, Y.; Hu, Z.; Zhang, X.; Wu, S.; Wang, R.; Zhu, Y. A Novel Electrodeposition route for fabrication of the superhydrophobic surface with unique self-cleaning, mechanical abrasion and corrosion resistance properties. *Chem. Eng. J.* **2016**, *303*, 37–47. [CrossRef]
26. Ding, Y.H.; Floren, M.; Tan, W. Mussel-inspired polydopamine for bio-surface functionalization. *Biosurface Biotribol.* **2016**, *2*, 121–136. [CrossRef] [PubMed]
27. Liebscher, J. Chemistry of polydopamine—scope, variation, and limitation. *Eur J. Org. Chem.* **2019**, *2019*, 4976–4994. [CrossRef]
28. *ASTM G69-20, Standard Test Method for Measurement of Corrosion Potentials of Aluminum Alloys*; ASTM International: West Conshohocken, PA, USA, 2020.
29. *ASTM G71-81, Standard Guide for Conducting and Evaluating Galvanic Corrosion Tests in Electrolytes*; ASTM International: West Conshohocken, PA, USA, 2019.
30. Lakshminarayanan, R.; Madhavi, S.; Poh Choo Sim, C. *Oxidative polymerization of dopamine: A High-Definition Multifunctional Coatings for Electrospun Nanofibers In Dopamine—Health and Disease*; Yenisetti, S.C., Ed.; IntechOpen: London, UK, 2018. Available online: https://www.intechopen.com/chapters/63723 (accessed on 12 May 2021). [CrossRef]

31. Qiu, W.Z.; Yang, H.C.; Xu, Z.K. Dopamine-Assisted Co-Deposition: An emerging and promising strategy for surface modification. *Adv. Colloid Interface Sci.* **2018**, *256*, 111–125. [CrossRef]
32. Wang, N.; Wang, Y.; Shang, B.; Wen, P.; Peng, B.; Deng, Z. Bioinspired one-step construction of hierarchical superhydrophobic surfaces for oil/water separation. *J. Colloid Interface Sci.* **2018**, *531*, 300–310. [CrossRef]
33. Della Vecchia, N.F.; Luchini, A.; Napolitano, A.; Derrico, G.; Vitiello, G.; Szekely, N.; Dischia, M.; Paduano, L. Tris buffer modulates polydopamine growth, aggregation, and paramagnetic properties. *Langmuir* **2014**, *30*, 9811–9818. [CrossRef]
34. Feng, L.; Yang, M.; Shi, X.; Liu, Y.; Wang, Y.; Qiang, X. Colloids and surfaces A: Physicochemical and engineering aspects copper-based superhydrophobic materials with long-term durability, stability, regenerability, and self-cleaning property. *Physicochem. Eng. Asp.* **2016**, *508*, 39–47. [CrossRef]
35. Boscher, N.D.; Vaché, V.; Carminati, P.; Grysan, P.; Choquet, P. A Simple and scalable approach towards the preparation of superhydrophobic surfaces-importance of the surface roughness skewness. *J. Mater. Chem. A* **2014**, *2*, 5744–5750. [CrossRef]
36. Che, P.; Liu, W.; Chang, X.; Wang, A.; Han, Y. Multifunctional silver film with superhydrophobic and antibacterial properties. *Nano Res.* **2016**, *9*, 442–450. [CrossRef]
37. Liu, Y.; Zhang, K.; Son, Y.; Zhang, W.; Spindler, L.M.; Han, Z.; Ren, L. A smart switchable bioinspired copper foam responding to different ph droplets for reversible oil-water separation. *J. Mater. Chem. A* **2017**, *5*, 2603–2612. [CrossRef]
38. Zhou, W.; Li, G.; Wang, L.; Chen, Z.; Lin, Y. A Facile Method for the Fabrication of a superhydrophobic polydopamine-coated copper foam for oil/water separation. *Appl. Surf. Sci.* **2017**, *413*, 140–148. [CrossRef]
39. Li, P.; Chen, X.; Yang, G.; Yu, L.; Zhang, P. Preparation of silver-cuprous oxide/stearic acid composite coating with superhydrophobicity on copper substrate and evaluation of its friction-reducing and anticorrosion abilities. *Appl. Surf. Sci.* **2014**, *289*, 21–26. [CrossRef]
40. Safaee, A.; Sarkar, D.K.; Farzaneh, M. Superhydrophobic properties of silver-coated films on copper surface by galvanic exchange reaction. *Appl. Surf. Sci.* **2008**, *254*, 2493–2498. [CrossRef]
41. Qu, L.; Dai, L. Novel silver nanostructures from silver mirror reaction on reactive substrates. *J. Phys. Chem. B* **2005**, *109*, 13985–13990. [CrossRef]
42. Fang, J.; You, H.; Kong, P.; Yi, Y.; Song, X.; Ding, B. Dendritic silver nanostructure growth and evolution in replacement reaction. *Cryst. Growth Des.* **2007**, *7*, 864–867. [CrossRef]
43. Tu, S.H.; Wu, H.C.; Wu, C.J.; Cheng, S.L.; Sheng, Y.J.; Tsao, H.K. Growing hydrophobicity on a smooth copper oxide thin film at room temperature and reversible wettability transition. *Appl. Surf. Sci.* **2014**, *316*, 88–92. [CrossRef]
44. Lee, S.M.; Kim, K.S.; Pippel, E.; Kim, S.; Kim, J.H.; Lee, H.J. Facile route toward mechanically stable superhydrophobic copper using oxidation-reduction induced morphology changes. *J. Phys. Chem. C* **2012**, *116*, 2781–2790. [CrossRef]
45. Kear, G.; Barker, B.D.; Walsh, F.C. Electrochemical corrosion of unalloyed copper in chloride media—A critical review. *Corros. Sci.* **2004**, *46*, 109–135. [CrossRef]
46. Stergioudi, F.; Vogiatzis, C.A.; Pavlidou, E.; Skolianos, S.; Michailidis, N. Corrosion resistance of porous NiTi biomedical alloy in simulated body fluids. *Smart Mater. Struct.* **2016**, *25*, 095024. [CrossRef]
47. Stergioudi, F.; Vogiatzis, C.A.; Gkrekos, K.; Michailidis, N.; Skolianos, S.M. Electrochemical corrosion evaluation of pure, carbon-coated and anodized Al foams. *Corros. Sci.* **2015**, *91*, 151–159. [CrossRef]

Article

Assessment of Stone Protective Coatings with a Novel Eco-Friendly Encapsulated Biocide

Martina Zuena, Ludovica Ruggiero, Giulia Caneva, Flavia Bartoli, Giancarlo Della Ventura, Maria Antonietta Ricci and Armida Sodo *

Dipartimento di Scienze, Università degli Studi "Roma Tre", Via della Vasca Navale 84, 00146 Roma, Italy; martina.zuena@uniroma3.it (M.Z.); ludovica.ruggiero@uniroma3.it (L.R.); giulia.caneva@uniroma3.it (G.C.); flavia.bartoli@uniroma3.it (F.B.); giancarlo.dellaventura@uniroma3.it (G.D.V.); mariaantonietta.ricci@uniroma3.it (M.A.R.)
* Correspondence: armida.sodo@uniroma3.it

Abstract: The conservation of stone monuments is a constant concern due to their continuous weathering, in which biofouling plays a relevant role. To enhance the effectiveness of biocidal treatments and to avoid environmental issues related to their possible toxicity, this research aims at formulating and characterizing a coating charged with an eco-friendly biocide and showing hydrophobic properties. For this purpose, zosteric sodium salt—a natural biocide product—has been encapsulated into two silica nanocontainers and dispersed into a tetraethoxysilane-based (TEOS) coating also containing TiO_2 nanoparticles. The coatings were applied on four different types of stone: brick, mortar, travertine, and Carrara marble. The effectiveness of the coating formulations and their compatibility concerning the properties of coated stones were assessed. The results showed that all coatings conferred a hydrophobic character to the substrate, as demonstrated by the increase of the static contact angle and the reduction in the capillary water absorption coefficient. The transmission of water vapor of the natural stones was preserved as well as their natural aspect. Furthermore, the coatings were homogeneously distributed on the surface and crack-free. Therefore, the protective capability of the coatings was successfully demonstrated.

Keywords: stone protection; monument biodeterioration; Si nanocontainers; antifouling; zosteric sodium salt; TiO_2 nanoparticles

Citation: Zuena, M.; Ruggiero, L.; Caneva, G.; Bartoli, F.; Della Ventura, G.; Ricci, M.A.; Sodo, A. Assessment of Stone Protective Coatings with a Novel Eco-Friendly Encapsulated Biocide. *Coatings* **2021**, *11*, 1109. https://doi.org/10.3390/coatings11091109

Academic Editor: Giorgos Skordaris

Received: 28 July 2021
Accepted: 12 September 2021
Published: 14 September 2021

Publisher's Note: MDPI stays neutral with regard to jurisdictional claims in published maps and institutional affiliations.

Copyright: © 2021 by the authors. Licensee MDPI, Basel, Switzerland. This article is an open access article distributed under the terms and conditions of the Creative Commons Attribution (CC BY) license (https://creativecommons.org/licenses/by/4.0/).

1. Introduction

Stone materials are widely used in architecture and artistic objects due to their durability, esthetics, availability, and easy manufacturing. Nevertheless, weathering processes may determine mechanical and esthetical concerns over time.

The most important weathering processes are connected to water penetrating the stone by the capillary rise or rainfall [1–4]. In particular, the presence of water may promote microorganism proliferation. This phenomenon is known as biodeterioration in the field of cultural heritage [4,5] and more widely as fouling. Its incidence depends on the structural and textural features of the stones, as well as on environmental factors [5–9].

Biodeterioration phenomena are usually treated by applying biocidal solutions on the stone, which can be potentially harmful since they often involve toxic or polluting substances [8,10]. The research on less toxic compounds and natural biocides has been widely developed in the last few years [11–15]. A common approach is the preservation of the artifacts by applying a coating that inhibits the microorganism colonization. Several biocides have been tested so far after dispersion into coating formulations. However, the main issue of these attempts is their short-lived antifouling action due to quick release and possible deterioration of the active compound. Furthermore, the use of a large amount of biocide, required to preserve the biocide function for longer time, is dangerous for the environment and human health. To overcome these issues, we studied the formulation of a novel coating based on the encapsulation of an eco-friendly biocide into proper nanocontainers. This strategy helped to reduce the biocide quantity and protracting its efficacy over time, thanks to a controlled release. The proposed coating combined hydrophobic and biocide properties. This was achieved by adopting a natural biocide product—the zosteric sodium salt (ZS) in order to increase the coating eco-compatibility—and dispersing it into a water repellent coating [16–18].

The zosteric acid or p-(sulfo-oxy) cinnamic acid (ZS, Figure 1), naturally found in *Zostera marina*, or eelgrass, presents an antifouling capability due to the sulfate ester group [19–24]. This antifoulant does not kill microorganisms but avoids their adhesion on cell surfaces at non-toxic concentrations.

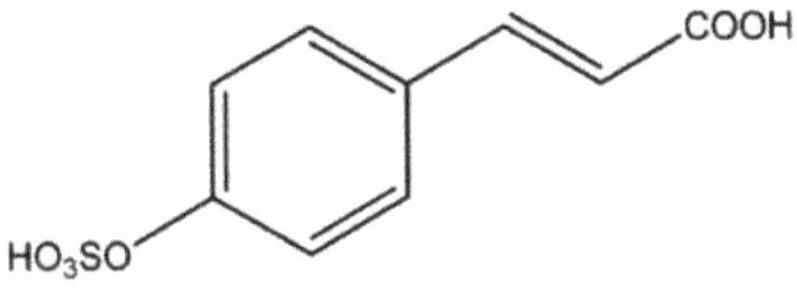

Figure 1. Structure formula of zosteric acid.

In our previous work, we encapsulated ZS into two different silica nanocontainers and we studied their release properties [25,26]. A preliminary in-vitro efficacy of encapsulated ZS against some common biodeteriogens has already been evaluated [26,27] and other experiments are currently ongoing, directly on site. Afterward, we dispersed the loaded nanodevices in a multifunctional TEOS-based coating also containing TiO_2 nanoparticles as photocatalytic agents. This formulation was based on our previous experience, where a commercial biocide, 2-mercaptobenzothiazole, was employed. We designed and characterized the composite coating on a glass slide, to evaluate its best composition in terms of optical properties and lack of cracking [28]. Subsequently, we applied this coating on stone supports [29]. In the present paper, we report laboratory tests on the same coating formulation, loaded with a different green biocide, to investigate its influence on the microstructural properties and visual aspect of brick, mortar, travertine, and Carrara marble, which are widely used in the Cultural Heritage field.

2. Materials and Methods

2.1. Substrates

Experiments were carried out on both natural and man-made stones used in buildings and monuments; namely, red fired brick, a natural hydraulic lime-based mortar, travertine, and Carrara marble. All samples were obtained with the dimensions of $5 \times 5 \times 1$ cm^3. The red-fired brick (BR) was acquired from a local wholesale, which uses traditional methods. The lime-based mortar (MO) was obtained from the same local wholesale by using a mixture of 1:2 of natural hydraulic lime, NHL 5 (Saint Astier), and standard river sand as aggregate (size < 4 mm). Water was added in a proper quantity to obtain a good workability of the mixture. The mortar samples were cured for 28 days at room conditions (RH% = 50%, 20 °C). Carrara marble (MA) and travertine (TR)—extracted in quarries located in Carrara and Tivoli, respectively—underwent an aging process to increase their porosity. The adopted procedure [30] required two heating-cooling cycles: heating to 600 °C, followed by storage in water overnight at room conditions (RH% = 50%, 20 °C).

2.2. Composition and Application of Coatings

Two different silica nanocontainers, namely silica nanocapsules (NC) and silica mesoporous nanoparticles (MNP) loaded with ZS, were synthesized according to the procedure reported in [31]. Acetyltrimethylammonium bromide (CTAB, Aldrich, St. Louis, MO, USA), ammonia solution (NH$_3$ aq. 30%, Aldrich, St. Louis, MO, USA), diethyl ether (Et$_2$O, Aldrich, St. Louis, MO, USA), tetraethoxysilane (TEOS, Aldrich, St. Louis, MO, USA) were used without any further purification. ZS was synthesized from trans-4-hydroxycinnamic and the sulfur trioxide pyridine complex, according to the procedure reported in [22]. Some properties of the obtained loaded nanocontainers are reported in Table 1.

Table 1. Properties of silica nanocontainers loaded with ZS.

Nanocontainer	Morphology	Size (nm)	Loading Capability (Weight %)
NC	Spherical	170 ± 20 (diameter)	2.1
MNP	Rods	100–1000 (length)	7.8

These nanoparticles were dispersed in a coating, prepared following [32] and [28] by using TEOS, ethanol (Carlo Erba Reagents S.r.l., Cornaredo, Italy), poly(dimethylsiloxane) hydroxyl-terminated (PDMS-OH, Sigma-Aldrich, St. Louis, MO, USA), and n-octylamine (Sigma-Aldrich, St. Louis, MO, USA) as a non-ionic surfactant, without any further purification. The synthesis protocol required the following mole ratio: 1TEOS/16Ethanol/10H_2O/0.04PDMS-OH/0.004 n-octylamine. The addition of PDMS-OH and n-octylamine to the original recipe reduced coating cracking and improved its adhesion to the stone surface as previously reported [28,29,33–35]. In fact, since TEOS is a silicate product, it is mostly compatible with silicate stones (brick in the present study) since it converts into amorphous silica once the polymerization process has been completed. In the case of carbonate stones (mortar, travertine and Carrara marble), the bonding between the carbonate stone and the silica gel is only physical [36,37].

In addition to the biocide-loaded nanocontainers, TiO_2 nanoparticles (anatase, Sigma-Aldrich) were also added. A total nanoparticle concentration of 0.1% w/w was used to obtain optimal results in terms of optical properties and lack of cracking, according to previous studies [28].

A coating without nanoparticles (Si_Control) was also applied as control, in addition to the investigated TEOS-based coatings containing TiO_2 nanoparticles and loaded with silica nanocapsules (Si_TiO_2–NC) or mesoporous nanoparticles (Si_TiO_2–MNP).

An overview of the analyzed samples and the relative characterization techniques is reported in Table 2.

Table 2. Analyzed samples and applied characterization techniques.

Sample	Description	Techniques
NT	Untreated samples	SCA, CM, WAC, WVP, OSR, SEM, PT
Si_Control	Coating without nanoparticles	SCA, CM, WAC, WVP, OSR, SEM, PT
Si_TiO_2–NC	Coating with TiO_2 nanoparticles and loaded silica nanocapsules	SCA, CM, WAC, WVP, OSR, SEM, PT
Si_TiO_2–MNP	Coating with TiO_2 nanoparticles and loaded silica mesoporous nanoparticles	SCA, CM, WAC, WVP, OSR, SEM, PT
Si_TiO_2	Coating with only titanium nanoparticles	PT

The coatings were applied by brush until saturation, i.e., the condition for which the stone surface remains wet for more than 1 min, to replicate a real application procedure on monuments. A total of 16 brushes were adopted for bricks and mortars, 4 for travertine, and 6 for Carrara marble, due to their different porosity and consequent absorption rate. The coatings were applied on the largest surface (5 × 5 cm^2) for all laboratory tests except absorption of water through capillarity, for which a smaller surface was treated (5 × 1 cm^2). About 1 week was required for coating solidification through polymerization and water evaporation, leaving the samples at laboratory conditions (RH% = 50%, 20 °C). For all stones and treatments, the amount of the applied product—that is, the sample weight difference before and right after the coating application—and the amount of retained product—that is, the sample weight difference before the coating application and one week after—are reported in Tables S1 and S2 in Supplementary Materials, respectively.

2.3. Assessment of Coated Stone Performance and Compatibility

2.3.1. Static Contact Angle (SCA)

The wettability of the stone surface, as well as the hydrophobic effect of coatings, were evaluated by measuring the static contact angle before and after the application of the coating on the same sample. A total of 10 water droplets (~5 µL each) were deposited on the stone surface at room temperature. The right and left angles observed between the water droplets and the stone surface after 10 s were captured by a high-resolution camera and then elaborated by computer analysis by using the AnalySIS Pro® 32 software (Soft Imaging System GmbH, Münster, Germany). The results are expressed as the average of all droplets on each surface.

2.3.2. Colorimetric Measurements (CM)

The color of the selected stone specimen before and after the application of the coatings was measured by using an Eoptis CLM-194 portable colorimeter (Trento, Italy) following NORMAL 43/93 [38]. The obtained values are expressed in the CIEL*a*b* color space, where three parameters determine the color location: L* indicates the lightness (0 = absolute black, 100 = absolute white), and a* and b* are the red/green and yellow/blue coordinates, respectively, with a* < 0 red and a* > 0 green, b* < 0 blue and b* > 0 yellow. The total color difference (ΔE^*) between untreated and treated samples was calculated according to the Equation (1):

$$\Delta E = \sqrt{\Delta L^{*2} + \Delta a^{*2} + \Delta b^{*2}} \tag{1}$$

For each stone, three samples were analyzed before and after the application of the coatings. Three points for each specimen were measured.

2.3.3. Water Absorption through Capillarity (WAC)

The water absorption through capillarity was measured by following the normative UNI EN 15801:2010 [39]. The studied samples were previously dried at 60 °C until a constant mass, m_0, was observed. Then, they were placed on a filter paper saturated with distilled water and their weight, m_i, was monitored at specific time intervals. Both the amount of absorbed water (Q) and the capillary water absorption coefficient (CWAC) were calculated. The first parameter, expressed as (kg/m^2), was obtained by the following Equation (2):

$$Q = [(m_i - m_0)/A] \tag{2}$$

where A is the surface in contact with the filter paper. The CWAC was calculated by linear fitting on the initial slope of Q versus the square root of time. The analysis was performed on the same sample before and after treatment. Three samples were used for each stone.

2.3.4. Water Vapor Permeability (WVP)

The water vapor permeability, expressed as the water vapor permeability coefficient (g), was measured by following the standard DIN 52615 [40]. Stone specimens were placed on a metal cup containing a saturated aqueous solution of KNO_3 in order to reach an internal humidity of 93% [41], and weighted every 24 h, until weight stabilization. The test was performed on the same sample, treated and untreated. Triplicates of specimens were used for all types of stone, and the average results are reported.

2.3.5. Optical Surface Roughness (OSR)

Measurements of optical surface roughness were performed to evaluate changes in the surface morphology by using a confocal Leica DCM 3D optical profilometer (Leica Microsystems, Wetzlar, Germany). For this analysis, the roughness parameter Rz was considered; it refers to the average differences, within a sample length, between the largest peak height and the largest peak valley depth [42,43]. For each sample, three datasets were analyzed on both treated and untreated surfaces. A total of 500 profiles were acquired in the Z direction within an analyzed area of 5 × 5 mm^2 with a resolution along the axis of 2.5 µm.

2.3.6. Scanning Electron Microscopy (SEM) Combined to Energy Dispersive Spectroscopy (SEM-EDS)

A Zeiss Sigma 300 SEM (Oberkochen, Germany) coupled with an ETSE (Everhart-Thornley secondary electron) detector were used to acquire SEM images from fragments of $1 \times 1 \times 1$ cm^3 from both untreated and treated samples. Combined with OSR, this analysis allows us to evaluate changes in the morphology of the surface and also the distribution of the coating, both on the surface and in-depth. All specimens were previously sputter-coated with gold and placed on a double-sided adhesive carbon tape. The operating voltages and working distances were set according to the analyzed samples. To evaluate the distribution of the coating both on the surface and in-depth, X-ray fluorescence chemical maps of Ti for brick and mortar, and Si for travertine and Carrara marble, were acquired with a 60 mm^2 Bruker high-resolution EDS detector (energy dispersive X-ray, Berlin, Germany). The investigated area was ~2400 m^2. Since the matrix of brick and mortar is rich in Si, this element was not investigated for these two materials.

2.3.7. Photocatalysis Testing (PT)

The photocatalytic activity of the TiO$_2$ nanoparticles inserted in the coatings was evaluated under laboratory conditions. Stone samples were covered with a solution of 1 mM of methyl orange (Riedel-de Haën, Seelze, Germany) used as a staining compound, which was diluted in ethanol to allow quick evaporation of the liquid once applied on the stones. Due to the different absorption of the stones, 750 µL were used for brick and mortar and 500 µL for travertine and Carrara marble. Moreover, also untreated samples (NT) and those treated with the coating loaded with only titanium nanoparticles (Si_TiO$_2$) were investigated as control. The stone specimens were placed in a ventilated chamber under a 365 nm UV light (Osram vitalux, Berlin, Germany), located at a distance of 20 cm, for 72 h [35]. The same colorimeter adopted to evaluate colorimetric changes on the surface due to the application of the coatings was used to estimate the photocatalytic degradation of methyl orange. The reference at time zero for the discoloration curves corresponds with the surfaces without methyl orange. The obtained data are reported after normalization to 100.

3. Results and Discussion

To assess the protective capability of the treatment, we first evaluated its hydrophobicity and possible color changes of the stone surface after application.

Information on the hydrophobicity of the coated stone surface was obtained by the SCA evaluation. The minimum acceptable SCA for stone protection is 90° [44]; the measured values for our treatments are reported in Table 3. The results concerning untreated stones are not reported, since the drops were absorbed too fast by the stone surfaces, due to their hydrophilic behavior. The obtained data demonstrate a good hydrophobization of the surfaces after treatment with all coatings. This confirms the hydrophobic behavior of the formulation, also shown in our previous work [29]. The addition of TiO$_2$ nanoparticles and silica nanocontainers lightly increased the hydrophobic effect for all stones, except for mortar. Indeed, in the latter case, the SCA did not sensibly change within the standard deviation. No relevant differences were recorded between the coatings charged with NC and MNP applied on all stones.

Colorimetric measurements must be performed to verify that no visible color change is observed after the application of coatings and that the natural aspect of the stone is preserved [45]. This means that the total color variation ΔE^* should not be >5 [46]. Data registered after coatings application are shown in Figure 2, while the results concerning the individual parameters, namely brightness L^*, red–green chromatic component a^*, and yellow–blue chromatic component b^* are reported in Table S3, in Supplementary Materials. All the investigated coatings showed a $\Delta E^* < 5$, with no relevant differences among empty coatings and charged ones. The presence of MNP caused a higher color change of the surface with respect to NC for brick and mortar. However, since the obtained values remained below the limit of $\Delta E^* = 5$, this difference can be considered irrelevant. Therefore, we can assume that the presence of nanoparticles did not influence the visual aspect of the treated stones.

Table 3. SCA values for all treated stones.

Sample	Average SCA (°)		
	Si_Control	Si_TiO$_2$-NC	Si_TiO$_2$-MNP
Brick	125.7 ± 4.7	138.2 ± 5.1	137.6 ± 4.5
Mortar	127.7 ± 3.5	130.1 ± 5.5	124.1 ± 4.3
Travertine	118.9 ± 5.1	120.2 ± 5.1	130.3 ± 1.9
Carrara Marble	125.5 ± 4.2	142.1 ± 3.3	134.2 ± 5.1

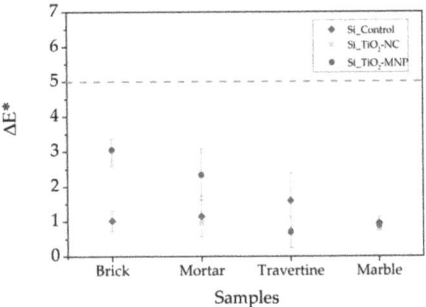

Figure 2. Mean global color variation ΔE^* between treated and untreated specimens in the CIEL*a*b* color space. The thin vertical bars represent the mean standard deviation.

Transfer properties of liquid water and vapor were assessed by capillarity and water vapor permeability measurements, to investigate efficacy and harmlessness of the treatments. In fact, a proper protective coating should reduce the penetration of liquid water from outside and, likewise, should not prevent the outflow of water vapor towards the external side of the stone [47]. The capillary absorption kinetics of the stones, reported as a function of the square root of time ($s^{1/2}$), before and after the application of the coatings, are displayed in Figure 3. The slope of the first part of each curve refers to the capillary water absorption coefficient (CWAC), while the second part measures the reached saturation. The reduction in CWAC with respect to untreated stones is reported in Table 4. All untreated stones reached a different level of water saturation, within a distinctive time interval, according to their different porosity. In fact, the quantity of water absorbed by brick was higher (~17 Kg/m^2) than mortar (~10 Kg/m^2). At the same time, travertine and Carrara marble absorbed a similar amount of water (~2.3 Kg/m^2 and ~1.7 Kg/m^2, respectively), which is lower with respect to the previous supports. Moving to the treated samples, all coatings showed a considerable reduction in water absorption with respect to the bare stones. Notice that, in the case of brick and mortar, data for the treated samples refer to the right vertical scale in Figure 3a–d, as Q changes by about one order of magnitude after treatment. For brick, a higher reduction of the initial slope of the curve is visible with Si_TiO$_2$-MNP and Si_TiO$_2$-NC compared with the Si_Control (Figure 3b). No differences are visible in the case of mortar (Figure 3d).

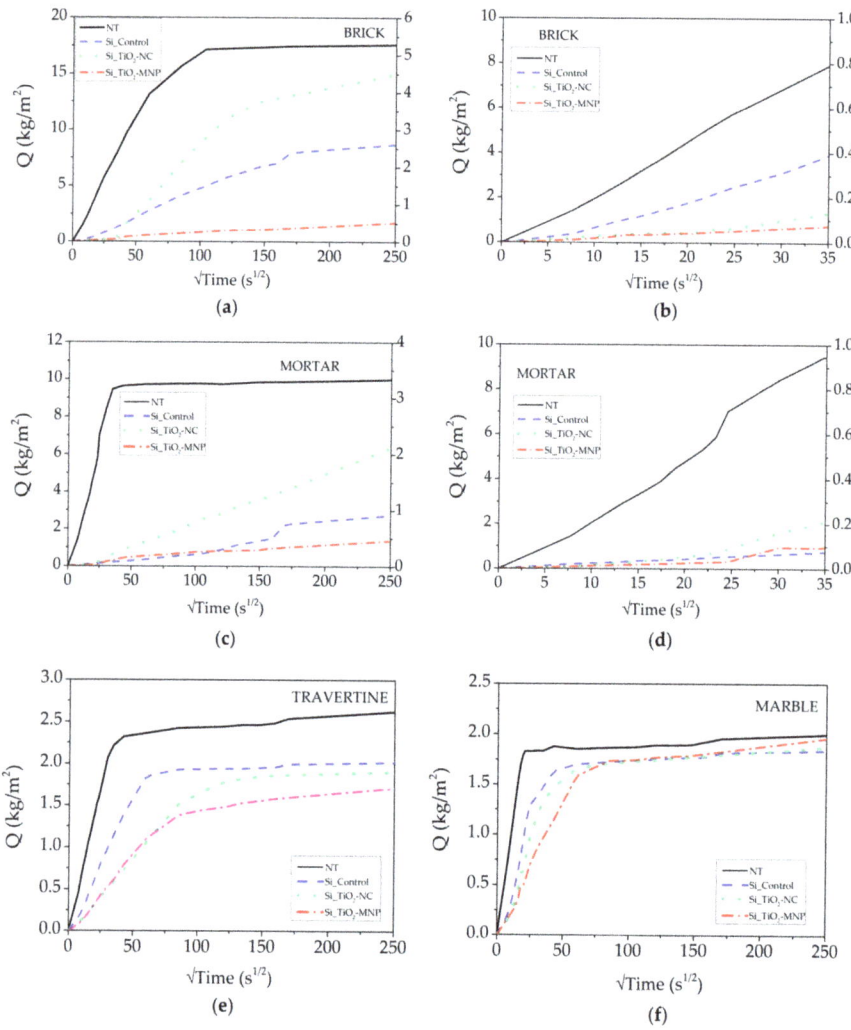

Figure 3. Plot of water absorption by capillarity test as a function of time: (**a**) brick, (**b**) brick (focus 0–35 $s^{1/2}$), (**c**) mortar, (**d**) mortar (focus 0–35 $s^{1/2}$), (**e**) travertine and (**f**) Carrara marble.

Table 4. Reduction (%) of the capillary water absorption coefficient (CWAC), due to the coating.

Sample	Reduction in CWAC (%)		
	Si_Control	Si_TiO$_2$–NC	Si_TiO$_2$–MNP
Brick	96.49	97.8	99.1
Mortar	98.97	98.9	99.5
Travertine	10.52	70.0	52.5
Carrara marble	46.25	47.8	68.8

In the case of brick and mortar, the relative reduction of CWAC was higher than 95% when compared to the untreated specimens (Table 4). A lower reduction was observed in the case of travertine and Carrara marble, possibly due to the lower number of applications (4 and 6, respectively). For travertine, the presence of nanoparticles had a higher effect on the reduction of CWAC. This may have been due to the size ratio between nanoparticles and stone porosity.

The measured water vapor permeability coefficient (g) for all treated specimens did not sensibly change with respect to the case of the untreated samples, within the standard deviation (Figure 4). No relevant differences were registered between the presence of NC and MNP. Therefore, our results indicate that the tested coatings preserved the original water permeability of the materials.

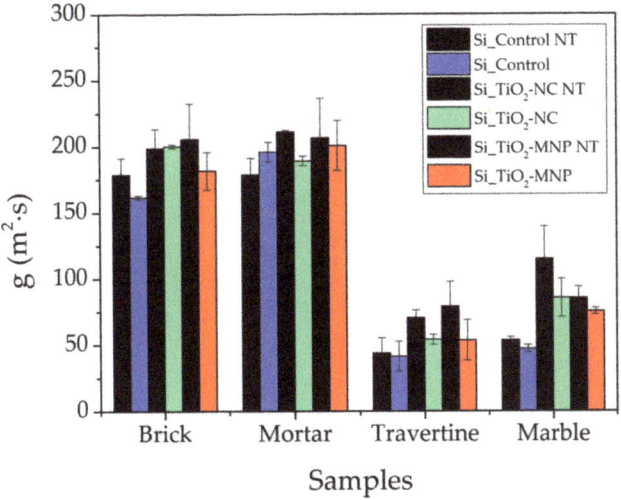

Figure 4. Mean water vapor permeability coefficient (g) shown for each stone before (black line on the left) and after (colored lines on the right) coatings application. The thin vertical black bar represents the mean standard deviation.

Roughness evaluations are reported in Figure S1, in Supplementary Materials. The results did not evidence particular trends of the roughness values after the application of the protective coatings, taking into account the significant standard deviations of Rz.

SEM observations on coated stones showed compact films and no cracks for both empty coatings (Figure S2b,d,f,h, in Supplementary Materials) and nanoparticles loaded ones (Figure 5). The behavior concerning the presence of NC and MNP was similar, thus confirming the results obtained in our previous work [29]. To evaluate the distribution of the coating preparation and their depth of penetration, we acquired Ti distribution maps from brick and mortar, and Si ones from travertine and Carrara marble. These maps (Figures S3 and S4, in Supplementary Materials) demonstrated a homogeneous distribution of the coatings on all stone surfaces. Moreover, the protective characteristic of the coatings was demonstrated by the lack of penetration in depth evidenced by the absence of Ti and Si in cross-section measurements.

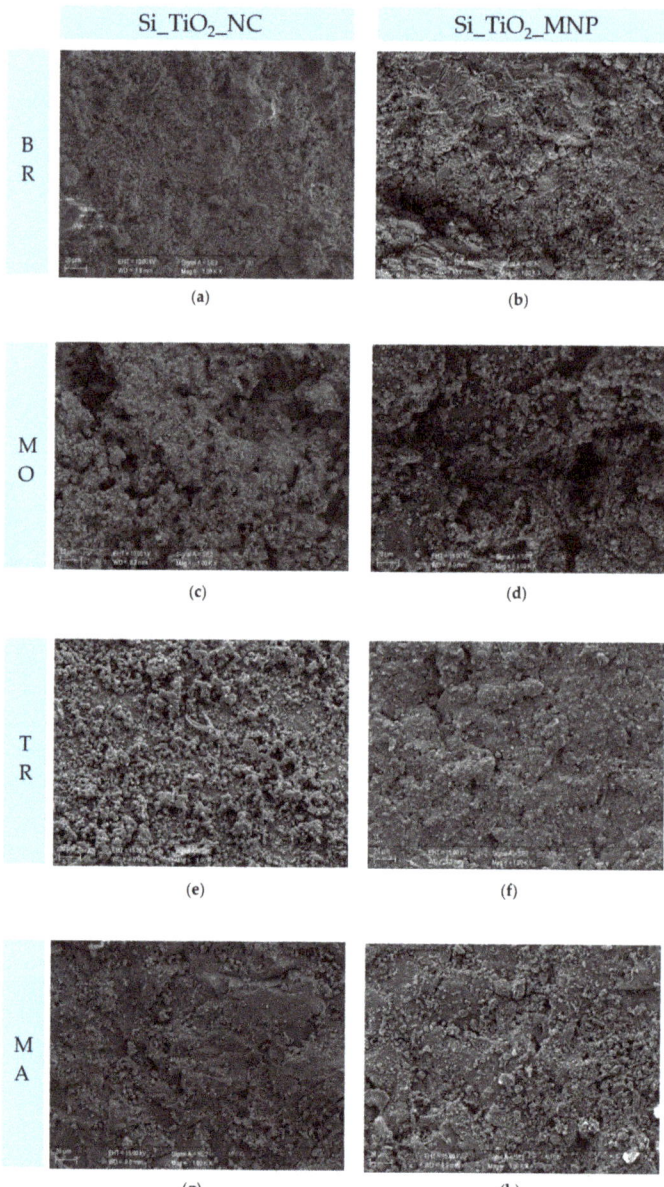

Figure 5. SEM images (1000×) of coated stones treated with Si-TiO$_2$-NC and Si-TiO$_2$-MNP: (**a,b**) brick (BR), (**c,d**), mortar (MO), (**e,f**) travertine (TR) and (**g,h**) Carrara marble (MA).

Aiming at evaluating the photocatalytic property of the TiO$_2$ dispersed in the coating formulation, the degradation of methyl orange over time was examined under laboratory conditions. The discoloration of stones was reported as ΔE^* variation (Figure 6). In the case of brick, the color of the used stained compound was very similar to that of substrate. For this reason, the color differences among the consecutive measurements had larger uncertainties compared to those performed on the other supports.

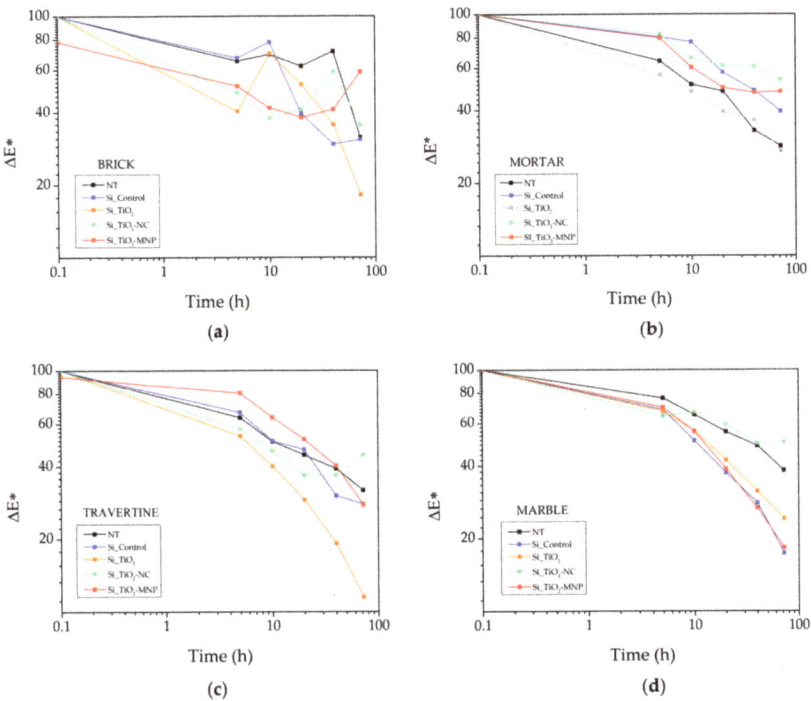

Figure 6. ΔE* variation connected to the discoloration of treated and untreated stones previously stained with methyl orange and then irradiated with UV light for 72 h: (**a**) brick, (**b**) mortar, (**c**) travertine and (**d**) Carrara marble.

The results show that only for travertine, for which a lower number of coating application was performed, there was a higher photocatalytic activity of Si_TiO$_2$ with respect to untreated samples and Si_Control. Both samples treated with Si-TiO$_2$-NC and Si-TiO$_2$-MNP showed worst results with respect to Si_TiO$_2$, possibly because the presence of silica nanocontainers caused an entrapment of TiO$_2$ nanoparticles, avoiding their correct contact with methyl orange. We noticed that the time evolution of ΔE* was not exponential.

Such results indicate that the coatings presented in this work had mechanical and optical properties comparable between them and with the previous formulations containing a commercial harmful biocide [29]. These results were achieved despite the different particle dimensions and shape, and were consequent to the size and chemical properties of the encapsulated molecule. Further evaluation on a broad spectrum of activity against the potential occurring biodeteriogen are in progress.

4. Conclusions

The application of four different TEOS-based multifunctional coatings, containing silica nanocontainers loaded with an eco-friendly biocide ZS and TiO$_2$ nanoparticle, confirmed their compatibility and effectiveness on the four tested stones, both natural and artificial. The presence of loaded-nanocontainers did not significantly alter the tested coating performances, while providing additional biocide efficacy. The presence of the coatings enhanced the stones' hydrophobicity and reduced water absorption through capillarity, with no evident difference between the two loaded nanocontainers. Importantly, our formulations did not significantly influence water vapor permeability and optical appearance of the stones. Thus, all fundamental requirements recommended to establish the effectiveness of a protective product for monument conservation were fulfilled by all presented coatings. We can, therefore, assume that these formulations are promising protective coatings even when stones with different mechanical and structural properties, as those tested here, are considered.

Supplementary Materials: The following are available online at https://www.mdpi.com/article/10.3390/coatings11091109/s1, Table S1. Amount of product applied—the difference of sample weight before and right after coating application, Table S2. Amount of product retained—the difference of sample weight before and 1 week after coating application, Table S3. Brightness L^*, red–green chromatic component a^* and yellow–blue chromatic component b^* acquired before and after the coatings application on all stones, Figure S1. The roughness R_z estimated for both treated and untreated stones (NT), Figure S2. SEM images (500×) of untreated samples and treated with empty coating: brick (BR) (a,b), mortar (MO) (c,d), travertine (TR) (e,f) and Carrara marble (MA) (g,h). NT: untreated; Si_Control, coating without nanoparticles, Figure S3. EDS-XFR Ti mapping acquired from brick and mortar: (a) brick treated with Si_TiO$_2$-NC, (b) brick treated with Si_TiO$_2$-MNP, (c) mortar treated with Si_TiO$_2$-NC and (d) mortar treated with Si-TiO$_2$-MNP, Figure S4. EDS-XFR Si mapping acquired from travertine and Carrara marble: (a) travertine treated with Si_TiO$_2$-NC, (b) travertine treated with Si_TiO$_2$-MNP, (c) Carrara Marble treated with Si_TiO$_2$-NC and (d) Carrara marble treated with Si-TiO$_2$-MNP.

Author Contributions: M.Z. carried out all the experiments, elaborated the data and wrote the paper. M.Z. and L.R. organized the experimental section, synthetized the coatings and finalized the photocatalysis proofs. G.D.V. supervised the SEM measurements. G.C., F.B. selected and provided the biocide, and performed biocidal tests. M.A.R. and A.S. supervised the project. All authors have read and agreed to the published version of the manuscript.

Funding: The authors acknowledge funding from Regione Lazio, Italy within the "SUPERARE" project (n. F86C18000650005) financed in the call "Gruppi di Ricerca" and "GRAL- Green and Long-lasting stone conservation products" project (n. F85F21001710009) financed in the call "Progetto Gruppi di Ricerca 2020". The Grant of Excellence Departments, MIUR (ARTICOLO 1, COMMI 314—337 LEGGE 232/2016) is also greatly acknowledged.

Institutional Review Board Statement: Not applicable.

Informed Consent Statement: Not applicable.

Data Availability Statement: Data reported in this article are available on request.

Acknowledgments: LIME laboratory (Roma Tre University, Rome, Italy) is acknowledged for the possibility of carrying out some analyses. We acknowledge Sergio Lo Mastro for SEM analysis and Edoardo Bemporad, Daniele De Felicis, Riccardo Moscatelli, and Edoardo Rossi (Engineering department, Roma Tre University) for the technical support in the contact angle and optical surface roughness measurements. We acknowledge Laura Crociani for synthesizing the zosteric sodium salt (Institute of Condensed Matter Chemistry and Technologies for Energy (ICMATE), Consiglio Nazionale delle Ricerche (CNR), Padova, Italy).

Conflicts of Interest: The authors declare no conflict of interest.

References

1. Camuffo, D. *Chapter 7—Atmospheric Water, Capillary Rise, and Stone Weathering*; Elsevier: Amsterdam, The Netherlands, 2019.
2. Saad, A.; Guédon, S.; Martineau, F. Microstructural weathering of sedimentary rocks by freeze-thaw cycles: Experimental study of state and transfer parameters. *Comptes Rendus—Geosci.* **2010**, *342*, 197–203. [CrossRef]
3. Steiger, M.; Charola, E. Chapter 4—Weathering and deterioration. In *Stone in Architecture*; Siegesmund, R.S., Ed.; Springer: Berlin/Heidelberg, Germany, 2011; pp. 227–316.
4. Traversetti, L.; Bartoli, F.; Caneva, G. Wind-driven rain as a bioclimatic factor affecting the biological colonization at the archaeological site of Pompeii, Italy. *Int. Biodeterior. Biodegrad.* **2018**, *134*, 31–38. [CrossRef]
5. Caneva, G.; Bartoli, F.; Ceschin, S.; Salvadori, O.; Futagami, Y.; Salvati, L. Exploring ecological relationships in the biodeterioration patterns of Angkor temples (Cambodia) along a forest canopy gradient. *J. Cult. Herit.* **2015**, *16*, 728–735. [CrossRef]
6. Caneva, G.; Bartoli, F.; Savo, V.; Futagami, Y.; Strona, G. Combining statistical tools and ecological assessments in the study of biodeterioration patterns of stone temples in angkor (Cambodia). *Sci. Rep.* **2016**, *6*, 1–8. [CrossRef]
7. Sterflinger, K.; Piñar, G. Microbial deterioration of cultural heritage and works of art—Tilting at windmills? *Appl. Microbiol. Biotechnol.* **2013**, *97*, 9637–9646. [CrossRef] [PubMed]
8. Tiano, P. Biodeterioration of stone monuments a worldwide issue. *Open Conf. Proc. J.* **2016**, *7*, 29–38. [CrossRef]

9. Casanova Municchia, A.; Bartoli, F.; Taniguchi, Y.; Giordani, P.; Caneva, G. Evaluation of the biodeterioration activity of lichens in the Cave Church of Üzümlü (Cappadocia, Turkey). *Int. Biodeterior. Biodegrad.* **2018**, *127*, 160–169. [CrossRef]
10. Urzi, C.; De Leo, F.; Krakova, L.; Pangallo, D.; Bruno, L. Effects of biocide treatments on the biofilm community in Domitilla's catacombs in Rome. *Sci. Total Environ.* **2016**, *572*, 252–262. [CrossRef] [PubMed]
11. Jeong, S.H.; Lee, H.J.; Kim, D.W.; Chung, Y.J. New biocide for eco-friendly biofilm removal on outdoor stone monuments. *Int. Biodeterior. Biodegrad.* **2018**, *131*, 19–28. [CrossRef]
12. Toreno, G.; Isola, D.; Meloni, P.; Carcangiu, G.; Selbmann, L.; Onofri, S.; Caneva, G.; Zucconi, L. Biological colonization on stone monuments: A new low impact cleaning method. *J. Cult. Herit.* **2018**, *30*, 100–109. [CrossRef]
13. Young, M.E.; Alakomi, H.L.; Fortune, I.; Gorbushina, A.A.; Krumbein, W.E.; Maxwell, I.; McCullagh, C.; Robertson, P.; Saarela, M.; Valero, J.; et al. Development of a biocidal treatment regime to inhibit biological growths on cultural heritage: BIODAM. *Environ. Geol.* **2008**, *56*, 631–641. [CrossRef]
14. Fidanza, M.R.; Caneva, G. Natural biocides for the conservation of stone cultural heritage: A review. *J. Cult. Herit.* **2019**, *38*, 271–286. [CrossRef]
15. Pinna, D. *Coping with Biological Growth on Stone Heritage Objects: Methods, Products, Applications, and Perspectives*; Apple Academic Press: Palm Bay, FL, USA, 2017; ISBN 978-354-077-3-405.
16. Colangiuli, D.; Calia, A.; Bianco, N. Novel multifunctional coatings with photocatalytic and hydrophobic properties for the preservation of the stone building heritage. *Constr. Build. Mater.* **2015**, *93*, 189–196. [CrossRef]
17. Ruffolo, S.A.; Ricca, M.; Macchia, A.; La Russa, M.F. Antifouling coatings for underwater archaeological stone materials. *Prog. Org. Coat.* **2017**, *104*, 64–71. [CrossRef]
18. Pinna, D.; Salvadori, B.; Galeotti, M. Monitoring the performance of innovative and traditional biocides mixed with consolidants and water-repellents for the prevention of biological growth on stone. *Sci. Total Environ.* **2012**, *423*, 132–141. [CrossRef]
19. Geiger, T.; Delavy, P.; Hany, R.; Schleuniger, J. Encapsulated Zosteric Acid Embedded in Poly [3- hydroxyalkanoate] Coatings—Protection against Biofouling. *Polym. Bull.* **2004**, *52*, 65–72. [CrossRef]
20. Boopalan, M.; Sasikumar, A. Studies on Biocide Free and Biocide Loaded Zeolite Hybrid Polymer Coatings on Zinc Phosphated Mild Steel for the Protection of Ships Hulls from Biofouling and Corrosion. *Silicon* **2011**, *3*, 207–214. [CrossRef]
21. Laabir, M.; Grignon-Dubois, M.; Masseret, E.; Rezzonico, B.; Soteras, G.; Rouquette, M.; Rieuvilleneuve, F.; Cecchi, P. Algicidal effects of *Zostera marina* L. and *Zostera noltii* Hornem. extracts on the neuro-toxic bloom-forming dinoflagellate *Alexandrium catenella*. *Aquat. Bot.* **2013**, *111*, 16–25. [CrossRef]
22. Villa, F.; Pitts, B.; Stewart, P.S.; Giussani, B.; Roncoroni, S.; Albanese, D.; Giordano, C.; Tunesi, M.; Cappitelli, F. Efficacy of Zosteric Acid Sodium Salt on the Yeast Biofilm Model Candida albicans. *Microb. Ecol.* **2011**, *62*, 584–598. [CrossRef]
23. Almeida, J.R.; Moreira, J.; Pereira, D.; Pereira, S.; Antunes, J.; Palmeira, A.; Vasconcelos, V.; Pinto, M.; Correia-da-Silva, M.; Cidade, H. Potential of synthetic chalcone derivatives to prevent marine biofouling. *Sci. Total Environ.* **2018**, *643*, 98–106. [CrossRef]
24. Cattò, C.; Dell'Orto, S.; Villa, F.; Villa, S.; Gelain, A.; Vitali, A.; Marzano, V.; Baroni, S.; Forlani, F.; Cappitelli, F. Unravelling the structural and molecular basis responsible for the anti-biofilm activity of zosteric acid. *PLoS ONE* **2015**, *10*, e0131519. [CrossRef] [PubMed]
25. Ruggiero, L.; Bartoli, F.; Fidanza, M.R.; Zurlo, F.; Marconi, E.; Gasperi, T.; Tuti, S.; Crociani, L.; Di Bartolomeo, E.; Caneva, G.; et al. Encapsulation of environmentally-friendly biocides in silica nanosystems for multifunctional coatings. *Appl. Surf. Sci.* **2020**, *514*, 145908. [CrossRef]
26. Ruggiero, L.; Crociani, L.; Zendri, E.; El Habra, N.; Guerriero, P. Incorporation of the zosteric sodium salt in silica nanocapsules: Synthesis and characterization of new fillers for antifouling coatings. *Appl. Surf. Sci.* **2018**, *439*, 705–711. [CrossRef]
27. Bartoli, F.; Zuena, M.; Sodo, A.; Caneva, G. The efficiency of biocidal silica nanosystems for the conservation of stone monuments: Comparative in vitro tests against epilithic green algae. *Appl. Sci.* **2021**, *11*, 6804. [CrossRef]
28. Ruggiero, L.; Fidanza, M.R.; Iorio, M.; Tortora, L.; Caneva, G.; Ricci, M.A.; Sodo, A. Synthesis and characterization of TEOS coating added with innovative antifouling silica nanocontainers and TiO_2 nanoparticles. *Frontiers* **2020**, *7*, 1–11. [CrossRef]
29. Zuena, M.; Ruggiero, L.; Della Ventura, G.; Bemporad, E.; Ricci, M.A.; Sodo, A. Effectiveness and compatibility of nanoparticle based multifunctional coatings on natural and man-made stones. *Coatings* **2021**, *11*, 480. [CrossRef]
30. Haake, S.; Simon, S.; Favaro, M. The bologna cocktail-evaluation of consolidation treatments on monuments in france and italy after 20 years on natural aging. In Proceedings of the 10th International Congress on Deterioration and Conservation of Stone, Stockholm, Sweden, 27 June–2 July 2004; pp. 423–430.
31. Ruggiero, L.; Di Bartolomeo, E.; Gasperi, T.; Luisetto, I.; Talone, A.; Zurlo, F.; Peddis, D.; Ricci, M.A.; Sodo, A. Silica nanosystems for active antifouling protection: Nanocapsules and mesoporous nanoparticles in controlled release applications. *J. Alloys Compd.* **2019**, *798*, 144–148. [CrossRef]
32. Xu, F.; Li, D.; Zhang, Q.; Zhang, H.; Xu, J. Effects of addition of colloidal silica particles on TEOS-based stone protection using n-octylamine as a catalyst. *Prog. Org. Coat.* **2012**, *75*, 429–434. [CrossRef]
33. Kapridaki, C.; Maravelaki-Kalaitzaki, P. TiO2-SiO2-PDMS nano-composite hydrophobic coating with self-cleaning properties for marble protection. *Prog. Org. Coat.* **2013**, *76*, 400–410. [CrossRef]
34. Pinho, L.; Elhaddad, F.; Facio, D.S.; Mosquera, M.J. A novel TiO_2—SiO_2 nanocomposite converts a very friable stone into a self-cleaning building material. *Appl. Surf. Sci.* **2013**, *275*, 389–396. [CrossRef]

35. Pinho, L.; Mosquera, M.J. Photocatalytic activity of TiO_2-SiO_2 nanocomposites applied to buildings: Influence of particle size and loading. *Appl. Catal. B Environ.* **2013**, *134–135*, 205–221. [CrossRef]
36. Scherer, G.W.; Wheeler, G.S. Silicate consolidants for stone. *Key Eng. Mater.* **2008**, *391*, 1–25. [CrossRef]
37. Sassoni, E.; Franzoni, E.; Pigino, B.; Scherer, G.W.; Naidu, S. Consolidation of calcareous and siliceous sandstones by hydroxyapatite: Comparison with a TEOS-based consolidant. *J. Cult. Herit.* **2013**, *14*, e103–e108. [CrossRef]
38. NORMAL 43/93 *Misure Colorimetriche di Superfici Opache (Italian Normative on Stone Material—Colorimetric Measurement of Opaque Surfaces)*; Commissione Beni Culturali UNI NORMAL: Roma, Italy, 1993.
39. EN UNI 15801:2010 Conservation of Cultural Property—Test Methods—Determination of Water Absorption by Capillarity. 2010. Available online: https://standards.iteh.ai/catalog/standards/cen/06e6ae49-1d9a-4318-b79b-eae91a02091c/en-15801-2009 (accessed on 9 December 2009).
40. DIN 52 615 Testing of Thermal Insulating Material. In *Determination of Water Vapour Permeability of Construction and Insulating Materials*; Deutsches Institut fur Normung E.V. (DIN): Berlin, Germany, 1987.
41. Greenspan, L. Humidity fixed points of binary saturated aqueous solutions. *J. Res. Natl. Bur. Stand. Sect. A Phys. Chem.* **1977**, *81A*, 89. [CrossRef]
42. Hobson, T. *Parameters and Definitions for Roundness and Surface Measurements*; Taylor Hobson Ltd.: Leicester, UK, 2000.
43. Della Volpe, C.; Penati, A.; Peruzzi, R.; Siboni, S.; Toniolo, L.; Colombo, C. Combined effect of roughness and heterogeneity on contact angles: The case of polymer coating for stone protection. *J. Adhes. Sci. Technol.* **2000**, *14*, 273–299. [CrossRef]
44. Alvarez de Buergo Ballester, M.; Fort González, R. Basic methodology for the assessment and selection of water-repellent treatments applied on carbonatic materials. *Prog. Org. Coat.* **2001**, *43*, 258–266. [CrossRef]
45. Sasse, H.R.; Snethlage, R. Methods for the evaluation of stone conservation treatments. In *Saving Our Cultural Heritage: The Conservation of Historic Stone Structure, Dahlem Workshop Reports*; Baer, N.S., Snethlage, R., Eds.; Wiley & Sons: Chichester, UK, 1997; pp. 223–243.
46. Delgado Rodrigues, J.; Grossi, A. Indicators and ratings for the compatibility assessment of conservation actions. *J. Cult. Herit.* **2007**, *8*, 32–43. [CrossRef]
47. Tsakalof, A.; Manoudis, P.; Karapanagiotis, I.; Chryssoulakis, I.; Panayiotou, C. Assessment of synthetic polymeric coatings for the protection and preservation of stone monuments. *J. Cult. Herit.* **2007**, *8*, 69–72. [CrossRef]

Article

Preparation of Sintered Brick with Aluminum Dross and Optimization of Process Parameters

Yu Zhang [1], Hongjun Ni [1,2,*], Shuaishuai Lv [1], Xingxing Wang [1], Songyuan Li [1] and Jiaqiao Zhang [1]

[1] School of Mechanical Engineering, Nantong University, Nantong 226019, China; 18501470690@163.com (Y.Z.); lvshuaishuai@ntu.edu.cn (S.L.); wangxx@ntu.edu.cn (X.W.); songyuanli2022@163.com (S.L.); 1810310038@stmail.ntu.edu.cn (J.Z.)

[2] Jiangsu Province Engineering Research Center of Aluminum Dross Solid Waste Harmless Treatment and Resource Utilization, Nantong 226019, China

* Correspondence: ni.hj@ntu.edu.cn

Abstract: Aluminum dross is produced in the process of industrial production and regeneration of aluminum. Currently, the main way to deal with aluminum dross is stacking and landfilling, which aggravates environmental pollution and resource waste. In order to find a green and environmental protection method for the comprehensive utilization, the aluminum dross was used as raw materials to prepare sintered brick. Firstly, the raw material ratio, molding pressure and sintering process were determined by single factor test and orthogonal test, and the mechanism of obvious change of mechanical strength of sintered brick was studied by XRD and SEM. The experimental results show that, the optimal formula of sintered brick is 50% aluminum dross, 37.50% engineering soil and 12.50% coal gangue. The optimum process parameters are molding pressure 10 MPa, heating rate 8 °C/min, sintering temperature 800 °C, holding time 60 min. The samples prepared under the above formula and process parameters present outstanding performance, and the compressive strength, flexural strength and water absorption rate are 16.21 MPa, 3.42 MPa and 17.12% respectively.

Keywords: aluminum dross; sintered brick; mechanical strength; orthogonal test; process parameter

1. Introduction

Globalization and population growth have increased the consumption of natural resources, resulting in a huge amount of waste. Taking the aluminum industry as an example, a large amount of industrial waste is produced every year, in which aluminum dross accounts for a large part. Some scholars have studied the influence of various factors on the harmless treatment of aluminum dross and adopted the aluminum dross after harmless treatment to prepare non-fired bricks [1], while some scholars have investigated the use of aluminum dross to prepare inorganic flocculants [2], coating material [3,4], refractory materials [5], steelmaking deoxidizers [6], concrete [7] and other products. However, the existing resource utilization methods have a small amount of disposal and cannot effectively use a large number of aluminum dross. The solid waste disposal method, which uses aluminum dross as raw material for the preparation of sintered brick, has the advantages of large disposal volume and easy to obtain solidified materials. This is an excellent method of solid waste disposal. In current urban construction, concrete is a relatively common building material. It is estimated that 11 billion tons of concrete are consumed every year [8], which will inevitably lead to a huge consumption of its raw materials, especially cement. The excellent properties of concrete cannot conceal its shortcomings of causing damage to the environment. According to reports, for every ton of ordinary cement produced, about 900 kg of CO_2 is produced, accounting for 7% of the total global CO_2 emissions [9]. Therefore, it is necessary to find a green and environmentally friendly material to deal with global warming and waste disposal problems.

Sintered brick has a long history. After thousands of years of development, different types of sintered brick have been derived. However, the traditional method of making sintered brick destroys land and consumes resources. Therefore, some experts and scholars have studied the use of coal gangue, fly ash, red mud, gold mine tailings, iron ore tailings and other raw materials to prepare sintered brick [10–14], thus realizing the comprehensive utilization of solid waste. The three raw materials used in this study are all solid wastes, and the preparation process will not cause harm to the surrounding environment.

In 2019, China's electrolytic aluminum output was 35.75 million tons, and imported bauxite was 100 million tons. Aluminum dross is a kind of industrial waste produced in the process of melting aluminum such as aluminum electrolysis, aluminum alloy production and waste aluminum regeneration [15,16]. The aluminum–silicon ratio reaches 20–50. It is a very important high-quality aluminum resource with high recycling value. As the AlN in aluminum dross is unstable at room temperature, it is easy to be damp and then produce ammonia [17]. As a result, the application of aluminum dross has been restricted in terms of resource utilization. However, AlN is oxidized by air during the sintering process to generate Al_2O_3 and other nitrogen oxides [18], so as to realize the harmless treatment of aluminum dross. The harmless treatment process of aluminum dross mainly includes two treatment methods: the pyrometallurgical process and hydrometallurgical process. However, the hydrometallurgical process is not as environmentally friendly as the pyrometallurgical process in terms of waste liquid and exhaust gas emissions [19]. In addition to aluminum dross, engineering soil and coal gangue are also used as raw materials for making sintered brick. As most countries began to carry out large scale infrastructure construction, a large amount of engineering soil was produced. The main phase of engineering soil is the quartz, and its chemical composition is similar to clay. It can effectively make up for the lack of SiO_2 content in aluminum dross. Coal gangue is generated in the coal mining process. At present, China's coal gangue stockpile has exceeded 5 billion tons, and about 160,000 tons of gangue are discharged for every 1 million tons of coal Because of its own combustible characteristics, it can increase the sintering temperature inside the bricks when preparing sintered brick. Therefore, it is feasible to prepare sintered bricks with aluminum dross as the main component, supplemented by engineering soil and coal gangue.

This article studies the effects of raw material ratio, molding pressure and sintering process on the properties of sintered brick. After obtaining the best process parameters for preparing sintered brick, the experiment was carried out for phase composition analysis and microscopic scanning analysis, and the formation mechanism of the mechanical strength of sintered brick was studied.

2. Materials and Methods

2.1. Raw Material

The aluminum dross selected for this experiment is from Jiangsu Haiguang Metal Co., Ltd. SuQian, China, specifically the final dross obtained after one-time aluminum dross recovery of part of the metal aluminum; the engineering soil is taken from Haian Zhengcheng New Wall Material Co., Ltd. NanTong, China; coal gangue is taken from Haian Zhengcheng New Wall Material Co., Ltd. NanTong, China; Water: deionized water.

The main components of aluminum dross are Al, Al_2O_3, AlN and a small amount of SiO_2. At high temperatures, Al and AlN can be oxidized to Al_2O_3 by air, and Al_2O_3 is the effective component of sintered brick. Aluminosilicate can be formed by Al_2O_3 combining with SiO_2 at a certain temperature to improve the mechanical strength of sintered brick. As the main substance in the engineering soil is SiO_2, the engineering soil must be mixed to effectively make up for the deficiency of SiO_2 in the aluminum dross. Coal gangue is added to sintered bricks as a kind of combustion-supporting agent. On the one hand, it can increase the content of SiO_2 in the raw materials; on the other hand, because of its own combustible characteristics, it effectively reduces the sintering temperature and saves energy. The XRD pattern of each raw material is shown in Figure 1. The mixed three raw materials in proportion for particle size analysis are shown in Figure 2.

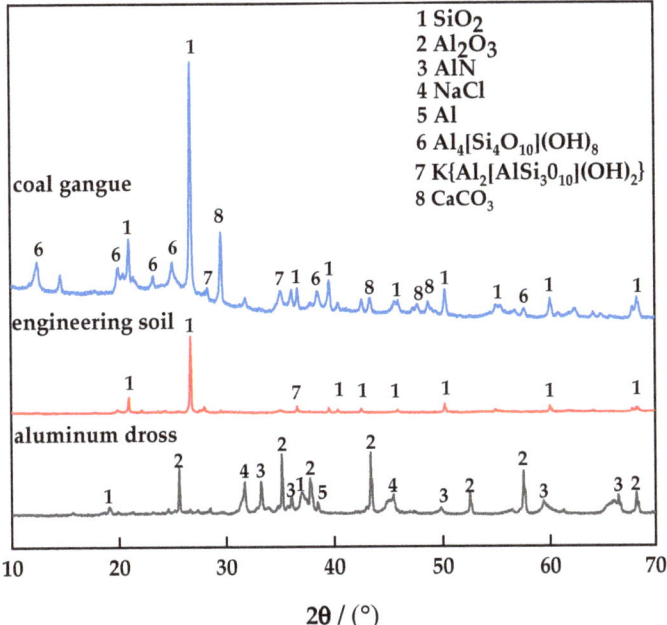

Figure 1. XRD pattern of the raw material.

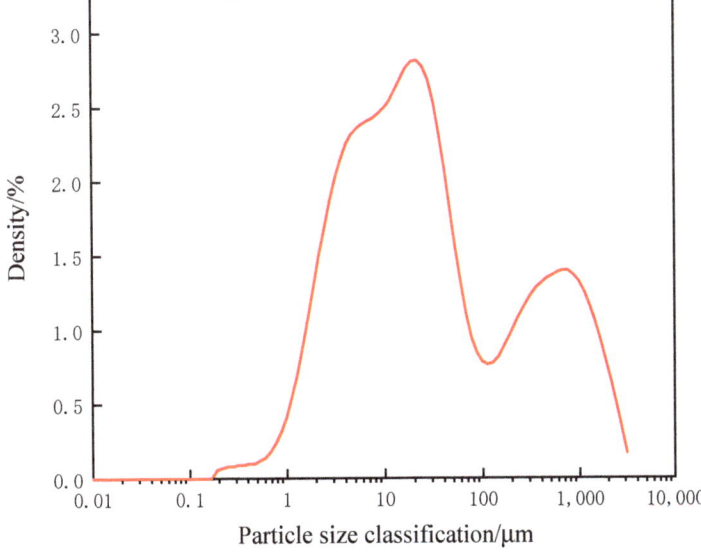

Figure 2. Raw material particle size.

2.2. Sample Preparation

Firstly, aluminum slag is added according to 30%–70%, engineering soil is added according to 20.73%–55.29%, and coal gangue is added according to 9.27%–14.71%. Then water is added, and the amount of water is 12% of the mass of brick. After sealing, the mixture is placed in a dark place and aged for 72 h to make the water evenly diffuse into the raw material. Coal gangue is added according to the calorific value of aging materials. According to the production practice of brick making enterprises, each kilogram of aging material usually needs about 837 kJ of heat, while the calorific value of coal gangue is about 2000 kJ/kg. The self-made mold is used for compression molding under 6 MPa–14 MPa, and the size of the brick is $40 \times 40 \times 160$ mm^3. After natural drying for 15 h, it was dried at 105 °C for 6 h to constant weight. The heating rate is 6~8 °C/min, the roasting temperature is 750~850 °C, and the temperature is maintained for 60~180 min to room temperature. The process flow of sintered brick preparation is shown in Figure 3. Five bricks were prepared for each experimental factor. The final data of compressive strength and flexural strength are the average values after removing maximum and minimum values. The sample is shown in Figure 4. The left side is the unsintered brick and the right side is the sintered brick. In terms of color, the unsintered brick has a darker color, while the color of the sintered brick has changed. In terms of appearance, the shape of unsintered brick is regular without surface damage, while the appearance of sintered brick is basically intact.

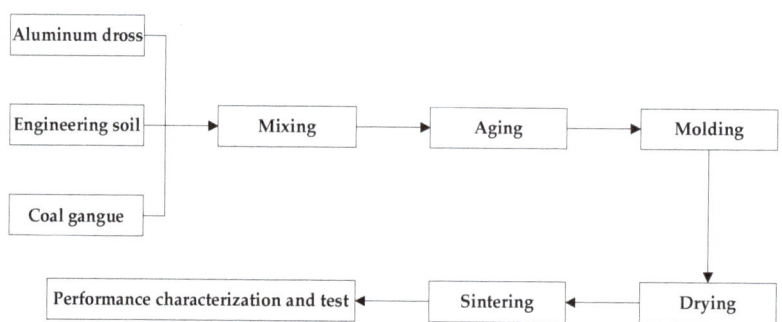

Figure 3. The process flow of sintered brick preparation.

Figure 4. Bricks before and after sintering: (a) Unsintered bricks, (b) Sintered brick.

2.3. Specimen Characterization

The measurement of the compressive strength of sintered brick refers to GB/T 5101-2017 "Sintered Ordinary Bricks" [20], and the flexural strength and water absorption refers to GB/T 2542-2012 "Testing Methods for Wall Bricks" [21]. The force loading speed of the compressive strength test is 2.4 KN/s, and the force loading speed of the flexural strength test is 0.05 KN/s. The microscopic morphology of the sample was observed with a field emission scanning electron microscope system (ZEISS Gemini SEM 300, Carl Zeiss AG,

Oberkochen, Germany), and the phase composition of the raw materials and the sample was analyzed and tested with an X-ray diffractometer manufactured by Rigaku, Tokyo, Japan. The XRD patterns of raw materials and products are analyzed by Jade6.0, drawn by origin2018, and the mechanical strength change graph is drawn by origin2018.

3. Results and Discussion

3.1. The Influence of the Mixing Amount of Aluminum Dross on the Properties of Sintered Brick

The range of aluminum dross is 30%–70%, and the other raw material admixtures are shown in Table 1.

Table 1. Raw material composition.

NO.	Aluminum Dross	Engineering Soil	Coal Gangue
1	30	55.29	14.71
2	40	46.68	13.32
3	50	37.50	12.50
4	60	29.74	10.26
5	70	20.73	9.27

The sintered bricks were naturally placed for 24 h and then tested for mechanical properties and water absorption. The experimental results are shown in Figures 5 and 6.

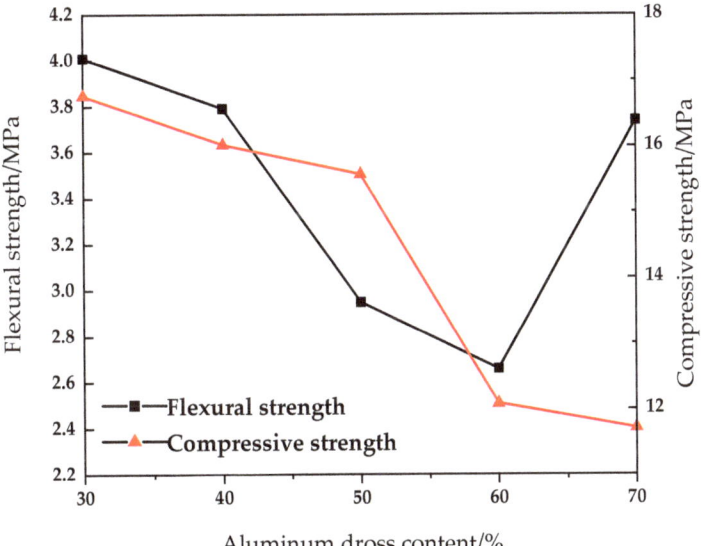

Figure 5. Effect of aluminum dross content on compressive and flexural properties of sintered brick.

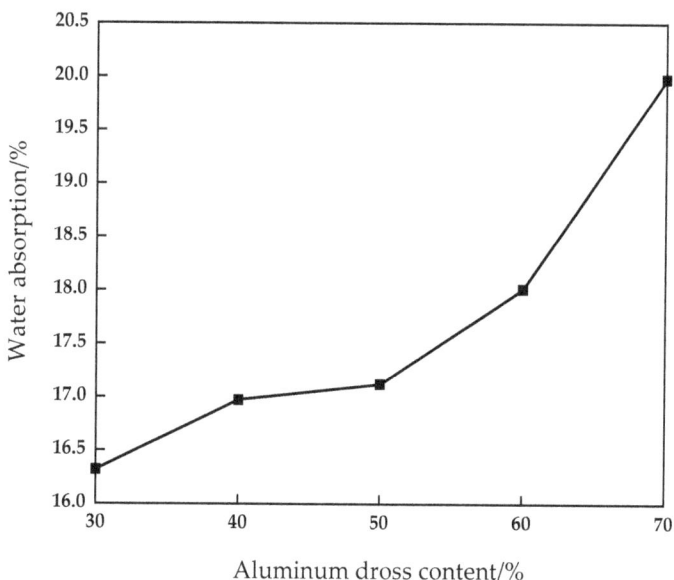

Figure 6. Effect of aluminum dross content on water absorption of fired brick.

It can be seen from Figures 5 and 6 that with the increase of aluminum dross addition, the compressive strength and flexural strength of the sintered brick decrease, and the water absorption increases. When the mixing amount of aluminum dross is 30%, 40% and 50%, the compressive strength of the brick is 16.76 MPa, 16.02 MPa and 15.57 MPa, respectively. The compressive strength decreases significantly after the aluminum dross content exceeds 50%. When the content of aluminum dross increased from 30% to 50%, the flexural strength decreased from 4.01 MPa to 2.95 MPa. When the aluminum dross content was 70%, the flexural strength increased sharply to 3.74 MPa. The water absorption rate increased from 16.32% to 19.98%, and green brick fluctuated from 9.15% to 11.28%. When the additional amount of aluminum dross was continuously increased, the SiO_2 content in the brick decreased, and it was difficult to combine with Al_2O_3 to form aluminosilicate in the roasting process. Aluminosilicate is the main structural framework of the sintered brick, so the mechanical properties were reduced. However, Al_2O_3 is also an effective component for the formation of mechanical strength of brick. Therefore, when the amount of aluminum dross is increased, it can be found that the downward trend of compressive strength slows down and the flexural strength increases.

The aluminum dross contains a small amount of aluminum. In the process of sintering, the aluminum is melted and filled between various phases, which can improve the overall density and mechanical properties of brick. However, with the increase of aluminum dross content, the content of SiO_2 in the product decreases, which is lower than the optimal range of SiO_2 content of 55–70%; at the same time, it increases the content of Al_2O_3. Both SiO_2 and Al_2O_3 are the main components for the mechanical strength of the brick. Therefore, when the aluminum dross content is 70%, the compressive strength decreases slowly and the flexural strength increases significantly. The main reason for the obvious increase of water absorption is that the content of AlN in the brick increases with the increase of aluminum dross addition. During the high temperature sintering process, AlN is converted into nitrogen oxides such as Al_2O_3 by oxygen. Therefore, there appear pores in the brick, which increases the water absorption. For the purpose of maximizing the utilization of aluminum dross, it is considered that the content of aluminum dross should be 50%, which meets the MU15 grade requirements of GB/T 5101-2017 "Sintered Ordinary Bricks" [19].

3.2. Influence of Molding Pressure on Properties of Sintered Brick

The content of aluminum dross is 50%. The molding pressure is set to 6 MPa, 8 MPa, 10 MPa, 12 MPa, 14 MPa, and the experimental results are shown in Figures 7 and 8.

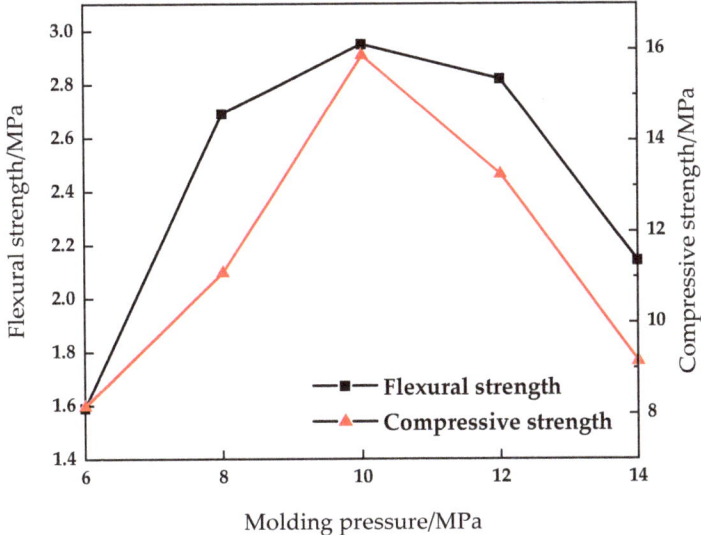

Figure 7. Effect of molding pressure on compressive and flexural properties of sintered brick.

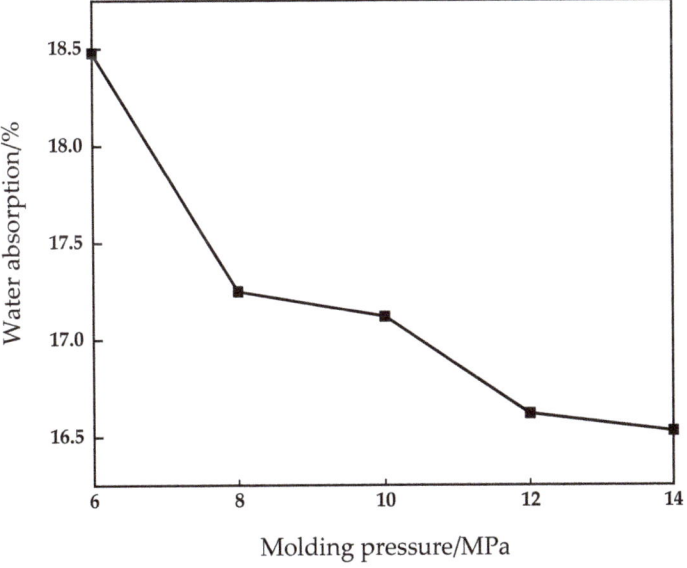

Figure 8. Effect of molding pressure on water absorption of fired brick.

It can be seen from Figures 7 and 8 that when the molding pressure reaches 10 MPa, the compressive strength of the sintered brick reaches the maximum value of 15.88 MPa. At the same time, the flexural strength of the brick reaches the maximum value of 2.95 MPa. When the molding pressure is 6 MPa and 14 MPa, the compressive strength is 8.16 MPa and 9.15 MPa respectively, which fails to meet the requirements of the MU10 compressive strength grade. When the molding pressure is 8 MPa and 12 MPa, the compressive strength is 11.1 MPa and 13.06 MPa respectively, which meets the requirements of mu10 compressive strength grade, but cannot meet the requirements of Mu15 compressive strength grade. Water absorption of brick is reduced from 18.48% to 16.53%, and that of green brick is reduced from 13.21% to 10.73%. As the molding pressure increases from 6 Mpa to 14 MPa, the density of brick increases from 1.93 g/cm^3 to 2.09 g/cm^3. As the molding pressure increases, both the compressive strength and the flexural strength of sintered bricks first increase and then decrease, while the water absorption rate decreases. This is mainly because when the molding pressure is low, the contact between the raw material particles of the brick is not close enough to form a larger pore, and its cohesive force and bite force are small. Therefore, good mechanics cannot be formed during the sintering process. When the forming pressure is high, the contact between the particles is closer, which causes the problem that the gas produced in the sintering process cannot be released in time. In this case, the volume of the brick expands and obvious cracks appear on the surface and inside, which reduces the mechanical properties of the brick.

3.3. The Effect of Firing System on the Performance of Sintered Brick

During the firing of the bricks, a series of complex physical and chemical changes take place in the various components inside the brick, which mainly include dehydration, isomorphic phase transformation, decomposition and the formation of new crystalline phases. Under high temperature conditions, Al and AlN in aluminum dross are oxidized by air, and the following Equations (1)–(6) occur.

$$Al + 3/4O_2 = 1/2Al_2O_3 \tag{1}$$

$$AlN + 3/4O_2 = 1/2Al_2O_3 + 1/2N_2 \tag{2}$$

$$AlN + 5/4O_2 = 1/2Al_2O_3 + NO \tag{3}$$

$$AlN + 7/4O_2 = 1/2Al_2O_3 + NO_2 \tag{4}$$

$$AlN + O_2 = 1/2Al_2O_3 + 1/2N_2O \tag{5}$$

$$AlN + 2O_2 = 1/2Al_2O_3 + 1/2N_2O_5 \tag{6}$$

The thermodynamic calculations of Equations (2)–(6) are shown in Figure 9. It can be seen from Figure 9 that the ΔG values of Equations (2)–(6) at 300–1100 °C are all less than 0. AlN in the aluminum dross can convert nitrogen and a variety of nitrogen oxides during the sintering process, and the tendency to convert to nitrogen is even greater. Nitrogen is the main component in the air and is harmless to the environment. Therefore, the preparation of sintered brick with aluminum dross as the main raw material is in line with the concept of green environmental protection.

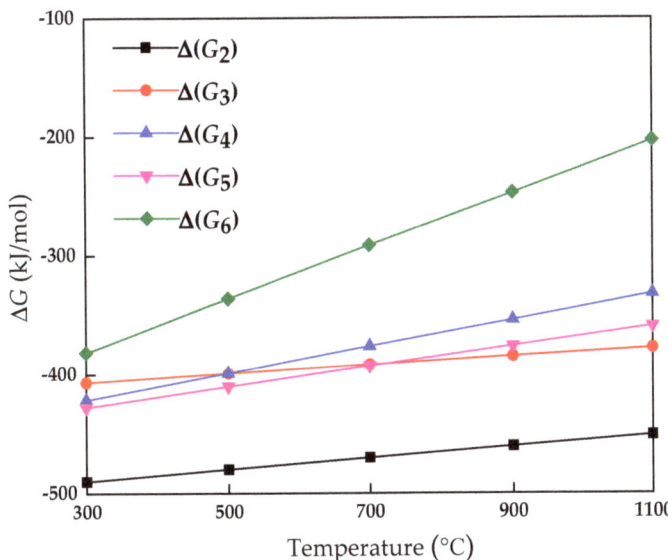

Figure 9. ΔG changes with temperature.

During the sintering experiment, orthogonal experiments were performed to analyze the significant influence of heating rate, sintering temperature, and holding time on the compressive strength of sintered brick. The best technological parameters of sintering brick were further found. In this orthogonal experiment, the above influencing factors are selected as the orthogonal experiment factors, and the three factors and three levels orthogonal experiment is carried out. Preliminary exploratory test results show that too fast heating speed leads to local burst of brick in furnace cavity. Therefore, considering various factors, the heating rate is 6 °C/min, 7 °C/min and 8 °C/min. A temperature gradient is set for every 100 °C increase and the temperature is maintained for 20 min. The selection of the sintering temperature is mainly based on the temperature of AlN oxidation [13] and the sintering temperature of traditional sintered brick. Therefore, the sintering temperature is set at 750 °C, 800 °C and 850 °C. The selection and level of each factor are shown in Table 2, the orthogonal experiment results are displayed in Table 3, and the analysis of variance is presented in Table 4.

Table 2. Orthogonal factors and standard selection table.

Factors	Heating Rate (°C/min)	Sintering Temperature (°C)	Holding Time (min)
Level 1	6	750	60
Level 2	7	800	120
Level 3	8	850	180

Table 3. The orthogonal pilot programs and experimental results.

No.	Heating Rate (°C/min)	Sintering Temperature (°C)	Holding Time (min)	Error	Test Results Compressive Strength (MPa)
	A	B	C	D	
1	1 (6)	1 (750)	1 (60)	1	9.48
2	1 (6)	2 (800)	2 (120)	2	10.75
3	1 (6)	3 (850)	3 (180)	3	9.32
4	2 (7)	1 (750)	2 (120)	3	9.63
5	2 (7)	2 (800)	3 (180)	1	13.78
6	2 (7)	3 (850)	1 (60)	2	10.63
7	3 (8)	1 (750)	3 (180)	2	15.57
8	3 (8)	2 (800)	1 (60)	3	16.21
9	3 (8)	3 (850)	2 (120)	1	12.31
K_1	29.55	34.68	36.32	35.57	-
K_2	34.04	40.74	32.69	36.95	-
K_3	44.09	32.26	38.67	35.16	-
R	14.54	8.48	5.98	1.79	-

Table 4. Variance analysis.

Factors	SS	df	MS	F	Significance
1	36.95	2	18.475	62.63	*
2	12.72	2	6.36	21.56	*
3	6.05	2	3.025	10.25	-
Errore	0.59	2	0.295	-	-
Sum	56.31	8	-	-	-

* Indicates whether the factor is significant.

The mechanical strength of the sintered brick is formed in the process of sintering, and its strength mainly comes from the silicate phase formed by the combination of SiO_2 and Al_2O_3 in different forms. When the sintering temperature is too low or holding time is too short, underfired brick will be produced. If the sintering temperature is too high, the brick will burn excessively, which will seriously reduce the quality of the brick. If the heating rate is too fast, the gas inside the brick will not have time to overflow, which will cause the volume expansion of the product and produce cracks on the surface or inside of the brick. And in severe cases, it will burst in the furnace cavity.

F test is performed on the data obtained in Table 4, and the query threshold value is $f_{0.05}(2,2) = 19$, $f_{0.01}(2,2) = 99$. Therefore, it can be seen that the heating rate and the calcination temperature have significant impact on the experimental results. Without considering the interaction, the optimal plan should take the level corresponding to the maximum K value of each factor, namely, A3B2. Since the holding time has no significant effect on the compressive strength of the sintered brick. When the heating rate is 8 °C/min and the sintering temperature is 800 °C, the holding time is 60, 120 and 180 min, respectively. It is found that different holding time has little effect on the mechanical strength of sintered brick. Therefore, from the perspective of energy saving, it is reasonable to set the heat preservation time as 60 min. Finally, A3B2C1 is the best solution. The heating rate was 8 °C/min, the sintering temperature was 800 °C and the holding time was 60 min. The results show that the compressive strength, flexural strength and water absorption of the sintered brick are 16.21 MPa, 3.42 MPa and 17.12%, respectively.

3.4. Microanalysis

The samples of the best group in the orthogonal test were detected by XRD, and analyze the phase composition inside the sintered bricks. The XRD pattern of the best experimental group of orthogonal test is shown in Figure 10.

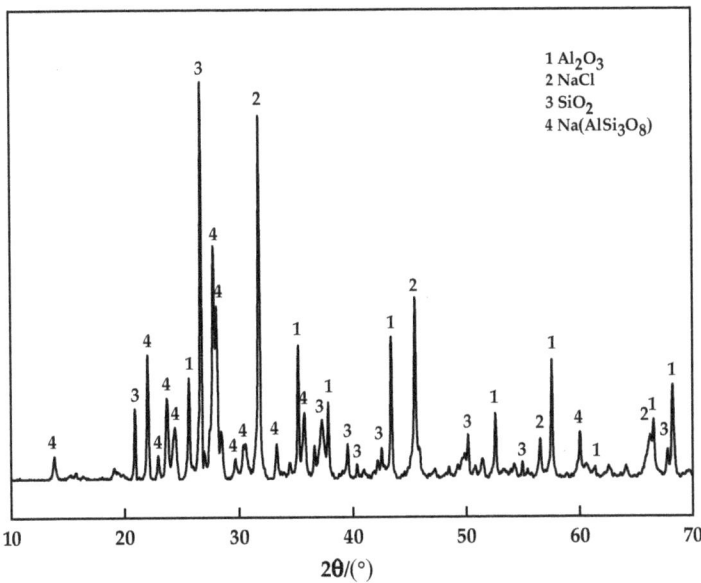

Figure 10. XRD pattern of the sample.

It is believed that the main materials of sintered brick after sintering are quartz, alumina, albite and soluble salts. Compared with the XRD patterns of raw materials, it can be seen that the AlN and Al in the aluminum dross disappear and transform into corresponding oxides after sintering. More importantly, albite is formed during the sintering process. At high temperatures, albite can form eutectic with quartz and aluminosilicate, fill between the crystal grains of the brick, and draw the particles closer under the effect of surface tension, making the brick more compact. The main crystalline phases are alumina, quartz and albite. These crystalline phases constitute the main framework of the sintered brick. The formation of new crystalline phase albite plays a key role in promoting sintering.

Figure 11a–c show the micro morphology of samples containing 30%, 50% and 70% aluminum dross, and Figure 11d shows the micro morphology of samples without aluminum dross. Then, it can be clearly seen that a glass phase appears inside the sintered brick without aluminum dross, which causes the particles to begin to melt and bond. It can be seen from Figure 11d that the glass phase is evenly distributed on the surface of the sample and connected with each other, which improves the mechanical strength of the sintered brick. Figure 11a–c show that with the increase of aluminum dross content, the glass phase inside the sintered brick becomes less and the porosity increases. This leads to a significant decrease in the mechanical properties of sintered bricks. However, with the addition of aluminum dross, the shape of crystal particles in the brick becomes more regular. As shown in Figure 11c, crystal particles mostly exist in the form of rods, strips and flakes. These are intertwined with each other, which helps to improve the mechanical properties of bricks to a certain extent.

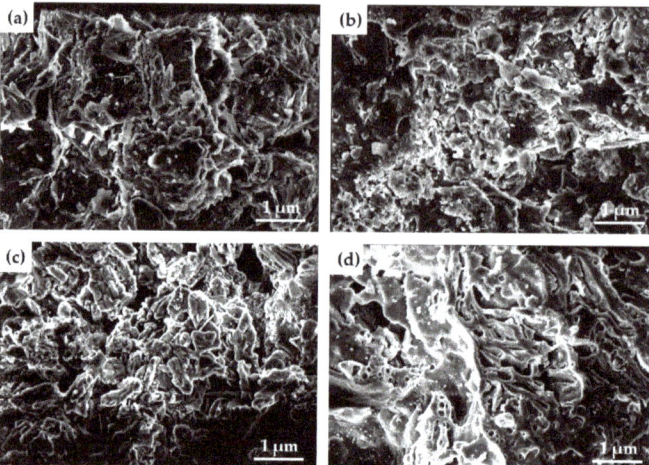

Figure 11. SEM images. (**a**) 30% aluminum dross (**b**) 50% aluminum dross (**c**) 70% aluminum dross (**d**) No aluminum dross.

3.5. TG Analysis

The samples with 40%, 50% and 60% aluminum dross content were selected for thermogravimetric analysis to study the variation of sample weight during the sintering process of sintered brick, as shown in Figure 12. It can be seen from Figure 11 that with the increase of temperature, the weight of the sample decreases. The weight change of 40% and 50% aluminum slag content is basically the same. However, when the aluminum dross content is 60%, the weight of the sample increases at low temperature. The reason is that the content of aluminum and AlN in the sample increases with the increase of aluminum dross content, and the weight of the sample increases during the oxidation process. When the temperature increases, the weight of the sample decreases in varying degrees, which is mainly due to the evaporation of the adsorbed water on the sample surface or the decomposition of other surface materials and impurities.

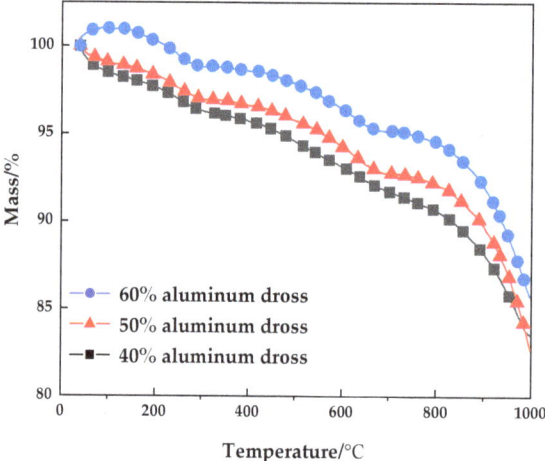

Figure 12. TG analysis.

4. Conclusions

(1) The best raw material ratio for the preparation of sintered brick is 50% aluminum dross, 37.5% engineering soil, 12.5% coal gangue. The best molding pressure is 10 MPa, the best sintering process: the heating rate is 8 °C/min, the temperature is 800 °C and the holding time is 60 min. According to the best preparation process, the sintered brick meets the requirements of Mu15 grade in GB/T 5101-2017 "fired ordinary brick".

(2) The sintering process of sintered brick is mainly affected by sintering temperature, holding time and heating rate. From the orthogonal test results, the heating rate and sintering temperature have a more significant impact on the overall mechanical strength of the sintered brick.

(3) After sintering, AlN and Al disappear and albite phase is formed. At high temperature, albite forms grain boundary with quartz and aluminosilicate, which is filled between the grains of the brick and improves the mechanical properties of the brick.

(4) Using aluminum dross, engineering soil and coal gangue to prepare sintered brick does not produce waste gas and waste liquid. This not only makes comprehensive use of solid waste, but also does not affect the surrounding environment in the production process, realizing the concept of green environmental protection and sustainable development.

Author Contributions: Conceptualization, H.N. and X.W.; methodology, Y.Z.; validation, Y.Z.; formal analysis, H.N. and S.L. (Shuaishuai Lv); investigation, Y.Z. and X.W.; resources, H.N. and S.L. (Shuaishuai Lv); writing—original draft preparation, Y.Z.; writing—review and editing, H.N. and S.L. (Songyuan Li); visualization, Y.Z. and J.Z.; supervision, H.N. All authors have read and agreed to the published version of the manuscript.

Funding: This research was Supported by the Priority Academic Program Development of Jiangsu Higher Education Institutions, grant number PAPD.

Institutional Review Board Statement: Not applicable.

Informed Consent Statement: Not applicable.

Data Availability Statement: Not applicable.

Conflicts of Interest: The authors declare no conflict of interest.

References

1. Wang, J.H.; Zhong, Y.Q.; Tong, Y.; Xu, X.L.; Lin, G.Y. Removal of AlN from secondary aluminum dross by pyrometallurgical treatment. *J. Cent. South. Univ.* **2021**, *28*, 386–397. [CrossRef]
2. Lin, Q.T.; Peng, H.L.; Zhong, S.X.; Xiang, J.X. Synthesis, characterization and secondary sludge dewatering performance of a novel combined silicon–aluminum–iron–starch flocculant. *J. Hazard. Mater.* **2015**, *285*, 199–206. [CrossRef] [PubMed]
3. Lv, S.S.; Zhang, J.Q.; Ni, H.J.; Wang, X.X.; Zhu, Y.; Gu, T. Study on Preparation of Aluminum Ash Coating Based on Plasma Spray. *Appl. Sci.* **2019**, *9*, 4980. [CrossRef]
4. Ni, H.J.; Zhang, J.Q.; Lv, S.S.; Gu, T.; Wang, X.X. Performance Optimization of Original Aluminum Ash Coating. *Coatings* **2020**, *10*, 831. [CrossRef]
5. Yoshimura, H.N.; Abreu, A.P.; Molisani, A.L.; de Camargo, A.C.; Portela, J.C.S.; Narita, N.E. Evaluation of aluminum dross waste as raw material for refractories. *Ceram. Int.* **2006**, *34*, 581–591. [CrossRef]
6. Zhan, D.P.; Zhang, H.S.; Jiang, Z.H. Effects of AlMnCa and AlMnFe Alloys on Deoxidization of Low Carbon and Low Silicon Aluminum Killed Steels. *J. Iron Steel Res. Int.* **2008**, *15*, 15–18. [CrossRef]
7. Elseknidy, M.H.; Salmiaton, A.; Nor, S.I.; Saad, A.H. A Study on Mechanical Properties of Concrete Incorporating Aluminum Dross, Fly Ash, and Quarry Dust. *Sustainability* **2020**, *12*, 9230. [CrossRef]
8. Kizito, P.M.; Revocatus, L.M.; Yusufu, A.C.J. Effect of Elevated Temperature on Compressive Strength and Physical Properties of Neem Seed Husk Ash Concrete. *Materials* **2020**, *13*, 1198.
9. Turner, L.K.; Collins, F.G. Carbon dioxide equivalent (CO_2-e) emissions: A comparison between geopolymer and OPC cement concrete. *Constr. Build. Mater.* **2013**, *43*, 125–130. [CrossRef]
10. Wei, Z.A.; Zhao, J.K.; Wang, W.S.; Yang, Y.H.; Zhuang, S.N.; Lu, T.; Hou, Z.K. Utilizing gold mine tailings to produce sintered bricks. *Constr. Build. Mater.* **2021**, *282*, 122655. [CrossRef]

11. Luo, L.Q.; Li, K.Y.; Weng, F.; Liu, C.; Yang, S.Y. Preparation, characteristics and mechanisms of the composite sintered bricks produced from shale, sewage sludge, coal gangue powder and iron ore tailings. *Constr. Build. Mater.* **2020**, *232*, 117250. [CrossRef]
12. Chiraporn, A.; Ryan, C.; McCuiston, T.; Prasartseree, P.; Pungpipat, S.O. Properties of Sintered Brick Containing Lignite Bottom Ash Substitutions. Mech. *Corros. Prop.* **2015**, *4122*, 138–142.
13. Ran, F.Z.; Gao, X.D.; Le, F.M.; Wei, J.G.; Jing, H.L. Preparation and Properties of Sintering Brick from Iron Tailings. *Mech. Corros. Prop.* **2012**, *1809*, 1023–1027.
14. Yang, H.L.; Liu, H.M. Performance Analysis on the Raw Material of Nantong Silt Sintered Brick. *Adv. Mater. Res.* **2012**, *1616*, 743–748. [CrossRef]
15. He, L.Q.; Shi, L.; Huang, Q.Z.; Hayat, W.; Shang, Z.B.; Ma, T.F.; Wang, M.; Yao, W.D.; Huang, H.Y.; Chen, R. Extraction of alumina from aluminum dross by a non-hazardous alkaline sintering process: Dissolution kinetics of alumina and silica from calcined materials. *Sci. Total Env.* **2021**, *777*, 146123. [CrossRef] [PubMed]
16. Meshram, A.; Jain, A.; Gautam, D.; Singh, K.K. Synthesis and characterization of tamarugite from aluminium dross: Part, I. *J. Environ. Manag.* **2019**, *232*, 978–984. [CrossRef] [PubMed]
17. Kristoffer, K.; Tomaž, K. Protection of AlN powder against hydrolysis using aluminum dihydrogen phosphate. *J. Eur. Ceram. Soc.* **2001**, *21*, 2075–2079.
18. Mostafa, M.; Ali, A. Hazardous aluminum dross characterization and recycling strategies: A critical review. *J. Environ. Manag.* **2018**, *223*, 452–468.
19. Shen, H.L.; Liu, B.; Ekberg, C.; Zhang, S.G. Harmless disposal and resource utilization for secondary aluminum dross: A review. *Sci. Total Env.* **2021**, *760*.
20. GB/T 5101-2017. *Fired Common Bricks*; General Administration of Quality Supervision, Inspection and Quarantine of the People's Republic of China: Beijing, China, 2017. (In Chinese)
21. GB/T 2542-2012. *Test Methods for Wall Bricks*; General Administration of Quality Supervision, Inspection and Quarantine of the People's Republic of China: Beijing, China, 2012. (In Chinese)

Article

Preparation and Characterization of PU/PET Matrix Gradient Composites with Microwave-Absorbing Function

Wenyan Gu [1], Rong Zhan [1], Rui Li [1], Jiaxin Liu [1] and Jiaqiao Zhang [2,3,*]

[1] School of Textile and Clothing, Nantong University, Nantong 226019, China; gu.wy@ntu.edu.cn (W.G.); 2012320005@stmail.ntu.edu.cn (R.Z.); 1715051043@stmail.ntu.edu.cn (R.L.); 1615052001@stmail.ntu.edu.cn (J.L.)
[2] School of Mechanical Engineering, Southeast University, Nanjing 211189, China
[3] School of Mechanical Engineering, Nantong University, Nantong 226019, China
* Correspondence: 1810310038@stmail.ntu.edu.cn

Abstract: In the field of microwave-absorbing materials, functional powder has always been the focus of research. In order to fabricate lightweight and flexible garment materials with microwave-absorbing function, the current work was carried out. Firstly, the general properties of polyurethane (PU) matrix composites reinforced with various microwave-absorbing powders were studied, and the carbon nanotubes (CNTs)/Fe_3O_4/PU film was proven to have the best general properties. Secondly, the needle-punched polyester (PET) nonwoven fabrics in 1 mm-thickness were impregnated into PU resin with the same composition of raw material as Fe_3O_4/CNTs/PU film, thereby the microwave-absorbing nonwovens with gradient structure were prepared. Moreover, the absorbing properties of the CNTs/Fe_3O_4/PU/PET gradient composites were tested and analyzed. Finally, the relationship between the mass ratio of CNTs and Fe_3O_4, and the microwave-absorbing properties was studied. The results show that the mass ratio of CNTs/Fe_3O_4 has a significant effect on the microwave-absorbing property of CNTs/Fe_3O_4/PU/PET. When the mass ratio of CNTs/Fe_3O_4 is 1:1, the prepared CNTs/Fe_3O_4/PU/PET gradient composite can achieve effective reflection loss in the range of more than 2 GHz in Ku-band (12–18 GHz), and the minimum reflection loss reaches −17.19 dB.

Keywords: microwave-absorbing; gradient composites; nonwoven; reflection loss

Citation: Gu, W.; Zhan, R.; Li, R.; Liu, J.; Zhang, J. Preparation and Characterization of PU/PET Matrix Gradient Composites with Microwave-Absorbing Function. *Coatings* **2021**, *11*, 982. http://doi.org/10.3390/coatings11080982

Academic Editor: Giorgos Skordaris

Received: 29 July 2021
Accepted: 17 August 2021
Published: 18 August 2021

Publisher's Note: MDPI stays neutral with regard to jurisdictional claims in published maps and institutional affiliations.

Copyright: © 2021 by the authors. Licensee MDPI, Basel, Switzerland. This article is an open access article distributed under the terms and conditions of the Creative Commons Attribution (CC BY) license (https://creativecommons.org/licenses/by/4.0/).

1. Introduction

At present, the rapid diffusion of electronic products produces a great number of radiations of electromagnetic waves around human beings [1,2]. Therefore, it is urgently needed to develop flexible microwave-absorbing textile with excellent microwave-absorbing properties to resist electromagnetic wave radiation and protect human health. A microwave-absorbing material is a kind of functional material that can absorb or attenuate the incident microwaves. It can convert or interfere with incident waves on the material surface by employing its inherent characteristics, to reduce the harm of electromagnetic wave radiation [3]. Microwave-absorbing functional powder has always been the focus of research in the field of microwave-absorbing materials. However, in the application of microwave absorbing functional powders, most of them are guided by the quarter theory, as while as taking into account the electromagnetic parameters of microwave absorbing materials [4]. As a result, the materials used for microwave absorption are often heavy and difficult to cut.

In recent years, due to the characteristics of low density, high conductivity, and high surface area, the carbon nanotube has gradually become a hot research spot in the field of microwave-absorbing composites [5–7]. However, microwave-absorbing composites filled with carbon nanotubes have obvious impedance mismatch problems, with relatively narrow effective absorption bandwidth. In order to solve the impedance mismatch problem, carbon nanotube is usually mixed with other fillers, such as Fe_3O_4 powder [8], graphite [9] or carbon fiber (CF) [10], which can not only obtain more balanced electromagnetic parameters, but also improve the microwave-absorbing properties. For example, Li et al. [11] enhanced the microwave-absorbing property of carbon nanotubes by adding Fe_3O_4 magnetic nanoparticle. The reflection loss of the optimized materials was less than -10 dB. Sandeep et al. [12] composed composite fillers of graphite and metal oxides, and prepared resin-based microwave-absorbing materials to improve the microwave-absorbing property of X-band. Wang et al. [13] prepared a flexible microwave-absorbing film based on graphene oxide/carbon nanotubes and Fe_3O_4 nanoparticles, which has been proven to have excellent microwave-absorbing properties in the range of 2–18 GHz.

Due to its special properties, Fe_3O_4 is often used as a filler to make composite materials with microwave absorption or electromagnetic shielding functions. Jacobo et al. [14] prepared a polyaniline (PANI)/Fe_3O_4 film composite with high conductivity, which indicated that the original performance of the materials can be improved after the Fe_3O_4 particles are filled. Yuvchenko et al. [15] studied the magnetic impedance of structured films in the presence of magnetic nanoparticles, which is helpful to the development of sensors for biomagnetic detection. The prepared non-woven fabric materials in our job with microwave absorption function can be used for human body wear, reduce the harm of electromagnetic waves to the human body, and has high application value in the medical field. The remarkable multi-modal function of magnetic nanoparticles used in our experiment, such as Fe_3O_4, is given by their size and morphology, which is very important for solving the challenge of slowing down the development of nano-biotechnology [16]. Apheshteguy et al. [17] used Fe_3O_4 to prepare medical magnetite magnetic nanoparticles and studied microwave resonant and zero-field absorption. Ansari et al. [18] summarized previous studies and believed that the magnetic iron oxide nanoparticles can be applied in the central nervous system. Kaczmarek et al. [19] used the magnetic nanoparticles for ultrasonic hyperthermia, which doubled the specific absorption rate.

It is well known that nonwovens are light, soft, and easy to process. As such, nonwoven fabrics have gradually come into the sight of researchers of microwave absorption. Bi etc. [20] assembled carbonyl iron and graphene aerogel onto nonwoven fabric, of which the widest efficient bandwidth covered 2.91–5.1 GHz and 10.99–18 GHz at the thickness of 6.0 mm, and the maximum RL is 22.3 dB. Egami etc. [21] coated polypyrrole on nonwoven fabric, and the results showed nonwoven sheets with extremely high frequencies absorption. However, the contribution of needle-punched fabric in the development of microwave absorption material has been rarely flagged up. Needle-punched fabric prosses irregular 3D pores formed by randomly tangled staple fibers, which endows them with various microwave absorption performance when loaded with different functional fillers.

In current study, the graphite, Fe_3O_4, carbonyl iron powder (CIP), carbon fiber, and carbon nanotubes were selected as the microwave-absorbing powders, and the PU was selected as the matrix to construct the microwave-absorbing materials. Through the comparison of properties such as tensile strength, adhesion, and antistatic ability, we found that the CNTs/Fe_3O_4/PU film has the best general performance. Furthermore, CNTs/Fe_3O_4/PU/PET gradient composites with the framework of needle-punched PET nonwoven fabric were prepared. In addition, its microwave-absorbing properties on the X-band and Ku-band were tested. The main contributions of our work are as follows. First, we proposed the preparation methods of microwave absorbing film and PU/PET matrix gradient composites. Moreover, the experimental tests proved that Fe_3O_4 and CNTs have outstanding general performance, suitable for the preparation of wearable non-woven fabrics with microwave-absorbing function. Finally, the microwave-absorbing performance of the gradient composite was tested, and it was proven that both the Fe_3O_4 content and the gradient structure have a great influence on the microwave absorption effect.

2. Materials and Methods

2.1. Materials and Specimen Preparation

2.1.1. Microwave-Absorbing Film

The preparation scheme of the microwave-absorbing film used in the general performance test is shown in Table 1, in which the thickness of the prepared microwave-absorbing wet film is 1 mm. In the single factor process of microwave-absorbing film, the waterborne polyurethane (PU, 601C type, Hefei Tairuike New Material Technology CO., Ltd., Hefei, China), thickener agent (Hefei Tairuike New Material Technology CO., Ltd., Hefei, China), and defoamer agent (X-690 type, Guangzhou Hongtai New Material CO., Ltd., Guangzhou, China) was used as matrix, and the graphite (20–80 mesh, Lingshou Zhanteng Mineral Products Processing Factory, Shijiazhuang, China), carbon fiber (CF, 500 mesh, Changzhou Hengfeng Nano Technology Co., Ltd., Changzhou, China), Fe_3O_4 (500 nm, with irregular shape, Qinghe Tuopu Metal Material Co., Ltd., Xingtai, China), and carbonyl iron powder (CIP, C913576 type, Shanghai Macklin Biochemical Co., Ltd., Shanghai, China) were used as the microwave-absorbing powders, respectively. The fillers mentioned above all have the functions of microwave absorption and electromagnetic shielding, which are often used in the related experiments. In the two-factors process of microwave-absorbing film, the combination of carbon nanotube (CNT, length > 5 μm, Changzhou Hengfeng Nano Technology Co., Ltd., Changzhou, China) and the absorbing powders in single factor process group were used as the microwave-absorbing powder, respectively. As contrast, pure PU film was processed as the control group.

Table 1. Preparation scheme of microwave-absorbing film for the general performance test.

Experiment Number	Group	Absorbing Powder 1/(wt.%)	Absorbing Powder 2/(wt.%)	PU/(wt.%)
F1	Control group	-	-	100
F2	Single factor process group	Graphite/1.5	-	100
F3		CFs/1.5	-	100
F4		Fe_3O_4/1.5	-	100
F5		CIP/1.5	-	100
F6	Two-factors process group	Graphite/1.5	CNTs/1.5	100
F7		CFs/1.5	CNTs/1.5	100
F8		Fe_3O_4/1.5	CNTs/1.5	100
F9		CIP/1.5	CNTs/1.5	100

The preparation process of microwave-absorbing film is shown in Figure 1. First, the 100 wt.% waterborne PU resin and 0.3 wt.% thickener were added to the beaker and stirred at a speed of 500 r/min for 10 min. Then, the microwave-absorbing powders were added to the waterborne PU and stirred at a speed of 800 r/min for 60 min, until the absorbing powders were dispersed. Next, the defoamer was added to the beaker and stirred at 500 r/min for 5 min to disperse the defoamer uniformly, and then let it stand for 30 min to defoam. After the foam disappears, it was poured into a glass mold with 1 mm depth. After air-dried for 48 h, the deionized water was added. After being soaked for 12 h, the coagulated film was removed and flattened with a pair of plates to obtain a microwave-absorbing film.

2.1.2. Microwave-Absorbing Impregnated Nonwoven Fabric

Employing needle-punched polyester (PET) nonwoven fabric in 1 mm-thickness (120 g/m^2, Yiwu Piccolo Electronic Commerce Co., Ltd., Yiwu, China) as a framework, the microwave-absorbing impregnated nonwoven fabric with gradient structure was prepared.

The preparation process of microwave-absorbing impregnated nonwoven fabric is shown in Figure 2. The preparation process of the absorbing solution was consistent with the preparation process of the microwave-absorbing film. After the foam disappeared, the absorbing solution was poured into a glass mold. The non-woven fabric substrate with a size of 30 cm × 30 cm was gently rinsed with deionized water to avoid errors due to sample deformation. Then, the non-woven fabric substrate was put into a constant temperature drying oven, and dried at a temperature of 65 °C for 24 h. After it was completely dried, it was taken out and naturally regained moisture for 24 h. Moreover, the bottom of the PET needle-punched nonwoven fabric was brought into contact with the absorbing solution in the glass mold to perform impregnation treatment. After standing for 24 h, a microwave-absorbing impregnated cloth with gradient structure was obtained.

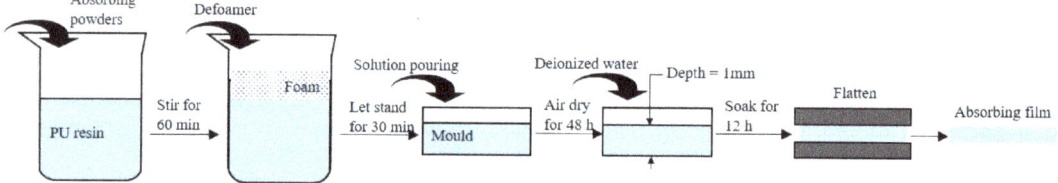

Figure 1. The preparation process of the microwave-absorbing film.

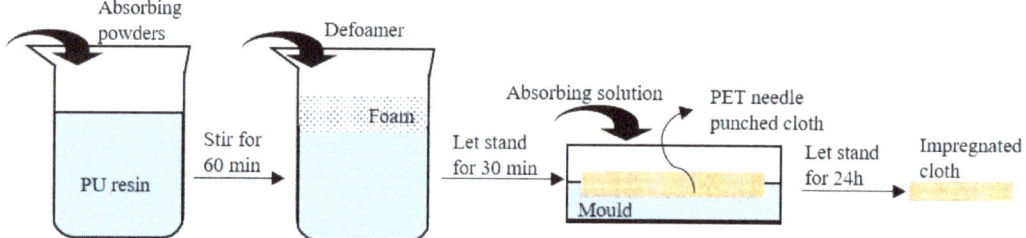

Figure 2. Preparation process of microwave-absorbing impregnated nonwoven fabric.

The physical pictures of the prepared microwave-absorbing impregnated nonwoven fabric are shown in Figure 3.

Figure 3. Physical images: (**a**) PET needle punched nonwoven fabric; (**b**) Microwave-absorbing impregnated nonwoven fabric.

2.2. Test Equipment and Methods

Desktop scanning electron microscopy (ZEISS Gemini SEM 300 type, Carl Zeiss AG, Oberkochen, Germany) was used to observe the samples' microstructures. The sample of the microwave-absorbing film was cut into 5 mm × 5 mm rectangles.

Material testing machine (3119-609 type, Instron company, Boston, MA, USA) was used to measure the tensile property of microwave-absorbing film. According to the standard (ISO 527-3:1995) [22], the film was cut into 100 mm × 15 mm rectangles. In the tensile test, the fracture at a tensile rate of 50 mm/min had the maximum tensile force and elongation. The thickness of the film can be measured by fabric thickness tester (YG(B)141D type, Wenzhou Darong Textile Instrument Co., Ltd., Wenzhou, China).

The prepared uniform coating was scraped on the cleaned and dried glass plate with 14-coating rod at a constant speed to produce a 32 μm thick coating. QFH-A Cross-Cut suit film adhesion tester is adopted in accordance with ASTM D3359-09 standard test method [23]. By cutting and penetrating the lattice image of coating, 100 small squares with a size of 2 mm × 2 mm were obtained, and then glued with 3M600 pressure-sensitive adhesive paper. Then, one end of the tape was quickly torn off from the direction of 90°. The magnifying glass was used to observe the area of falling squares. The percentage of the area of falling squares to the total surface area was evaluated by coating adhesion level, which was divided into 0 to 5, with 5 as the worst.

According to the standards JJG920-2017 [24], optical density meter (LS117 type, Shenzhen Linshang Technology Co., Ltd., Shenzhen, China) was used to test the optical density and transmissivity of microwave-absorbing film.

The surface resistance tester (VICTOR 385 type, Shenzhen Yisheng Shengli Technology Co., Ltd., Shenzhen, China) was used to test the resistance and impedance of microwave-absorbing film. During the test, the sensor meter was placed on the surface of microwave-absorbing film and cannot be in contact with other objects. It should be noted that the tests were performed multiple times.

According to the bow reflection method the measurement methods for the reflectivity of radar absorbing material (GJB 2038A-2011) [25], the vector network analyzer (AV3672C, China Electronics Technology Instrument Co., Ltd., Qingdao, China) was used to measure the microwave-absorbing property of the materials. The test bands were X-band and Ku-band. The frequency ranges were 8.2–12.4 GHz and 12–18 GHz. The sample size was 30 cm × 30 cm.

3. Experiment Results and Discussion

3.1. General Performance

The general performance of films has a great effect on the stability of the matrix. Therefore, single-factor experiments on general performance of functionalized PU films and pure PU were conducted to pick out a kind of PU film, with good general performance, for the process of microwave-absorbing compound materials.

3.1.1. SEM Images

The SEM images of different microwave-absorbing films are shown in Figure 4. In the process of taking the SEM images, the extra high tension (EHT) of PU film was set to 2.0 kV. In order to obtain clear images, the working distance (WD) was continuously changed with the test pieces. Figure 4a shows a pure PU film as a control group, with a flat surface. Figure 4b presents a PU film doped with graphite. Due to the sheet structure of graphite, it is easy to see some swellings formed by the lapping of sheets on the surface of the film. Since the graphite sheets are very easy to slip, smaller particles will also appear during the preparation process. Figure 4c shows a PU film mixed with CFs. It is obvious that the directions along the length of the CFs in the film distribute at random in three-dimensional space. Figure 4d shows a PU film doped with Fe_3O_4 powder. Some bumps result from irregularly shaped Fe_3O_4 appearing on the surface of the film. Figure 4e presents a PU film mixed with CIP. Except for some micro-powders, there are also dozens of microns holes on the surface of the film, indicating that the CIP causes exothermic reaction, during the high-speed mechanical stirring process, generating gas and heat. This is because CIP powder has great activity and can catalyze the exothermic reaction between CO and CO_2, which will further affect the stability of CIP in the later film formation process. The CIP in the film formation process is affected by heat and continues to release CO, thereby forming many micro-holes in the film. Figure 4f is a PU film doped with graphite and CNTs. The three-dimensional lap shape formed by the graphite flake structure is still obvious, but the size of such three-dimensional lap structure is significantly reduced, in the situation of coexisting of graphite flake and CNTs, comparing with Figure 4b. Figure 4g shows a PU film mixed with CFs and CNTs. The CFs are scattered in a straight or diagonal manner, and the curled CNTs are randomly distributed both on the surface of the CFs and in the film. Therefore, the distribution of CNTs can be approximately regarded as the approximate linear distribution along the CFs and the three-dimensional distribution in the film. Compared with the surface of other functional PU film, part of the three-dimensional space of the CNTs is lost, which is unfavorable for the loss and absorption of microwaves. Figure 4h shows a PU film doped with Fe_3O_4 powders and CNTs. Irregular Fe_3O_4 powders and tubular CNTs can be found randomly distributed in the three-dimensional space of the film, which are conducive to the absorption of microwaves. Figure 4i shows a PU film doped with CIP and CNTs. There are no obvious micropores on the surface of the film. It is considered that the presence of CNTs inhibits the exothermic reaction of the CIP, so there are no obvious pores on the surface.

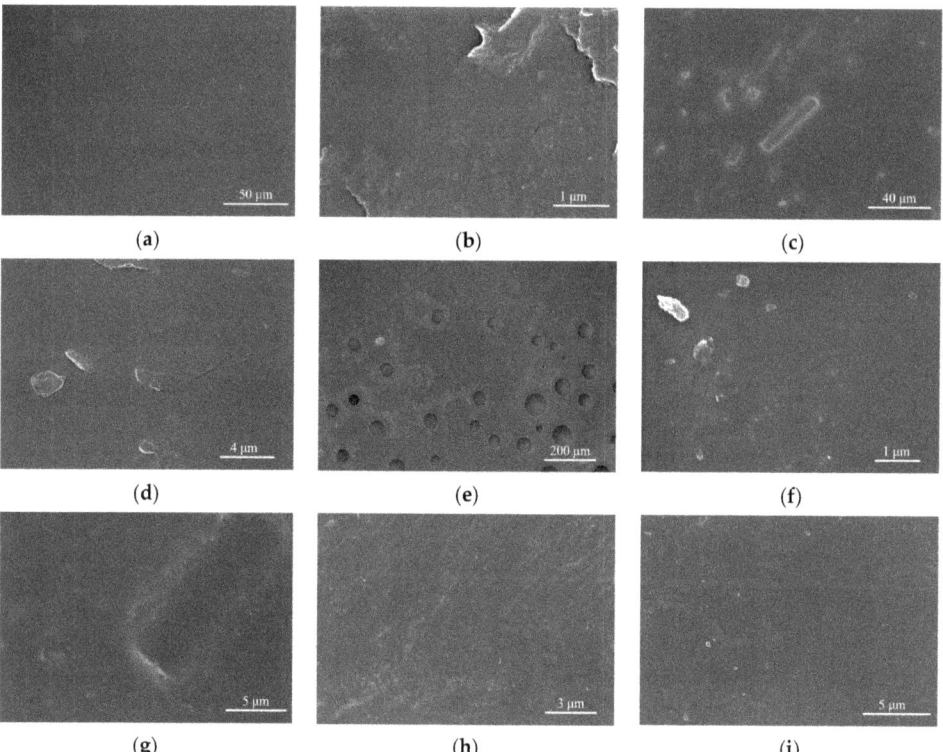

Figure 4. The SEM images of pure and functional PU films: (**a**) F1 (WD = 8.2 mm, EHT = 2.0 kV); (**b**) F2 (WD = 8.2 mm, EHT = 2.0 kV); (**c**) F3 (WD = 8.3 mm, EHT = 2.0 kV); (**d**) F4 (WD = 8.1 mm, EHT = 2.0 kV); (**e**) F5 (WD = 8.2 mm, EHT = 2.0 kV); (**f**) F6 (WD = 8.1 mm, EHT = 2.0 kV); (**g**) F7 (WD = 7.6 mm, EHT = 2.0 kV); (**h**) F8 (WD = 8.2 mm, EHT = 2.0 kV); (**i**) F9 (WD = 8.2 mm, EHT = 2.0 kV).

3.1.2. Tensile Property

Functional PU films can not only bear forces, but also protect the functional phases from external force damage, resulting in a stable function performance. Therefore, thin film materials with good tensile properties are the prerequisite for the process of microwave-absorbing impregnated nonwoven fabrics with excellent microwave-absorbing function.

Figure 5 shows the tensile strength and breaking elongation rate of pure and functional PU films filled with different fillers. In control group F1, the tensile strength and elongation of pure PU is 3.15 MPa and 12.29%, respectively. Single-factor process groups F2–F5 are the tensile strength and elongation of the functional PU films loading graphite, CFs, nano Fe_3O_4 powder and CIP, respectively. It can be found that the loading of graphite, CFs and nano Fe_3O_4 powder in 1.5 wt.%, keeping the PU resin in 100 wt.%, can improve the tensile strength of PU film. The reason is that graphite, CFs and nano Fe_3O_4 powder have stable performance in high-speed mechanical stirring process in water, and can induce better crystallization of PU, thereby enhancing the tensile strength of PU film. However, the loading of CIP reduces the tensile strength of CIP/PU film. Also, the CIP has been found to be so unstable in high-speed mechanical stirring process in water that it releasees a lot of heat, which impairs the formation of PU film and eventually leads to a significant decrease in tensile strength. It is worth mentioning that the loading of graphite can significantly improve the tensile strength and breaking elongation of PU film to 3.15 MPa and 12.29%, respectively. In contrast, although the addition of CFs can significantly increase tensile strength, it cannot significantly improve breaking elongation.

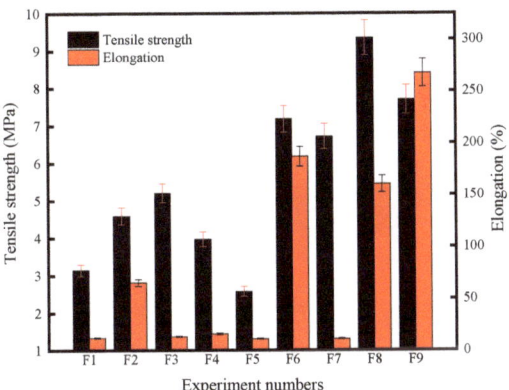

Figure 5. Tensile strength and elongation of pure and functional PU films loaded with different fillers.

The two-factors process groups F6–F9 come from the single-factor process groups F2–F5, into which CNTs are incorporated. The experimental results show that the tensile strength of the two-factors process groups is more than twice that of the control group and is nearly double that of the corresponding sample in the single-factor process groups, which indicates that the loading of CNTs improves the tensile strength of functional PU film. Especially, when Fe_3O_4 combined with CNTs dispersed in PU film, the tensile strength of the film reaches a maximum of 9.35 MPa. The breaking elongations of almost all two-factors process groups increase by about 2–20 times higher than that of the control group, but that of CIP is almost the same as that of the control group. In experiment F9, the combination of CIP and CNTs in PU film makes the breaking elongation of the film achieve a maximum of 267.46%.

The fractograph of pure and functional PU films loaded with varied fillers is observed with SEM images, as shown in Figure 6. As can be seen from the figure, the cross-sections of pure PU film (Figure 6a), graphite/CNTs/PU film (Figure 6f) and CIP/CNT/PU film (Figure 6i) are relatively similar and are all uneven, containing a great number of fragments. It is believed that these films fracture slowly, and a large deformation occurs. Furthermore, the cross-sections of graphite/PU film (Figure 6b), CFs/PU film (Figure 6c), Fe_3O_4/PU film (Figure 6d), CFs/CNTs/PU film (Figure 6g), and Fe_3O_4/CNTs/PU film (Figure 6h) look neat, indicating that brittle fractures have occurred, and the fillers make a significant contribution to the enhancement of the strength of the PU films. The CIP/PU film (Figure 6e) is special. Its fracture is relatively neat, scattered with holes. On the fracture, the marks of multiple brittle fracture are very obvious. It is thought that in the fabrication of CIP/PU film, a lot of gas and heat is generated, which results in stress concentration along the film. It is harmful to the tensile performance.

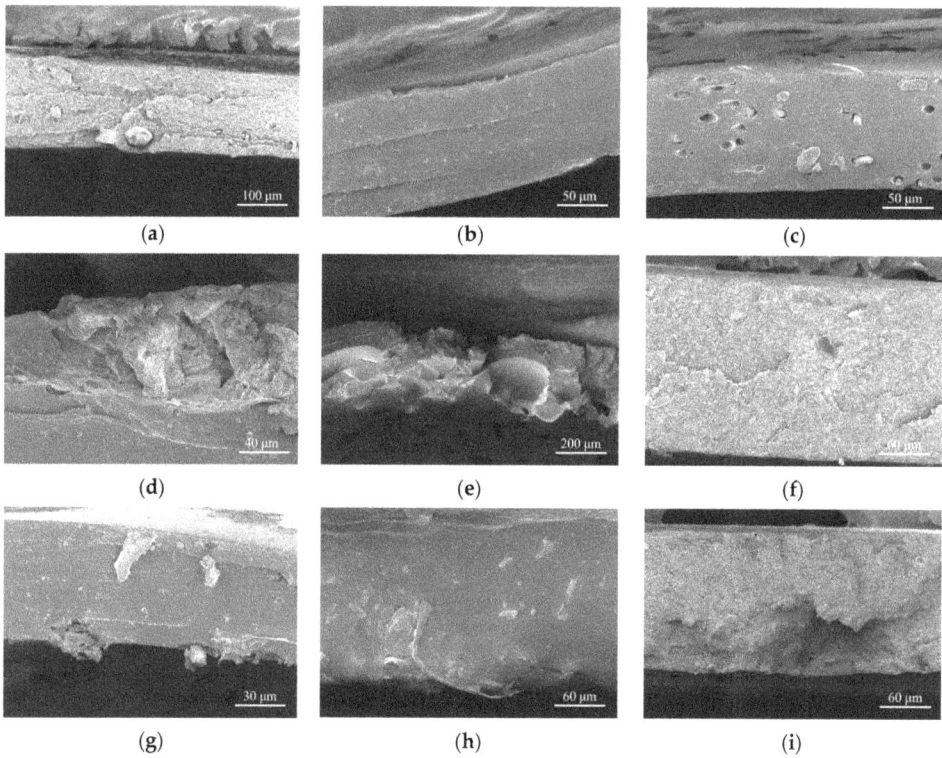

Figure 6. The SEM images of fractograph of pure and functional PU films: (**a**) F1 (WD = 5.7 mm, EHT = 2.0 kV); (**b**) F2 (WD = 6.9 mm, EHT = 2.0 kV); (**c**) F3 (WD = 5.9 mm, EHT = 2.0 kV); (**d**) F4 (WD = 7.0 mm, EHT = 2.0 kV); (**e**) F5 (WD = 6.9 mm, EHT = 2.0 kV); (**f**) F6 (WD = 5.9 mm, EHT = 2.0 kV); (**g**) F7 (WD = 6.1 mm, EHT = 2.0 kV); (**h**) F8 (WD = 7.1 mm, EHT = 2.0 kV); (**i**) F9 (WD = 7.0 mm, EHT = 2.0 kV).

3.1.3. Adhesion

The adhesion can examine the performance of coating film. Only when a microwave-absorbing film and its substrate have good adhesion do they not fall off easily and exert their microwave-absorbing properties on the basis of a complete film coverage.

As can be seen from Table 2, the adhesion level of control group F1 is level 3. In the single-factor process group, the adhesion levels of F2 and F3 are both level 1, and the adhesion levels of F4 and F5 are both level 2. The adhesion levels in the two-factors process group are greater than or equal to level 3. When the amount of PU is 100%, part of the coating will fall off. After adding a single microwave-absorbing filler such as graphite, CFs, nano Fe_3O_4 powder and CIP, the adhesion of coating is improved. Especially, the loading of graphite or CFs will help the adhesion of coating achieve the best level. When the CNTs are added, the adhesion level of coating film increases. Moreover, large patches of films peel off. Through comparison, it can be seen that the adhesion levels of the coating films loaded with CNTs is higher than or equal to those without CNTs. Considering the high strength of CNTs, the length dispersion is basically along the plane of the film. The effect of surface tension will weaken the dispersion of functional particles, such as CNTs, micro powders, microfibers, and so on, at the interface, resulting in the weakening of adhesion performance. When the CNTs meet flake graphite, the dispersion characteristics of CNTs will affect the adhesion performance of the flake graphite, resulting in the decline of the overall adhesion performance. When the CNTs meet CIP powder, the presence of CNTs inhibits the exothermic reaction of the CIP, so there are no obvious pores both on the surface and in the film. CNTs and CIP powder are uniformly dispersed in the film, but due to the effect of surface tension, there are few opportunities for them to emerge, resulting in a decrease in the adhesion performance.

Table 2. Test results of the adhesion of the pure and functional PU films.

Experiment Number	Level
F1	3
F2	1
F3	1
F4	2
F5	2
F6	4
F7	3
F8	3
F9	4

3.1.4. Optical Density and Transmittance

During the experiment, we found that the addition of fillers will reduce the transparency of the film. The microwave absorption film prepared by us is mainly used for human body wear, and the change of transparency will have a certain impact on the use occasion of the composite. Therefore, we performed qualitative measurements on the parameters of optical density (OD) and transmittance. The test results of OD and transmittance of pure and functional PU films are shown in Figure 7. The LS117 optical density meter used 380–760 nm whole white light for the test of OD and transmittance, which complied with the international commission on illumination (CIE) photopic function standard.

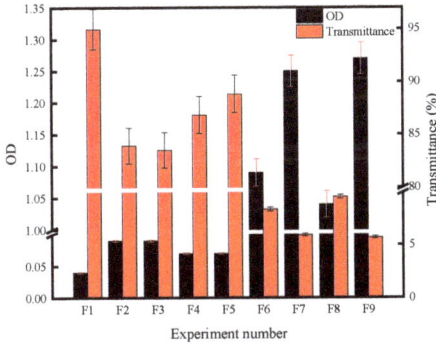

Figure 7. Test results of optical density and transmittance of pure and of functional PU films.

It is observed from the Figure 7 that the optical density of the pure PU film is the smallest (0.04), and the transmittance is the largest (94.97). When such functional phases as graphite, carbon fiber, nano Fe_3O_4 powder, and carbonyl iron are loaded in the films in one-factor process groups, respectively (experiment number F2–F5), the optical densities of the films increase slightly, remaining at a low level (less than 0.1), while the transmittances of those one-factor process group films decrease to 83%–90%. In the two-factors process groups, a certain amount of carbon nanotubes is loaded in the films, combining with the unchanged amount of powder filler in one-factor process group films. This results in the rapid rise of optical density of functional PU films, which are all above 1, while the transmittance drops sharply. In other words, the addition of CNTs will seriously reduce the transparency of the microwave-absorbing film, because the SWCNTs inside the CNTs will absorb light [26].

3.1.5. Antistatic Property

The fillers in this experiment play an important role in absorbing and shielding electromagnetic waves [27,28]. The results of surface resistance test reflect the influence of the fillers on the antistatic property of the functional PU films, as shown in Table 3. The surface resistance of control group F1 is in the order of 10^9 Ω. F2–F5 are the single-factor process groups. Among them, the surface resistance of F2 with graphite is the smallest, which is in the order of 10^9 Ω, while the surface resistances of F3, F4 and F5 all reach the order of 10^{11} Ω. The antistatic properties of these four kinds of films in the single-factor process groups are between conductors and insulators, reaching the order of a semiconductor, which is similar to pure PU film. In the two-factors process groups F6–F9, incorporating of CNTs into the single-factor process groups significantly reduces the order of magnitude of surface resistance, achieving 10^4–10^5 Ω. The antistatic properties of these four kinds of films in two-factors process groups all reach the level of a conductor. The CNTs possess excellent conductivity [29], showing good electrical properties even in a semiconductor matrix.

Table 3. Test results of the impedance and surface resistance of the pure and functional PU films.

Experiment Number	Surface Resistance (Ω)	Antistatic Property
F1	10^9	Semiconductor
F2	10^9	Semiconductor
F3	10^{11}	Semiconductor
F4	10^{11}	Semiconductor
F5	10^{11}	Semiconductor
F6	10^4	Conductor
F7	10^5	Conductor
F8	10^5	Conductor
F9	10^4	Conductor

3.1.6. Selection of Filling Scheme

Based on the analysis of the above experimental results, it can be found that the combination of Fe_3O_4 and CNTs can effectively improve the mechanical properties of microwave-absorbing film. For example, the tensile strength reaches 9.35 MPa, which is 193.15% higher than that of control group F1, and 79.46% higher than that of single-factor process group F3. In addition, Fe_3O_4/CNTs/PU film also performs best in the properties of adhesion, transmission and antistatic in the two-factors process group.

A good general performance is the basis of microwave-absorbing performance. Therefore, Fe_3O_4/CNTs/PU film (F8) with excellent general performance is selected to impregnate nonwoven fabric in 1 mm thickness to carry out further research on microwave-absorbing properties.

3.2. Microwave-Absorbing Property

3.2.1. Gradient Absorbing Structure

The cross-section is observed with SEM, of the Fe_3O_4/CNTs/PU/PET microwave-absorbing impregnated nonwoven fabric. As shown in Figure 8a, in the normal direction of the impregnated nonwoven fabric surface, the resin wrapped densely on the lower surface of the fiber web. At the same time, the resin on the upper surface is so rare that several single fibers can be easily made out.

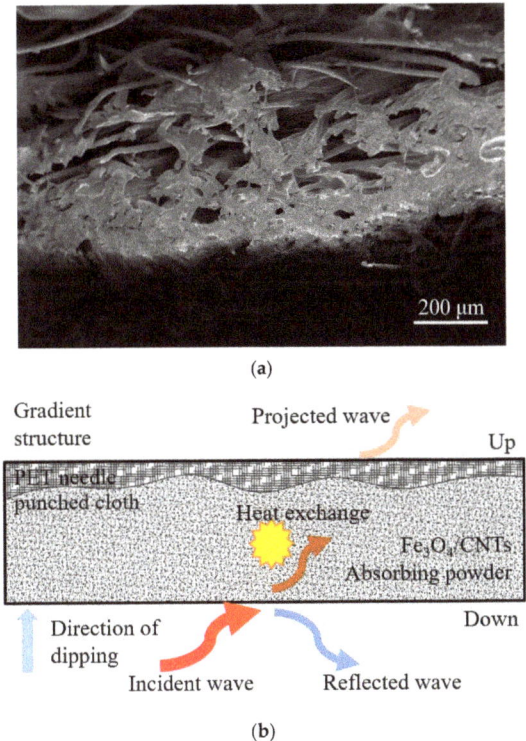

Figure 8. Structure and principle of PU/PET matrix microwave-absorbing gradient composites: (**a**) SEM image of cross-section of PU/PET matrix microwave-absorbing gradient composites (WD = 7.4 mm, EHT = 2.0 kV); (**b**) Principle of PU/PET matrix microwave-absorbing gradient composites.

In the lower surface of the fiber web, that is, the side contacting the impregnating liquid, the resin is densely distributed, and the fibers are wrapped by the resin.

In the normal direction of the impregnated non-woven fabric, the resin distributes in a gradient manner, which is thick and weighty at the bottom, while thin and insubstantial on the top. It is undoubtedly that such a gradient structure could enhance the penetration of microwaves on the outer surface, and the absorption and loss inside the matrix, as shown in Figure 8b.

3.2.2. Influence of Mass Ratio of Fe_3O_4/CNTs on Microwave-Absorbing Property

With regard to the fact that Fe_3O_4/CNTs/PU film (F8) shows excellent general performance, the PET nonwoven fabric is impregnated in the PU slurry with the same composition of raw material as the Fe_3O_4/CNTs/PU film. Also, considering the mass ratio of Fe_3O_4 and CNTs maybe have a great influence on microwave-absorbing property of Fe_3O_4/CNTs/PU/PET, another two impregnated PET nonwoven fabric with varied mass ratios of Fe_3O_4 and CNTs were investigated, as shown in Table 4. During the experiment, the content of CNTs remained unchanged, and the content of Fe_3O_4 increased exponentially to 0.75, 1.5, and 2.25 wt.%, respectively. Figure 9 shows the SEM micrograph and particle size distribution of Fe_3O_4 powders. It can be seen from Figure 9 that the Fe_3O_4 particles have irregular shapes, but most of the particles are spherical. The diameter of the Fe_3O_4 particles is mainly concentrated below 500 nm, accounting for 96.25% of the total number of particles, and the average diameter of the particles is 341.82 nm. Figure 10 shows the XRD patterns of Fe_3O_4. The diffraction peaks appear at 2θ of 18.34°, 30.14°, 35.50°, 37.10°, 43.12°,

53.48°, 57.10°, 62.58°, and 74.01°. Calculated from Bragg's Law [17], the corresponding crystal face index is (111), (220), (311), (222), (400), (422), (511), (440), and (533). It is found that the XRD pattern is basically consistent with the characteristic of peak position and peak intensity of the cubic crystal system Fe_3O_4 (JCPDS 99-0073) standard card.

Table 4. Experimental scheme for the ratio of Fe_3O_4/CNTs.

Experiment Number	CNTs/(wt.%)	Fe_3O_4/(wt.%)	PU/(wt.%)
F10	1.5	2.25	100
F11	1.5	1.5	100
F12	1.5	0.75	100

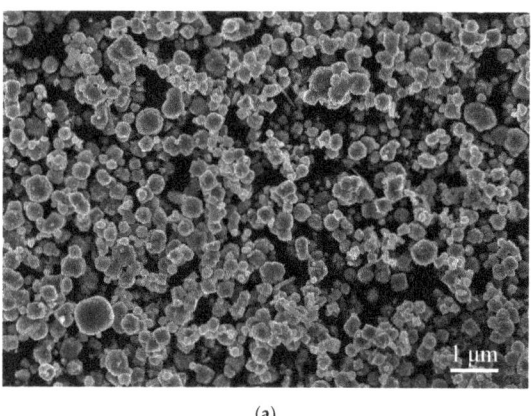
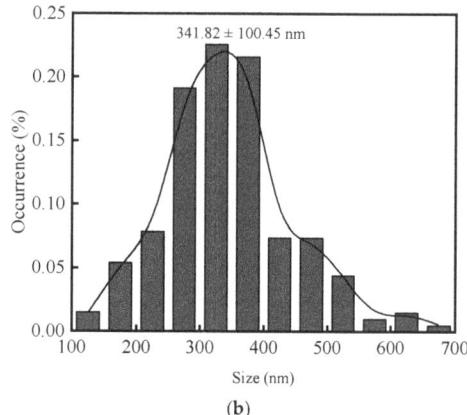

Figure 9. The SEM micrograph and particle size distribution of Fe_3O_4 powders (**a**) SEM image (WD = 9.2 mm, EHT = 5.0 kV); (**b**) particle size distribution.

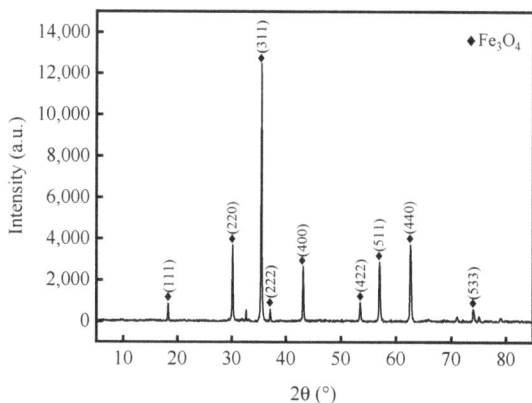

Figure 10. The XRD pattern of Fe_3O_4.

The bow reflection method is employed to observe the microwave-absorbing performances of Fe_3O_4/CNTs/PU/PET matrix gradient composites, with varied mass ratio of Fe_3O_4 and CNTs, in the ranges of 8.2–12.4 GHz and 12–18 GHz.

Figure 11 shows the results of the absorbing performance test. The microwave-absorbing performance of these three samples face down in the frequency range of 8.2 to 12.4 GHz are as follows. First, Fe_3O_4/CNTs/PU/PET matrix gradient composite made from 1.5 wt.% CNTs, 1.5 wt.% Fe_3O_4, and 100 wt.% PU shows the highest maximum reflection loss, which is −2.12 dB. Secondly, Fe_3O_4/CNTs/PU/PET matrix gradient composite composed of 1.5 wt.% CNTs, 2.25 wt.% Fe_3O_4, and 100 wt.% PU resin ranks second, with the maximum reflection loss of −1.62B. Third, Fe_3O_4/CNTs/PU/PET matrix gradient composite containing 1.5 wt.% CNTs, 0.75 wt.% Fe_3O_4 and 100 wt.% PU resin has the lowest maximum reflection loss, which is −1.00 dB.

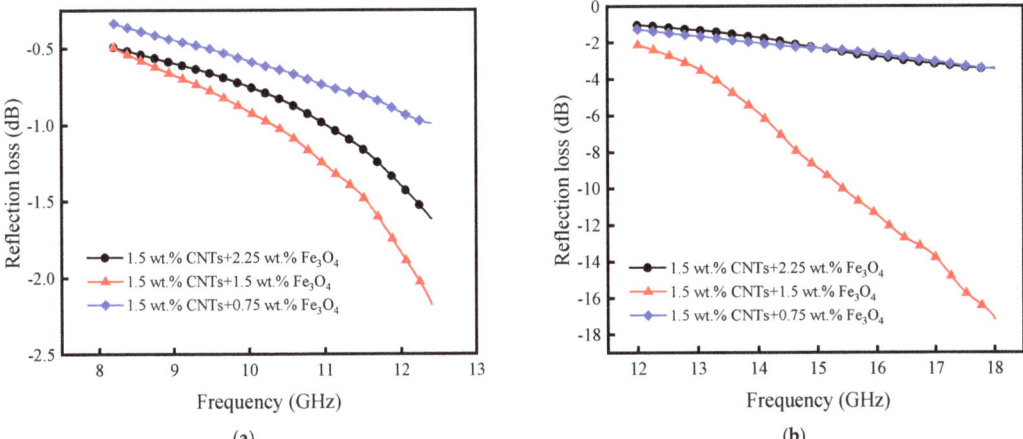

Figure 11. The influence of different mass ratio of CNTs and Fe_3O_4 on microwave-absorbing performance of Fe_3O_4/CNTs/PU/PET gradient composites, with composites face down: (**a**) 8.2–12.4 GHz; (**b**) 12–18 GHz.

The microwave-absorbing performance of these three samples in the 12–18 GHz frequency band are as follows. The Fe_3O_4/CNTs/PU/PET matrix gradient composite composed of 1.5 wt.% CNTs, 1.5 wt.% Fe_3O_4 and 100 wt.% PU resin has a maximum reflection loss of −17.19 dB. In addition, the microwave-absorbing performance of the Fe_3O_4/CNTs/PU/PET matrix gradient composite containing 1.5 wt.% CNTs, 2.25 wt.% Fe_3O_4 and 100 wt.% PU resin and that of 1.5 wt.% CNTs, 0.75 wt.% Fe_3O_4 and 100 wt.% PU resin are both −3 dB, with minute differences.

The results suggest that the reflection loss of Fe_3O_4/ CNTs/ PU/ PET gradient composites can be improved, when carbon nanotubes and Fe_3O_4 are in the mass ratio of 1:1. Particularly, a satisfied efficient reflection loss can be achieved in the frequency range of more than 2 GHz in Ku-band, and the minimum reflection loss reaches −17.19 dB. If the mass ratio is too large or too small, no efficient microwave-absorbing performance is shown. It is considered that impedance match is achieved when the mass ratio of CNTs and Fe_3O_4 is 1:1, the microwave-absorbing reagent can consume most of the electromagnetic waves. If the mass ratio of dielectric and magnetic medium is too large or too small, the impedance mismatches, resulting in poor microwave-absorbing performance. It seems that the key in the development of the thin and lightness of microwave-absorbing materials is just to solve the impedance matching problem among conductor, dielectric and magnetic medium.

To verify this speculation, the microwave-absorbing performance of these three samples face up in the frequency bands of both X-band and Ku-band are also displayed, seen in Figure 12a,b. Both figures show no efficient reflection loss. However, with the mass ratio of Fe_3O_4 increasing from 0.75 to 2.25 wt.%, the reflection loss is improved. So, keeping the absorbers the same, variation of the injected direction of electromagnetic wave onto a gradient composite leads to a significant difference in their electromagnetic absorbing performance. It is obvious that the structural parameters of micro objects also play a key role in impedance matching. It can be inferred that impedance matching, which impacts greatly on microwave-absorbing materials, depends not only on the electromagnetic performances of conductors, dielectric and magnetic mediums, but also on their macro structures' parameters.

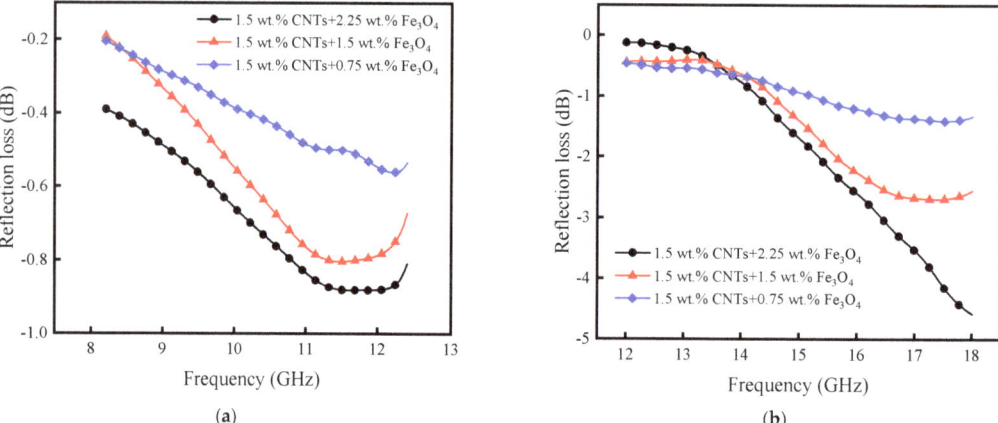

Figure 12. The influence of different mass ratio of CNTs and Fe$_3$O$_4$ on microwave-absorbing performance of Fe$_3$O$_4$/CNTs/PU/PET gradient composites, with composites face up: (**a**) 8.2–12.4 GHz; (**b**) 12–18 GHz.

4. Conclusions

- Microwave-absorbing powder can significantly improve a film's tensile property. When Fe$_3$O$_4$ and CNTs are combined in usage of a filler, the tensile strength of the film reaches a maximum value of 9.35 MPa. When CIP and CNTs are mixed as fillers, the breaking elongation of the film reaches the maximum, which is 267.46%. CNTs can increase the adhesion level of coating film, while the other microwave-absorbing powders will help enhance the coating adhesion. The antistatic property of films in the single-factor process groups is similar to that of pure PU film, which is similar to semiconductor. The loading of CNTs in PU film significantly reduces the surface resistance, so that the functional PU films loaded with CNTs is similar to conductor.
- The combination of Fe$_3$O$_4$ and CNTs can effectively improve the mechanical properties of PU film. For example, the tensile strength is 193.15% higher than that of the control group F1, and 79.46% higher than that of the single factor process group F3.
- The mass ratio of CNTs and Fe$_3$O$_4$ has a significant impact on the microwave-absorbing properties of PU/PET matrix gradient composites. In the Ku-band, when the mass ratio of CNTs and Fe$_3$O$_4$ is 1:1, the PU/PET matrix microwave-absorbing composite in the thickness of 1 mm has a maximum reflection loss of −17.19 dB. When the mass ratio of CNTs and Fe$_3$O$_4$ is more than or less than 1:1, PU/PET matrix gradient composites show poor microwave-absorbing performance.
- The Fe$_3$O$_4$/CNTs/PU/PET matrix gradient composite was easily fabricated with needle punched nonwoven and shown notable electromagnetic absorbing performance, indicating the great prospects of gradient nonwovens in electromagnetic absorbing fields.
- The Fe$_3$O$_4$/CNTs/PU/PET matrix gradient composite was easily fabricated with needle punched nonwoven and shown notable absorbing performance, indicating the great prospects of gradient nonwovens in electromagnetic absorbing fields. Our current work tries to clarify the macro performance of different PU/PET matrix gradient composites with microwave-absorbing function. Further research involving the relationship between different structure forms at micro-level, meso-level and macro-level will be conducted in the future.

Author Contributions: Conceptualization, W.G.; methodology, R.Z. and R.L.; validation, J.Z.; formal analysis, J.Z.; resources, W.G.; writing-original draft, R.Z. and J.L.; writing-review and editing, W.G. and J.Z.; investigation, R.L.; visualization, J.Z.; supervision, W.G. All authors have read and agreed to the published version of the manuscript.

Funding: This research was funded by Jiangsu Overseas Research and Training Program for University Prominent Young and Middle-aged Teachers and Presidents (Grant No. 148), and the Innovative Project of Nantong University Students (Grant No. 143).

Institutional Review Board Statement: Not applicable.

Informed Consent Statement: Not applicable.

Data Availability Statement: Not applicable.

Conflicts of Interest: The authors declare no conflict of interest.

References

1. Wanasinghe, D.; Aslani, F.; Ma, G. Electromagnetic shielding properties of carbon fibre reinforced cementitious composites. *Constr. Build. Mater.* **2020**, *260*, 120439. [CrossRef]
2. Yang, J.; Liao, X.; Wang, G.; Chen, J.; Guo, F.; Tang, W.; Wang, W.; Yan, Z.; Li, G. Gradient structure design of lightweight and flexible silicone rubber nanocomposite foam for efficient electromagnetic interference shielding. *Chem. Eng. J.* **2020**, *390*, 124589. [CrossRef]
3. Jia, Z.; Lan, D.; Lin, K.; Qin, M.; Kou, K.; Wu, G.; Wu, H. Progress in low-frequency microwave absorbing materials. *J. Mater. Sci. Mater. Electron.* **2018**, *29*, 17122–17136. [CrossRef]
4. Zhan, J.; Yao, Y.; Zhang, C.; Li, C. Synthesis and microwave absorbing properties of quasione-dimensional mesoporous NiCo2O4 nanostructure. *J. Alloys Compd.* **2014**, *585*, 240–244. [CrossRef]
5. Gao, Y.; Gao, X.; Li, J.; Guo, S. Improved microwave absorbing property provided by the filler's alternating lamellar distribution of carbon nanotube/carbonyl iron/poly (vinyl chloride) composites. *Compos. Sci. Technol.* **2018**, *158*, 175–185. [CrossRef]
6. Zheng, X.; Hu, Q.; Wang, Z.; Nie, W.; Wang, P.; Li, C. Roll-to-roll layer-by-layer assembly bark-shaped carbon nanotube/Ti3C2Tx MXene textiles for wearable electronics. *J. Colloid Interface Sci.* **2021**, *602*, 680–688. [CrossRef]
7. Savi, P.; Giorcelli, M.; Quaranta, S. Multi-walled carbon nanotubes composites for microwave absorbing applications. *Appl. Sci.* **2019**, *9*, 851. [CrossRef]
8. Song, X.; Li, X.; Yan, H. Preparation and microwave absorption properties of MWCNTs/Fe3O4/NBR composites. *Diam. Relat. Mater.* **2019**, *100*, 107573. [CrossRef]
9. Zhao, T.; Jin, W.; Ji, X.; Yan, H.; Jiang, Y.; Dong, Y.; Yang, Y.; Dang, A.; Li, H.; Li, T.; et al. Synthesis of sandwich microstructured expanded graphite/barium ferrite connected with carbon nanotube composite and its electromagnetic wave absorbing properties. *J. Alloys Compd.* **2017**, *712*, 59–68. [CrossRef]
10. Singh, S.K.; Akhtar, M.J.; Kar, K.K. Hierarchical carbon nanotube-coated carbon fiber: Ultra, lightweight, thin, and highly efficient microwave absorber. *ACS Appl. Mater. Interfaces* **2018**, *10*, 24816–24828. [CrossRef]
11. Li, N.; Huang, G.; Li, Y.; Xiao, H.; Feng, Q.; Hu, N.; Fu, S. Enhanced microwave absorption performance of coated carbon nanotubes by optimizing the Fe3O4 nanocoating structure. *ACS Appl. Mater. Interfaces* **2017**, *9*, 2973–2983. [CrossRef]
12. Singh, S.K.; Akhtar, M.J.; Kar, K.K. Impact of Al2O3, TiO2, ZnO and BaTiO3 on the microwave absorption properties of exfoliated graphite/epoxy composites at X-band frequencies. *Compos. Part B Eng.* **2019**, *167*, 135–146. [CrossRef]
13. Wang, L.; Jia, X.; Li, Y.; Yang, F.; Zhang, L.; Liu, L.; Ren, X.; Yang, H. Synthesis and microwave absorption property of flexible magnetic film based on graphene oxide/carbon nanotubes and Fe3O4 nanoparticles. *J. Mater. Chem. A* **2014**, *2*, 14940–14946. [CrossRef]
14. Jacobo, S.E.; Aphesteguy, J.C.; Anton, R.L.; Schegoleva, N.N.; Kurlyandskay, G.V. Influence of the preparation procedure on the properties of polyaniline based magnetic composites. *Eur. Polym. J.* **2007**, *43*, 1333–1346. [CrossRef]
15. Yuvchenko, A.A.; Lepalovskii, V.N.; Vas'kovskii, V.O.; Safronov, A.P.; Volchkov, S.O.; Kurlyandskaya, G.V. Magnetic impedance of structured film meanders in the presence of magnetic micro- and nanoparticles. *Tech. Phys.* **2014**, *59*, 230–236. [CrossRef]
16. Zamani Kouhpanji, M.R.; Stadler, B.J.H. A guideline for effectively synthesizing and characterizing magnetic nanoparticles for advancing nanobiotechnology: A review. *Sensors* **2020**, *20*, 2554. [CrossRef] [PubMed]
17. Aphesteguy, J.C.; Jacobo, S.E.; Lezama, L.; Kurlyandskaya, G.V.; Schegoleva, N.N. Microwave resonant and zero-field absorption study of doped magnetite prepared by a co-precipitation method. *Molecules* **2014**, *19*, 8387–8401. [CrossRef]
18. Ansari, S.A.M.K.; Ficiarà, E.; Ruffinatti, F.A.; Stura, I.; Argenziano, M.; Abollino, O.; Cavalli, R.; Guiot, C.; D'Agata, F. Magnetic iron oxide nanoparticles: Synthesis, characterization and functionalization for biomedical applications in the central nervous system. *Materials* **2019**, *12*, 465. [CrossRef]
19. Kaczmarek, K.; Hornowski, T.; Dobosz, B.; Józefczak, A. Influence of magnetic nanoparticles on the focused ultrasound hyperthermia. *Materials* **2018**, *11*, 1607. [CrossRef]
20. Bi, S.; Tang, J.; Wang, D.J.; Su, Z.A.; Hou, G.L.; Li, H.; Li, J. Lightweight non-woven fabric graphene aerogel composite matrices for assembling carbonyl iron as flexible microwave absorbing textiles. *J. Mater. Sci. Mater. Electron.* **2019**, *30*, 17137–17144. [CrossRef]

21. Egami, Y.; Yamamoto, T.; Suzuki, K.; Yasuhara, T.; Higuchi, E.; Inoue, H. Stacked polypyrrole-coated non-woven fabric sheets for absorbing electromagnetic waves with extremely high frequencies. *J. Mater. Sci.* **2012**, *47*, 382–390. [CrossRef]
22. ISO 527-3:1995. *Plastics-Determination of Tensile Properties Part 3: Test Conditions for Films and Sheets*; International Organization for Standardization: Geneva, Switzerland, 1995.
23. ASTM D3359-09. *Standard Test Methods for Measuring Adhesion by Tape Test*; American Society for Testing and Materials: West Conshohocken, PA, USA, 2009.
24. JJG920-2017. *Diffuse Transmission Visual Densitometers*; National Optical Metrology Technical Committee: Beijing, China, 2017. (In Chinese)
25. GJB 2038A-2011. *The measurement Methods for Reflectivity of Radar Absorbing Material*; General Assembly Electronic Information Basic Department: Beijing, China, 2011. (In Chinese)
26. Jiang, S.; Hou, P.; Chen, M.; Wang, B.; Sun, D.; Tang, D.; Jin, Q.; Guo, Q.; Zhang, D.; Du, J.; et al. Ultrahigh-performance transparent conductive films of carbon-welded isolated single-wall carbon nanotubes. *Sci. Adv.* **2018**, *4*, eaap9264. [CrossRef] [PubMed]
27. Wang, D.; Wu, Z.; Li, F.; Gan, X.; Tao, J.; Yi, J.; Liu, Y. A combination of enhanced mechanical and electromagnetic shielding properties of carbon nanotubes reinforced Cu-Ni composite foams. *Nanomaterials* **2021**, *11*, 1772. [CrossRef] [PubMed]
28. Kim, H.-J.; Kim, S.-H.; Park, S. Effects of the carbon fiber-carbon microcoil hybrid formation on the effectiveness of electromagnetic wave shielding on carbon fibers-based fabrics. *Materials* **2018**, *11*, 2344. [CrossRef] [PubMed]
29. Ulloa-Castillo, N.A.; Martínez-Romero, O.; Hernandez-Maya, R.; Segura-Cárdenas, E.; Elías-Zúñiga, A. Spark plasma sintering of aluminum-based powders reinforced with carbon nanotubes: Investigation of electrical conductivity and hardness properties. *Materials* **2021**, *14*, 373. [CrossRef]

Article

Experimental Study Regarding the Possibility of Blocking the Diffusion of Sulfur at Casting-Mold Interface in Ductile Iron Castings

Denisa Anca, Iuliana Stan *, Mihai Chisamera, Iulian Riposan and Stelian Stan

Materials Science and Engineering Faculty, Politehnica University of Bucharest, 313 Spl. Independentei, 060042 Bucharest, Romania; denisa_elena.anca@upb.ro (D.A.); chisameramihai@gmail.com (M.C.); iulian.riposan@upb.ro (I.R.); constantin.stan@upb.ro (S.S.)
* Correspondence: iuliana.stan@upb.ro

Abstract: The main objective of this work is to investigate the mechanism of sulfur diffusion from the mold (sand resin P-toluol sulfonic acid mold, sulfur-containing acid) in liquid cast iron in order to limit the graphite degeneration in the surface layer of iron castings. A pyramid trunk with square section samples was cast. On the opposite side of the feed canal of the samples, steel sheets with different thicknesses (0.5, 1, and 3 mm) were inserted with the intention of blocking the diffusion of sulfur from the mold into the cast sample during solidification. The structure evaluation (graphite and matrix) in the surface layer and the casting body was recorded. The experimental results revealed that by blocking the direct diffusion of sulfur at the mold-casting interface, a decrease of the demodified layer thickness (for 0.5 mm steel sheet thickness) is obtained until its disappearance (for steel sheet thicknesses of more than 1 mm). The paper contains data that may be useful in elucidating the mechanism of graphite degeneration in the superficial layer of ductile iron castings. Based on the obtained results, we recommend using such barriers on the metal-mold interface, which are able to limit sulfur diffusion from the mold/core materials into the iron castings, in order to limit or even cease graphite degeneration in the Mg-treated surface iron casting layer. The paper presents additional data related to the interaction of sulfur at the ductile iron casting-mold interface previously analyzed.

Keywords: ductile cast iron; sulfur diffusion; structure; graphite degeneration; surface layer

Citation: Anca, D.; Stan, I.; Chisamera, M.; Riposan, I.; Stan, S. Experimental Study Regarding the Possibility of Blocking the Diffusion of Sulfur at Casting-Mold Interface in Ductile Iron Castings. *Coatings* **2021**, *11*, 673. https://doi.org/10.3390/coatings11060673

Academic Editor: Giorgos Skordaris

Received: 19 April 2021
Accepted: 29 May 2021
Published: 1 June 2021

Publisher's Note: MDPI stays neutral with regard to jurisdictional claims in published maps and institutional affiliations.

Copyright: © 2021 by the authors. Licensee MDPI, Basel, Switzerland. This article is an open access article distributed under the terms and conditions of the Creative Commons Attribution (CC BY) license (https://creativecommons.org/licenses/by/4.0/).

1. Introduction

Although the graphite degeneration phenomenon was highlighted in the first iron castings, with compact graphite shapes obtained, not enough is known about the phenomenon at present. Many papers have identified its surface degenerated structure as an effect of sulfur diffusion from the mold into the Mg-treated iron melt [1–8].

The analysis of the demodifying graphite structure in the superficial layer of compact graphite iron castings requires the most accurate knowledge of physico-chemical processes at the liquid metal-mold interface. It is important especially in the cooling phase of liquid metal when it is in direct contact with the mold. During solidification, contact between the liquid and the mold is made by means of a solidified crust, which implies a slower diffusion and an interruption of the contact of the alloy with the atmosphere of the mold.

Previous works [1,6–10] have pointed out that decreasing the residual magnesium content aggravates the surface graphite degeneration five times more in mold including sulfur, compared with no-sulfur mold media. If the sulfur contribution of the mold is diminished, such as by blocking its transfer to the iron melt, the graphite degeneration in the casting surface layer can be avoided or, at least, diminished. Graphite degeneration in the casting surface layer is promoted by a S-bearing coating, or conversely, it is possible to limit the surface layer thickness using desulfurization type coatings (Al_2O_3, $CaCO_3$, Basic slag, CaF_2, Talc, Mg), with Mg-bearing coatings as performance [1,6–8]. It has also been found that the negative role of oxygen and especially of sulfur in graphite degeneration in the surface casting layer can be counteracted by the addition of materials able to block the diffusion of these elements into the iron melt, such as carbonic material (more efficient to limit oxygen effect) and iron powder (especially to limit sulfur negative effects) [9].

Recently, it was found that in the presence of a thin steel sheet at the metal-mold interface, the thickness of the surface layer decreases or is just avoided. It was suggested that it acts as a barrier, blocking sulfur diffusion into the iron melt [10]. The present paper aims to elucidate some aspects of the demodifiyng process related to the mechanism of the transition of sulfur from the mold into liquid cast iron through the prism of the diffusion blocking effect at the interface.

2. Materials and Methods

The base iron melt was prepared using the induction furnace with acid liner (100 kg capacity, 2400 Hz frequency, Inductro, Bucharest, Romania). A total of 100 wt.% cast iron scrap, carbonic material (0.37 wt.% addition, >98 wt.% C) and CaC_2 (0.43 wt.% addition) were used as charge materials, and 2 wt.% FeSiCaMgRE (6 wt.% Mg) alloy was used as nodulizer (Tundish Cover treatment technique, Politehnica University, Bucharest, Romania), followed by ladle inoculation (CaBaFeSi, 0.5 wt.% addition). The thermal regime of iron melt processing was as follows: superheating temperature: $Ts = 1550\ °C$; Mg-treatment temperature: $Tm = 1530\ °C$; pouring temperature: $Tp = 1350\ °C$. The inoculated Mg-treated iron was poured in a pyramid trunk with square section samples and with steel sheets (0.5, 1, and 3 mm thickness) inserted at the casting-mold interface (Figure 1). For comparison, a sample without a steel sheet was cast (for more details, see paper [10]).

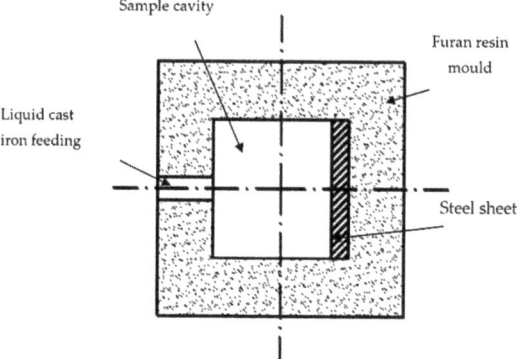

Figure 1. The inserted steel sheet position in mold (transversal section).

The structure evaluation in the surface layer and the casting body was made according to Figure 2. Graphite morphology analysis [10] was conducted by using a professional automat image analyzer (OMNIMET ENTERPRISE and analySIS® FIVE Digital Imaging Solutions software version 5.0, Waukegan, IL, USA), using both the standard cast iron modulus and particle analysis software.

The structure analysis in the casting body was recorded on the three parallel directions on a 5 mm casting thickness, starting at a 3 mm distance from the casting surface (Figure 2a). Nine analysis points, with a 0.66 mm distance between them in each direction, were used. The average value of the 27 analyses was considered.

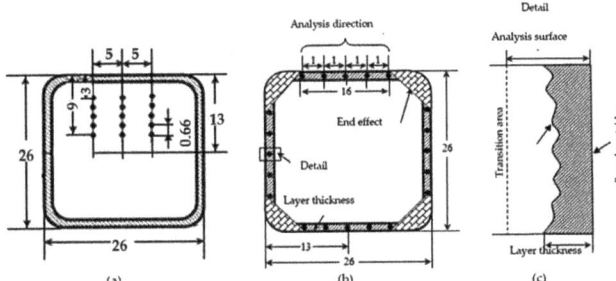

Figure 2. The analysis procedure to evaluate the structure parameters (graphite and metal matrix) in the casting body (**a**) and in the surface layer and skin (surface layer) thickness (**b**,**c**) (Units-mm).

Figure 2b,c shows the procedure of structure analysis on the section of the surface layer. In order to avoid the end effect, the structure analysis was recorded at a distance of 16 mm, with 17 analysis points each at a 1mm distance between them. The degenerated graphite layer was measured by metallographic microscope analysis (100:1 magnification) as a visible layer thickness up to prevalent nodular graphite morphology presence, in un-etched and Nital etched conditions. For this purpose, with the assistance of the automatic image analyzer, images of the degenerate layer were captured on the un-etched and Nital etched samples, on which measurements of the thickness of the degenerate layer were made, after the previous calibration of the instrument. The degenerate surface layer was delimited based on the visible decrease of graphite compactness, on the un-etched samples, and by measuring the thickness of the predominantly ferritic area determined by the decrease of graphite compactness. There are some differences between the two measurement variants because in the un-etched state the degenerate layer was more difficult to delimit, while the boundary pearlite/ferrite had a more visible contrast. The measurement points' layer thickness are at 100 μm between these points, with an average value consideration. Graphite nodularity (according to the relation (1)) on the casting section, starting from the casting surface, is also used to evaluate the skin thickness.

3. Results and Discussion

3.1. Chemical Composition

The chemical composition was evaluated for the base cast iron (wt.%: 3.37 C, 1.42 Si, 0.54 Mn, 0.021 S, 3.83 CE (carbon equivalent)), for the cast iron after the Mg-treatment (wt.%: 3.44 C, 2.56 Si, 0.62 Mn, 0.013 S, 0.059 Mg, 0.02 Ce, 0.0086 La, 4.24 CE), and inoculation (wt.%: 3.37 C, 2.93 Si, 0.62 Mn, 0.015 S, 0.048 Mg, 0.015 Ce, 0.006 La, 4.2 CE). The resulting ouble treated cast irons (Mg-treatment + Inoculation) are characterized by an eutectic position in the Fe-C diagram and a good nodulizing potential (Mg-Ce-La residual contents). The residual elements content is below the limits considered critical for the shape of graphite (wt.%: 0.006–0.012 Al, 0.005–0.006 Ti, 0.045–0.05 Cr, 0.036–0.037 Ni, 0.04–0.05 Cu, 0.008–0.009 Mo, 0.005–0.006 Co, 0.002–0.003 V, 0.0002–0.0004 Pb, 0.0004–0.0005 Sn, 0.0055–0.006 As, <0.001 Sb, <0.001 Te, 0.0009–001 B, 0.0002–0.0005 Zn, 0.008–0.01 N, <0.0005 Bi).

3.2. Structure Characteristics

Nodular graphite morphology characterizes the structure of the casting body, with more than 80% graphite nodularity (Equation (1) in all of the tested variants (Figure 3)).

$$N = [(\sum A_{NG} + 1/2 \sum A_{IG})/\sum A_{tot}] \times 100\% \tag{1}$$

where

A_{NG}—area of particles classified as nodules (RSF = 0.625–1.0);
A_{IG}—area of particles classified as intermediates (RSF = 0.525–0.625);

A_{tot}—area of all of the graphite particles;
RSF—roundness shape factor = $4A/\pi l^2$;
A—area of the graphite particle in question;
l—maximum axis length of the graphite particle in question (maximum distance between two points on the graphite particle perimeter).

The presence of the steel sheet at the mold–cavity interface led to graphite nodularity improvement (5%) and the nodule count increasing (1.5 time), with the presence of the fine nodules (max. 30 µm, Size 7, ISO 945) 1.75 time higher, favoring the ferrite amount increasing in the casting body. Under steel sheet influence, acting as a chilling agent (higher cooling rate), a limited amount of carbides resulted (2%–4%).

By blocking direct diffusion at the interface with the help of steel plates, a decrease in the thickness of the demodified layer (0.5 mm sheet thickness) was obtained until its disappearance (at 1 and 3 mm sheet thicknesses) (Figure 4).

At 0.5 mm sheet thickness, a discontinuous layer of degenerate graphite appeared. Its appearance may have been due to the rapid overheating of the steel sheet and the layer's arrival in a semi-viscous state, which favored the diffusion of sulfur from the mold in liquid cast iron. It was found that the layer thicknesses obtained on the samples etched with Nital 2% were higher than those obtained by measuring on un-etching samples, which confirms that the demodifying effect is manifested inside the samples, inducing changes in the metal matrix.

It is suggested that in the first stage of solidification, the diffusion of sulfur from the furan resin mold in steel sheet occurred. By pouring the iron melt into the furan resin mold, under the water vapors and the temperature action, P-toluol sulfonic acid (PTS) dissociated with sulfuric acid forming, according to the Equation (2):

$$CH_3C_6H_4SO_3H \text{ (PTD acid)} + H_2O \text{ (water steam)} \rightarrow C_6H_5CH_3 \text{(toluene)} + H_2SO_4 \quad (2)$$

The dissociation of sulfuric acid and aromatic sulfonic acids from the catalyst composition (which ensures furan resin ionic polymerization and the mold hardening) resulted in the SO_2 migrating to the interface and creating a concentration of sulfur on the metal surface following the Equation (3):

$$3\langle Fe \rangle + \{SO_2\} = \langle FeS \rangle + 2\langle FeO \rangle \quad (3)$$

The reaction was similar to the sulfurizing process of the solid metal charge in the cupola of gaseous atmospheres containing SO_2.

From the ΔG relation ($\Delta G = -73500 + 25.33T$, J), it resulted that with the temperature increase, the reaction was carried out to the left, in the sense of sulfur oxidation. However, experiments [11] performed by maintaining a sample of steel in the SO_2 atmosphere have shown that as the temperature increases the sulfurization process intensifies, especially above the temperature of 800 °C [12]. This has been attributed to the formation of sulfur eutectics that pass into the liquid state (FeO-FeS, at 940 °C and Fe-FeS, at 975 °C), favouring the adsorption of sulfur from gases. At the same time, the temperature increase determines the diffusion intensification process of sulfur through the solid metal, reducing its concentration in the superficial area, thus favoring the reaction to the right.

The formed iron sulfide determined the concentration of sulfur in the superficial layer of the steel sheet and created the conditions for diffusion through the steel sheet in which the concentration of sulfur was lower (less than 0.05 wt.% S).

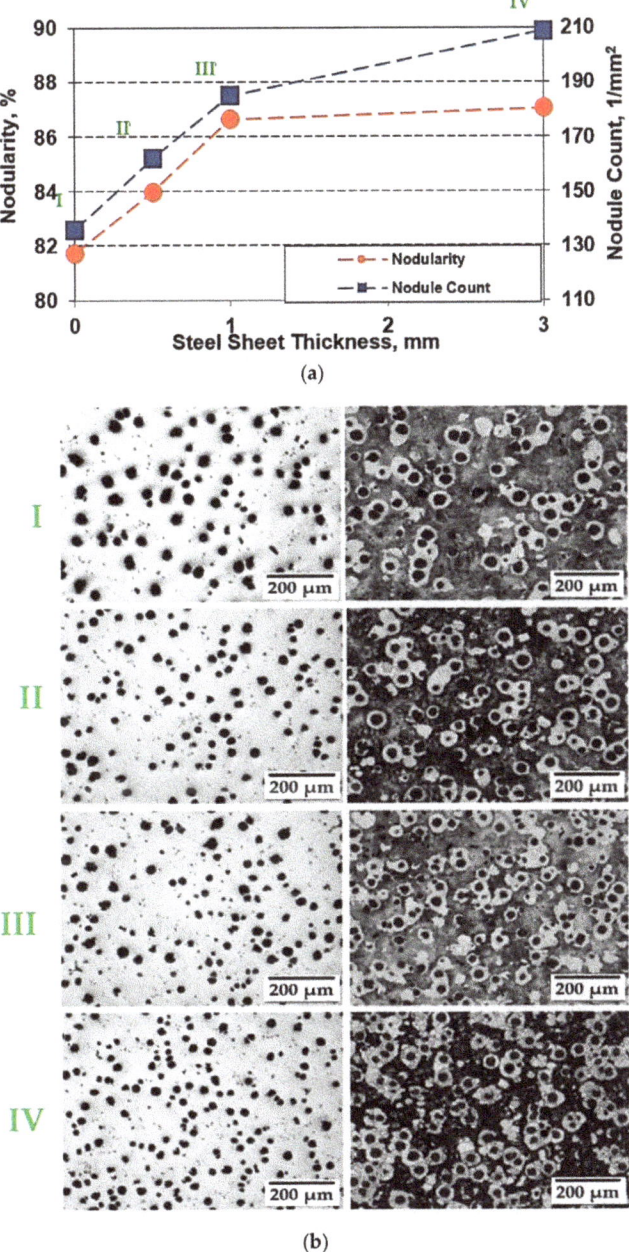

Figure 3. Graphite Nodularity and Nodule Count (**a**) and Structure (**b**) in the casting body as steel sheet thickness influence in un-etching and etching conditions, respectively (**I**-sample without steel sheet, **II**-sample with 0.5 mm steel sheet thickness, **III**-sample with 1 mm steel sheet thickness, **IV**-sample with 3 mm steel sheet thickness).

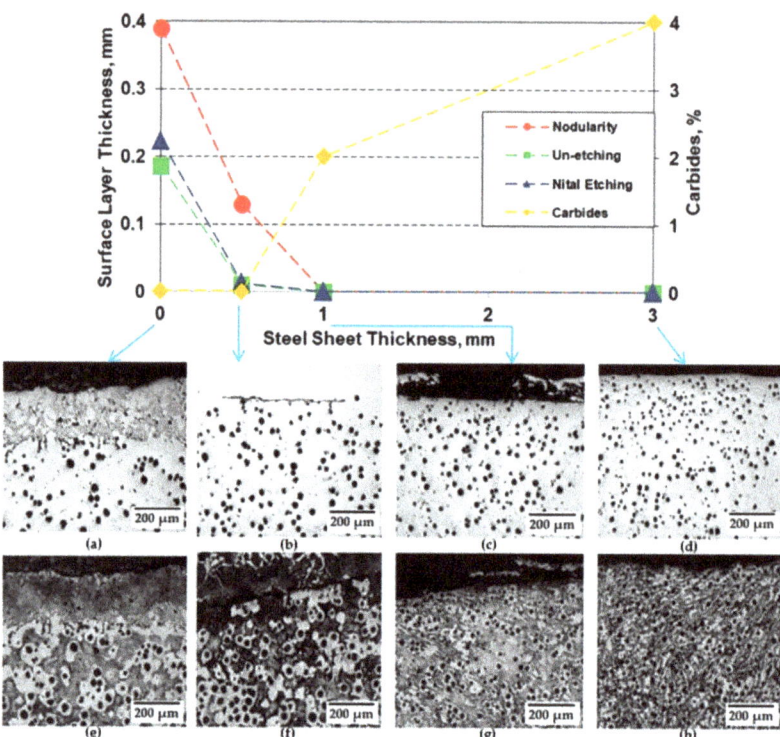

Figure 4. Graphite degenerated layer thickness as steel sheet thickness influence in un-etching and Nital etching conditions, respectively. Superficial layer structure as reference (**a,e**), 0.5 mm steel sheet influence (**b,f**), 1 mm steel sheet influence (**c,g**) and 3 mm steel sheet influence respectively; **a, b, c, d**-un-etched samples; **e, f, g, h**-Nital etched samples.

In the second stage, the diffusion of sulfur from the steel sheet in the liquid cast iron sample occurred.

In the case of steel sheet barriers introduction, the time of solid crust formation was considerably reduced due to an increase of the solidification constant.

In the first stage, the diffusion of sulfur from the furan resin mold and the steel sheet occurred.

The diffusion of sulfur from the gas phase through the steel sheet is governed by Fick's second law, whose analytically transposed solution for flat finite media and short durations is:

$$F = \frac{C_m - C_0}{C_s - C_0} = 1.128 \times \frac{\sqrt{D \times t}}{1} \tag{4}$$

where

F is the sulfur saturation fraction;

C_m—the average concentration of the broadcast substance (in our case, sulfur) at time t > 0;

C_0—the initial concentration of the broadcast substance at time t = 0 (in our case, the sulfur concentration in the steel sheet, $S_0 = 0.05\%$–0.06%) is admitted: $S_0 = 0.055$ wt.%;

C_s—the concentration of the broadcast substance at the interface (the sulfur content in the boundary layer created by the reaction between Fe and SO_2); for the calculation, the sulfur concentration at the interface is considered 1 wt.%;

D—the diffusion coefficient of sulfur in the solid steel sheet (at a temperature of 1000 °C, $D_S = (2...7) \times 10^{-10}$ cm^2/s [11]);

l—the distance on the sheet depth.

For a steel sheet with 0.5 mm thickness, at t = 10 s, it results:

$$F = \frac{1.128\sqrt{4.5 \cdot 10^{-10} \cdot 10}}{0.5 \cdot 10^{-1}} \simeq 0.0015$$

Result:
$$\frac{C_m - C_0}{C_s - C_0} = \frac{C_m - 0.055}{1 - 0.055} = 0.0015$$

C_m = 0.0564 wt.%, which means an increase of sulfur in the steel sheet of about 2.6 wt.% after 10 s.

For steel sheets of 1 mm and 3 mm, respectively, the concentrations C_m = 0.0550 wt.% S result, which means that at thicknesses greater than 1 mm there is no noticeable concentration of sulfur in the sheets ($C_m = C_0$).

Next, the diffusion of sulfur from the steel sheet into the liquid cast iron from the sample:

- The diffusion coefficient of sulfur in liquid cast iron at 1400 °C is [11]: $D_S^{1400} = 3.50 \times 10^{-5}$ cm^2/s;
- The sulfur concentration in the steel sheet after the diffusion of sulfur from the mold, C_S = 0.0564 wt.% (calculated);
- Sulfur concentration in modified cast iron, C_0 = 0.015 wt.%;
- Diffusion layer thickness is considered equal to the demodified layer average thickness (for the present example, on the 0.5 mm steel sheet, a demodified layer determined by means of the shape factors, δ_{str} = 0.13 mm, resulted [10]);

The sulfur saturation fraction according to Equation (4) is:

$$F = \frac{1.128\sqrt{3.5 \times 10^{-5} \times t}}{0.13 \times 10^{-1}} = 0.51 \times \sqrt{t} = \frac{C_m - C_0}{C_s - C_0} = \frac{C_m - 0.015}{0.0564 - 0.015}$$

The average concentration of sulfur in the broadcast layer results,

$$C_m = 0.015 + 0.0211 \times \sqrt{t} \qquad (5)$$

Example: For t = 10 s, C_m = 0.0817 wt.% S.

This amount of sulfur can annihilate 0.0817 · 24/32 = 0.061 wt.% Mg; thus, more than the residual Mg content of cast iron (0.048 wt.%). This explains the appearance of a small degenerated layer in the sample.

The diffusion of sulfur in liquid cast iron was limited by the solidified cast iron crust formation at which the diffusion coefficient is much lower (D_S = (2...7) × 10^{-10} cm^2/s, at a temperature of 1000 °C) [11]. In this context, it would have been useful to determine the moment of solid crust formation at the interface.

For the samples analyzed in this paper, a law of solidified crust thickness variation (δ) resulted like: $\delta = 0.908 \times 10^{-3} \times \sqrt{t} = 3.12 \times 10^{-3}$ m whence it results for δ = 0, $\sqrt{t} \geq \frac{3.12}{0.908} \geq 11.8$ s.

This means that after about 12 s from the casting, the solid crust formation began for the samples cast in the furan resin mold, without steel sheets.

During this time, reactions at the interface between the liquid iron during cooling and the mold developed.

In the case of the steel sheet barriers' introduction, the solid crust formation duration was considerably reduced due to the increase of the solidification constant. For example, at an estimated value of the solidification constant in the case of casting in steel molds, k = 2.8 × 10^{-3} m/s$^{\frac{1}{2}}$, a duration of solidified crust formation of about 1.2 s results. This is also the reason why the demodified layer thickness in the ductile cast iron samples, in which steel sheets at the interface were inserted, was practically zero compared to the sample without protection sheets in which the demodified layer thickness exceeded 200 µm. Moreover, as resulted from the calculations, at steel sheet thicknesses greater than 1 mm, this became an effective barrier against the diffusion of sulfur from the mold into the cast piece.

Besides the main effect of degeneration of the graphite, other effects of sulfur diffusion were recorded, considered secondary but with implications on the quality of the surface layer:

- The chill effect (free carbides formation) occurred in the white cast iron layer formation (ledeburite + cementite + pearlite) in the interface areas where liquid cast iron infiltrated into the pores of the mold (Figure 5a,b). In these areas, due to the large contact surface between the liquid iron and the mold, the diffusion of sulfur was favoured, leading to concentrations of over 0.3 wt.%–0.4 wt.% S at which its free carbides formation effect was visible. This is only a supposition. Future investigation must elucidate this aspect.
- Formation of complex compounds in the surface layer of the samples and in the slag film on their surface (Figure 5c,d). They were evaluated only by morphologic aspect. Energy-dispersive X-ray analysis (EDAX)examinations are needed for quantification.
- Formation of graphite particles with an advanced degree of degeneration, indicating changes in growth directions and Mg/S ratio as a result of changes in sulfur concentration (Figure 5e).

The steel sheets could not be fully melted due to their high melting temperature, but they created effects at the interface due to the diffusion of carbon from cast iron into steel, increasing the cooling rate of liquid cast iron at the interface and limiting the transition of sulfur from mold to cast iron due to the much lower diffusion rate in the solid state compared to the liquid state. These effects were manifested by:

- Formation of mixed graphite structures at the interface (Figure 6a);
- Modification of the matrix structure in steel sheets from ferritic to ferrite-pearlitic and even hypereutectoid steel (pearlite + secondary cementite), due to the strong diffusion of carbon from the liquid cast iron into the steel sheets (Figure 6b);
- Formation of metastable cast iron structures (pearlite + cementite + graphite) (Figure 6c).

These effects, visible in the steel sheet with a thickness of 0.5 mm, faded in those with greater thicknesses (1–3 mm), where it took the cooling effect to slow down the diffusion processes, and these effects favored the rapid formation of a solid cast iron crust in which the demodification effect no longer manifested.

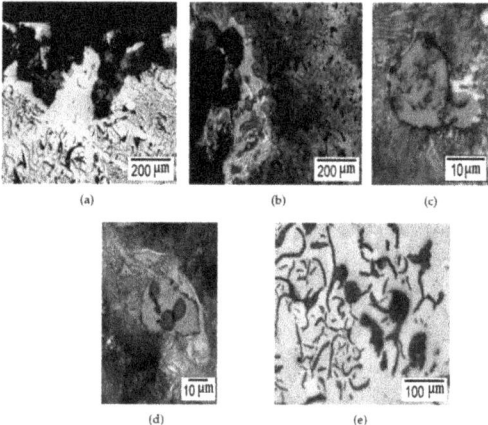

Figure 5. Effects of sulfur in the surface layer of cast samples: (**a**,**b**)-free carbides formation; (**c**,**d**)-formation of complex compounds; (**e**)-complex graphite shapes.

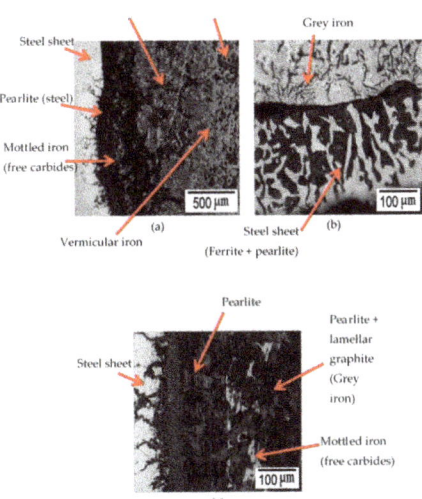

Figure 6. The steel sheets' effects on the interface liquid ductile iron-furan resin mold: (**a**)-mixed structures at the interface (steel-cast iron); (**b**)-Widmannstaten structure in the steel sheet; (**c**)-mottled cast iron structure in the surface layer (details from photo (**a**)).

4. Conclusions

The introduction of barriers to the liquid cast iron-mold interface, in the form of steel sheets, has generated effects depending on their ability to change the thermal balance of solidification.

In the case of using steel sheets with a 0.5 mm thickness, the presence of degenerate graphite was observed in the surface layer, which means that there was a diffusion of sulfur.

The steel sheet with a 0.5 mm thickness, although not completely assimilated, did not eliminate the effect of superficial demodification, but caused secondary effects due to a change in the thermal balance at the interface, given that the diffusion processes, especially of C and S, generated the formation of complex structures specific to cast iron–steel border areas.

The steel sheets with 1–3 mm thickness also stopped the superficial demodification process, by accelerating surface cooling and favoring the early formation of a solidified cast iron crust in which the demodification effect does not occur. It can be said that at higher thicknesses, the steel sheets show the effect of external coolers. The role of coolers is highlighted by the finishing of graphite in cast iron samples and the appearance of carbides in the metal matrix.

Theoretical calculus shows the possibility of sulfur diffusion limitations by the use of protective steel sheets in the liquid-mold interface. The thickness of these sheets must be calculated as dependent on casting solidification parameters (casting modulus, purring temperature, eutecticity degree of ductile iron, etc.). Future investigations are necessary in order to establish the concrete conditions of practical applications.

Author Contributions: D.A., I.S., M.C., I.R., and S.S. contributed equally in conceiving, designing and performing the experiments, analyzing the data, and writing the paper. All authors have read and agreed to the published version of the manuscript.

Funding: This research received no external funding.

Institutional Review Board Statement: Not applicable.

Informed Consent Statement: Not applicable.

Data Availability Statement: Data available on request due to restrictions.

Conflicts of Interest: The authors declare no conflict of interest.

References

1. Ivan, N.; Chisamera, M.; Riposan, I. Influence of magnesium content and coating type on graphite degeneration in surface layer of iron castings in resin sand—PTSA molds. *ISIJ Int.* **2012**, *52*, 1848–1855. [CrossRef]
2. Boonmee, S.; Stefanescu, D.M. Occurrence and effect of casting skin in compacted graphite iron. *Int. J. Cast. Met. Res.* **2016**, *29*, 47–54. [CrossRef]
3. Stefanescu, D.M.; Wills, S.; Massone, J.; Duncan, F. Quantification of casting skin in ductile and compacted graphite irons and its effect on tensile properties. *Int. J. Met.* **2008**, *2*, 7–28. [CrossRef]
4. Labrecque, C.; Gagne, M.; Cabanne, P.; Francois, C.; Beret, C.; Hoffmann, F. Comparative study of fatigue endurance limit for 4 and 6 mm thin wall ductile iron castings. *Int. J. Met.* **2008**, *2*, 7–17. [CrossRef]
5. Danko, R.; Gorny, M.; Holtzer, M.; Zymankowska-Kumon, S. Effect of the quality of furan molding sand on the skin layer of ductile iron castings. *ISIJ Int.* **2015**, *54*, 1288–1293. [CrossRef]
6. Chisamera, M.; Ivan, N.; Riposan, I.; Stan, S. Iron casting skin management in no-bake mold—Effects of magnesium residual level and mold coating. *China Foundry* **2015**, *12*, 222–230.
7. Ivan, N.; Chisamera, M.; Riposan, I.; Stan, S. Control of graphite degeneration in the surface layer of Mg-treated iron castings in resin sand—P-Toluol Sulphonic Acid (PTSA) molds. *AFS Trans.* **2013**, *121*, 379–390.
8. Anca, D.; Chisamera, M.; Stan, S.; Riposan, I. Graphite degeneration in high Si, Mg-treated iron castings: Sulfur and oxygen addition effects. *Int. J. Met.* **2020**, *14*, 663–671. [CrossRef]
9. Anca, D.; Chisamera, M.; Stan, S.; Stan, I.; Riposan, I. Sulfur and oxygen effects on high-Si ductile iron casting skin formation. *Coatings* **2020**, *10*, 618.
10. Anca, D.; Stan, I.; Chisamera, M.; Riposan, I.; Stan, S. Control of the Mg-treated iron casting skin formation by S-diffusion blocking at the metal—Mold interface. *Coatings* **2020**, *10*, 680. [CrossRef]
11. Sofroni, L.; Riposan, I.; Brabie, V.; Chisamera, M. *Cast Iron*; EDP: Bucharest, Romania, 1985.
12. Anca, D. Research on the Loss Modifyung Effect in the Superficial Layer Phenomenon of the Iron Castings. Ph.D. Thesis, Politehnica University of Bucharest, Bucharest, Romania, 2014.

Review

MoS$_2$-Based Substrates for Surface-Enhanced Raman Scattering: Fundamentals, Progress and Perspective

Yuan Yin [1], Chen Li [1], Yinuo Yan [1], Weiwei Xiong [1], Jingke Ren [2] and Wen Luo [1,*]

[1] School of Science, Wuhan University of Technology, Wuhan 430070, China; yinyuan@whut.edu.cn (Y.Y.); lichenlc@whut.edu.cn (C.L.); yanyinuo@whut.edu.cn (Y.Y.); vivianx@whut.edu.cn (W.X.)
[2] School of Materials Science and Engineering, Wuhan University of Technology, Wuhan 430070, China; renjingke518@163.com
* Correspondence: luowen_1991@whut.edu.cn

Abstract: Surface-enhanced Raman scattering (SERS), as an important tool for interface research, occupies a place in the field of molecular detection and analysis due to its extremely high detection sensitivity and fingerprint characteristics. Substantial efforts have been put into the improvement of the enhancement factor (EF) by way of modifying SERS substrates. Recently, MoS$_2$ has emerged as one of the most promising substrates for SERS, which is also exploited as a complementary platform on the conventional metal SERS substrates to optimize the properties. In this minireview, the fundamentals of MoS$_2$-related SERS are first explicated. Then, the synthesis, advances and applications of MoS$_2$-based substrates are illustrated with special emphasis on their practical applications in food safety, biomedical sensing and environmental monitoring, together with the corresponding challenges. This review is expected to arouse broad interest in nonplasmonic MoS$_2$-related materials along with their mechanisms, and to promote the development of SERS studies.

Keywords: surface-enhanced Raman spectroscopy; MoS$_2$; two-dimensional materials; charge transfer

1. Introduction

SERS is an effective spectroscopic technique for the detection and analysis of molecules adsorbed on rough metal surfaces or other nanostructures with high sensitivity and accuracy [1]. It was first observed due to the miraculous increase in signals of surface-enhanced Raman spectra of pyridine adsorbed on electrochemically rough silver in 1973, 45 years after the first discovery of Raman scattering [2]. Since then, the scope of SERS has been gradually broadened due to its characteristics such as high sensitivity and selectivity, rapid measurement, fingerprint characteristics, nondestructive examination, and good biocompatibility (Figure 1a). Numerous experiments have been carried out to explain the mechanism of enormous signal enhancement. It was not until the proposal of electromagnetic mechanism (EM) and chemical mechanism (CM) in 1977 that the reinforcement mechanism of SERS was basically formed and well acknowledged [3,4]. Later, as more newly fabricated materials were applied as SERS substrates, the reinforcement mechanism was gradually refined and has become the important focus of theoretical research on SERS [5,6].

The conventional substrates for SERS were some surface-roughened noble metals such as gold, silver and copper [7,8]. There are two main ideas to improve detection accuracy and expand the application scope of metal substrate for SERS: (1) moving concentration from precious metals to transition metal nanoarrays [9–12] and (2) changing the configuration and shape of Au/Ag nanoparticles (Au/Ag NPs) [13,14], such as Au nanocube array [15], concave–convex nanostructure array [16], checkerboard nanostructure [17], among others [18,19] (Figure 1b).

Another research idea is to combine different kinds of materials to exploit the advantages of diverse materials jointly and achieve better performance in various aspects [20,21]. At present, with two-dimensional (2D) materials flourishing, researchers are attempting to

---

Citation: Yin, Y.; Li, C.; Yan, Y.; Xiong, W.; Ren, J.; Luo, W. MoS$_2$-Based Substrates for Surface-Enhanced Raman Scattering: Fundamentals, Progress and Perspective. *Coatings* **2022**, *12*, 360. https://doi.org/10.3390/coatings12030360

Academic Editor: Alessio Lamperti

Received: 30 January 2022
Accepted: 5 March 2022
Published: 8 March 2022

Publisher's Note: MDPI stays neutral with regard to jurisdictional claims in published maps and institutional affiliations.

Copyright: © 2022 by the authors. Licensee MDPI, Basel, Switzerland. This article is an open access article distributed under the terms and conditions of the Creative Commons Attribution (CC BY) license (https://creativecommons.org/licenses/by/4.0/).

utilize these materials as substrates for SERS. Some of these 2D semiconductor materials are emphasized for their high chemical stability, good biocompatibility and controllability during fabrication [22–24]. Among them, MoS_2 emerges as one of the most promising materials as a new platform for SERS research owing to its extraordinary adsorption capacity and fluorescence quenching ability [25,26]. Moreover, MoS_2 has its apparent merits in photoelectric devices, electrochemistry and biosensors [27]. The band gap of MoS_2 can be adjusted from 1.29 to 1.9 eV as the number of layers increases, demonstrating its flexibility and light absorption [28]. Although the SERS enhancement factor of a single pristine MoS_2 monolayer is relatively small, its combination with metal nanoparticles can overcome the weak adsorption capacity and aggregated oxidation of a single metal substrate [29,30].

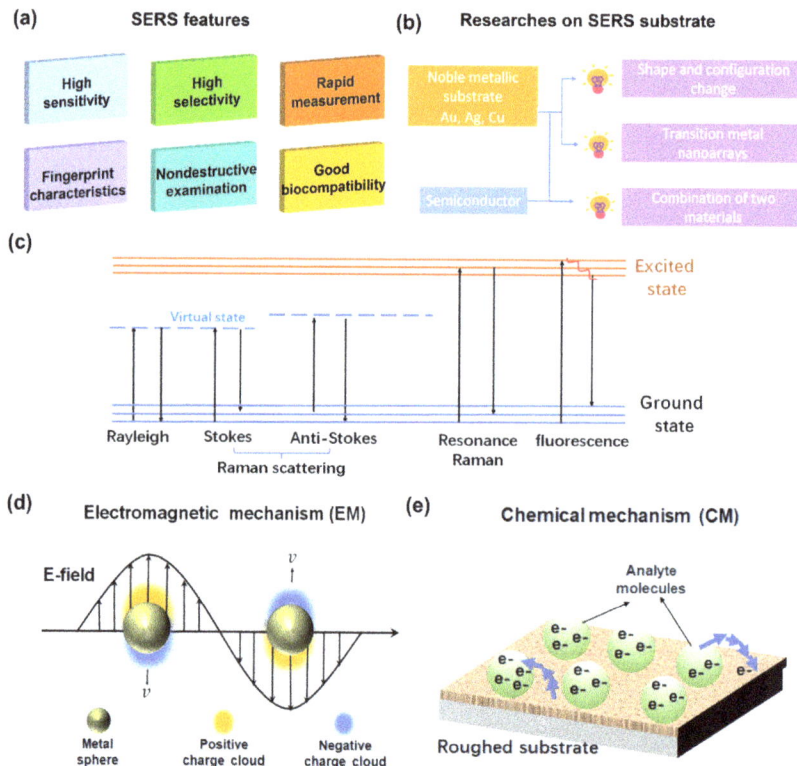

Figure 1. (a) Features of SERS; (b) development of research on SERS substrate; (c) schematic of Raman scattering; (d) EM: surface plasmon resonance on metal surface; (e) CM: charge transfer between analyte molecules and roughed substrate.

In this review, we begin by introducing the fundamentals of SERS, including two commonly accepted SERS mechanisms and MoS_2-related SERS mechanisms. Then, several basic synthesis methods, advances and multitudinous applications of MoS_2-based substrates are discussed. Finally, we summarize the current situation of SERS centered on MoS_2 material and look forward to its development trend. We hope this review encourages broad interest and sheds light on the synthesis and application for SERS with high sensitivity.

2. Fundamentals of SERS
2.1. SERS Mechanism

Raman scattering was first discovered in 1928 and refers to the change in frequency of a light wave when it is scattered. When light is scattered from an atom or molecule, most photons are elastic scattering, also known as Rayleigh scattering, meaning that the photons have the same frequency before and after scattering [31]. However, a small part (about 10^{-10}–10^{-6}) of the photons changes in frequency, namely Raman scattering (Figure 1c), and SERS is based on Raman scattering.

EM is considered the leading cause of the SERS phenomenon, represented by the surface plasmon resonance (SPR) model of the metal (Figure 1d), which has a long-range effect and has little to do with the type of adsorbent molecules [32–35]. Surface plasmons were first observed in the spectrum of light diffracted on a metallic diffraction grating. It was soon proven that the anomaly was associated with the excitation of electromagnetic surface waves on the surface of the diffraction grating [36]. It was recognized that under the action of the photoelectric field, electrons near the metal surface would produce dense vibrations, causing excitation of the plasmons, especially on a rough surface [37]. Once the incident light frequency matches the oscillation frequency of electrons near the metal surface, a local surface plasma resonance occurs. Then, the electromagnetic field in the vicinity of the nanoparticles is significantly enhanced [38]. The electromagnetic enhancement also includes the image field enhancement [39] and tip lightning rod effect [40], though their contribution is smaller than the SPR mentioned above.

CM refers to the charge resonance transition generated by the interaction between molecule and substrate (Figure 1e), as well as charge transfer between analyte molecules. Unlike EM, CM has a short-range effect and mainly depends on the type of adsorbent molecules and the interplay between detected molecules and the substrate [41,42]. Because of the complexity of the substrate and the detection of molecular species, CM is far more complicated than EM [43].

CM primarily suggests that electrons in the metal are excited to the charge-transfer state under laser irradiation of appropriate wavelength, which causes relaxation of the molecular nuclear skeleton. Therefore, when the electron returns to the metal, the photon emitted is one less vibrational quantum of energy than the incoming light. The enhancement is caused by the resonance of the scattering process with the charge-transfer state [44].

2.2. MoS$_2$-Related SERS Mechanism

It was proven that the EF of 2D monolayer MoS$_2$ as SERS substrate to detect 4-Mercaptopyridine could reach 10^5, much higher than the previous observation on 2D graphene and boron nitride [45]. This considerable enhancement in the SERS signal is probably attributed to interface dipole interaction and the enhanced charge transfer from 2D MoS$_2$ to organic molecules when in resonance [46]. Although the enhancement mechanism of the metal/MoS$_2$ composite substrate is not well understood, theorists generally believe that both EM and CM make contributions.

One of the more plausible explanations is that since the Fermi level of MoS$_2$ is higher than that of Au nanoparticles, Au nanoparticles act as the electron capture centers in the conduction band of MoS$_2$, resulting in the transfer of electrons from MoS$_2$ to Au. Therefore, potential changes and the formation of the Schottky barrier occur on the surface of MoS$_2$ (Figure 2a,b). This process, together with the electromagnetic enhancement of Au nanoparticles, remarkably enhances the Raman signal [47].

Another explanation takes the more general interaction between semiconductors and metals into account and suggests that light exposure plays a crucial role in enhancing the Raman signal. In this course, two conditions may contribute to signal enhancement. First, owing to the localized surface plasmon resonance (LSPR) of metal nanocrystals, local enhanced electromagnetic fields are generated on the surface (the gray areas in Figure 2c,d). When the metal nanocrystals are close enough to MoS$_2$, the local electromagnetic field and the absorption spectrum of MoS$_2$ overlap, which promotes the electron transfer from the valence band to the MoS$_2$ conduction band and generates electron-hole pairs (Figure 2c). Because of the strong field of the metal nanocrystals, the intensity of this process is increased by several orders of magnitude relative to light alone. During their interaction, the enhancement effect of the Raman signal is further amplified. The other functioning mechanism is related to plasmonic sensitization. In simple terms, the laser-induced plasmonic sensitization excites electrons in the conduction band of metal nanocrystals to overcome the Schottky barrier and jump into the conduction band of MoS_2 (Figure 2d). The electromagnetic and chemical enhancements are amplified by plasma excitation and electron transfer during the entire process.

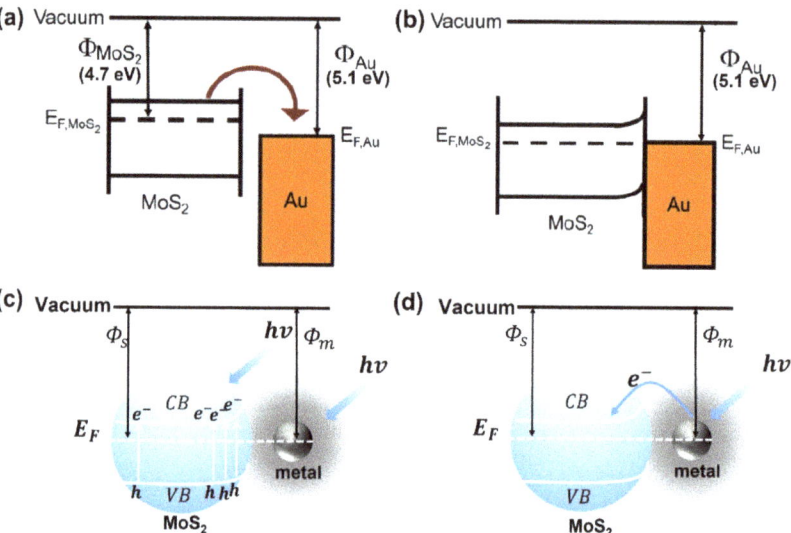

Figure 2. (a,b) Schematic diagram of Fermi level movement of Au/MoS_2 hybrid SERS substrate. Reprinted with permission from ref. [48]. Copyright 2014 Springer Nature. (c) Mechanism for plasmonic enhancement of light absorption. (d) Mechanism for plasmonic sensitization and electron excitation from the metal nanocrystal to MoS_2.

3. Hybrid SERS Nanostructures Based on MoS_2

3.1. Synthesis of MoS_2-Related SERS Substrates

Different research groups have prepared MoS_2-based SERS substrate using diverse methods. They have found differences in the final enhancement effect through comparison, which implies that preparation technology can influence the efficiency and detection sensitivity of SERS to some extent [49–54]. The preparation methods of SERS substrates have been constantly updated and developed with the development of preparation technology. In the following, we describe some classical preparation methods and discuss the effect of impurities and defects introduced during the preparation process on the detection limits.

3.1.1. Synthesis of MoS_2

Currently, there is a wide variety of preparation technologies for SERS substrates to MoS_2. The following are some traditional approaches, namely the hydrothermal/solvothermal method, chemical vapor deposition (CVD) method and mechanical force stripping method.

Hydrothermal/Solvothermal Method

The hydrothermal/solvothermal method uses molybdenum-containing compounds as molybdenum sources and high-purity sulfur compounds as sulfur sources and surfactants. These two are mixed by reaction, and the liquid sample mixture is acquired after complete stirring. The mixture is then dried, heated and molded through a closed-kettle under high temperature (Figure 3a). In the sealed heating process, MoS_2 substrates with different morphologies can be obtained by controlling the reaction time, temperature and the number of reaction reagents. For instance, Jiang et al. [55] used $Na_2MoO_4 \cdot 2H_2O$ (molybdenum sources), CH$_3$CSNH$_2$ (sulfur sources) and H$_4$[Si(W$_3$O$_{10}$)$_4$]·xH$_2$O to collect deposited MoS$_2$ in the autoclave. It was employed as SERS substrate to detect carbohydrate antigen 19-9 in serum directly, and the final minimum detection concentration reached 10^{-14} mol·L^{-1} level.

Chemical Vapor Deposition (CVD)

The CVD method is one of the traditional methods for the preparation of large-area nanofilm materials [56]. After decades of technical innovation, it is considered a mature technology for preparing 2D nanofilm materials [57,58]. The preparation process involves placing the growth base in a CVD tubular furnace, passing it through the precursor gas and allowing it to react on the surface of the substrate [59]. In preparing MoS$_2$ nanosheet films by the CVD method, Mo is first sputtered on SiO$_2$ substrate, then MoS$_2$ nanometer-thin films are grown on the surface through the reaction between Mo and sulfur vapor in the furnace. The size and thickness of the MoS$_2$ substrate can be modulated artificially by altering the thickness of Mo metal films. Zhan and colleagues [60] used this method to deposit Mo on the SiO$_2$ substrate surface and fabricated a MoS$_2$ thin-film layer by heating sulfur powder and reacting with Mo at a high temperature. Zheng et al. [61] used electrochemically oxidized Mo foil as a growth material to achieve layer-by-layer growth of MoS$_2$ by rapid sulfidation of Mo oxides in the gas phase (Figure 3b).

Mechanical Stripping Method

The mechanical stripping method applies the viscosity of special tape to act on the surface of MoS$_2$ material to weaken Van der Waals forces of MoS$_2$ between layers (Figure 3c). Without breaking the covalent bond, MoS$_2$ layered structure or even a single layer structure can be obtained. The thin layer is attached to the SiO$_2$/Si substrate surface to form a MoS$_2$ substrate. Yan et al. [62] obtained MoS$_2$ substrate for Rh6G molecule detection with a minimum detection limit of 10^{-8} mol·L^{-1} by the mechanical stripping method combined with heating and annealing treatments.

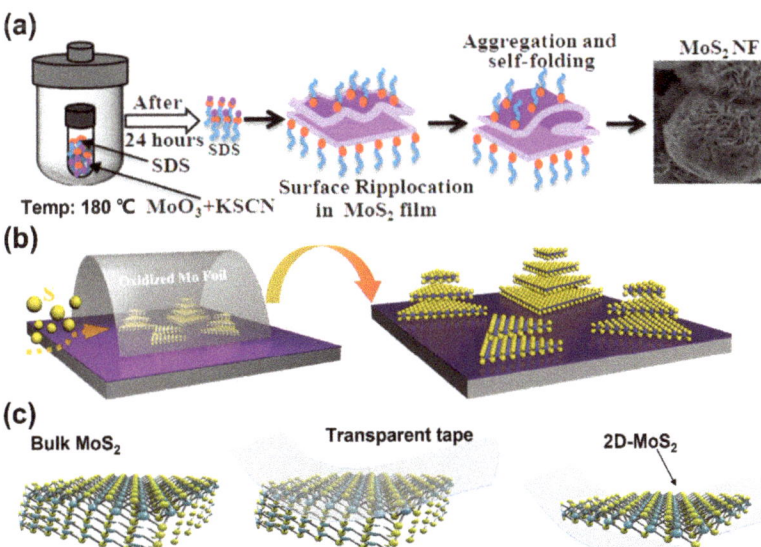

Figure 3. (a) Schematic of hydrothermal method to fabricate MoS$_2$ nanoflowers. Reprinted with permission from ref. [63]. Copyright 2018 Elsevier. (b) CVD growth of MoS$_2$ flakes using arched oxidized Mo foil as precursor. Reprinted with permission from ref. [61]. Copyright 2017 John Wiley and Sons. (c) The flowchart of the mechanical stripping method to prepare 2D MoS$_2$.

3.1.2. Synthesis of Metal/MoS$_2$ Hybrid Substrate

Recent approaches to prepare metal/MoS$_2$ hybrids substrates for SERS can be organized into three categories: (1) physical methods: physically depositing specific metal on MoS$_2$ or placing ready-made metal nanoparticles on MoS$_2$ directly; (2) chemical methods: spontaneous reduction method, self-assembly technology, thermal reduction method (including hydrothermal method, solvothermal method, microwave-assisted hydrothermal method), among others; (3) nanoetching methods: plasma etching, electron beam lithography (EBL) and photoetching. We emphatically describe the latter two approaches in this part.

Spontaneous Reduction Method

The spontaneous reduction method refers to the initiative reduction reaction between the prepared MoS$_2$ film and the precursor solution of metal nanoparticles such as HAuCl$_4$ solution (the precursor of gold NPs), to obtain directly the metal/MoS$_2$ composite substrate. This spontaneous reduction can occur at room temperature or more uniformly and rapidly through auxiliary means such as heating [50,64].

For example, in a typical preparation method, the concentrations of HAuCl$_4$ significantly influence the character of the Au NPs-loaded MoS$_2$ surface and eventually the Raman EF. As the concentration of the precursor HAuCl$_4$ increases, the Au NPs on the AuNPs@MoS$_2$ show more hotspots and more aggregation, and the detection limit of AuNPs@MoS$_2$ for Rh6G decreases and then increases [50]. Therefore, too high or too low precursor concentration is not conducive to the SERS enhancement effect of the hybrid substrates, and the regulation of selecting the appropriate HAuCl$_4$ concentration has become one of the main concerns of the experimentalists.

Hydrothermal Reduction Method

Hydrothermal reduction is the process of reducing metal cations in solution under different conditions while adding MoS$_2$ material. The cations are attracted by unsaturated sulfur on the MoS$_2$ surface to form chemical bonds, and eventually the metal/MoS$_2$ composite SERS substrate is obtained. Singha's group [63] adopted the hydrothermal method to modify MoS$_2$ with Au NPs and detected free bilirubin in human blood, which showed high sensitivity, stability and good reproducibility. However, compared with the traditional hydrothermal method, the microwave-assisted hydrothermal method is more frequently used to prepare nanomaterials [50,65,66]. Microwaves are utilized as a heating tool to realize stirring on the molecular level. It overcomes the shortcoming of uneven heating in the hydrothermal vessel, thus shortening reaction time and improving efficiency [67–69]. Kim and coworkers [70] reported this facile method and observed that the gold nanoparticles tend to grow at defective sites, mainly at the edges and the line defects in the basal planes.

During a hydrothermal reaction, flowing high purity argon is usually used to avoid oxidation during the reaction, which greatly affects the SERS sensitivity of the final substrate [63]. According to Kim's work, the chemical intercalation–exfoliation process in the hydrothermal method created more defects in the substrate surface of MoS$_2$ flakes than its single-crystal counterpart when preparing MoS$_2$, which facilitated the deposition of higher density gold nanoparticles [70]. The Au NPs@MoS$_2$ obtained by this method ultimately exhibited better enhancement and lower detection limits.

Nanoetching Method

Nanoetching technology was first applied in the integrated process and had irreplaceable advantages in micrographics [71]. Its advantages such as fast processing speed, high precision, minor damage to substrates and no pollution make it a popular technique.

The typical representative of electron beam processing is electron beam lithography (EBL), which mainly uses electron beams to induce surface reaction beams for microprocessing. The reaction between the atoms on the substrate surface and the adsorbed molecules or ions is facilitated by irradiating the specimen by the electron beam, and the designed pattern is finally obtained on the substrate by the liftoff technique. Zhai et al. [72] applied EBL to fabricate a Au nanoarray on the monolayer MoS_2 film, which was used as a SERS substrate to realize CV detection of 10^{-6}–10^{-15} M. They considered it to be combined with the separation technology to form a sensor that can quickly detect trace molecules in a natural environment.

The focused laser beam can locally transform the MoS_2 film into microscopic patterns with active nucleation sites. When the modified film is in complete contact with the reaction substance, selective modification can be achieved at specific locations to flexibly prepare a thin layer of MoS_2 decorated by metal NPs. Lu and coworkers [73] employed this technology to realize self-designed pattern preparation of Au NPs decorating MoS_2. They controlled the localized modification of the materials by changing the laser power, MoS_2 film thickness and reaction time. It was proven that the prepared hybrid substrate can detect aromatic organic molecules with outstanding performance.

The femtosecond laser is another technique that is widely adopted to modify MoS_2 with metal NPs [51,53]. It can induce photoelectrons generated on the film surface and greatly promote the interaction between metal cations and photoelectrons on the film surface. Then, the reduction and in situ deposition of metal NPs on MoS_2 nanosheets formed the metal/MoS_2 hybrid substrates for SERS.

Nanoetching technology possesses unique advantages in terms of precise tuning. The roughness of the laser-treated MoS_2 film is about three times greater than that of the pristine, which facilitates the deposition of metal nanoparticles [73]. The hybrid substrates prepared by this method show stronger SERS activity, whose detection limit can reach as low as 1 fM for CV detection. In addition, the power of the laser also affects the Raman intensity at the same concentration of the analytes [72].

3.2. Advances in MoS_2-Based SERS Substrate

3.2.1. MoS_2 Substrates

With the development of chemical mechanisms, SERS substrate material is not confined to metal, and MoS_2 emerges as a promising substrate material owing to its distinct merits shown in SERS studies. For single MoS_2 material research, researchers have placed more focus on 2D material, which can be roughly divided into two directions: (1) special treatment of MoS_2 material, such as plasma treatment and usage of the femtosecond laser to induce defect sites on the surface to enhance the charge transfer; and (2) stacking the single-layer 2D MoS_2 material according to the set angle to obtain double-layer MoS_2.

For the former research direction, it was found that the plasma-processed MoS_2 nanosheets can perform better in SERS. The Raman intensities of Rh6G on MoS_2 nanoflakes were enhanced more than tenfold after oxygen-plasma and argon-plasma treatments [74]. Other external treatments such as pressure and femtosecond laser were verified that they could reinforce charge transfer between the substrate and molecules to induce MoS_2 defect sites and realize pressure or photoluminescence control [75–77]. Sun et al. [77] found that there are more transferred charges between the substrate and analytes with increasing applied pressure (Figure 4a), which also leads to an increase in the enhancement factor. For the latter, Xia and coworkers [78] studied prominent resonance Raman and photoluminescence spectroscopic differences between AB (60°, Figure 4c) and AA' (0°, Figure 4d) stacked bilayer MoS_2, and considered that the 0° stacked MoS_2 bilayer was superior to the 60° stacked one in interlayer electron coupling, hence its Raman enhancement effect was more outstanding (Figure 4b).

Compared with metal, MoS_2 possesses unique adsorption capacity, especially for some aromatic molecules, because the π bonds of MoS_2 interplay with those of aromatic molecules [73,79]. Furthermore, MoS_2 has good fluorescence quenching ability and can quench background fluorescence, which is conducive to detection and substrate stability at low concentration [26,80]. However, its electromagnetic enhancement is extremely weak, and chemical enhancement alone can hardly contribute significantly to sensitivity.

In addition, owing to the selectivity and complexity of chemical enhancement for the detection of molecules, these substrates can only be implemented for some particular organic molecules such as aromatic molecules.

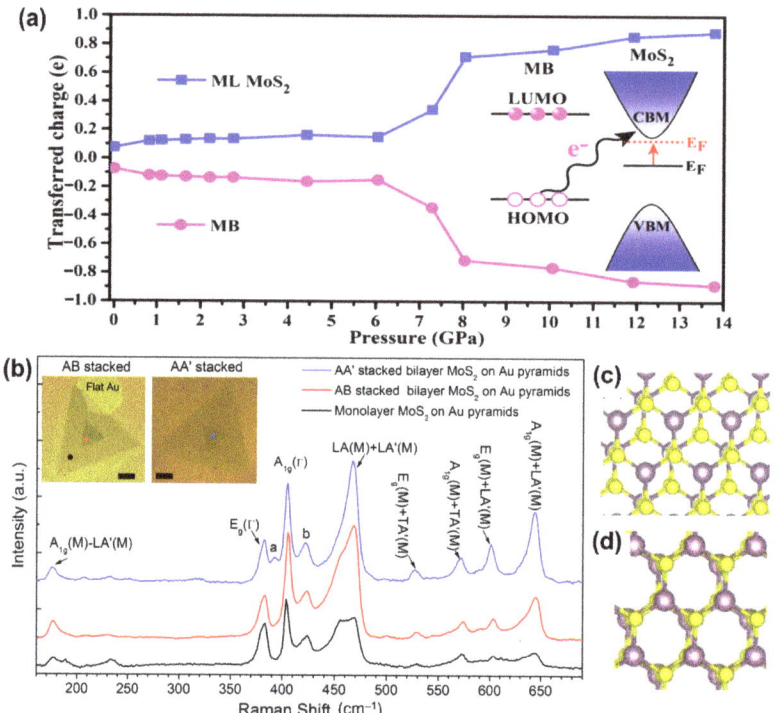

Figure 4. (**a**) Charge transfer at the monolayer MoS$_2$ (ML-MoS$_2$)/methylene blue (MB) interface varies with pressure. The charge gained by ML-MoS$_2$ and the charge lost by MB molecules are set to be positive and negative, respectively. Reprinted with permission from ref. [76]. Copyright 2021 American Chemical Society. (**b**) Resonance Raman spectra of monolayer and bilayer MoS$_2$ on Au nanopyramids. Reprinted with permission from ref. [78]. Copyright 2015 American Chemical Society. Atomic structure diagram of (**c**) AB and (**d**) AA' staked bilayer MoS$_2$ (purple balls represent Mo atoms, and yellow balls represent S atoms). Reprinted with permission from ref. [78]. Copyright 2015 American Chemical Society.

3.2.2. Metal/MoS$_2$ Hybrid Substrates

Metal/MoS$_2$ hybrid substrates are considered admirable SERS substrates, with EFs that can reach 10^8 and even up to 10^{12} after some special processing such as changing shape and metal nanoparticles configuration [81]. Because of the prominent enhancement effect of this substrate, experimentalists have conducted various studies on it, making this kind of substrate become one of the important research topics of SERS in recent years.

One was on Au NPs /MoS$_2$, and the researchers found that these composite substrates enhanced significantly better than either single Au or single MoS$_2$. Subsequently, Rani et al. [82] used low-power focused laser cutting to carve artificial edges on the MoS$_2$ monolayer. The intensive accumulation of Au NPs along the artificial edges led to the aggregation of SERS hotspots in the same places, which made it possible to generate SERS hotspots with ideal location and geometry shape in a controllable way on a large-area substrate (Figure 5). Liang's group [83] prepared 3D MoS$_2$-nanospheres, 3D MoS$_2$-nanospheres @Au seeds and 3D MoS$_2$-nanospheres @Ag-NPs hybrids structures, and calculated their enhancement factors as 500, 7.5×10^6 and 1.2×10^8, respectively.

Through experiments, they believed that silver nanoparticles were more suitable than gold nanoparticles as modification materials for MoS_2. This is because silver nanoparticles can be closer to each other and have higher coverage, leading to more hotspots on the surface and stronger signal enhancement.

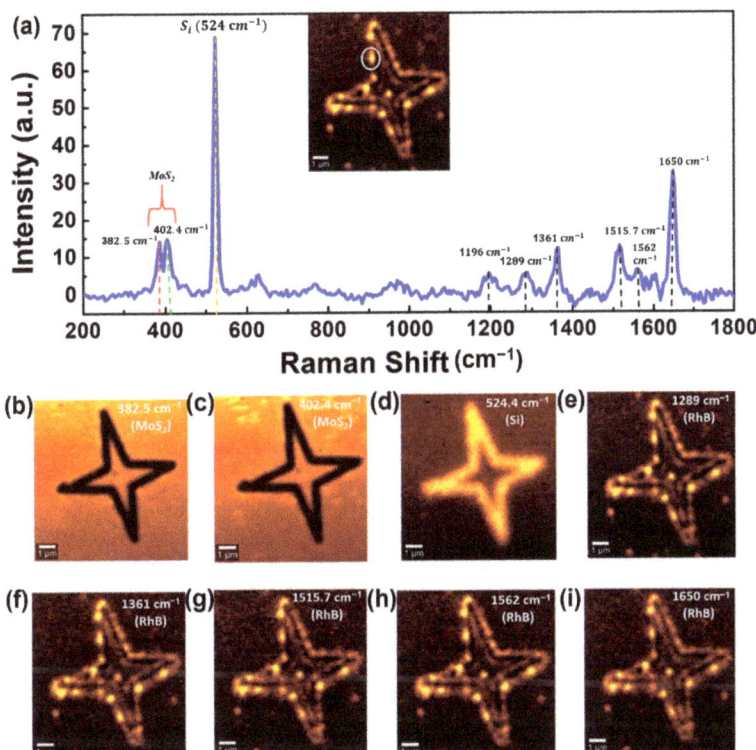

Figure 5. (**a**) Raman peaks of Rhodamine B (black dotted line) and signature signals of MoS_2 (red and green dotted line) and Si (yellow dotted line), obtained at the edge of laser-etched MoS_2 surface modified with Au NPs. (**b**–**i**) Micromapping images of a star-shaped feature at characteristic peaks of MoS_2, Si and Rhodamine B, illustrating localized hotspots generated along the factitious edges of the star-shaped nanostructure. Reprinted with permission from ref. [82]. Copyright 2020 American Chemical Society.

Compared with the substrates mentioned above, the metal/MoS_2 hybrid substrate concentrated the advantages of metal and MoS_2, so it shows better adsorption effect, higher sensitivity, stronger stability and lower detection limit, and has gradually become a new platform in SERS research. Because of the modification of MoS_2 by metal nanoparticles, the composite substrate exhibits not only stronger electromagnetic enhancement, but also better chemical enhancement effect and fluorescence quenching effect, which can greatly reduce external interference. However, owing to the unclear mechanisms of the interaction between metal and MoS_2 and the complicated production process, further development of this kind of substrate has been limited to some extent.

3.3. Practical Applications

Most organics have SERS activity and their molecules are very close in size to the analytes in the plasmonic structure, making SERS very suitable for the detection of these small molecules. Furthermore, SERS is promising to become a viable alternative to mass spectrometry and chromatographic-based techniques, owing to its potential for high sensitivity, specificity and capability of rapid measurements. In this section, we mainly focus on SERS technique applications such as food safety detection, biomedical sensing and environmental monitoring, particularly examining MoS_2-based SERS.

3.3.1. Food Safety Detection

Food issues have always been a concern, and there have been numerous reports of excessive additives found in food. Therefore, a sensitive and credible approach for detection techniques is imperative. Since most illegal additives have Raman activity, the SERS technique can examine their contents, ensuring powerful food supervision [84–86]. Li et al. [87] prepared a 3D flexible Ag NPs@MoS_2/pyramidal polymer structure, which not only had a large surface area but also could generate dense hotspots. The minimum detection limits of the structure reached 10^{-13} and 10^{-14} M for Rh6G and CV as probe molecules, respectively, which showed the ultrasensitivity of the structure. Long-term repeated use experiments showed the stability and reproducibility of the structure. In addition, they used the structure to achieve ultralow in situ detection of melamine in milk with a measured detection limit of 10^{-6} M, which was found to be within the safe range according to FDA regulations.

The detection of pesticide and antimicrobial residues on food is also the main interest for SERS in food safety. Therefore, some researchers have detected the residues by MoS_2-based SERS. Chen and coworkers [88] developed an Ag NPs-MoS_2 composite substrate with striking SERS activity and photocatalytic efficiency. They established two calibration curves with ultralow detection limits of 6.4×10^{-7} and 9.8×10^{-7} mg/mL for the standard solutions of tetramethylthiuram disulfide (TMTD) and methyl parathion (MP). Finally, they successfully achieved a recyclable detection of single and mixed residues of TMTD and MP on eggplant, Chinese cabbage, grape and strawberry by using different monitoring protocols depending on the size and level of surface roughness (Figure 6). Zhai et al. [72] prepared Au nanodisk array-monolayer MoS_2 (ADAM) composites material by EBL technique. They experimentally demonstrated the good stability of the composite at different laser frequencies and temperatures. According to their research, the ADAM composite material was highly sensitive as a SERS substrate, enabling CV detection in the range of 10^{-6}–10^{-15} M with detection limits as low as 1 fM. They also used this active substrate continuously for the detection of antimicrobial residues in aquatic products and found it under the safety limits of the EU directive.

However, it is still challenging for SERS to achieve in situ analysis of toxic residues in real foods due to the complexity of real foods and the low concentration of contaminants. Therefore, how to further improve the sensitivity of SERS detection and realize the combination with rapid separation techniques needs to be further explored.

3.3.2. Biomedical Sensing

SERS promises to be a viable alternative in the field of bioanalytical sensing due to its numerous merits mentioned above and its potential to be integrated into small packages for measurement at the point of care, which can be used for clinical testing to cure some intractable diseases [89–92]. At present, SERS is mainly applied for three aspects in biomedical sensing: (1) analyzing the properties of biomolecular components, (2) effectively detecting target substances in various mixtures and (3) cell imaging.

For biomolecular analysis, Guerrini et al. [93] performed label-free analysis of unmodified dsDNA by SERS using positively charged silver colloids. The electrostatic adhesion of DNA promoted the aggregation of nanoparticles into stable clusters, producing intense and reproducible SERS spectra at the nanoscale. Based on this, they reported the quantitative identification of hybridization substances, along with SERS identification of single base mismatches and base methylation (5-methylated cytosine and N6-methylated adenine) in duplexes. Moreover, when a SERS probe is scaled down to a quantum scale, it is possible to study epigenetic features of cancer stem cells and gene expression aberrations in genomic

DNA. Ganesh et al. [94] performed experiments based on this principle, pointing out differences in genomic DNA between cancer and noncancer cells, and achieved tracking of both genes. For the tracking detection aspect, Singha and coworkers [63] used Au NPs/MoS$_2$ nanoflower hybrid as a SERS substrate, which used Rh6G as a probe molecule with detection limits as low as 10^{-12} M. This SERS biosensor detected free bilirubin under the interference of key interfering factors such as glucose, cholesterol and phosphate in human serum, showing good selectivity and reliability, as well as potential for clinical diagnosis. For medical imaging, Fei et al. [64] fabricated gold nanoparticles@MoS$_2$ quantum dots (Au NPs@MoS$_2$ QDs) core-shell nanocomposite. The pinhole-free, chemically inert and ultrathin MoS$_2$ QDs shell protected the Au core from the chemical environment and probe molecules. The detection limit of this hybrid substrate for crystal violet could reach 0.5 nM. In addition, the hybrid was also used as a nanoprobe for label-free near-infrared SERS imaging of 4T1 cells. Finally, they successfully obtained distinguishable SERS images from 4T1 cells.

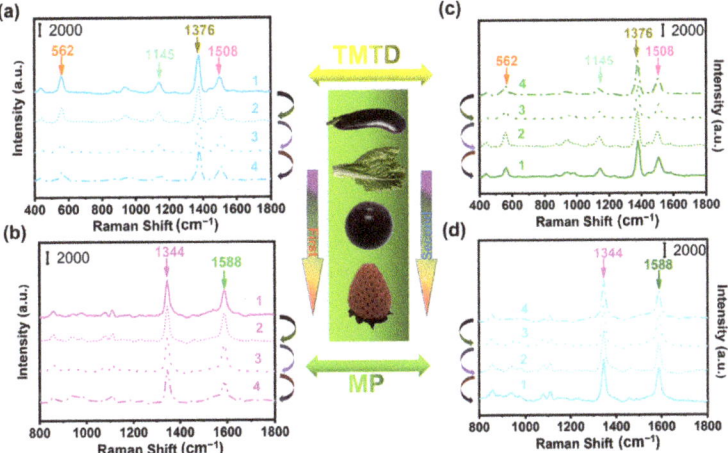

Figure 6. Recyclable SERS-based detection on eggplant (denoted as 1), Chinese cabbage (2), grape (3) and strawberry (4); first cycle for (**a**) TMTD and (**b**) MP and second cycle for (**c**) TMTD and (**d**) MP. Reprinted with permission from ref. [88]. Copyright 2020 American Chemical Society.

However, there are certain challenges in the application of SERS in biomedicine. Because of the complex structure of biological macromolecules, elastic scattering may occur in all directions from all parts of the cell. Elastic scattering generates severe background signals on the SERS spectra, which seriously interferes with the analysis and detection. Therefore, how to optimize the SERS probe, reduce the risk of signal interference in SERS detection and enhance its screening capability will be the focus of research for this application.

3.3.3. Environmental Monitoring

With the development of modern industry, environmental pollution has become a problem that cannot be ignored. Some organic pollutants can enter our food chain through water pollution and soil pollution, posing a severe threat to our health and adversely affecting the balance of the ecosystem. Therefore, using simple and accurate measurement methods to monitor the environment has become the focus of our attention. SERS has been applied to detect the water environment and soil owing to its various merits.

Zhao et al. [95] prepared a PSi/MoS$_2$/Au NPs MSC (pyramidal Si/MoS$_2$/Au NPs multiscale cavity), and they used this hybrid substrate to detect CV alcohol solutions from 10^{-5} to 10^{-10} M, and found that the main characteristic peaks were still evident when the concentration reached 10^{-10} M. They used PSi/MoS$_2$/Au MSC and PSi/MoS$_2$

MSC to determine the hydrophilic and hydrophobicity of the mixtures with different concentrations, respectively, to compare and analyze the superiority of PSi/MoS$_2$/Au MSC compared to PSi/MoS$_2$ MSC (Figure 7a,b). Through experiments, they reported that PSi/MoS$_2$/Au MSC can achieve targeted monitoring of organic contaminants and act as a visible-light self-cleaning SERS substrate with good recovery properties. In addition, a Cu/CuO @Ag nanowire complex that transforms from hydrophilic to hydrophobic under infrared light was prepared as a multifunctional SERS substrate [96]. The substrates have controlled wettability and can self-separate in multiphase solutions and adsorb to the two terminals of the substrate. Malachite green and formalin were used as two probe molecules at two different terminals to obtain the lowest detection limits of 10^{-9} M and 10^{-5} M. The substrate, after complete hydrophobic modification, can also be used to extract the organic phase (Figure 7c) in "oil/water" mixtures and as a probe for in situ detection.

However, there are still some difficulties. Because of the reality that wastewater is a heterogeneous solution, organic contaminants are often heterogeneously distributed in it, which leads to inaccurate collection of Raman signals. In addition, for some heavy metal ions in the environment, the SERS technique cannot be used directly for detection, but requires further modification of the probe to enable indirect detection. These difficulties need to be further refined.

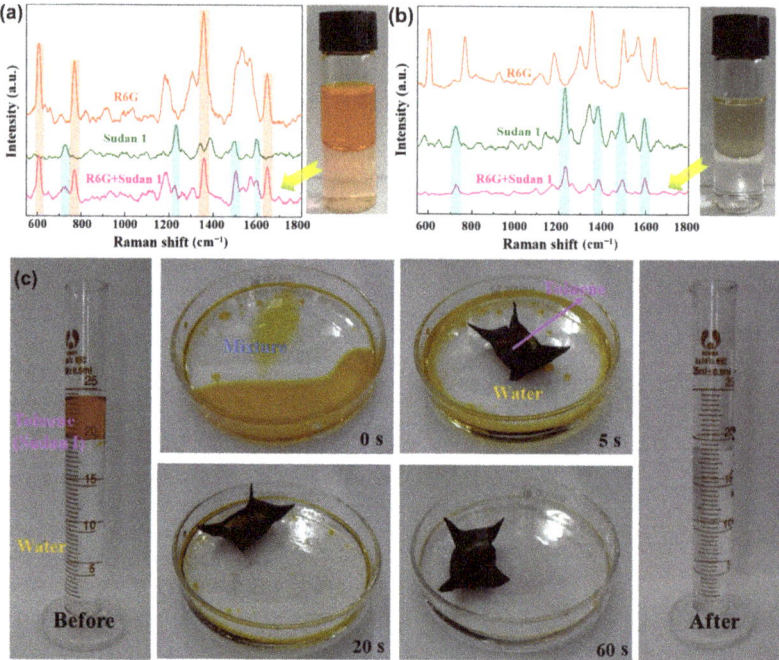

Figure 7. (**a**) The SERS spectra of Rh6G aqueous solution (10^{-5} M), Sudan 1 toluene solution (10^{-3} M) and their mixture detected from PSi/MoS$_2$ MSC. (**b**) The SERS spectra of Rh6G aqueous solution (10^{-9} M), Sudan 1 toluene solution (10^{-5} M), and their mixture detected from PSi/MoS$_2$/Au MSC [95]. (**c**) Extract toluene from the mixed "water/toluene" solution by hydrophobic Cu/CuO @Ag substrate. Reprinted with permission from ref. [96]. Copyright 2020 Elsevier.

4. Conclusions

In this minireview, the superiority of MoS$_2$-based substrate is emphasized, because these kinds of substrates possess characteristics such as solid fluorescence quenching effects and adsorption abilities. These advantages can make up for the deficiency of roughened metal. The probable mechanisms of MoS$_2$-based SERS are depicted in detail, mainly attributed to the enhanced charge transfer. Then, we introduce the synthesis, advances and practical applications of MoS$_2$ and metal/MoS$_2$ hybrid substrate in sequence. The search or modification of SERS substrates, such as changing the shape, nanoparticle configuration and unique surface treatment, to improve EF and lower detection limits has been the focus of experiments. Refinement of existing enhancement mechanisms can contribute to establishing the direction of SERS substrate pursuit and modification. It is worth mentioning that although EM and CM are sufficient to explain most of the existing phenomena, they still need further improvement, which will become the emphasis of future research.

In recent years, the combination with a wide range of technologies such as rapid separation techniques has been essential to broaden the scope of SERS applications. However, SERS substrates prepared by conventional methods may exhibit problems such as heterogeneity and instability, which may limit the development of this technology and its widespread application. Thus, how to design an optimal synthesis method to achieve high reproducibility and mass production also becomes a focus in the future. In terms of application, it is evident that the application of SERS technology in the field of bioscience has become a general trend. In addition to the detection and analysis of biomarkers mentioned above, the technology can also be used in assisted tumor location, protein analysis, etc., which will greatly benefit humanity. In conclusion, MoS$_2$ has served as a new platform for SERS research, but the future development and application prospect of this material deserve further exploration.

Funding: This work was funded by National Innovation and Entrepreneurship Training Program for College Students, China (Grant Nos.: S202110497144, 3120400002304).

Institutional Review Board Statement: Not applicable.

Informed Consent Statement: Not applicable.

Conflicts of Interest: The authors declare no conflict of interest.

References

1. Barbillon, G. Fabrication and SERS Performances of metal/Si and metal/ZnO Nanosensors: A Review. *Coatings* **2019**, *9*, 86. [CrossRef]
2. Fleischmann, M.; Hendra, P.J.; McQuillan, A.J. Raman spectra of pyridine adsorbed at a silver electrode. *Chem. Phys. Lett.* **1974**, *26*, 163–166. [CrossRef]
3. Jeanmaire, D.L.; Van Duyne, R.P. Surface Raman spectroelectrochemistry Part I. Heterocyclic, aromatic, and aliphatic amines adsorbed on the anodized silver electrode. *J. Electroanal. Chem.* **1977**, *84*, 1–20. [CrossRef]
4. Albrecht, M.G.; Creighton, J.A. Anomalously intense Raman spectra of pyridine at a silver electrode. *J. Am. Chem. Soc.* **1977**, *99*, 5215–5217. [CrossRef]
5. Lee, Y.; Kim, H.; Lee, J.; Yu, S.H.; Hwang, E.; Lee, C.; Ahn, J.-H.; Cho, J.H. Enhanced Raman scattering of rhodamine 6g films on two-dimensional transition metal dichalcogenides correlated to photoinduced charge transfer. *Chem. Mater.* **2016**, *28*, 180–187. [CrossRef]
6. Jiang, R.; Li, B.; Fang, C.; Wang, J. Metal/semiconductor hybrid nanostructures for plasmon-enhanced applications. *Adv. Mater.* **2014**, *26*, 5274–5309. [CrossRef] [PubMed]
7. Baibarac, M.; Cochet, M.; Łapkowski, M.; Mihut, L.; Lefrant, S.; Baltog, I. SERS spectra of polyaniline thin films deposited on rough Ag, Au and Cu. Polymer film thickness and roughness parameter dependence of SERS spectra. *Synth. Met.* **1998**, *96*, 63–70. [CrossRef]
8. Wu, D.Y.; Liu, X.M.; Duan, S.; Xu, X.; Ren, B.; Lin, S.H.; Tian, Z.Q. Chemical enhancement effects in SERS spectra: A quantum chemical study of pyridine interacting with copper, silver, gold and platinum metals. *J. Phys. Chem. C* **2008**, *112*, 4195–4204. [CrossRef]
9. Tian, Z.-Q.; Ren, B.; Wu, D.-Y. Surface-enhanced Raman scattering: From noble to transition metals and from rough surfaces to ordered nanostructures. *J. Phys. Chem. B* **2002**, *106*, 9463–9483. [CrossRef]
10. Kim, K.; Lee, J.W.; Shin, K.S. Cyanide SERS as a platform for detection of volatile organic compounds and hazardous transition metal ions. *Analyst* **2013**, *138*, 2988–2994. [CrossRef]
11. Bezerra, A.G.; Machado, T.N.; Woiski, T.D.; Turchetti, D.A.; Lenz, J.A.; Akcelrud, L.; Schreiner, W.H. Plasmonics and SERS activity of post-transition metal nanoparticles. *J. Nanopart. Res.* **2018**, *20*, 142. [CrossRef]
12. Kong, K.V.; Lam, Z.; Lau, W.K.O.; Leong, W.K.; Olivo, M. A transition metal carbonyl probe for use in a highly specific and sensitive SERS-based assay for glucose. *J. Am. Chem. Soc.* **2013**, *135*, 18028–18031. [CrossRef]

13. Zhou, H.; Yang, D.; Ivleva, N.P.; Mircescu, N.E.; Niessner, R.; Haisch, C. SERS detection of bacteria in water by in situ coating with Ag nanoparticles. *Anal. Chem.* **2014**, *86*, 1525–1533. [CrossRef] [PubMed]
14. Zhong, L.-B.; Yin, J.; Zheng, Y.-M.; Liu, Q.; Cheng, X.-X.; Luo, F.-H. Self-assembly of Au nanoparticles on PMMA template as flexible, transparent, and highly active SERS substrates. *Anal. Chem.* **2014**, *86*, 6262–6267. [CrossRef] [PubMed]
15. Chirumamilla, M.; Das, G.; Toma, A.; Gopalakrishnan, A.; Zaccaria, R.P.; Liberale, C.; De Angelis, F.; Di Fabrizio, E. Optimization and characterization of Au cuboid nanostructures as a SERS device for sensing applications. *Microelectron. Eng.* **2012**, *97*, 189–192. [CrossRef]
16. Zenidaka, A.; Tanaka, Y.; Miyanishi, T.; Terakawa, M.; Obara, M. Comparison of two-dimensional periodic arrays of convex and concave nanostructures for efficient SERS templates. *Appl. Phys. A* **2011**, *103*, 225–231. [CrossRef]
17. Chen, S.; Han, L.; Schülzgen, A.; Li, H.; Li, L.; Moloney, J.V.; Peyghambarian, N. Local electric field enhancement and polarization effects in a surface-enhanced Raman scattering fiber sensor with chessboard nanostructure. *Opt. Express* **2008**, *16*, 13016–13023. [CrossRef] [PubMed]
18. Yan, J.; Su, S.; He, S.; He, Y.; Zhao, B.; Wang, D.; Zhang, H.; Huang, Q.; Song, S.; Fan, C. Nano rolling-circle amplification for enhanced SERS hot spots in protein microarray analysis. *Anal. Chem.* **2012**, *84*, 9139–9145. [CrossRef]
19. Pallares, R.M.; Su, X.; Lim, S.H.; Thanh, N.T.K. Fine-tuning of gold nanorod dimensions and plasmonic properties using the Hofmeister effects. *J. Mater. Chem. C* **2016**, *4*, 53–61. [CrossRef]
20. Zhang, H.; Zhang, W.; Gao, X.; Man, P.; Sun, Y.; Liu, C.; Li, Z.; Xu, Y.; Man, B.; Yang, C. Formation of the AuNPs/GO@MoS$_2$/AuNPs nanostructures for the SERS application. *Sens. Actuators B Chem.* **2019**, *282*, 809–817. [CrossRef]
21. Chen, J.; Liu, G.; Zhu, Y.-Z.; Su, M.; Yin, P.; Wu, X.-J.; Lu, Q.; Tan, C.; Zhao, M.; Liu, Z.; et al. Ag@MoS$_2$ core–shell heterostructure as SERS platform to reveal the hydrogen evolution active sites of single-layer MoS$_2$. *J. Am. Chem. Soc.* **2020**, *142*, 7161–7167. [CrossRef] [PubMed]
22. Zheng, Z.; Cong, S.; Gong, W.; Xuan, J.; Li, G.; Lu, W.; Geng, F.; Zhao, Z. Semiconductor SERS enhancement enabled by oxygen incorporation. *Nat. Commun.* **2017**, *8*, 1993. [CrossRef] [PubMed]
23. Yang, B.; Jin, S.; Guo, S.; Park, Y.; Chen, L.; Zhao, B.; Jung, Y.M. Recent development of SERS technology: Semiconductor-based study. *ACS Omega* **2019**, *4*, 20101–20108. [CrossRef]
24. Wang, X.; Shi, W.; She, G.; Mu, L. Surface-Enhanced Raman Scattering (SERS) on transition metal and semiconductor nanostructures. *Phys. Chem. Chem. Phys.* **2012**, *14*, 5891–5901. [CrossRef] [PubMed]
25. Lan, L.; Chen, D.; Yao, Y.; Peng, X.; Wu, J.; Li, Y.; Ping, J.; Ying, Y. Phase-dependent fluorescence quenching efficiency of MoS$_2$ nanosheets and their applications in multiplex target biosensing. *ACS Appl. Mater. Interfaces* **2018**, *10*, 42009–42017. [CrossRef]
26. Tegegne, W.A.; Su, W.-N.; Tsai, M.-C.; Beyene, A.B.; Hwang, B.-J. Ag nanocubes decorated 1T-MoS$_2$ nanosheets SERS substrate for reliable and ultrasensitive detection of pesticides. *Appl. Mater. Today* **2020**, *21*, 100871. [CrossRef]
27. Liu, J.; Chen, X.; Wang, Q.; Xiao, M.; Zhong, D.; Sun, W.; Zhang, G.; Zhang, Z. Ultrasensitive monolayer MoS$_2$ field-effect transistor based DNA sensors for screening of down syndrome. *Nano Lett.* **2019**, *19*, 1437–1444. [CrossRef]
28. Mak, K.F.; Lee, C.; Hone, J.; Shan, J.; Heinz, T.F. Atomically thin MoS$_2$: A new direct-gap semiconductor. *Phys. Rev. Lett.* **2010**, *105*, 136805. [CrossRef]
29. Alamri, M.; Sakidja, R.; Goul, R.; Ghopry, S.; Wu, J.Z. Plasmonic Au nanoparticles on 2D MoS$_2$/Graphene van der Waals heterostructures for high-sensitivity surface-enhanced Raman spectroscopy. *ACS Appl. Nano Mater.* **2019**, *2*, 1412–1420. [CrossRef]
30. Xie, L.; Lu, J.-L.; Liu, T.; Chen, G.-Y.; Liu, G.; Ren, B.; Tian, Z.-Q. Key Role of direct adsorption on SERS sensitivity: Synergistic effect among target, aggregating agent, and surface with au or ag colloid as surface-enhanced Raman spectroscopy substrate. *J. Phys. Chem. Lett.* **2020**, *11*, 1022–1029. [CrossRef]
31. Yu, Z.; Brus, L. Rayleigh and Raman scattering from individual carbon nanotube bundles. *J. Phys. Chem. B* **2001**, *105*, 1123–1134. [CrossRef]
32. Otto, A. Theory of first layer and single molecule Surface Enhanced Raman Scattering (SERS). *Phys. Status Solidi (a)* **2001**, *188*, 1455–1470. [CrossRef]
33. Tian, X.; Chen, L.; Xu, H.; Sun, M. Ascertaining genuine SERS spectra of p-aminothiophenol. *RSC Adv.* **2012**, *2*, 8289–8292. [CrossRef]
34. Willets, K.A.; Van Duyne, R.P. Localized surface plasmon resonance spectroscopy and sensing. *Annu. Rev. Phys. Chem.* **2007**, *58*, 267–297. [CrossRef] [PubMed]
35. Meyer, S.A.; Le Ru, E.C.; Etchegoin, P.G. Combining Surface Plasmon Resonance (SPR) Spectroscopy with Surface-Enhanced Raman Scattering (SERS). *Anal. Chem.* **2011**, *83*, 2337–2344. [CrossRef] [PubMed]
36. Homola, J. Electromagnetic Theory of Surface Plasmons. In *Surface Plasmon Resonance Based Sensors*; Homola, J., Ed.; Springer: Berlin/Heidelberg, Germany, 2006; Volume 4, pp. 3–44. [CrossRef]
37. Zhao, J.; Zhang, C.; Lu, Y.; Wu, Q.; Yuan, Y.; Xu, M.; Yao, J. Surface-enhanced Raman spectroscopic investigation on surface plasmon resonance and electrochemical catalysis on surface coupling reaction of pyridine at Au/TiO$_2$ junction electrodes. *J. Raman Spectrosc.* **2020**, *51*, 2199–2207. [CrossRef]
38. Schwartzberg, A.M.; Zhang, J.Z. Novel optical properties and emerging applications of metal nanostructures. *J. Phys. Chem. C* **2008**, *112*, 10323–10337. [CrossRef]
39. David, C.; Richter, M.; Knorr, A.; Weidinger, I.; Hildebrandt, P. Image dipoles approach to the local field enhancement in nanostructured Ag–Au hybrid devices. *J. Chem. Phys.* **2010**, *132*, 24712. [CrossRef]

40. Liao, P.F.; Wokaun, A. Lightning rod effect in surface enhanced Raman scattering. *J. Chem. Phys.* **1982**, *76*, 751–752. [CrossRef]
41. Schlücker, S. Surface-Enhanced Raman Spectroscopy: Concepts and chemical applications. *Angew. Chem. Int. Ed.* **2014**, *53*, 4756–4795. [CrossRef]
42. Kovacs, G.J.; Loutfy, R.O.; Vincett, P.S.; Jennings, C.; Aroca, R. Distance dependence of SERS enhancement factor from Langmuir-Blodgett monolayers on metal island films: Evidence for the electromagnetic mechanism. *Langmuir* **1986**, *2*, 689–694. [CrossRef]
43. Ma, N.; Zhang, X.-Y.; Fan, W.; Han, B.; Jin, S.; Park, Y.; Chen, L.; Zhang, Y.; Liu, Y.; Yang, J.; et al. Controllable preparation of SERS-active Ag-FeS Substrates by a cosputtering technique. *Molecules* **2019**, *24*, 551. [CrossRef] [PubMed]
44. Moore, J.E.; Morton, S.M.; Jensen, L. Importance of correctly describing charge-transfer excitations for understanding the chemical effect in SERS. *J. Phys. Chem. Lett.* **2012**, *3*, 2470–2475. [CrossRef] [PubMed]
45. Muehlethaler, C.; Considine, C.R.; Menon, V.; Lin, W.-C.; Lee, Y.-H.; Lombardi, J.R. Ultrahigh Raman enhancement on monolayer MoS_2. *ACS Photon.* **2016**, *3*, 1164–1169. [CrossRef]
46. Xia, M. 2D materials-coated plasmonic structures for sers applications. *Coatings* **2018**, *8*, 137. [CrossRef]
47. Shakya, J.; Patel, A.S.; Singh, F.; Mohanty, T. Composition dependent Fermi level shifting of Au decorated MoS_2 nanosheets. *Appl. Phys. Lett.* **2016**, *108*, 013103. [CrossRef]
48. Bhanu, U.; Islam, M.R.; Tetard, L.; Khondaker, S.I. Photoluminescence quenching in gold—MoS_2 hybrid nanoflakes. *Sci. Rep.* **2015**, *4*, 5575. [CrossRef]
49. Sha, P.; Su, Q.; Dong, P.; Wang, T.; Zhu, C.; Gao, W.; Wu, X. Fabrication of Ag@Au (core@shell) nanorods as a SERS substrate by the oblique angle deposition process and sputtering technology. *RSC Adv.* **2021**, *11*, 27107–27114. [CrossRef]
50. Su, S.; Zhang, C.; Yuwen, L.; Chao, J.; Zuo, X.; Liu, X.; Song, C.; Fan, C.; Wang, L. Creating SERS hot spots on MoS_2 nanosheets with in situ grown gold nanoparticles. *ACS Appl. Mater. Interfaces* **2014**, *6*, 18735–18741. [CrossRef]
51. Pan, C.; Song, J.; Sun, J.; Wang, Q.; Wang, F.; Tao, W.; Jiang, L. One-step fabrication method of MoS_2 for high-performance surface-enhanced Raman scattering. *J. Phys. Chem. C* **2021**, *125*, 24550–24556. [CrossRef]
52. Miao, P.; Ma, Y.; Sun, M.; Li, J.; Xu, P. Tuning the SERS activity and plasmon-driven reduction of p-nitrothiophenol on a Ag@MoS_2 film. *Faraday Discuss.* **2019**, *214*, 297–307. [CrossRef] [PubMed]
53. Zuo, P.; Jiang, L.; Li, X.; Li, B.; Ran, P.; Li, X.; Qu, L.; Lu, Y.F. Metal (Ag, Pt)–MoS_2 hybrids greenly prepared through photochemical reduction of femtosecond laser pulses for SERS and HER. *ACS Sustain. Chem. Eng.* **2018**, *6*, 7704–7714. [CrossRef]
54. Er, E.; Hou, H.-L.; Criado, A.; Langer, J.; Möller, M.; Erk, N.; Liz-Marzán, L.M.; Prato, M. High-yield preparation of exfoliated 1T-MoS_2 with SERS activity. *Chem. Mater.* **2019**, *31*, 5725–5734. [CrossRef]
55. Jiang, J.; Shen, Q.; Xue, J.; Qi, H.; Wu, Y.; Teng, Y.; Zhang, Y.; Liu, Y.; Zhao, X.; Liu, X. A highly sensitive and stable SERS sensor for malachite green detection based on Ag nanoparticles in situ generated on 3D MoS_2 nanoflowers. *ChemistrySelect* **2020**, *5*, 354–359. [CrossRef]
56. Lee, Y.-H.; Zhang, X.-Q.; Zhang, W.; Chang, M.-T.; Lin, C.-T.; Chang, K.-D.; Yu, Y.-C.; Wang, J.T.-W.; Chang, C.-S.; Li, L.-J.; et al. Synthesis of large-area MoS_2 atomic layers with chemical vapor deposition. *Adv. Mater.* **2012**, *24*, 2320–2325. [CrossRef]
57. Liu, H.F.; Wong, S.L.; Chi, D.Z. CVD growth of MoS_2-based two-dimensional materials. *Chem. Vap. Depos.* **2015**, *21*, 241–259. [CrossRef]
58. Jeon, J.; Jang, S.K.; Jeon, S.M.; Yoo, G.; Jang, Y.H.; Park, J.-H.; Lee, S. Layer-controlled CVD growth of large-area two-dimensional MoS_2 films. *Nanoscale* **2015**, *7*, 1688–1695. [CrossRef]
59. Kumar, M.; Ando, Y. Chemical vapor deposition of carbon nanotubes: A review on growth mechanism and mass production. *J. Nanosci. Nanotechnol.* **2010**, *10*, 3739–3758. [CrossRef]
60. Zhan, Y.; Liu, Z.; Najmaei, S.; Ajayan, P.M.; Lou, J. Large-area vapor-phase growth and characterization of MoS_2 atomic layers on a SiO_2 Substrate. *Small* **2012**, *8*, 966–971. [CrossRef]
61. Zheng, J.; Yan, X.; Lu, Z.; Qiu, H.; Xu, G.; Zhou, X.; Wang, P.; Pan, X.; Liu, K.; Jiao, L. High-mobility multilayered MoS_2 flakes with low contact resistance grown by chemical vapor deposition. *Adv. Mater.* **2017**, *29*, 1604540. [CrossRef]
62. Yan, D.; Qiu, W.; Chen, X.; Liu, L.; Lai, Y.; Meng, Z.; Song, J.; Liu, Y.; Liu, X.-Y.; Zhan, D. Achieving high-performance surface-enhanced Raman scattering through one-step thermal treatment of bulk MoS_2. *J. Phys. Chem. C* **2018**, *122*, 14467–14473. [CrossRef]
63. Singha, S.S.; Mondal, S.; Bhattacharya, T.S.; Das, L.; Sen, K.; Satpati, B.; Das, K.; Singha, A. Au nanoparticles functionalized 3D-MoS_2 nanoflower: An efficient SERS matrix for biomolecule sensing. *Biosens. Bioelectron.* **2018**, *119*, 10–17. [CrossRef] [PubMed]
64. Fei, X.; Liu, Z.; Hou, Y.; Li, Y.; Yang, G.; Su, C.; Wang, Z.; Zhong, H.; Zhuang, Z.; Guo, Z. Synthesis of Au NP@MoS_2 quantum dots Core@Shell nanocomposites for SERS bio-analysis and label-free bio-imaging. *Materials* **2017**, *10*, 650. [CrossRef] [PubMed]
65. Liang, X.; Zhang, X.-J.; You, T.-T.; Yang, N.; Wang, G.-S.; Yin, P.-G. Three-dimensional MoS_2-NS@Au-NPs hybrids as SERS sensor for quantitative and ultrasensitive detection of melamine in milk. *J. Raman Spectrosc.* **2018**, *49*, 245–255. [CrossRef]
66. Vattikuti, S.P.; Nagajyothi, P.; Devarayapalli, K.; Yoo, K.; Nam, N.D.; Shim, J. Hybrid Ag/MoS_2 nanosheets for efficient electrocatalytic oxygen reduction. *Appl. Surf. Sci.* **2020**, *526*, 146751. [CrossRef]
67. Yu, H.-P.; Zhu, Y.-J.; Lu, B.-Q. Highly efficient and environmentally friendly microwave-assisted hydrothermal rapid synthesis of ultralong hydroxyapatite nanowires. *Ceram. Int.* **2018**, *44*, 12352–12356. [CrossRef]
68. Kharisov, B.I.; Kharissova, O.V.; Ortiz-Mendez, U. (Eds.) Microwaves: Microwave-Assisted Hydrothermal Synthesis of Nanoparticles. In *CRC Concise Encyclopedia of Nanotechnology*; CRC Press: Boca Raton, FL, USA, 2016; pp. 588–599. [CrossRef]

69. Yu, X.; Zhao, Z.; Sun, D.; Ren, N.; Yu, J.; Yang, R.; Liu, H. Microwave-assisted hydrothermal synthesis of Sn_3O_4 nanosheet/rGO planar heterostructure for efficient photocatalytic hydrogen generation. *Appl. Catal. B Environ.* **2018**, *227*, 470–476. [CrossRef]
70. Kim, J.; Byun, S.; Smith, A.J.; Yu, J.; Huang, J. Enhanced electrocatalytic properties of transition-metal dichalcogenides sheets by spontaneous gold nanoparticle decoration. *J. Phys. Chem. Lett.* **2013**, *4*, 1227–1232. [CrossRef]
71. Kang, S. Replication technology for micro/nano optical components. *Jpn. J. Appl. Phys.* **2004**, *43*, 5706–5716. [CrossRef]
72. Zhai, Y.; Yang, H.; Zhang, S.; Li, J.; Shi, K.; Jin, F. Controllable preparation of Au-MoS_2 nano-array composite: Optical properties study and SERS application. *J. Mater. Chem. C* **2021**, *9*, 6823–6833. [CrossRef]
73. Lu, J.; Lu, J.H.; Liu, H.; Liu, B.; Gong, L.; Tok, E.S.; Loh, K.P.; Sow, C.H. Microlandscaping of Au Nanoparticles on few-layer MoS_2 films for chemical sensing. *Small* **2015**, *11*, 1792–1800. [CrossRef] [PubMed]
74. Sun, L.; Hu, H.; Zhan, D.; Yan, J.; Liu, L.; Teguh, J.S.; Yeow, E.K.L.; Lee, P.S.; Shen, Z. Plasma modified MoS_2 nanoflakes for surface enhanced Raman scattering. *Small* **2014**, *10*, 1090–1095. [CrossRef] [PubMed]
75. Zuo, P.; Jiang, L.; Li, X.; Ran, P.; Li, B.; Song, A.; Tian, M.; Ma, T.; Guo, B.; Qu, L.; et al. Enhancing charge transfer with foreign molecules through femtosecond laser induced MoS_2 defect sites for photoluminescence control and SERS enhancement. *Nanoscale* **2019**, *11*, 485–494. [CrossRef] [PubMed]
76. Sun, H.; Yao, M.; Liu, S.; Song, Y.; Shen, F.; Dong, J.; Yao, Z.; Zhao, B.; Liu, B. SERS selective enhancement on monolayer MoS_2 enabled by a pressure-induced shift from resonance to charge transfer. *ACS Appl. Mater. Interfaces* **2021**, *13*, 26551–26560. [CrossRef]
77. Sun, H.; Yao, M.; Song, Y.; Zhu, L.; Dong, J.; Liu, R.; Li, P.; Zhao, B.; Liu, B. Pressure-induced SERS enhancement in a MoS_2/Au/R6G system by a two-step charge transfer process. *Nanoscale* **2019**, *11*, 21493–21501. [CrossRef]
78. Xia, M.; Li, B.; Yin, K.; Capellini, G.; Niu, G.; Gong, Y.; Zhou, W.; Ajayan, P.M.; Xie, Y.-H. Spectroscopic signatures of AA' and AB stacking of chemical vapor deposited bilayer MoS_2. *ACS Nano* **2015**, *9*, 12246–12254. [CrossRef]
79. Lai, H.; Ma, G.; Shang, W.; Chen, D.; Yun, Y.; Peng, X.; Xu, F. Multifunctional magnetic sphere-MoS_2@Au hybrid for surface-enhanced Raman scattering detection and visible light photo-Fenton degradation of aromatic dyes. *Chemosphere* **2019**, *223*, 465–473. [CrossRef]
80. Li, Z.; Jiang, S.; Xu, S.; Zhang, C.; Qiu, H.; Li, C.; Sheng, Y.; Huo, Y.; Yang, C.; Man, B. Few-layer MoS_2-encapsulated Cu nanoparticle hybrids fabricated by two-step annealing process for surface enhanced Raman scattering. *Sens. Actuators B Chem.* **2016**, *230*, 645–652. [CrossRef]
81. Zeng, Z.; Tang, D.; Liu, L.; Wang, Y.; Zhou, Q.; Su, S.; Hu, D.; Han, B.; Jin, M.; Ao, X.; et al. Highly reproducible surface-enhanced Raman scattering substrate for detection of phenolic pollutants. *Nanotechnology* **2016**, *27*, 455301. [CrossRef]
82. Rani, R.; Yoshimura, A.; Das, S.; Sahoo, M.R.; Kundu, A.; Sahu, K.K.; Meunier, V.; Nayak, S.K.; Koratkar, N.; Hazra, K.S. Sculpting artificial edges in monolayer MoS_2 for controlled formation of surface-enhanced Raman hotspots. *ACS Nano* **2020**, *14*, 6258–6268. [CrossRef]
83. Liang, X.; Wang, Y.-S.; You, T.-T.; Zhang, X.-J.; Yang, N.; Wang, G.-S.; Yin, P.-G. Interfacial synthesis of a three-dimensional hierarchical MoS_2-NS@Ag-NP nanocomposite as a SERS nanosensor for ultrasensitive thiram detection. *Nanoscale* **2017**, *9*, 8879–8888. [CrossRef] [PubMed]
84. Wu, Y.; Yu, W.; Yang, B.; Li, P. Self-assembled two-dimensional gold nanoparticle film for sensitive nontargeted analysis of food additives with surface-enhanced Raman spectroscopy. *Analyst* **2018**, *143*, 2363–2368. [CrossRef] [PubMed]
85. Neng, J.; Zhang, Q.; Sun, P. Application of surface-enhanced Raman spectroscopy in fast detection of toxic and harmful substances in food. *Biosens. Bioelectron.* **2020**, *167*, 112480. [CrossRef] [PubMed]
86. Zhang, D.; You, H.; Yuan, L.; Hao, R.; Li, T.; Fang, J. Hydrophobic slippery surface-based surface-enhanced Raman spectroscopy platform for ultrasensitive detection in food safety applications. *Anal. Chem.* **2019**, *91*, 4687–4695. [CrossRef] [PubMed]
87. Li, C.; Yu, J.; Xu, S.; Jiang, S.; Xiu, X.; Chen, C.; Liu, A.; Wu, T.; Man, B.; Zhang, C. Constructing 3D and flexible plasmonic structure for high-performance SERS application. *Adv. Mater. Technol.* **2018**, *3*, 1800174. [CrossRef]
88. Chen, Y.; Liu, H.; Tian, Y.; Du, Y.; Ma, Y.; Zeng, S.; Gu, C.; Jiang, T.; Zhou, J. In Situ recyclable surface-enhanced Raman scattering-based detection of multicomponent pesticide residues on fruits and vegetables by the flower-like MoS_2@Ag hybrid substrate. *ACS Appl. Mater. Interfaces* **2020**, *12*, 14386–14399. [CrossRef]
89. Marks, H.; Schechinger, M.; Garza, J.; Locke, A.; Coté, G. Surface enhanced Raman spectroscopy (SERS) for in vitro diagnostic testing at the point of care. *Nanophotonics* **2017**, *6*, 681–701. [CrossRef]
90. Tran, V.; Walkenfort, B.; König, M.; Salehi, M.; Schlücker, S. Rapid, Quantitative, and ultrasensitive point-of-care testing: A portable SERS reader for lateral flow assays in clinical chemistry. *Angew. Chem. Int. Ed.* **2019**, *58*, 442–446. [CrossRef]
91. Granger, J.H.; Schlotter, N.E.; Crawford, A.C.; Porter, M.D. Prospects for point-of-care pathogen diagnostics using surface-enhanced Raman scattering (SERS). *Chem. Soc. Rev.* **2016**, *45*, 3865–3882. [CrossRef]
92. Gao, X.; Zhang, H.; Fan, X.; Zhang, C.; Sun, Y.; Liu, C.; Li, Z.; Jiang, S.; Man, B.; Yang, C. Toward the highly sensitive SERS detection of bio-molecules: The formation of a 3D self-assembled structure with a uniform GO mesh between Ag nanoparticles and Au nanoparticles. *Opt. Express* **2019**, *27*, 25091–25106. [CrossRef]
93. Guerrini, L.; Krpetić, Željka; Van Lierop, D.; Alvarez-Puebla, R.A.; Graham, D. Direct surface-enhanced Raman scattering analysis of DNA duplexes. *Angew. Chem.* **2015**, *127*, 1160–1164. [CrossRef]
94. Ganesh, S.; Venkatakrishnan, K.; Tan, B. Quantum scale organic semiconductors for SERS detection of DNA methylation and gene expression. *Nat. Commun.* **2020**, *11*, 1135. [CrossRef] [PubMed]

95. Zhao, X.; Liu, C.; Yu, J.; Li, Z.; Liu, L.; Li, C.; Xu, S.; Li, W.; Man, B.; Zhang, C. Hydrophobic multiscale cavities for high-performance and self-cleaning surface-enhanced Raman spectroscopy (SERS) sensing. *Nanophotonics* **2020**, *9*, 4761–4773. [CrossRef]
96. Liu, C.; Yang, M.; Yu, J.; Lei, F.; Wei, Y.; Peng, Q.; Li, C.; Li, Z.; Zhang, C.; Man, B. Fast multiphase analysis: Self-separation of mixed solution by a wettability-controlled CuO@Ag SERS substrate and its applications in pollutant detection. *Sens. Actuators B Chem.* **2020**, *307*, 127663. [CrossRef]

MDPI
St. Alban-Anlage 66
4052 Basel
Switzerland
www.mdpi.com

Coatings Editorial Office
E-mail: coatings@mdpi.com
www.mdpi.com/journal/coatings

Disclaimer/Publisher's Note: The statements, opinions and data contained in all publications are solely those of the individual author(s) and contributor(s) and not of MDPI and/or the editor(s). MDPI and/or the editor(s) disclaim responsibility for any injury to people or property resulting from any ideas, methods, instructions or products referred to in the content.

www.ingramcontent.com/pod-product-compliance
Lightning Source LLC
LaVergne TN
LVHW070247100526
838202LV00015B/2189